Thorsten M. Buzug
Computed Tomography

Thorsten M. Buzug

Computed Tomography

From Photon Statistics to Modern Cone-Beam CT

With 475 Figures and 10 Tables

 Springer

Thorsten M. Buzug, Prof. Dr.
Institut für Medizintechnik
Universität zu Lübeck
Ratzeburger Allee 160
23538 Lübeck
Germany
E-mail: buzug@imt.uni-luebeck.de

ISBN 978-3-642-07257-4   e-ISBN 978-3-540-39408-2

DOI 10.1007/978-3-540-39408-2

Cover design: Frido Steinen-Broo, eStudio Calamar, Spain

Printed on acid-free paper

9 8 7 6 5 4 3 2 1

springer.com

Preface

This book provides an overview of X-ray technology, the historic developmental milestones of modern CT systems, and gives a comprehensive insight into the main reconstruction methods used in computed tomography. The basis of reconstruction is, undoubtedly, mathematics. However, the beauty of computed tomography cannot be understood without a detailed knowledge of X-ray generation, photon–matter interaction, X-ray detection, photon statistics, as well as fundamental signal processing concepts and dedicated measurement systems. Therefore, the reader will find a number of references to these basic disciplines together with a brief introduction to the underlying principles of CT.

This book is structured to cover the basics of CT: from photon statistics to modern cone-beam systems. However, the main focus of the book is concerned with detailed derivations of reconstruction algorithms in 2D and modern 3D cone-beam systems. A thorough analysis of CT artifacts and a discussion of practical issues, such as dose considerations, provide further insight into modern CT systems. While mainly written for graduate students in biomedical engineering, medical engineering science, medical physics, medicine (radiology), mathematics, electrical engineering, and physics, experienced practitioners in these fields will benefit from this book as well.

The didactic approach is based on consistent notation. For example, the notation of computed tomography is used in the signal processing chapter. Therefore, contrary to many other signal processing books, which use time-dependent values, this book uses spatial variables in one, two or three dimensions. This facilitates the application of the mathematics and physics learned from the earlier chapters to detector array signal processing, which is described in the later chapters. Additionally, special attention has been paid to creating a text with detailed and richly discussed algorithm derivations rather than compact mathematical presentations. The concepts should give even undergraduate students the chance to understand the principal reconstruction theory of computed tomography. The text is supported by a large number of illustrations representing the geometry of the projection situation. Since the impact of cone-beam CT will undeniably increase in the future, three-dimensional reconstruction algorithms are illustrated and derived in detail.

This book attempts to close a gap. There are several excellent books on medical imaging technology that give a comprehensive overview of modern X-ray technology, computed tomography, magnetic resonance imaging, ultrasound, or nuclear medicine modalities like PET and SPECT. However, these books often do not go into the mathematical detail of signal processing theory. On the other hand, there are a number of in-depth mathematical books on computed tomography that do not discuss practical issues. The present book is based on the German book *Einführung in die Computertomographie*, which first appeared during the summer of 2004. Fortunately, since the book was used by many of my students in lectures on *Engineering in Radiology*, *Medical Engineering*, *Signals and Systems in Medicine*, and *Tomographic Methods*, I received a lot of feedback regarding improvements on the first edition. Therefore, the idea arose to publish an English version of the book, which is a corrected and extended follow-up.

I would like to thank Siemens Medical Solutions, General Electric Medical Systems, and Philips Medical Systems, who generously supported my laboratories in the field of computed tomography. In particular, I would like to thank my friend Dr. Michael Kuhn, former Director of Philips Research Hamburg. It was his initiative that made possible the first installation of CT in my labs in 1999. Additionally, I have to thank Mrs. Annette Halstrick and Dr. Hans-Dieter Nagel (Philips Medical Systems Hamburg), Leon de Vries (Philips Medical Systems Best), Doris Pischitz, Jürgen Greim and Robby Rokkitta (Siemens Medical Solutions Erlangen), Dieter Keidel and Jan Liedtke (General Electric Medical Systems) for many photos in this book. I would like to thank Wolfgang Härer (AXI CC, Siemens Medical Solutions), Dr. Gerhard Brunst, (General Electric Medical Systems), Dr. Armin H. Pfoh, Director of General Electric Research Munich, Dr. Wolfgang Niederlag (Hospital Dresden-Friedrichstadt), Prof. Dr. Heinz U. Lemke (Technical University Berlin), Dr. Henrik Turbell (Institute of Technology, Linköpings Universitet), and Prof. Dr. Dr. Jürgen Ruhlmann (Medical Center Bonn) for the courtesy to allow me to use their illustrations and photos. Further, I have to thank the Digital Collections and Archives of Tufts University, the Collection of Portraits of the Austrian Academy of Sciences, and the Röntgen-Kuratorium Würzburg e.V. for the courtesy to allow me to use their photos.

Additionally, I have to thank my friends, colleagues, and students for proof-reading and translating parts of the book. In alphabetical order I appreciated the help of:

Dr. Bernd David (Philips Research Laboratories Hamburg)
Katie Dechambre (Milwaukee School of Engineering)
Erin Fredericks (California Polytechnic State University, San Luis Obispo)
Sebastian Gollmer (University of Lübeck)
Dr. Franko Greiner (University of Kiel)
Tobias Knopp (University of Lübeck)
Dieter Lukhaup (Schriesheim-Altenbach)
Andreas Mang (University of Lübeck)
Prof. Dr.-Ing. Alfred Mertins (University of Lübeck)
Jan Müller (University of Lübeck)
Dr. Hans-Dieter Nagel (Philips Medical Systems Hamburg)
Susanne Preissler (RheinAhrCampus Remagen)

Tony Shepherd (University College London)
Vyara Tonkova (RheinAhrCampus Remagen)

Many special thanks go to Sebastian Gollmer, Andreas Mang, and Jan Müller, who did the copy editing of the complete manuscript. However, for the errors that remain, I alone am responsible and apologize in advance.

I would like to thank the production team at le-tex as well as Paula Francis for copy editing. Further, I have to thank Springer Publishing, especially Dr. Ute Heilmann and Wilma McHugh for their excellent cooperation over the last few years.

Finally, I would like to thank my wife Kerstin, who has supported and sustained my writing efforts over the last few years. Without her help, patience, and encouragement this book would not have been completed.

Lübeck, June 2008   Thorsten M. Buzug

Contents

1   Introduction 1
    1.1   Computed Tomography – State of the Art 1
    1.2   Inverse Problems 2
    1.3   Historical Perspective 4
    1.4   Some Examples 7
    1.5   Structure of the Book 11

2   Fundamentals of X-ray Physics 15
    2.1   Introduction 15
    2.2   X-ray Generation 15
         2.2.1   X-ray Cathode 16
         2.2.2   Electron–Matter Interaction 19
         2.2.3   Temperature Load 23
         2.2.4   X-ray Focus and Beam Quality 24
         2.2.5   Beam Filtering 28
         2.2.6   Special Tube Designs 30
    2.3   Photon–Matter Interaction 31
         2.3.1   Lambert–Beer's Law 32
         2.3.2   Mechanisms of Attenuation 34
    2.4   Problems with Lambert–Beer's Law 46
    2.5   X-ray Detection 48
         2.5.1   Gas Detectors 48
         2.5.2   Solid-State Scintillator Detectors 50
         2.5.3   Solid-State Flat-Panel Detectors 52
    2.6   X-ray Photon Statistics 59
         2.6.1   Statistical Properties of the X-ray Source 60
         2.6.2   Statistical Properties of the X-ray Detector 64
         2.6.3   Statistical Law of Attenuation 66
         2.6.4   Moments of the Poisson Distribution 68
         2.6.5   Distribution for a High Number of X-ray Quanta 70
         2.6.6   Non-Poisson Statistics 72

3 Milestones of Computed Tomography 75
 3.1 Introduction 75
 3.2 Tomosynthesis 76
 3.3 Rotation–Translation of a Pencil Beam (First Generation).......... 79
 3.4 Rotation–Translation of a Narrow Fan Beam (Second Generation) ... 83
 3.5 Rotation of a Wide Aperture Fan Beam (Third Generation) 84
 3.6 Rotation–Fix with Closed Detector Ring (Fourth Generation) 87
 3.7 Electron Beam Computerized Tomography 89
 3.8 Rotation in Spiral Path... 90
 3.9 Rotation in Cone-Beam Geometry 91
 3.10 Micro-CT 93
 3.11 PET-CT Combined Scanners 96
 3.12 Optical Reconstruction Techniques 98

4 Fundamentals of Signal Processing 101
 4.1 Introduction 102
 4.2 Signals... 102
 4.3 Fundamental Signals 102
 4.4 Systems.. 104
  4.4.1 Linearity 104
  4.4.2 Position or Translation Invariance 105
  4.4.3 Isotropy and Rotation Invariance........................ 105
  4.4.4 Causality 106
  4.4.5 Stability 106
 4.5 Signal Transmission 106
 4.6 Dirac's Delta Distribution 109
 4.7 Dirac Comb 112
 4.8 Impulse Response... 115
 4.9 Transfer Function 116
 4.10 Fourier Transform 118
 4.11 Convolution Theorem 124
 4.12 Rayleigh's Theorem... 125
 4.13 Power Theorem 125
 4.14 Filtering in the Frequency Domain 126
 4.15 Hankel Transform 128
 4.16 Abel Transform 132
 4.17 Hilbert Transform 133
 4.18 Sampling Theorem and Nyquist Criterion 135
 4.19 Wiener–Khintchine Theorem................................... 141
 4.20 Fourier Transform of Discrete Signals 144
 4.21 Finite Discrete Fourier Transform............................. 145
 4.22 z-Transform 147
 4.23 Chirp z-Transform 148

5   Two-Dimensional Fourier-Based Reconstruction Methods 151
     5.1   Introduction 151
     5.2   Radon Transformation 153
     5.3   Inverse Radon Transformation and Fourier Slice Theorem 163
     5.4   Implementation of the Direct Inverse Radon Transform 167
     5.5   Linogram Method 170
     5.6   Simple Backprojection 175
     5.7   Filtered Backprojection 179
     5.8   Comparison Between Backprojection and Filtered Backprojection ... 183
     5.9   Filtered Layergram: Deconvolution of the Simple Backprojection 187
     5.10   Filtered Backprojection and Radon's Solution 191
     5.11   Cormack Transform.. 194

6   Algebraic and Statistical Reconstruction Methods 201
     6.1   Introduction 201
     6.2   Solution with Singular Value Decomposition 207
     6.3   Iterative Reconstruction with ART 211
     6.4   Pixel Basis Functions and Calculation of the System Matrix 218
            6.4.1   Discretization of the Image: Pixels and Blobs 219
            6.4.2   Approximation of the System Matrix in the Case of Pixels 221
            6.4.3   Approximation of the System Matrix in the Case of Blobs 222
     6.5   Maximum Likelihood Method.................................. 223
            6.5.1   Maximum Likelihood Method for Emission Tomography 224
            6.5.2   Maximum Likelihood Method for Transmission CT 230
            6.5.3   Regularization of the Inverse Problem 235
            6.5.4   Approximation Through Weighted Least Squares........... 238

7   Technical Implementation 241
     7.1   Introduction 241
     7.2   Reconstruction with Real Signals 242
            7.2.1   Frequency Domain Windowing 244
            7.2.2   Convolution in the Spatial Domain 247
            7.2.3   Discretization of the Kernels........................... 252
     7.3   Practical Implementation of the Filtered Backprojection 255
            7.3.1   Filtering of the Projection Signal 255
            7.3.2   Implementation of the Backprojection 258
     7.4   Minimum Number of Detector Elements 258
     7.5   Minimum Number of Projections 259
     7.6   Geometry of the Fan-Beam System............................. 261
     7.7   Image Reconstruction for Fan-Beam Geometry................... 262
            7.7.1   Rebinning of the Fan Beams 265
            7.7.2   Complementary Rebinning............................. 270

	7.7.3	Filtered Backprojection for Curved Detector Arrays	272
	7.7.4	Filtered Backprojection for Linear Detector Arrays	280
	7.7.5	Discretization of Backprojection for Fan-Beam Geometry	286
7.8	Quarter-Detector Offset and Sampling Theorem		293

8   Three-Dimensional Fourier-Based Reconstruction Methods 303
- **8.1** Introduction 303
- **8.2** Secondary Reconstruction Based on 2D Stacks of Tomographic Slices 304
- **8.3** Spiral CT 309
- **8.4** Exact 3D Reconstruction in Parallel-Beam Geometry 321
  - **8.4.1** 3D Radon Transform and the Fourier Slice Theorem 321
  - **8.4.2** Three-Dimensional Filtered Backprojection 326
  - **8.4.3** Filtered Backprojection and Radon's Solution 327
  - **8.4.4** Central Section Theorem 329
  - **8.4.5** Orlov's Sufficiency Condition 335
- **8.5** Exact 3D Reconstruction in Cone-Beam Geometry 336
  - **8.5.1** Key Problem of Cone-Beam Geometry 339
  - **8.5.2** Method of Grangeat 341
  - **8.5.3** Computation of the First Derivative on the Detector 347
  - **8.5.4** Reconstruction with the Derivative of the Radon Transform .. 348
  - **8.5.5** Central Section Theorem and Grangeat's Solution 350
  - **8.5.6** Direct 3D Fourier Reconstruction with the Cone-Beam Geometry 354
  - **8.5.7** Exact Reconstruction using Filtered Backprojection 357
- **8.6** Approximate 3D Reconstructions in Cone-Beam Geometry 366
  - **8.6.1** Missing Data in the 3D Radon Space 366
  - **8.6.2** FDK Cone-Beam Reconstruction for Planar Detectors 371
  - **8.6.3** FDK Cone-Beam Reconstruction for Cylindrical Detectors ... 388
  - **8.6.4** Variations of the FDK Cone-Beam Reconstruction 390
- **8.7** Helical Cone-Beam Reconstruction Methods 394

9   Image Quality and Artifacts 403
- **9.1** Introduction 403
- **9.2** Modulation Transfer Function of the Imaging Process 404
- **9.3** Modulation Transfer Function and Point Spread Function 410
- **9.4** Modulation Transfer Function in Computed Tomography 412
- **9.5** SNR, DQE, and ROC 421
- **9.6** 2D Artifacts 423
  - **9.6.1** Partial Volume Artifacts 423
  - **9.6.2** Beam-Hardening Artifacts 425
  - **9.6.3** Motion Artifacts 432
  - **9.6.4** Sampling Artifacts 435
  - **9.6.5** Electronic Artifacts 435
  - **9.6.6** Detector Afterglow 437
  - **9.6.7** Metal Artifacts 438

9.6.8   Scattered Radiation Artifacts 443
9.7   3D Artifacts 445
    9.7.1   Partial Volume Artifacts 446
    9.7.2   Staircasing in Slice Stacks 448
    9.7.3   Motion Artifacts 450
    9.7.4   Shearing in Slice Stacks Due to Gantry Tilt 451
    9.7.5   Sampling Artifacts in Secondary Reconstruction 454
    9.7.6   Metal Artifacts in Slice Stacks 455
    9.7.7   Spiral CT Artifacts 456
    9.7.8   Cone-Beam Artifacts 458
    9.7.9   Segmentation and Triangulation Inaccuracies 459
9.8   Noise in Reconstructed Images 462
    9.8.1   Variance of the Radon Transform 462
    9.8.2   Variance of the Reconstruction 464
    9.8.3   Dose, Contrast, and Variance 467

10   Practical Aspects of Computed Tomography 471
10.1   Introduction 471
10.2   Scan Planning 471
10.3   Data Representation 475
    10.3.1   Hounsfield Units 475
    10.3.2   Window Width and Window Level 476
    10.3.3   Three-Dimensional Representation 479
10.4   Some Applications in Medicine 482

11   Dose 485
11.1   Introduction 485
11.2   Energy Dose, Equivalent Dose, and Effective Dose 486
11.3   Definition of Specific CT Dose Measures 487
11.4   Device-Related Measures for Dose Reduction 493
11.5   User-Related Measures for Dose Reduction 499

References 503

Subject Index 511

1 Introduction

Contents

1.1   Computed Tomography – State of the Art **1**
1.2   Inverse Problems **2**
1.3   Historical Perspective **4**
1.4   Some Examples **7**
1.5   Structure of the Book **11**

1.1
Computed Tomography – State of the Art

Computed tomography (CT) has evolved into an indispensable imaging method in clinical routine. It was the first method to non-invasively acquire images of the inside of the human body that were not biased by superposition of distinct anatomical structures. This is due to the projection of all the information into a two-dimensional imaging plane, as typically seen in planar X-ray fluoroscopy. Therefore, CT yields images of much higher contrast compared with conventional radiography. During the 1970s, this was an enormous step toward the advance of diagnostic possibilities in medicine.

However, research in the field of CT is still as exciting as at the beginning of its development during the 1960s and 1970s; however, several competing methods exist, the most important being magnetic resonance imaging (MRI). Since the invention of MRI during the 1980s, the phasing out of CT has been anticipated. Nevertheless, to date, the most widely used imaging technology in radiology departments is still CT. Although MRI and positron emission tomography (PET) have been widely installed in radiology and in nuclear medicine departments, the term tomography is clearly associated with X-ray computed tomography[1].

Some hospitals actually replace their conventional shock rooms with a CT-based virtual shock room. In this scenario, imaging and primary care of the patient takes place using a CT scanner equipped with anesthesia devices. In a situation

---

[1] In the United States computed tomography is also called CAT (computerized axial tomography).

where the fast three-dimensional imaging of a trauma patient is necessary (and it is unclear whether MRI is an adequate imaging method in terms of compatibility with this patient), computed tomography is the standard imaging modality. Additionally, due to its ease of use, clear interpretation in terms of physical attenuation values, progress in detector technology, reconstruction mathematics, and reduction of radiation exposure, computed tomography will maintain and expand its established position in the field of radiology.

Furthermore, the preoperatively acquired CT image stack can be used to synthetically compute projections for any given angulations. A surgeon can use this information in order to get an impression of the images that are taken intraoperatively by a C-arm image intensifier. Therefore, there is no need to acquire additional radiographs and the artificially generated projection images actually resemble conventional radiographs. Additionally, the German Employer's Liability Insurance Association insists on a CT examination in severe accidents that occur at work. Therefore, CT has advanced to become the standard diagnostic imaging modality in trauma clinics. Patients with heavy trauma, fractures, and luxations benefit greatly from the clarification provided by imaging techniques such as computed tomography.

Recently, interesting technical, anthropomorphic, forensic, and archeological (Thomsen et al. 2003) as well as paleontological (Pol et al. 2002) applications of computed tomography have been developed. These applications further strengthen the method as a generic diagnostic tool for non-destructive material testing and three-dimensional visualization beyond its medical use. Magnetic resonance imaging fails whenever the object to be examined is dehydrated. In these circumstances, computed tomography is the three-dimensional imaging method of choice.

1.2
Inverse Problems

The mathematics of CT image reconstruction has influenced other scientific fields and vice versa. The backprojection technique, for instance, is used in both geophysics and radar applications (Nilsson 1997). Clearly, the fundamental problem of computed tomography can be easily described: Reconstruct an object from its shadows or, more precisely, from its projections. An X-ray source with a fan- or cone-beam geometry penetrates the object to be examined as a patient in medical applications, a skull found in archeology or a specimen in nondestructive testing (NDT). In the so-called third generation scanners, the fan-shaped X-ray beam fully covers a slice section of the object to be examined.

Depending on the particular paths, the X-rays are attenuated at varying extents when running through the object; the local absorption is measured with a detector array. Of course, the shadow that is cast in only one direction is not an adequate basis for the determination of the spatial distribution of distinct structures inside a three-dimensional object. In order to determine this structure, it is necessary to irradiate the object from all directions. Figure 1.1 schematically illustrates this principle, where $p_{\gamma_i}(\xi)$ represents the attenuation profile of the beam versus the X-ray detector array coordinate ξ under a particular projection angle γ_i. If the different attenuation or absorption profiles are plotted over all angles of rotation γ_i of the sampling unit, a sinusoidal arrangement of the attenuation or projection integral values is obtained. In two dimensions, these data, $p_{\gamma_i}(\xi)$, represent the radon space of the object, which is essentially the set of raw data.

In a special CT acquisition protocol, the spiral measurement process, the X-ray tube is continuously rotated, with the examination table being moved linearly through the measuring field. This scan process produces data that take a spiral or helical path. This method offers the possibility of computing any number of slices retrospectively, so that an accurate three-dimensional rendering can be expected. To allow this acquisition technique, a slip ring transfer system has been developed. In such a system, power supply to the X-ray tube and signal transfer from the detector system is guaranteed, even though the imaging system continuously rotates.

From a mathematical point of view, image reconstruction in computed tomography is the task of computing the spatial structure of an object that casts shadows using these very shadows. The solution for this problem is complex and involves techniques in physics, mathematics, and computer science. The described scenario is referred to as the inverse problem in mathematics.

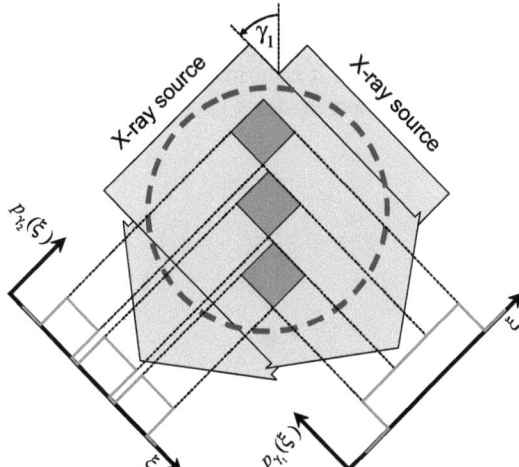

Fig. 1.1. Schematic illustration of computed tomography (CT). Three homogeneous objects with quadratic intersection areas are exposed with X-ray under the projection angles γ_1 and γ_2. Each projection angle produces a specific shadow, $p_\gamma(\xi)$, which, measured with the detector array represents the integral X-ray attenuation profile. The geometric shadow boundaries are indicated with *dashed lines*. However, analysis of the profile under the first projection angle, γ_1, on its own does not allow one to deduce an estimate of the number of separated objects

1.3
Historical Perspective

A particular kind of mathematical problem in CT became popular in the 1950s when the astrophysicist Bracewell proved that the resolution of telescopes could be significantly improved if spatially distributed telescopes are appropriately synchronized. However, in 1936, similar problems with the same mathematical basis had already been discussed (cf. Cormack [1982] and the references therein; many examples are collected by Deans [1983]).

One example can be found in the field of statistics: Given all marginal distributions of a probability distribution density, is it possible to deduce the probability density itself? In this example, the marginal distributions represent an equivalent example for the measured projections in computed tomography. Another example can be found in astrophysics: Looking from earth into a particular direction of the universe, only the radial velocity component of the stars can be obtained via the spectral Doppler red shift. This again represents the same inverse problem that has to be solved if the distribution of the actual three-dimensional velocity vector is to be reconstructed from the radial velocity components acquired from all available directions.

In computed tomography, the meaning of the mathematical term *inverse problem* is immediately apparent. In contrast to the situation shown in Fig. 1.1, the spatial distribution of the attenuating objects that produce the projection shadow is not known a priori. This, actually, is the reason for acquiring the projections along the rotating detector coordinate ξ over a projection angle interval of at least $180°$. Figure 1.2 illustrates this situation. It is an inversion of integral transforms. From a sequence of measured projection shadows $\{p_{\gamma_1}(\xi), p_{\gamma_2}(\xi), p_{\gamma_3}(\xi), \ldots\}$, the spa-

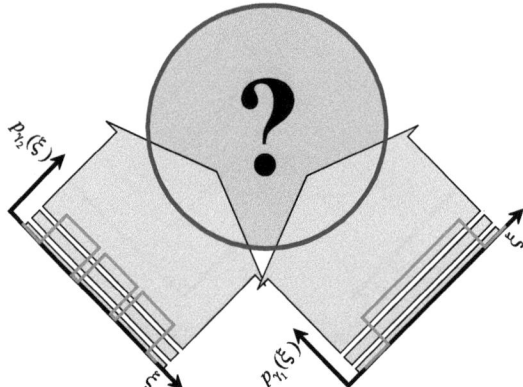

Fig. 1.2. Schematic illustration of the inverse problem posed by CT. Attenuation profiles, $p_\gamma(\xi)$, have been measured for a set of projection angulations, γ_1 and γ_2. The unknown geometry, or the object with its associated spatial distribution of attenuation coefficients, has to be calculated from a complete set of attenuation profiles $\{p_{\gamma_1}(\xi), p_{\gamma_2}(\xi), p_{\gamma_3}(\xi), \ldots\}$

Fig. 1.3. *Left*: Allan MacLeod Cormack (1924–1998) shortly after the official announcement of the Nobel Prizes for medicine in 1979 (courtesy of Tufts University, Digital Collections and Archives). *Right*: Sir Godfrey Hounsfield (1919–2004) in front of his first EMI CT scanner (courtesy of General Electric Medical Systems)

Fig. 1.4. *Left*: Johann Radon (1887–1956; courtesy of the Austrian Academy of Sciences [OAW], collection of portraits). *Right*: Wilhelm Conrad Röntgen (1845–1923; courtesy of Röntgen-Kuratorium Würzburg e.V.)

tial distribution of the objects, or more precisely, the spatial distribution of the attenuation coefficients within a chosen section through the patient, must be estimated.

In 1961, the solution to this problem was applied for the first time to a sequence of X-ray projections for which an anatomical object had been measured from different directions. Allen MacLeod Cormack (1924–1998) and Sir Godfrey Hounsfield (1919–2004) are pioneers of medical computed tomography and in 1979 received the Nobel Prize for Medicine for their epochal work during the 1960s and 1970s. Figure 1.3 shows a picture of the two Nobel Prize winners.

Table 1.1. Summary of historical CT milestones

Year	Milestone
1895	Röntgen discovers a new kind of radiation, which he named X-ray
1901	Röntgen receives the Nobel Prize for physics
1906	Bockwinkel employs the Lorentz's solution in the reconstruction of three-dimensional functions from two-dimensional area integrals
1917	Radon publishes his epochal work on the solution of the inverse problem of reconstruction
1925	Ehrenfest extends the solution of Lorentz to n dimensions using the Fourier transform
1936	Cramer and Wold solve the reconstruction problem in statistics in which the probability distribution is obtained from a complete set of marginal probability distributions
1936	Eddington solves the reconstruction problem in the field of astrophysics to calculate the distribution of star velocities from the distribution of their measured radial components
1956	Bracewell applies Fourier techniques for the solution of the inverse problem in radio astronomy
1958	The Ukrainian scientist Korenblyum develops an X-ray scanner and tries to measure thin slices through the patient with analogue reconstruction principles
1963	Cormack contributes the first mathematical implementations for tomographic reconstruction in South Africa
1969	Hounsfield shows proof of the principle with the first CT scanner based on a radioactive source at the EMI research laboratories
1972	Hounsfield and Ambrose publish the first clinical scans with an EMI head scanner
1975	Set-up of the first whole body scanner with a fan-beam system
1979	Hounsfield and Cormack receive the Nobel Prize for Medicine
1983	Demonstration of electron beam CT (EBCT)
1989	Kalender publishes the first clinical spiral-CT
1991	Demonstration of multi-slice CT (MSCT)

Cormack pointed out that previously the Dutch physicist H.A. Lorentz had found a solution to the three-dimensional problem in which the desired function had to be reconstructed from two-dimensional surface integrals (Cormack 1982). Lorentz himself did not publish the results and so, unfortunately, the context of his work is still unknown today. The result, however, is associated with Lorentz by H. Bockwinkel, who mentioned the work in a 1906 publication on light propagation in crystals.

A detailed, mathematical basis to the solution of the inverse problem in computed tomography was published by the Bohemian mathematician Johann Radon in 1917 (cf. Fig. 1.4, left) (Radon 1917). Due to the complexity and depth of the mathematical publication, however, the consequences of his ground-breaking results were revealed very late in the mid-20th century. Additionally, the paper was published in German, which hindered a wide distribution of the work. In ▸ Chap. 5, an excerpt of his original work is reprinted.

In Table 1.1, some of the historical milestones and development steps of computed tomography are summarized. The list undoubtedly has to start with Wilhelm Conrad Röntgen (1845–1923), who received the Nobel Prize for physics in 1901 (cf. Fig. 1.4, right). Before 1960, a significant number of mathematical contributions in the field of inverse problems were developed independently and are summarized here only retrospectively.

1.4
Some Examples

Figures 1.5 to 1.8 show several examples of computer tomographic images that illustrate different anatomical regions often used in clinical practice. Modern CT scanners yield images with an excellent soft tissue contrast. In Fig. 1.5, the slices are annotated with the relevant scan protocol parameter. The most important parameters are the acceleration voltage (which determines the energy of the X-ray quanta), the tube current (which determines the intensity of the radiation), the slice thickness (which is the axial thickness of the X-ray fan beam), and the gantry tilt (which is the angulation of the CT frame with respect to the axial axis). In spiral-CT, the pitch is an additional parameter that defines the table feed in units of slice thickness.

In clinical practice, besides choosing an appropriate set of scan parameters (cf. Figs. 1.5 and 1.6), it is necessary to have a planning step for accurate anatomical scanning before CT slice sequence acquisition. In this planning step, the slices must be adapted to the anatomical situation, and furthermore, the dose for sensitive organs must be minimized. The planning is accomplished on the basis of an overview scan that looks similar to simple projection radiography (cf. ▸ Chap. 10). Here, the exact position and orientation of the slice can be interactively defined.

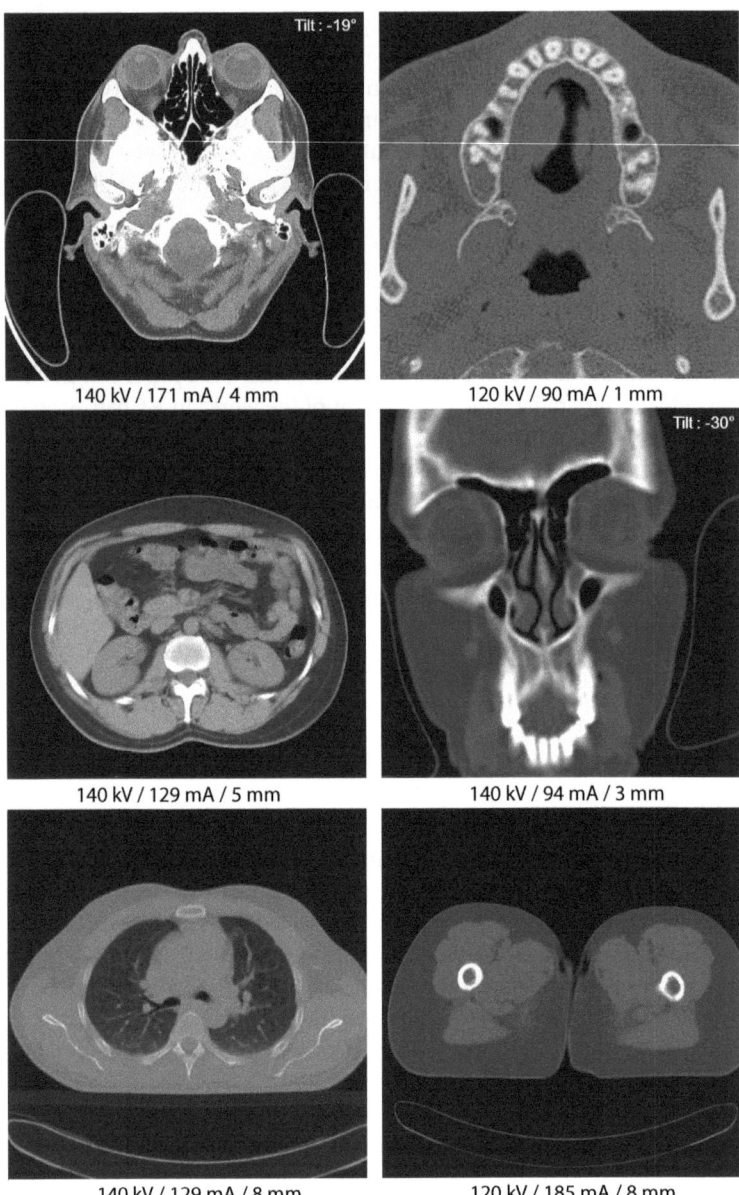

Fig. 1.5. Examples of CT images. Modern CT scanners produce images with excellent soft tissue contrast (courtesy of J. Ruhlmann)

Figures 1.7 and 1.8 illustrate that computed tomography is a three-dimensional modality. The geometrically precise slice stack can be constructed in a secondary reconstruction step to yield a virtual three-dimensional volume. In Fig. 1.7, five

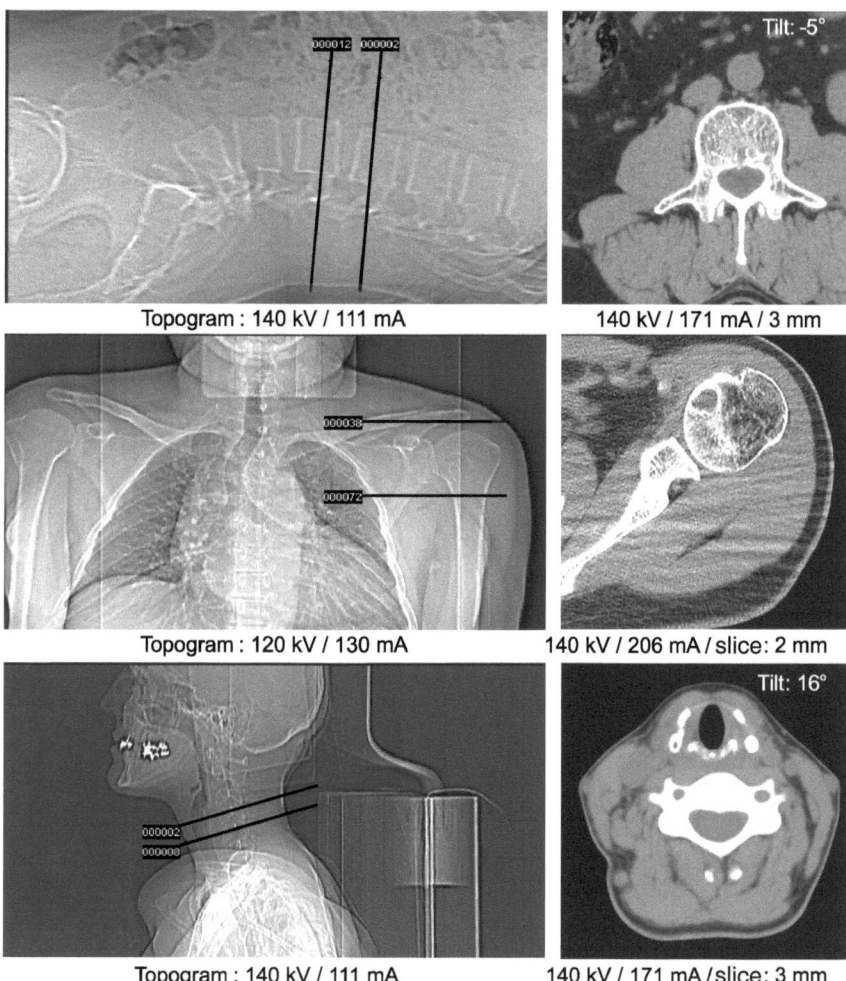

Fig. 1.6. A patient overview must be acquired for CT scan sequence planning. Depending on the manufacturer the overview scan is called a topogram, a scout view, a scanogram or a pilot view. The geometric scan interval and gantry tilt are determined interactively (courtesy of J. Ruhlmann)

slices are illustrated as an example. Additionally, the patient's skin and lung was segmented with a simple threshold and visualized using a surface-rendering procedure. For the same data set an alternative visualization is presented in Fig. 1.8. Multi-planar reformatting (MPR) is used to show angulated sections through the three-dimensional stack of slices. Typically, the principal sections (the sagittal, coronal, and axial slices) are presented to the radiologist. In Fig. 1.9, the principal slice directions are illustrated.

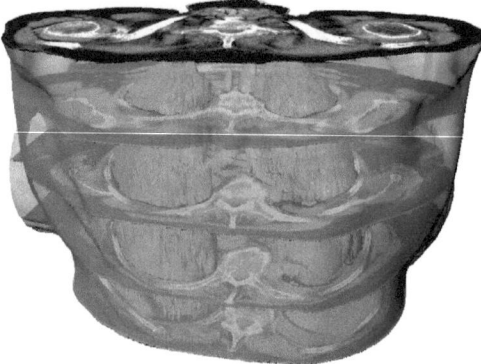

Fig. 1.7. Conventional CT produces two-dimensional slices. However, CT becomes a three-dimensional imaging modality if consecutive slices are arranged as axial stacks (courtesy of J. Ruhlmann)

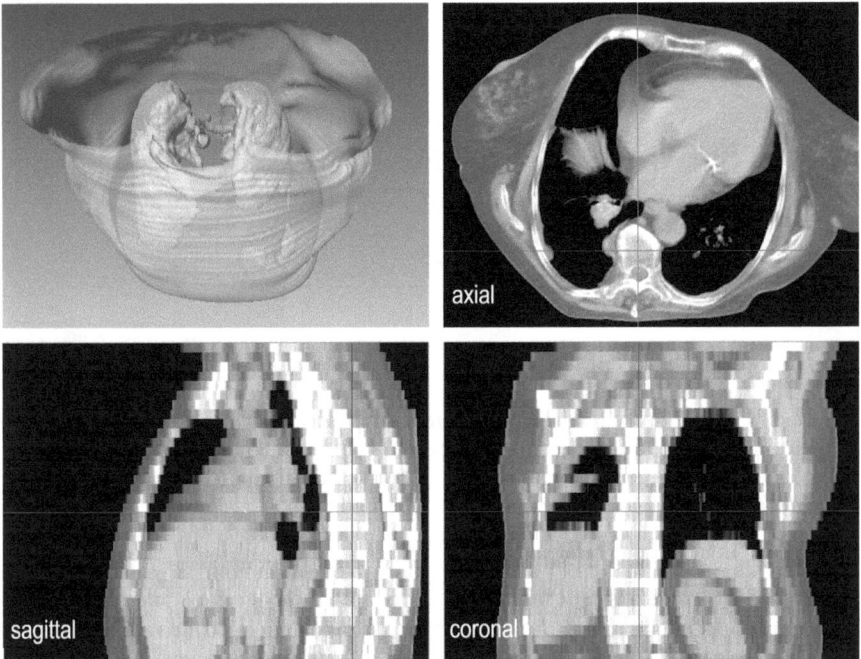

Fig. 1.8. The arrangement of a set of axial CT slices to build up a three-dimensional volume is called secondary reconstruction. This data representation allows deeper diagnostic insights. Typically, segmented organs of interest are displayed using either surface rendering or an approach in which the gray values are presented in an orthonormal reformatting consisting of the sagittal, coronal, and axial view (courtesy of J. Ruhlmann)

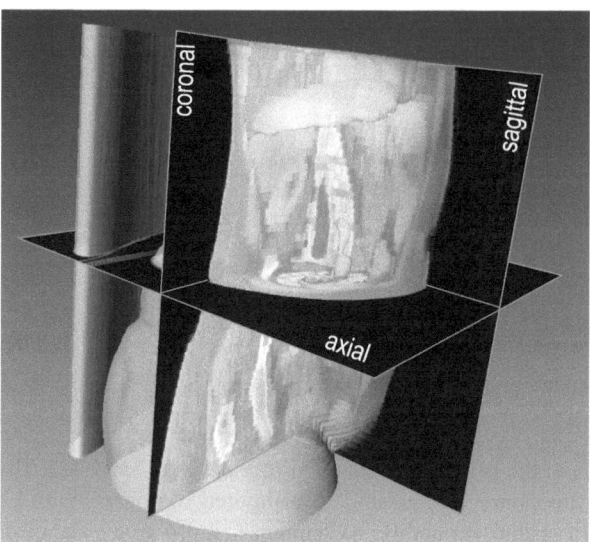

Fig. 1.9. Primarily, with CT, an axial slice sequence is acquired and reconstructed. Using interpolation, coronal and sagittal slices can be calculated from the stack. This procedure is called multi-planar reformatting (MPR)

1.5
Structure of the Book

This book gives a comprehensive overview of the main reconstruction methods in computed tomography. The basis of the reconstruction is undoubtedly mathematics. However, the beauty of computed tomography cannot be understood without a basic knowledge of X-ray physics, signal processing concepts and measurement systems. Therefore, the reader will find a number of references to these basic disciplines as well as a brief introduction to many of the underlying principles.

With respect to the subtitle of this book, it is structured to cover the basics of CT, from photon statistics to modern cone-beam systems. Without an elementary knowledge of X-ray physics, a number of the described imaging effects and artifacts cannot readily be understood. In ▸ Chap. 2, X-ray generation, photon–matter interaction, X-ray detection, and photon statistics are briefly summarized. In ▸ Chap. 3, a retrospective overview of the historical milestones on the road map of the technical developments in computed tomography is given. Starting with tomosynthesis in the 1920s and 1930s, the different types or generations of CT are characterized. The chapter concludes with motivation for the modern scanner concepts like electron-beam CT (EBCT), micro-CT, and especially helical cone-beam CT. Although remarkable advances in CT technology have been achieved, Fig. 1.10 shows that the appearance of the gantry has undergone only a slight change throughout the years.

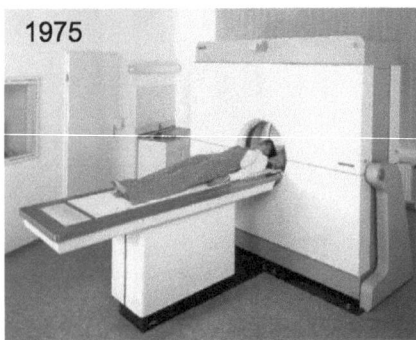 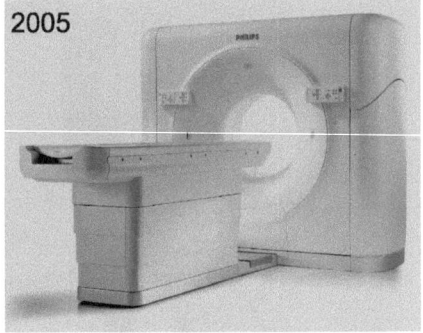

Fig. 1.10. Design of CT gantries in 1975 and 2005 (courtesy of Philips Medical Systems)

In ▸ Chap. 4, the principles of signal processing are reviewed. This chapter focuses on the necessary background of computed tomography and consequently uses signals of spatial variability. ▸ Chapters 5 and 6 give a detailed overview of two-dimensional reconstruction mathematics. The most important algorithms are derived step by step. In ▸ Chap. 5, the Fourier-based methods are collected. In ▸ Chap. 6, the algebraic and statistical approaches are explained.

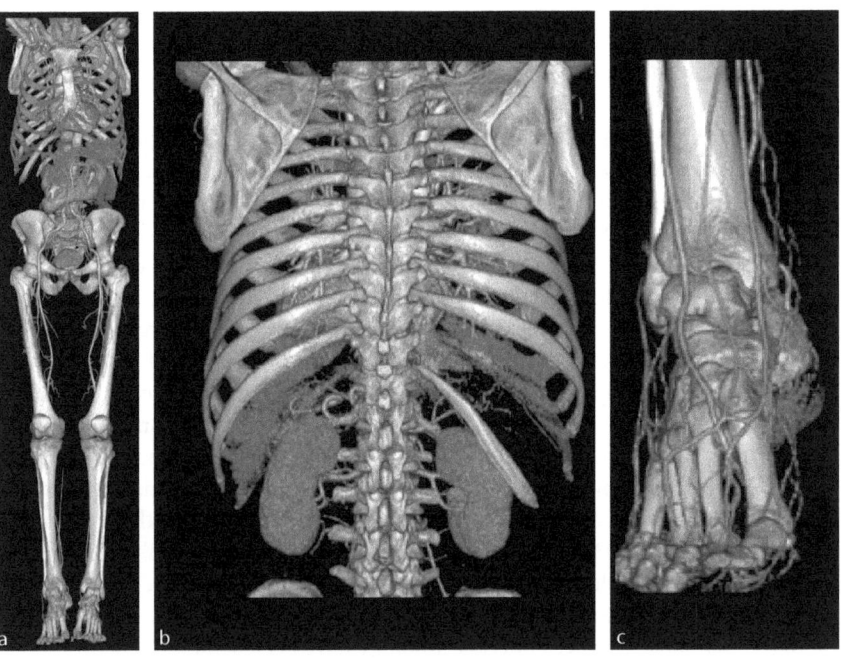

Fig. 1.11a–c. Whole body scans can be performed with the latest generation of CT systems, including a multi-slice detector system. Even very small vessels of the feet can be precisely visualized (courtesy of Philips Medical Systems)

In ▸ Chap. 7, the limitations of the practical implementation of the previously described methods are discussed. Specifically, the correspondence of the parallel pencil-beam and the fan-beam X-ray system are demonstrated.

In ▸ Chap. 8, the three-dimensional methods of CT image or volume reconstruction are reviewed. It is shown that some of the ideas are consequent extensions of the methods discussed in ▸ Chap. 5. The methods described in this chapter represent the basis of a highly active field of research. A description of the existing manifold algorithmic variations in the field of helical cone-beam methods, for instance, is beyond the scope of this book. However, in Fig. 1.11 an example of the impressive quality of the three-dimensional reconstruction results of modern multi-slice CT scanners is given.

In ▸ Chap. 9, an introduction to the methods of image quality evaluation is given. The chapter focuses on typical artifacts of computed tomography, whereby two-dimensional and three-dimensional artifacts are differentiated. Additionally, the important fourth power law is derived that describes the correspondence among signal-to-noise ratio, dose, and detector element size.

In ▸ Chap. 10, some practical aspects of computed tomography are described. This includes CT planning, which uses the overview scan mentioned previously, the mapping of the physical attenuation values to the Hounsfield scale, and a list of exemplary application fields of CT in practice. Finally, ▸ Chap. 11 concludes the book with a review of dose issues in clinical computed tomography.

2   Fundamentals of X-ray Physics

Contents

2.1	Introduction	**15**
2.2	X-ray Generation	**15**
2.3	Photon–Matter Interaction	**31**
2.4	Problems with Lambert–Beer's Law	**46**
2.5	X-ray Detection	**48**
2.6	X-ray Photon Statistics	**59**

2.1
Introduction

For the discovery of a new radiation capable of high levels of penetration, Wilhelm Conrad Röntgen was awarded with the first Nobel Prize for physics in 1901. In 1895, in experiments with accelerated electrons, he had discovered radiation with the ability to penetrate optically opaque objects, which he named X-rays. In this chapter, the generation of X-rays, photon–matter interaction, X-ray detection, and statistical properties of X-ray quanta will be described. However, the scope of this chapter is limited to physical principles that are relevant to computed tomography (CT). A more comprehensive description can be found in many physics text books, for example Demtröder (2000), and in overviews on radiological technology, for example Curry et al. (1990). One of the main reasons for the wide exploitation of Röntgen's radiation was the simple equipment required for X-ray generation and detection. Nevertheless, the development of robust, high-power X-ray tubes that are optimized for use in CT, is ongoing.

2.2
X-ray Generation

X-ray radiation is of electromagnetic nature; it is a natural part of the electromagnetic spectrum, with a range that includes radio waves, radar and microwaves, infrared, visible and ultraviolet light to X- and γ-rays. In electron-impact X-ray sources, the radiation is generated by the deceleration of fast electrons[1] entering a solid metal anode, and consists of waves with a range of wavelengths[2] roughly between 10^{-8} m and 10^{-13} m. Thus, the radiation energy depends on the electron velocity, ν, which in turn depends on the acceleration voltage, U_a, between cathode and anode so that with the simple conservation of energy[3]

$$eU_a = \frac{1}{2}m_e\nu^2 \tag{2.1}$$

the electron velocity can be determined.

2.2.1
X-ray Cathode

In medical diagnostics acceleration voltages are chosen between 25 kV and 150 kV, for radiation therapy they lie between 10 kV and 300 kV, and for material testing they can reach up to 500 kV. Figure 2.1 shows a schematic drawing of an X-ray

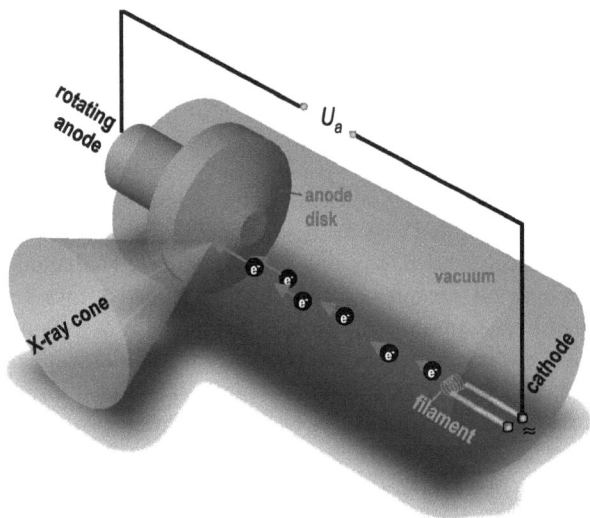

Fig. 2.1. Schematic drawing of an X-ray tube. Thermal electrons escape from a cathode filament that is directly heated to approximately 2,400 K. The electrons are accelerated in the electric field between cathode and anode. X-ray radiation emerges from the deceleration of the fast electrons following their entry into the anode material

---

[1] Charge of electrons: $e = 1.602 \cdot 10^{-19}$ C; mass of electrons: $m_e = 9.109 \cdot 10^{-31}$ kg.

[2] There is no clear definition of the X-ray wavelength interval. The range overlaps with ultraviolet and γ-radiation.

[3] Acceleration energy is measured in units called electron volts (eV). 1 eV is the energy that an electron will gain if it is accelerated by an electrical potential of one volt. The same unit is used to measure X-ray photon energy.

tube. Electrons are emitted from a filament, which is directly heated to approximately 2,400 K to overcome the binding energy of the electrons to the metal of the filament[4].

The binding energy, E_v, is due to two main effects (Bergmann and Schäfer 1999). The first effect is the formation of a dipole layer at the cathode surface. When a free inner electron moves toward the surface of the metal, electrostatic forces prevent the electron from escaping. However, before reversing its direction, the electron overshoots the outer metal ion layer, and, as a result, an electron is missing inside the metal for charge neutralization. The surplus of positive charge at the inner side of the surface layer, together with the negative electron outside the metal, forms an electric dipole layer. The electric field inside the dipole layer slows down electrons trying to leave the metal. This effect results in the part W_{Dipole} of the work function.

The other part originates from what is called a mirror-image force. Due to electrostatic influence, an electron above a metal surface causes a charge displacement inside the metal. The resulting electric field between the electron and the metal surface looks like the field between a charge, $-e$, above and a virtual mirror charge, $+e$, below the metal boundary at the same distance x from the boundary. To bring an electron from distance d above the surface to infinity, the work

$$W_{\text{mirror}} = \frac{e^2}{4\pi\varepsilon_0} \int_d^\infty \frac{\mathrm{d}x}{(2x)^2} = \frac{e^2}{16\pi\varepsilon_0 d} \tag{2.2}$$

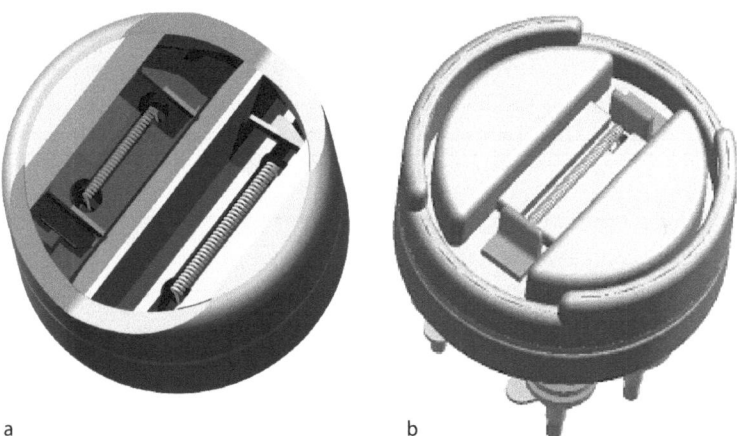

a   b

Fig. 2.2. The electron beam is controlled by a cylindrically shaped electrode, containing the cathode with opposite potential. This electron optics is called a *Wehnelt* cylinder, or is sometimes also called a focusing cup. In this way, the electrons are steered onto a small focal point on the anode. Shown in **a** is a dual-filament and in **b** a modern mono-filament; both are designed to produce focal spot sizes of 0.5 mm and 1.5 mm (courtesy of Philips Medical Systems)

---

[4] Filaments are usually made of thoriated tungsten with a melting point at 3,410 °C.

must be applied. Since the metal–vacuum interface is not an ideal mirror surface, it is sensible to start the integration at a distance of approximately one atom diameter ($d \approx 10^{-10}$ m), leading to a work of $W_{\mathrm{mirror}} \approx 3.6\,\mathrm{eV}$.

Due to their thermal energy, electrons are boiled off from the filament. This process is called thermionic emission. The temperature of the metal must be high enough to increase the kinetic energy, E_{kin}, of the electrons such that $E_{\mathrm{kin}} > E_v = W_{\mathrm{dipole}} + W_{\mathrm{mirror}}$. The emission current density, j_e, is essentially a func-

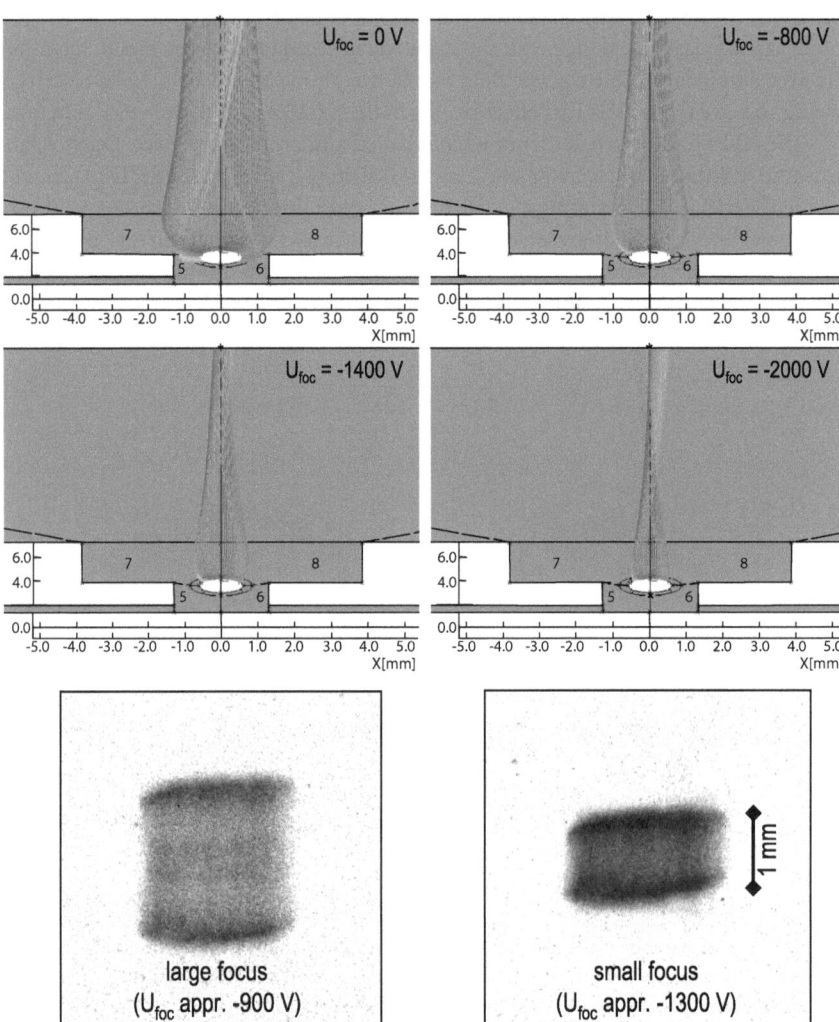

Fig. 2.3. Simulation of electron trajectories emitted from the filament and accelerated onto the anode. The potential at the *Wehnelt* cylinder controls the electron focus on the anode. *Below*: Shape and size of a large and a small X-ray focus (courtesy of Philips Medical Systems)

tion of the temperature and can be described by the *Richardson–Dushman* equation

$$j_e = C_{RD}\, T^2\, e^{-\frac{\varphi}{kT}}\,,\tag{2.3}$$

where C_{RD} is the *Richardson–Dushman*[5] constant

$$C_{RD} = \frac{4\pi m_e k^2 e}{h^3}\,,\tag{2.4}$$

k is the Boltzmann constant[6] and φ is the work function[7] defined as the difference between E_v and the Fermi energy edge. An electron cloud forms around the filament and these electrons are subsequently accelerated toward the anode. When the electrons reach the surface of the anode they will be stopped abruptly.

To produce a small electron focus on the anode, the trajectories of the accelerated electrons must be controlled by electron optics. The focusing device can be seen in Fig. 2.2. Basically, it is a cup-shaped electrode that forms the electric field near the filaments such that the electron current is directed to a small spot. In Fig. 2.3 it can be seen how the potential of the *Wehnelt* optics (frequently named *Wehnelt* cylinder) influences the electron trajectories. In this way, the cylinder can easily be controlled to produce a large or small X-ray focus. The effect that this has on imaging quality will be discussed in a later section.

2.2.2
Electron–Matter Interaction

With the entry of accelerated electrons into the anode, sometimes also called the *anticathode*, several processes take place close to the anode surface. Generally, the electrons are diffracted and slowed down by the Coulomb fields of the atoms in the anode material. The deceleration results from the interaction with the orbital electrons and the atomic nucleus. As known from classical electrodynamics, acceleration and deceleration of charged particles creates an electric dipole and electromagnetic waves are radiated. Usually, several photons emerge throughout the complete deceleration process of one single electron. Figure 2.4a illustrates two successive deceleration steps. It can happen, however, that the entire energy, eU_a, of an electron is transformed into a single photon. This limit defines the maximum energy of the X-ray radiation, which can be determined by

$$eU_a = h\nu_{max} = E_{max}\,.\tag{2.5}$$

The limit E_{max} corresponds to the minimum wavelength

$$\lambda_{min} = \frac{hc}{eU_a} = \frac{1.24\,\text{nm}}{U_a/\text{kV}}\,,\tag{2.6}$$

---

[5] For ideal metals $C_{RD} \approx 120\,\text{A cm}^{-2}\,\text{K}^{-2}$. However, in practice C_{RD} is material-dependent.
[6] Boltzmann constant: $k = 1.38 \cdot 10^{-23}\,\text{J K}^{-1}$.
[7] 4.5 eV for tungsten.

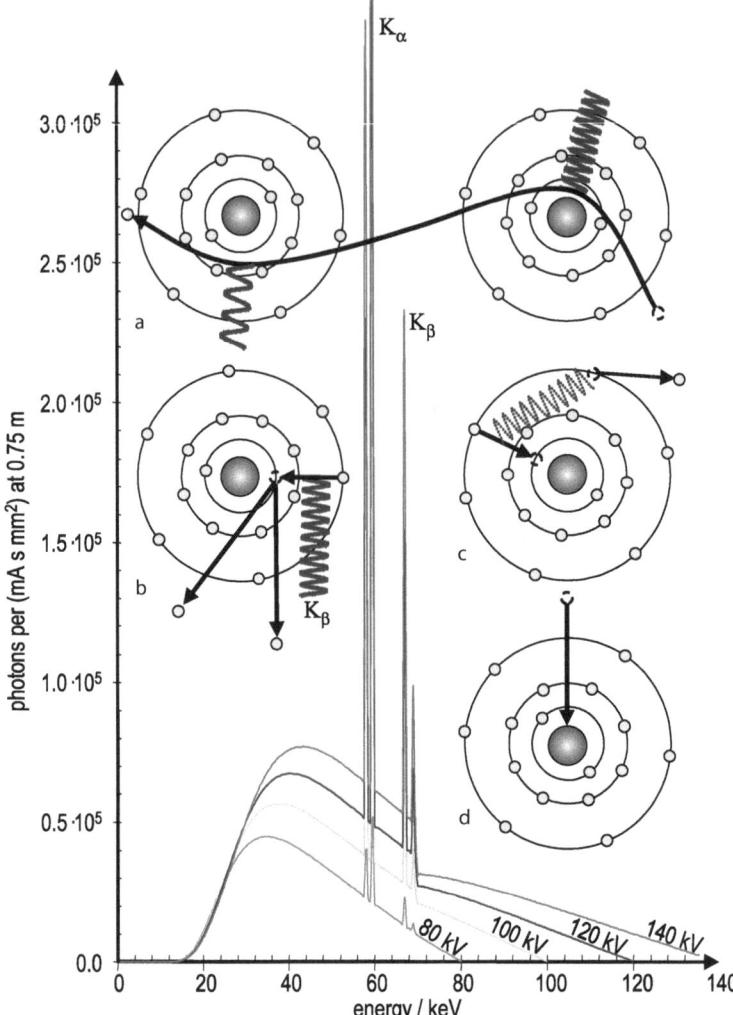

Fig. 2.4. X-ray spectrum of a tungsten anode at acceleration voltages in the range of U_a = 80–140 kV. The anode angle is 10° and 2 mm Al filtering has been applied. The intensity versus wavelength plot shows the characteristic line spectrum as well as the continuous *bremsstrahlung* (courtesy of B. David, Philips Research Labs). The minimum wavelength is determined by the total energy, eU_a, of the electron reaching the anode. Process illustrations: *a* *bremsstrahlung*, *b* characteristic emission, *c* Auger process and *d* direct electron-nucleus collision

where h is Planck's constant and c is the speed of light[8]. While the acceleration voltage determines the energy interval of the X-ray spectrum, the intensity of the generated X-ray spectrum or the number of X-ray quanta, is solely controlled by the anode current.

---

[8] Planck's constant: $h = 6.626 \cdot 10^{-34}$ Js; speed of light in vacuum: $c = 2.998 \cdot 10^{8}$ m/s.

Due to the fact that the slowing down of electrons in the anode material is a multi-process deceleration cascade, a continuous distribution of energies can be shown by the *bremsstrahlung* (cf. Fig. 2.4). Since the free electrons are unbound, their energy cannot be quantized. The energy balance can be described by the following equation

$$E_{\mathrm{kin}}(e^-)(+atom\ lattice) \rightarrow (atom\ lattice+)E_{\mathrm{kin}-h\nu}(e^-) + h\nu \ . \quad (2.7)$$

Unfortunately, the process of X-ray generation by electron deflection is a rare one. It has been shown (Agawal 1991) that the intensity of *bremsstrahlung* follows

$$I \propto Zh\left(\nu_{\mathrm{max}} - \nu\right) \ , \quad (2.8)$$

where Z is the atomic number of the anode material. The conversion efficiency from kinetic electron energy to *bremsstrahlung* energy can be described by

$$\eta = KZU_{\mathrm{a}} \ , \quad (2.9)$$

where K is a material constant that was found by Kramers (1923) to theoretically[9] be $K = 9.2 \cdot 10^{-7} \mathrm{kV}^{-1}$ when the acceleration voltage, U_{a}, is given in kV.

The continuous *bremsstrahlung* is superimposed by a characteristic line spectrum, which originates from direct interaction of fast electrons with the inner shell electrons of the anode material. If an electron on the K-shell or K-orbital is kicked out of the atom by a collision with a fast electron, i.e., the atom is ionized by the loss of an inner electron, an electron of one of the higher shells fills the vacant position on the K-shell. An example for the energy balance of the transition of an L-electron to the K-shell is given by

$$E_{\mathrm{K}}(Atom^+) \rightarrow E_{\mathrm{L}}(Atom^+) + h\nu_{\mathrm{KL}} \ . \quad (2.10)$$

As the inner shells represent states having a lower potential energy than the outer shells, this process is accompanied by the emission of a photon. Due to the high-energy difference between the inner shells, these photons, with a wavelength

$$\lambda = \frac{hc}{E_i - E_j} \quad (2.11)$$

are X-ray quanta. This process creates sharp lines in the X-ray spectrum that are characteristic finger prints for the anode material. The notation of these lines is agreed upon as follows: K_α, K_β, K_γ, ... denote the transition of an electron from the L-, M-, N-, ... shell to the K-shell and L_α, L_β, L_γ, ... denote the transition from the M-, N-, O-, ... shell to the L-shell, etc. Figure 2.4b illustrates a K_β emission.

The position of the characteristic K-line spectrum is given by Moseley's Law

$$\lambda = \frac{hc}{E_n - E_1} = \frac{hc}{13.6\,\mathrm{eV}(Z-1)^2(1 - 1/n^2)} \ , \quad (2.12)$$

where Z is again the atomic number of the anode material and n is the principal quantum number of the electron falling to the K-shell.

---

[9] Experiments are in good agreement with Kramers' result, but show a slight dependence on Z.

The result is the production of a large number of X-ray quanta at a few discrete energies. It can be seen in Fig. 2.4 that the probability of X-ray quanta emerging by the K_α process is higher than the probability of *bremsstrahlung* quanta at the same energy. However, in total the characteristic X-ray radiation contributes far less to the total intensity than the *bremsstrahlung*. Figure 2.4 shows a typical spectrum for a tungsten anode at acceleration voltages between $U_a = 80\,\text{kV}$ and $140\,\text{kV}$. In Fig. 2.4c a competing process, named the Auger process, is illustrated as well. Instead of emitting the K_β radiation the atom may absorb the photon by emitting another electron, called an Auger electron. This is seen as a non-radiative process. The emerging Auger electrons are mono-energetic.

The probability of the Auger process is constant for all elements. However, the probability of emitting characteristic radiation is given by the ratio

$$P = \frac{Z^4}{Z^4 + a} = \frac{1}{1 + \frac{a}{Z^4}} < 1 , \tag{2.13}$$

where a is a positive constant[10]. For elements with low atomic numbers, the Auger process dominates, whereas for heavy elements, characteristic emission dominates.

The direct collision of the fast electron with the nucleus of an anode atom is indicated in Fig. 2.4d. This interaction represents an ideal conversion of the entire kinetic energy of the electron to *bremsstrahlung* in one single deceleration process. Obviously, the collision contributes to the upper limit of the X-ray spectrum. However, from the X-ray intensity at the upper energy limit of the spectrum, it can be concluded that direct collision is a very rare interaction process. The mean energy lost of the electron in matter can quantitatively be described by

$$\left(\frac{dE}{dx}\right) = \left(\frac{dE}{dx}\right)_{bremsstrahlung} + \left(\frac{dE}{dx}\right)_{ionization} , \tag{2.14}$$

where the first term of (2.14) is given by quantum electrodynamics (QED)

$$\left(\frac{dE}{dx}\right)_{bremsstrahlung} = -4\alpha N_A \rho \frac{Z^2}{A}\left(\frac{e^2}{m_e c^2}\right)^2 E \ln\left(\frac{183}{Z^{1/3}}\right) , \tag{2.15}$$

where α is the fine-structure constant[11], Z is the atomic number, ρ is the density, A is the atomic weight of the material, and N_A is the Avogadro constant[12].

The second term of (2.14) is given by the Bethe–Bloch equation

$$\left(\frac{dE}{dx}\right)_{ionization} = -4\pi N_A \rho \frac{Z}{A}\left(\frac{e^4}{m_e c^2}\right)\frac{z^2}{\beta^2}\left[\frac{1}{2}\ln\left(\frac{2m_e c^2 \beta^2 \gamma^2 T_{max}}{I}\right) - \beta^2 - \frac{\delta}{2}\right] . \tag{2.16}$$

The electron velocity is given in units of the speed of light, i.e., $\beta = v/c$, and, here, the charge of the electron z is given in units of the elementary charge, therefore, $z = 1$. γ is the Lorentz factor, i.e., $\gamma = (1 - \beta^2)^{-1/2}$. T_{max} is the maximum kinetic energy to be transferred in a single collision. I is the mean ionization energy of the material and δ is a density correction of the ionization energy.

---

[10] For K-shell emission $a = 1.12 \cdot 10^6$ (Otendal 2006).

[11] Fine-structure constant: $\alpha = e^2/(4\pi\varepsilon_0 \hbar c) \approx 1/137.036$.

[12] Avogadro constant: $N_A = 6.023 \cdot 10^{23}\,\text{mol}^{-1}$.

2.2.3
Temperature Load

Using (2.9), it can be estimated that the quantum efficiency of the conversion from kinetic energy into X-ray radiation, within a tungsten anode (W, Z = 74), and working with an acceleration voltage of U_a = 140 kV, is roughly in the magnitude of η = 0.01. This means that 99% of the kinetic energy is transferred locally to the lattice, heating up the anode. As a result, CT X-ray tubes have serious heat problems. Since it is the energy deposition in the target volume that produces the heat load, the tube current and the duration of exposure or, more precisely, the product of current in milliamperes and exposure time in seconds, are two important parameters of the practical scan protocol the radiologist has to choose appropriately. The heat capacity of an X-ray tube is measured in *Heat Units*

$$HU = U \cdot I \cdot t . \qquad (2.17)$$

For several decades rotating anode disks have been used to distribute the thermal load over the entire anode. The anode target material is rotated about the central axis and therefore, new, cooler anode material is constantly rotated into position at the focal spot (Mudry 2000). In this way, the energy of the electron beam is spread out over a line, called the focal line, rather than being concentrated at one single

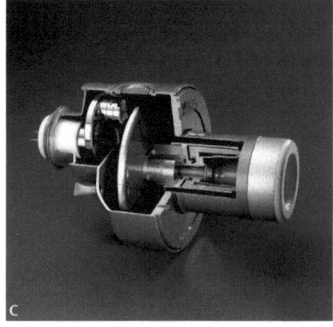
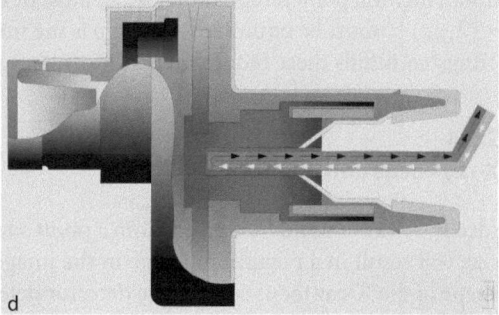

Fig. 2.5a–d. X-ray tubes through the years **a** 1978, **b** 1980, **c** 2002, and **d** schematic illustration of a modern X-ray tube with a rotating anode disk (courtesy of Philips Medical Systems)

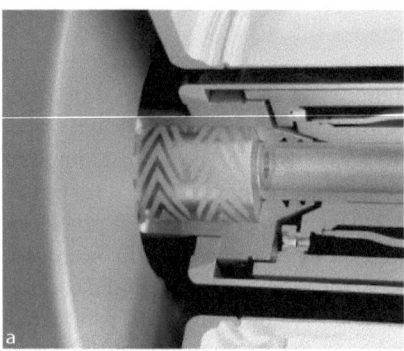

Fig. 2.6. a X-ray anode with **b** spiral-groove bearing (courtesy of Philips Medical Systems)

point. This line focus principle was invented in 1919, and in 1929 the idea of a rotating anode was realized (Otendal 2006). Figure 2.5 shows different types of CT X-ray tubes with rotating anode disks that have been used down the years. In Fig. 2.5a a tube from 1978 is shown. It was integrated into single-slice CT with a scan time of 3 s per slice. The heat capacity was 2–3 10^6 HU. In 1980, an X-ray tube with a heat capacity of 4–6 10^6 HU came onto the market (Fig. 2.5b). This tube was ready to be integrated into spiral-CT systems. In Fig. 2.5c a modern tube that is used today is shown. It has a heat capacity of 5–8 10^6 HU and has been integrated into recent multi-slice CT systems. A significant part of the heat (about 30%) is conducted via the bearing of the rotating anode as schematically drawn in Fig. 2.5d. The rest of the heat is transferred via radiation to the housing of the X-ray tube. The rotation frequency is very high, causing mechanical parts of the tube to be subjected to g-forces of up to 40 g. A liquid metal-filled spiral-groove bearing as shown in Fig. 2.6 allows very high continuous power compared with conventional ball bearings. Often, a heat exchanger is placed on the rotating acquisition disk to cool the anode.

For the design of an optimal X-ray tube anode it has been found that a material with high efficiency, i.e., a large Z, high thermal conductivity, λ_t, and a maximum melting point temperature, ϑ_{max}, must be chosen. For rotating anodes $Z\vartheta_{max} \times (\lambda_t \rho c)^{1/2}$ must be optimized, where ρ is the mass density and c the heat capacity. Tungsten fulfills these requirements (Morneburg 1995).

2.2.4
X-ray Focus and Beam Quality

Ideally, X-rays should be created from a point source, because an increase in source size will result in a penumbra fringe in the image of any object point. The size and shape of the X-ray focus seen by the detector determines the quality of the resulting image. The effective target area, called the optical focus, depends on the orientation of the anode surface that is angulated with respect to the electron beam. The projection of the focus shape onto the detector must be minimized to obtain a sharp image. However, this surface angle increases the tube power limit, because it allows the heat to be deposited across a relatively large spot while the apparent spot size at the detector will be smaller by a factor of the sine of the anode angle (Mudry 2000).

Figure 2.7 schematically illustrates that the image quality is degraded by a large focus diameter, due to significant partial shadow areas of each object point. The mathematical expression that measures the image quality is called the *modulation transfer function* (MTF). Mathematical measures of image quality will be discussed in detail in ▶ Chap. 9. A large angle between the incident electron beam and the anode surface normal is generally not desired because there is a certain probability that the electrons are elastically reflected from the surface and thus do not contribute to X-ray generation. The probability of this backscattering effect increases as the atomic number of the anode material increases and as the angle between the surface normal and the anode rotation axis decreases.

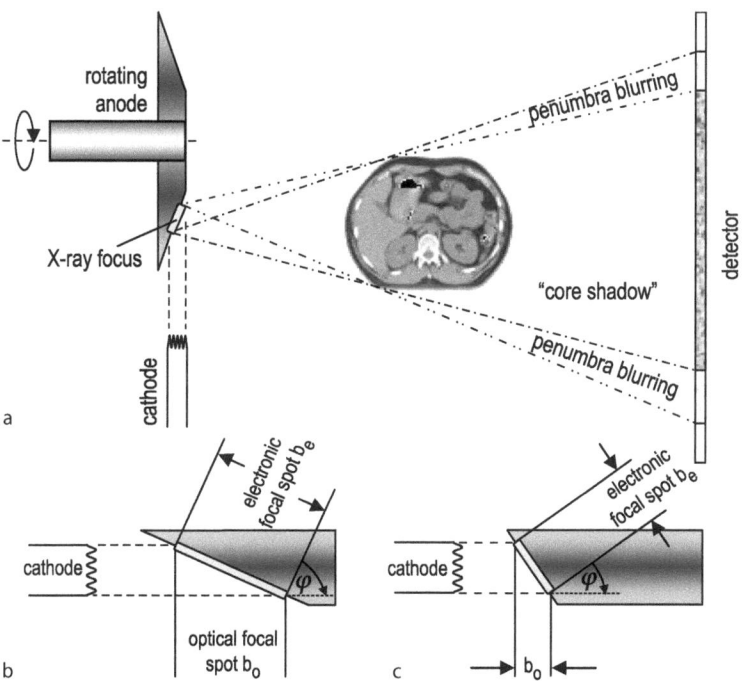

Fig. 2.7a–c. The size of the optical focal point is determined by the orientation of the anode surface normal relative to the electron beam. The larger the optical focal point, the more blurred the image becomes, since all points of the object are surrounded by a penumbra fringe

Typically, the X-ray focus diameter[13] for diagnostic tubes is found to be between 0.3 mm and 2 mm. The penetration depth of electrons into the anode depends on the kinetic energy and the anode material and can be found up to 30 µm. The radiologist may have the option of choosing between two focus sizes. This is technically realized by focusing the electron beam onto two target points with different anode orientations. Figure 2.8 illustrates that the spatial resolution of the detector can also be doubled by switching between two foci during data acquisition. This concept is called a flying focus.

The basic assumption of the imaging principles of fluoroscopy and of CT reconstruction is that the object to be scanned is illuminated homogeneously. Therefore, another factor that must be considered regarding image quality is the directional characteristic of the X-ray. Slight deviations from a homogeneous beam profile can be compensated for by specially formed filters mounted on the X-ray tube and by detector calibration. The next paragraphs give a brief explanation of the physical mechanism that leads to inhomogeneous object illumination.

Generally, two effects are responsible for a directional characteristic of the effective X-ray beam. As discussed above, *bremsstrahlung* is produced by dipole radiation that shows the specific *Hertzian* antenna characteristic. Fast electrons accelerated with $U_a = 120\,kV$ have a velocity of roughly 70% of the speed of light, meaning that relativistic effects emerge. In the case of dipole radiation, the relativistic model reveals an antenna characteristic that points into the same direction as the electron velocity. This relativistic effect grows with increasing kinetic energy of the electrons. In Fig. 2.9a the antenna characteristic for a relativistic dipole is shown. However, with respect to the total radiation characteristic of the X-ray tube, this effect is less important because, as mentioned above, *bremsstrahlung* is produced in a multiple deceleration cascade. Since the direction of the electron changes with each deceleration step, in practice, the radiation is homogeneous over 4π. For this reason, X-ray

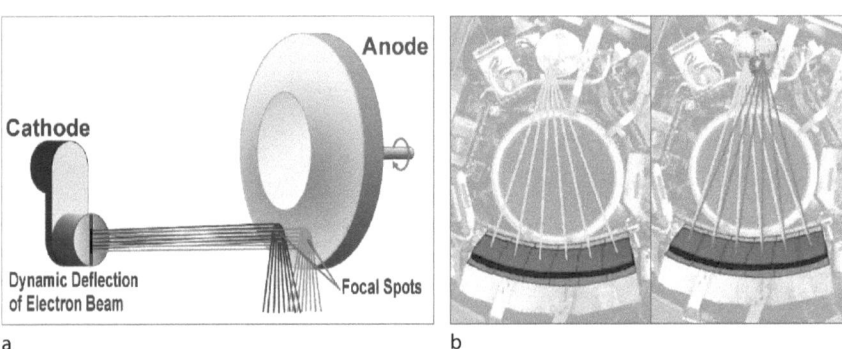

a   b

Fig. 2.8a,b. With a flying focus the spatial resolution of the sampling unit can be doubled

---

[13] The diameter of the X-ray focus depends on the diagnostic application (see for example Huda and Slone [1995]).

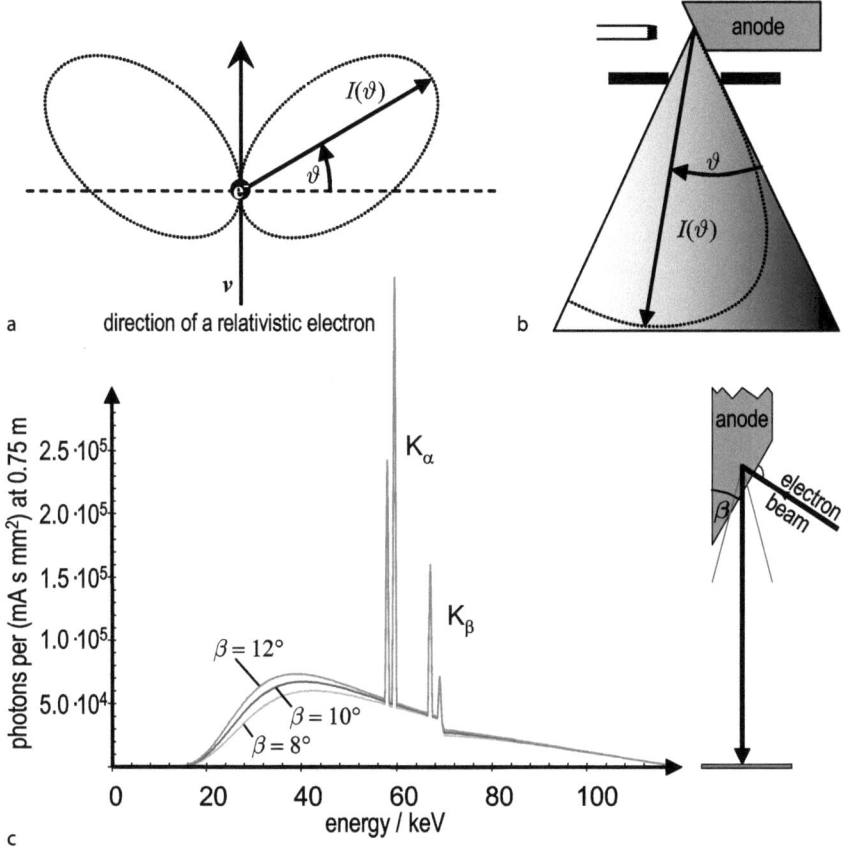

Fig. 2.9. a Radiation directionality of a Hertz dipole producing *bremsstrahlung* during relativistic electron deceleration. **b** *Heel* effect: Intensity reduction of X-ray due to self absorption at the anode surface. **c** X-ray spectra of a tungsten anode simulated for different anode angles at $U_a = 120\,\mathrm{kV}$ and 2 mm Al filtering (courtesy of B. David, Philips Research Labs)

tubes have a protective housing with a window for the useful beam; otherwise the 4π characteristic would cause the patient and the radiologist to be exposed to an unnecessary dosage.

More important than the relativistic effects are the surface effects of the anode. X-ray beams leaving the anode tangential to the anode surface are reduced in intensity when arriving at the detector. This intensity reduction is due to the self-absorption of photons by the anode, caused by the microscopic roughness of the anode surface. In Fig. 2.9b the characteristic of X-ray intensity versus radiation direction is schematically illustrated. The X-ray intensity decreases gradually with a reduction in the angle, ϑ, between beam and anode surface. This effect, which is called the *Heel* effect, grows during the lifetime of any X-ray tube, because the roughness of the anode surface increases (see Fig. 2.10) due to erosion by electron bombard-

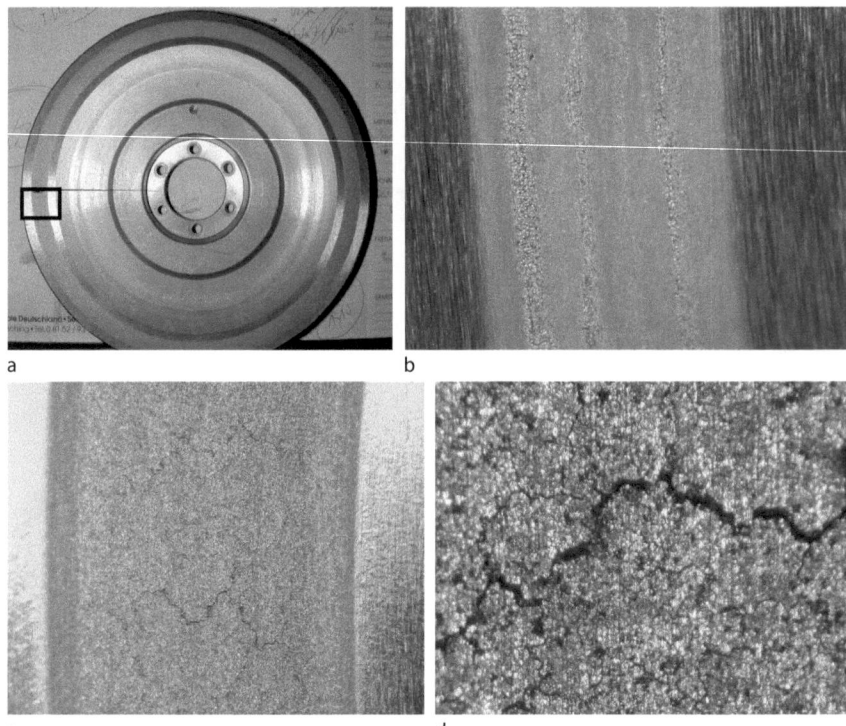

Fig. 2.10. a Appearance of a used anode disk after 200,000 scans. **b–d** Different magnifications of the focus lines. The surface shows 1- to 3-mm-deep cracks and a loss of grains (courtesy of Philips Medical Systems)

ment along the focus line (Heinzerling 1998). Figure 2.9c shows the resulting X-ray spectra simulated for a tungsten anode at an acceleration voltage of $U_a = 120\,\text{kV}$ and 2 mm Al beam filtering. It can be seen that the X-ray intensity decreases with the anode angle, β, especially for lower energy X-rays, due to the self absorption process.

However, in modern CT systems the cone of the utilized X-rays is becoming larger and larger. Therefore, it is important to know the radiation characteristics of the anode.

2.2.5
Beam Filtering

It should be noted at this point, that the image quality of CT specifically suffers from the fact that attenuation is a complicated function of the wavelength. The details of the functional behavior are given in the next section. Generally, one has to bear in mind that a low-energy X-ray, i.e., radiation with a larger wavelength, is more strongly attenuated when passing through matter than high-energy X-ray[14]. As a consequence, the center of the polychromatic X-ray is shifted to higher energies or harder radiation respectively. This is the origin of what is called beam hardening, which produces artifacts in the reconstructed images, because today it is standard to

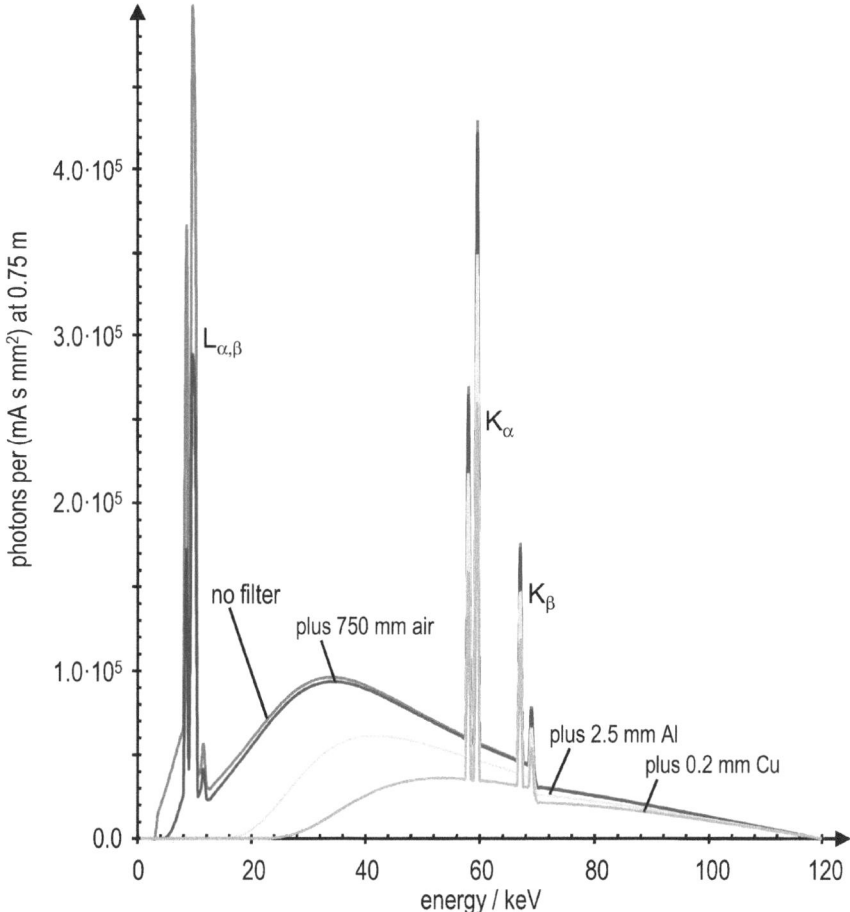

Fig. 2.11. Beam hardening of an X-ray spectrum produced by a tungsten anode (anode angle 10°, acceleration voltage $U_a = 120$ kV) due to filtering by a flat, 2.5-mm source side-mounted aluminum filter and a 0.2-mm copper filter respectively. The amount of intensity reduction depends on the wavelength. The intensity of the high-energy *bremsstrahlung* of the copper-filtered spectrum is even higher than the intensities of the characteristic K_α and K_β lines (courtesy of B. David, Philips Research Labs)

---

[14] Low energy quanta are generally undesired in X-ray imaging. They increase the dose to the patient, but do not contribute to imaging, because they are almost totally absorbed by the human body.

consider the X-ray to be monochromatic within the mathematical reconstruction process.

Artifacts due to beam hardening will be discussed in ▸ Sect. 9.6.2. In Fig. 2.11 the X-ray spectrum of a tungsten anode is shown. The beam hardening is demonstrated by a source side aluminum and copper filter respectively. Generally, a flat metal filter measuring a few millimeters is mounted to the X-ray tube. The filtering of the useful beam reduces the number of X-ray quanta while increasing the average energy of the radiation. This pre-hardening of the radiation reduces beam-hardening artifacts during image reconstruction and the dose to which the patient is exposed.

2.2.6
Special Tube Designs

Although much effort has been made to increase the heat capacity of X-ray tube anodes, the power of these tubes is still limited. Since there is a desire for higher power tubes that are also of lower weight, specially designed X-ray devices have recently been developed. One design abandons the solid-state principle of the anode. In these tubes a liquid metal jet is subjected to fast electrons. The main idea is very simple. Liquid metal, eutectics of SnPb, GaInSn, or PbBiInSn, turbulently streaming through a tube close to the cathode, is heated at the focal spot. While the heated

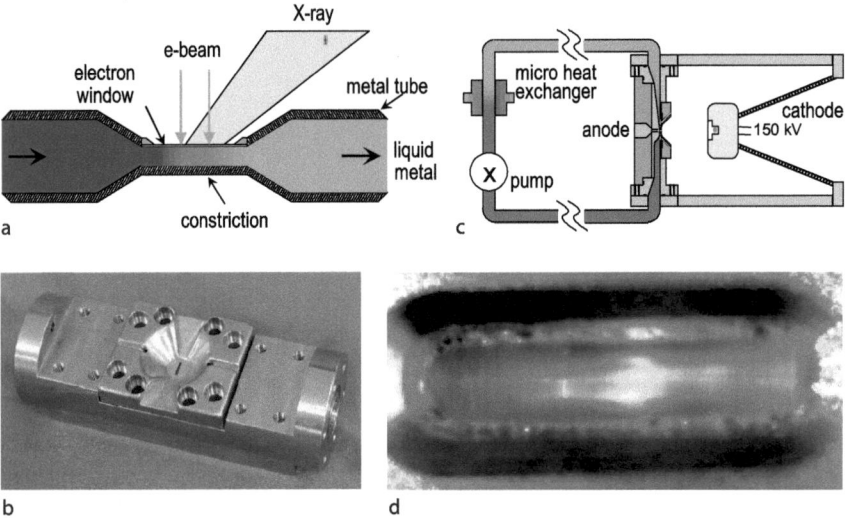

Fig. 2.12. a Basic principle of the liquid metal X-ray tube. **b** The anode module consists of a constricted fluid channel incorporating an exchangeable electron-beam window module. **c** The anode module itself is connected to a fluid circuit containing a displacement pump and a water-cooled cross-flow heat exchanger. **d** Infrared imaging of the focal spot temperature (courtesy of Philips Research, Hamburg [David et al. 2003, 2004])

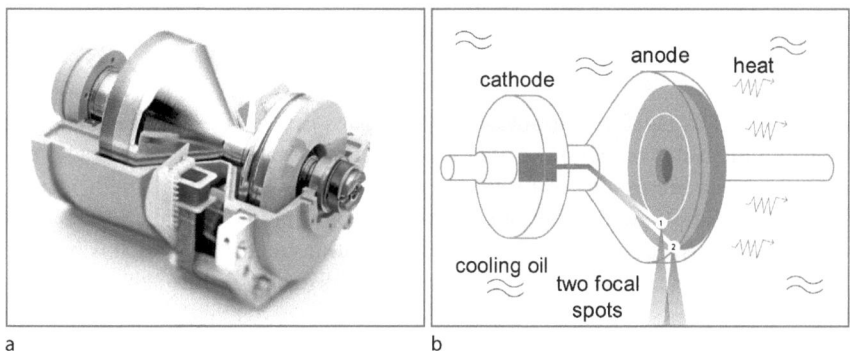

a   b

Fig. 2.13a,b. Straton design: Directly cooled Straton^{TM} X-ray tube with alternating focal spots (courtesy of Siemens Medical Solution)

material is transported through the tubing, cold metal enters the focal spot area. A schematic diagram of what is known as the LIMAX tube (Liquid Metal Anode X-ray) is shown in Fig. 2.12. The liquid metal is cooled effectively by circulation through a heat exchanger. The liquid metal is separated from the vacuum by a diamond, tungsten or molybdenum window of several microns in thickness (David et al. 2003, 2004). In comparison to stationary anode X-ray tubes, the LIMAX design has shown a significant improvement in its ability to be continuously loaded. However, in the current state of development, the peak power does not reach values, to the order of 150 kVA, that are required for the latest CT generation.

A different strategy for direct tube cooling is implemented by the Straton X-ray tube (see Fig. 2.13a,b). The direct anode cooling, realized by completely embedding the rotating housing in cooling oil, eliminates the need for heat storage capacity and therefore permits a smaller tube size. Thanks to its low weight, this kind of X-ray device has been integrated into the current state-of-the-art dual source generation of CT scanners.

2.3
Photon–Matter Interaction

The X-ray is known to have a very high, material-dependent capability of matter penetration. However, the number of photons, i.e., the radiation intensity, decreases exponentially while running through an object along the incident direction. This attenuation is due to absorption and scattering. The reason for an exponential reduction in photon number is that each photon is removed individually from the incident beam by an interaction. In this chapter, the most important physical mechanisms of photon–matter interaction (i.e., Rayleigh scattering, Compton scattering, photoelectric absorption, and pair production), as well as their mathematical models, will be explained briefly.

2.3.1
Lambert–Beer's Law

As illustrated in Fig. 2.14, all physical mechanisms that lead to the attenuation of radiation intensity (reduction of photons) measured by a detector behind a homogeneous object are usually to be subsumed by a single attenuation coefficient, μ. Within this simple model the total attenuation of a monochromatic X-ray beam can be calculated in the following way. The radiation intensity – which is proportional to the number of photons – after passing a distance $\Delta\eta$ through an object is determined by

$$I(\eta + \Delta\eta) = I(\eta) - \mu(\eta)I(\eta)\Delta\eta . \tag{2.18}$$

By simple reordering of (2.18) the quotient

$$\frac{I(\eta + \Delta\eta) - I(\eta)}{\Delta\eta} = -\mu(\eta)I(\eta) \tag{2.19}$$

can be obtained. Taking the limit of (2.19) leads to the differential quotient

$$\lim_{\Delta\eta \to 0} \frac{I(\eta + \Delta\eta) - I(\eta)}{\Delta\eta} = \frac{dI}{d\eta} = -\mu(\eta)I(\eta) . \tag{2.20}$$

In a first step, the medium is assumed to be homogeneous, i.e., in (2.20) the object can be described by a single attenuation constant $\mu(\eta) \equiv \mu$ along the entire length of penetration. This leads to an ordinary linear and homogeneous, first-order differential equation with constant coefficients. The solution is obtained by separation of variables. Consider the right-hand side of (2.20), which can be separated to

$$\frac{dI}{I(\eta)} = -\mu \, d\eta . \tag{2.21}$$

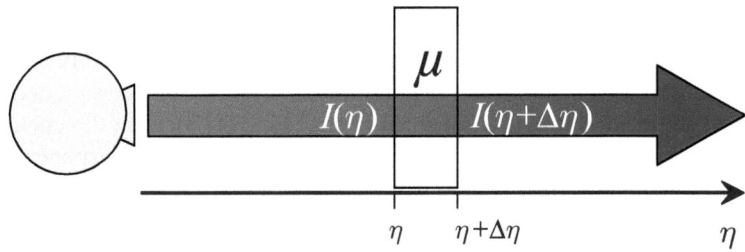

Fig. 2.14. Mathematical model of monochromatic X-ray attenuation. The photons are running through an object of thickness $\Delta\eta$ with a constant attenuation coefficient, μ. Equal parts of the same absorbing medium attenuate equal fractions of the radiation

Integration of both sides of (2.21)

$$\int \frac{dI}{I(\eta)} = -\mu \int d\eta \tag{2.22}$$

gives

$$\ln |I| = -\mu \eta + C . \tag{2.23}$$

Due to the physical fact that the intensity is defined as a positive quantity, the absolute sign, $|\cdot|$, is obsolete so that subsequent exponentiation leads to

$$I(\eta) = e^{-\mu\eta+C} . \tag{2.24}$$

With the initial condition $I(0) = I_0$, the special solution of the differential equation (2.20)

$$I(\eta) = I_0 e^{-\mu\eta} , \tag{2.25}$$

is obtained, known as *Beer's* law of attenuation[15]. The linear attenuation coefficient, μ, of (2.25) is an additive combination of a scatter coefficient, μ_s, and an absorption coefficient, α, i.e.,

$$\mu = \mu_s + \alpha . \tag{2.26}$$

The attenuation coefficient is measured in units of $[\mu] = \text{m}^{-1}$. *Beer's* law holds for pencil-beam geometry only, where the scattered radiation is completely removed from the main beam. With a wide beam, much of the scattered radiation carries on with a forward component to its direction (Barrett and Swindell 1981). Generally, the linear attenuation coefficient is given by

$$\mu = \frac{\rho N_A}{A} \cdot \sigma_{tot} = n \cdot \sigma_{tot} , \tag{2.27}$$

where ρ is again the density, A is the atomic weight of the material and N_A is the Avogadro constant. That means the attenuation coefficient is given by the number of target atoms per unit volume, n, times the total photon atomic cross-section, σ_{tot}, for either scattering or absorption.

By introducing the absorber density, ρ, the mass attenuation coefficient

$$\kappa = \mu/\rho \tag{2.28}$$

can be defined. The mass attenuation coefficient is measured in units of $[\kappa] = \text{m}^2/\text{kg}$. This coefficient is useful for estimating the mass of a material required to attenuate the primary beam in the design of X-ray shielding.

---

[15] This observation holds for monoenergetic X-ray. In the case of polychromatic X-ray one must integrate over all energies.

2.3.2
Mechanisms of Attenuation

In the following paragraphs the principal individual competing physical processes modeled by (2.25) will be explained briefly. Generally, photon interaction can result in a change in incoming photon energy and/or photon number and/or traveling direction. In Fig. 2.15, the single processes are summarized schematically.

2.3.2.1
Rayleigh or Thomson Scattering

Rayleigh[16] or Thomson[17] scattering is an elastic scattering event that can be observed if the diameter of the scattering nucleus is small compared with the wavelength of the incident radiation. The incident and the scattered X-ray have equal wavelength, but the directions of the scattered rays are different from those of the incident beam. Consequently, there is no energy transfer. The cross-section can be derived by the classical model in which the electric field of the incoming beam drives strongly bound electrons of an atom up and down. This makes the electrons radiate – again, as in the case of the emergence of *bremsstrahlung* discussed in ▸ Sect. 2.2.2, due to their acceleration – during oscillation, creating a dipole. The Thomson cross-section (Feynman 1966) is given by

$$\sigma_{\text{Thomson}} = \frac{8\pi r_e^2}{3} \frac{\omega^4}{\left(\omega^2 - \omega_0^2\right)^2} \tag{2.29}$$

where

$$r_e = \frac{1}{4\pi\varepsilon_0} \frac{e^2}{m_e c^2} \tag{2.30}$$

is the classical electron radius and, ε_0 is the permittivity of free space[18].

Principally, at low energies ($\omega < \omega_0$, with ω_0 being the natural frequency of the bounded atom electrons) elastic scattering is an important part of attenuation because the probability of scattering is proportional to the 4th power of the radiation frequency[19]. However, at very high frequencies $\omega \gg \omega_0$, the Thomson cross-section becomes constant and competing processes become dominant[20]. Thomson, or Rayleigh, scattering is a coherent process by which photons interact with atomic bound electrons, leaving the target atoms neither excited nor ionized. Obviously,

---

[16] The scattering process is called Rayleigh scattering if the momentum transferred by the photon is small compared with the momentum of the electron in the atom. In this case one must sum the amplitudes of the X-rays scattered by all electrons in the atom (see Sergé [1965]).

[17] The scattering processes is often called Thomson scattering if it is a photon–nucleus interaction.

[18] Permittivity of free space: $\varepsilon_0 = 8.854 \cdot 10^{-12}$ F/m.

[19] This is the reason why the sky appears blue.

[20] Therefore, coherent scattering is of less interest in classical CT. However, it has been shown recently that coherent scattering can be used for bone characterization.

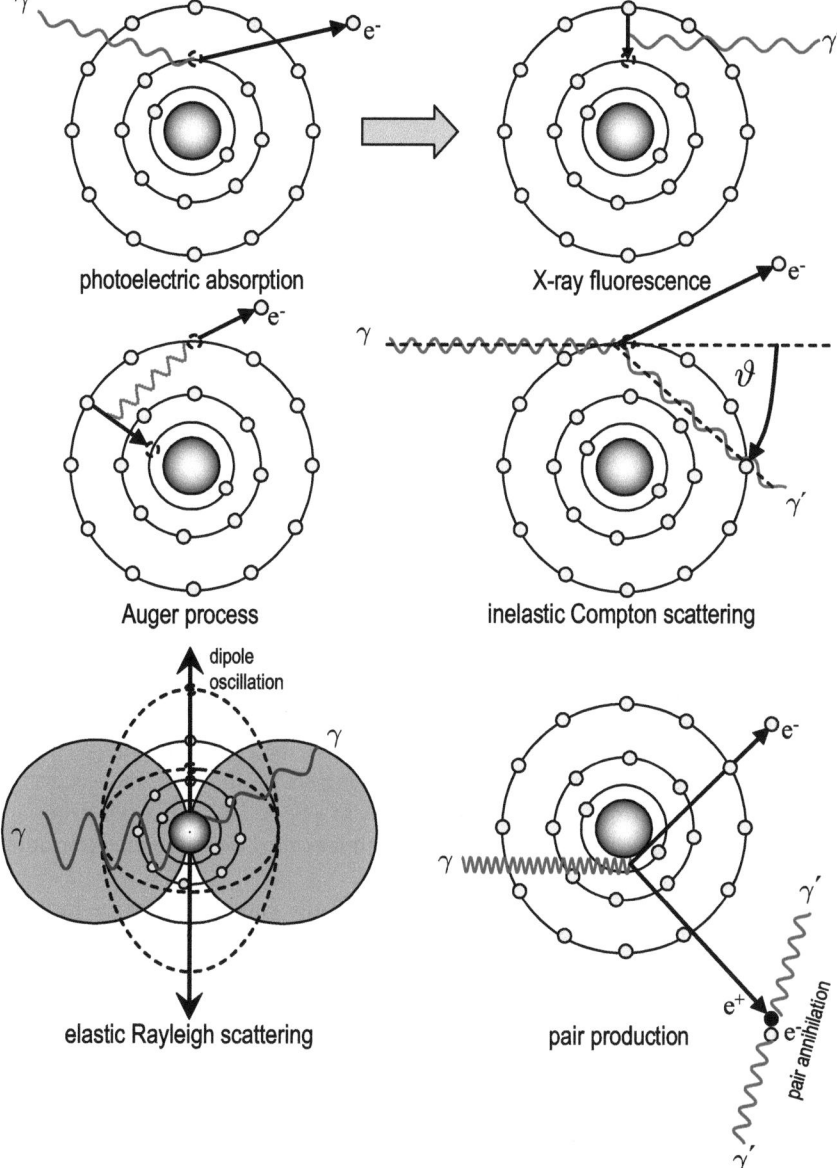

Fig. 2.15. Principles of photon–matter interaction. For the Rayleigh process the dipole antenna characteristic is illustrated. For pair production the successive process of pair annihilation is illustrated as well

the model does not take into account the quantum aspects of light and, therefore, for energies being considered here, this classical scattering model yields results in disagreement to what is found by experiments.

There are three further mechanisms of X-ray attenuation that take the quantum mechanical interaction principles into account. These are either pure photon absorption processes (photoelectric absorption and pair production) or a mixture of scattering and photo energy absorption (Compton scattering).

2.3.2.2
Photoelectric Absorption[21]

The entire energy of an X-ray photon, $h\nu$, can be absorbed by an atom, if the binding energies of atomic electrons are smaller than $h\nu$. The interacting electron is raised to a state of the continuous spectrum, i.e., the electron of a lower shell is kicked off the atom and travels through the material as a free photoelectron. The energy balance of this ionizing process is given by

$$h\nu \ (+ \ Atom) \ \rightarrow \ E_{ion}(Atom^+) \ + \ E_{kin}(e^-). \tag{2.31}$$

(2.31) means that the electron leaves the atom with a kinetic energy equal to the difference between the quantum energy of the incident photon and the binding energy of the electron. This energy difference is removed from the primary beam and transferred locally to the lattice in the form of heat. The vacancy left by the electron that was kicked out is filled by electrons from outer shells or, in the case of solids, by electrons from the band. As a result of recombination, characteristic X-ray fluorescence lines can be measured. If the radiation energy of the successive recombination process for the electron vacancy described by (2.31) is large enough to kick out another electron in the more outlying shells, the new free electron is called an Auger electron. As mentioned already in ▶ Sect. 2.2.2, Auger electrons are monoenergetic particles. This process is sometimes called radiation-free transition or internal conversion.

The linear absorption coefficient depends on the energy, $h\nu$, of the incident X-ray beam and the atomic number, Z, of the absorber material. It has been demonstrated experimentally that a useful rule of thumb is given by

$$\alpha = k\frac{\rho}{A}\frac{Z^4}{(h\nu)^3}, \tag{2.32}$$

where k is a constant that depends on the shell involved, ρ is the density, and A is the atomic weight of the material. Thus, the absorption is given by

$$\alpha \propto Z^4 \lambda^3. \tag{2.33}$$

This strong Z^4 dependence of the absorption coefficient is utilized within the choice of radio-opaque contrast media based on for example iodine (I, $Z = 53$) or barium (Ba, $Z = 56$) (Lauenberger and Lauenberger 1999). In Fig. 2.16 the attenuation

---

[21] For recovering the nature of the photoelectric absorption in 1905, *Albert Einstein* received the Nobel Prize for physics in 1921.

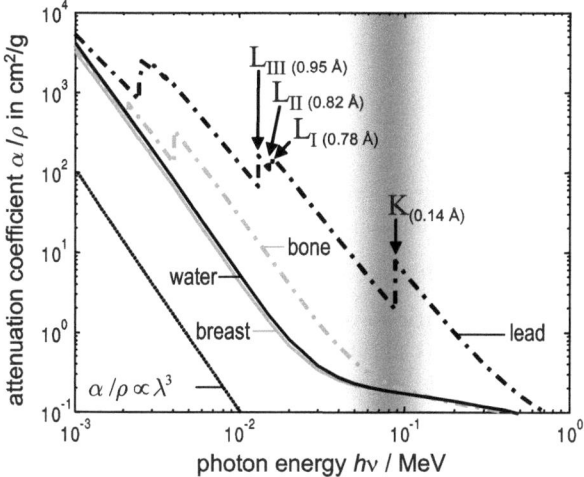

Fig. 2.16. Mass attenuation coefficient (α/ρ measured in units of cm^2/g) for lead and water as well as for the bio-tissues bone and soft tissue given versus the incident radiation energy. For absorption processes above the K-shell the curve shows a fine structure (compiled with data from the web database XCOM [Berger et al. 2004])

coefficients are given for lead (Pb, $Z = 82$), water (H_2O), and two bio-materials. It can be seen that (2.33) is a piecewise function for lead. However, for water and bio-materials the functional dependence of the attenuation coefficient in the diagnostically relevant photon-energy interval ($50\,keV \leq h\nu \leq 140\,keV$) becomes more complicated.

Due to (2.33), lead (Pb, $Z = 82$) shields X-ray radiation approximately 1,580 times better than aluminum (Al, $Z = 13$). It must be noted that the mass attenuation coefficient, κ, defined by (2.28) shows a $\kappa \propto Z^3$ dependency only, because the atomic weight, A, is proportional to the atomic number, Z (Demtröder 2000).

If the mass attenuation coefficient is considered as a function of the incident radiation wavelength, a complicated structure is revealed. In Fig. 2.16 it can be seen that the absorption curve for lead exhibits characteristic absorption edges.

To understand the absorption edges at λ_k ($k = 1, 2, \ldots$), excitation energies belonging to the wavelengths must be discussed. For wavelengths $\lambda \leq \lambda_k$, the energy of the X-ray beam is high enough to excite an electron from an inner shell, i.e., to lift up the electron to the energy continuum, or ionize the atom. However, for wavelengths $\lambda > \lambda_k$, the photon energy is too small to lift the electron to an excited, quantum mechanically allowed state. Therefore, at this very edge, the penetrated material suddenly becomes transparent. This principle is repeated when the wavelength is increased for electrons of the L-, M-, ... shell. For any shell above the K-shell a *fine structure*[22] appears in the absorption spectrum (3 for the L-shell, 5 for the M-shell, etc.). If the photon energy is higher than the binding energy of the

---

[22] The *fine structure* occurs due to L-S coupling and, for heavy atoms, jj coupling.

K-shell, more than 80% of the photoelectric absorption takes place in the emission of K-shell electrons.

2.3.2.3
Compton Scattering[23]

An X-ray photon with energy $h\nu$ sometimes collides with a quasi-free[24] electron. The energy balance of this collision process is given by

$$h\nu(+ e^-) \rightarrow E_{kin}(e^-) + h\nu' . \tag{2.34}$$

In contrast to the photoelectric absorption, the X-ray photon loses only a part of its energy during the Compton collision. Thus, the scattered photon is of lower energy when it continues its travel through the matter. The energy level of the scattered photon can be measured under the scatter angle ϑ, via the shift of wavelength

$$\Delta\lambda = \frac{h}{m_e c}(1 - \cos(\vartheta)) . \tag{2.35}$$

The complementary part of the energy is carried by the kinetic energy of the recoil electron that is kicked off the atom. This Compton electron is also called a secondary electron. Both the scattered photon and the Compton electron may have enough energy to undergo further ionizing interactions. (2.35) can be derived by considering the conservation of energy and momentum. It says that in the forward scattering direction no energy is transferred, whereas in the backward scattering direction most of the quantum energy is transferred. However, even at 180° of deflection the scattered X-ray quanta retain at least two-thirds of their initial energy (Bushong 2001). Sometimes Compton backscattering produces ghost images of the detector back board.

Overall, the probability of Compton scattering depends on the electron density n and not on the atomic number of the scattering medium. Since the electron density difference between distinct tissues is small, Compton scattering provides low-contrast information. The total cross-section for Compton scattering can be derived from the Klein–Nishina equation (Leroy and Rancoita 2004)

$$\sigma_{Compton} = 2\pi r_e^2 \left[\left(\frac{1+\mathcal{E}}{\mathcal{E}^2}\right) \left(2\frac{(1+\mathcal{E})}{1+2\mathcal{E}^2} - \frac{\ln(1+2\mathcal{E})}{\mathcal{E}}\right) + \frac{\ln(1+2\mathcal{E})}{2\mathcal{E}} - \frac{1+3\mathcal{E}}{(1+2\mathcal{E})^2} \right], \tag{2.36}$$

where $\mathcal{E} = h\nu/(m_e c^2)$ is the reduced energy of the incoming photon. Substituting (2.36) into (2.27) leads to

$$\mu_{Compton} = n \cdot \sigma_{Compton} , \tag{2.37}$$

the desired (material-independent) energy dependence of the attenuation coefficient due to Compton scattering.

---

[23] For recovering the nature of incoherent scattering *Arthur Holly Compton* received the Nobel Prize for physics in 1927.

[24] For instance a weakly bound valence electron in the outer shell.

2.3.2.4
Pair Production

For photon energies of $h\nu > 1.022\,\mathrm{MeV}$ the creation probability for an electron–positron pair raises continuously inside the Coulomb field of a nucleus or an electron[25]. The energy balance of this pair production process is given by

$$h\nu \to e^- + e^+ + 2E_{\mathrm{kin}} \,, \tag{2.38}$$

if *Einstein's* mass–energy equivalency

$$h\nu = 2m_e c^2 + 2E_{\mathrm{kin}} \tag{2.39}$$

is satisfied.

The positron, e^+, is the anti-particle of the electron, e^-. Therefore, it has the same physical properties with the exception that the electric charge of a positron is positive[26]. In Fig. 2.15 it is schematically shown that, in matter, the positron always meets an electron after a very short traveling length. When an electron and its anti-matter particle, e^+, have a collision, they disintegrate and two γ-rays emerge (each with the energy of $m_e c^2$ if the particles come together at rest). This annihilation process can be described by the following balance:

$$e^- + e^+ \to 2h\nu \,. \tag{2.40}$$

Pair annihilation plays a key role in positron emission tomography (PET). Due to the conservation of linear momentum, called annihilation radiation, the two photons travel apart in approximately[27] opposite directions. Since the entire electron and positron mass is transformed, each photon has an energy of $511\,\mathrm{keV}$. The annihilation radiation is measured by coincidence detection.

Besides the physical mechanisms described above, it is also possible that the photon is scattered or absorbed by the nucleus. The photonuclear cross-section is, however, negligible for the diagnostic energy window, because this interaction contributes only to 5–10% of the total attenuation within a narrow energy interval between a few MeV and a few tens of MeV (Leroy and Rancoita 2004). The total cross-section for pair production for photon energies in the interval

$$\left(m_e c^2\right) \ll h\nu \ll \alpha^{-1} Z^{-1/3} \left(m_e c^2\right)$$

can be approximated (Leroy and Rancoita 2004) via

$$\sigma_{\mathrm{pair\ production}} = \alpha r_e^2 Z^2 \left[\frac{28}{9}\ln(2\mathcal{E}) - \frac{218}{27} + \frac{6.45}{\mathcal{E}}\right] \,, \tag{2.41}$$

---

[25] Pair production in the field of an electron is sometimes called triplet production (Leroy and Rancoita 2004).

[26] Positrons were predicted by P.A.M. Dirac in 1927. C. Anderson proved their existence in 1932.

[27] Due to the conservation of linear momentum.

where α is again the fine-structure constant. Table 2.1 summarizes the photon inter-action principles.

The contributions of the four attenuation processes described in the paragraphs above to the total attenuation of X-ray in matter depends on the wavelength of the incident beam and the material that is penetrated. In Fig. 2.17 (top), the functional behavior of the total attenuation coefficient due to Rayleigh and Compton scattering, photoelectric absorption, and pair production, is given versus the quantum energy of the photons for lead (Pb). The data are obtained from the currently available web database XCOM (Berger et al. 2004).

For high energies, $h\nu > 10\,\text{MeV}$, pair production is the main contribution to the total attenuation. In the mid-energy range of $h\nu \approx 1\,\text{MeV}$, incoherent Compton scattering is the dominant process. However, in the diagnostic energy window for CT[28], $50\,\text{keV} \leq h\nu \leq 140\,\text{keV}$ as indicated in Fig. 2.17 (top), photoelectric absorption governs the overall behavior. For that reason, lead is an excellent X-ray shielding material. The contribution of the undesired Rayleigh and Compton scattering is more than one magnitude lower than the required X-ray photon absorption. Contrary to the behavior of lead, water (H_2O) would not be an appropriate shielding material. In Fig. 2.17 (bottom), the different contributions of attenuation processes are given versus the incident photon energy.

For high energies $h\nu > 30\,\text{MeV}$ pair production is the main contribution to the total attenuation of water. In a broad mid-energy range, $40\,\text{keV} \leq h\nu \leq 30\,\text{MeV}$, incoherent Compton scattering is the dominant process. Within the diagnostic energy window for CT, indicated in Fig. 2.17 (bottom), Compton scattering governs the overall behavior. Even Rayleigh scattering shows a higher probability than the photoelectric absorption. For typical X-ray tube voltages between $100\,\text{kV}$ and $140\,\text{kV}$, the contribution of Compton scattering to the total attenuation is more than two magnitudes higher than the contribution of photoelectric absorption. Since the human body consists mainly of water, this fact has its consequences for dose considerations in diagnostic imaging, because organs adjacent to the scanned region are subjected to radiation by the secondary X-ray photons, $h\nu'$, described by (2.34).

Table 2.1. X-ray photon interaction in matter (Tisson 2006)

	Orbital electrons	Nuclei	Electric field of the atom
Ideal absorption	Photoelectric absorption	Photo disintegration	Pair production
Elastic scattering	Rayleigh scattering	Thomson scattering	Delbrück scattering
Inelastic scattering	Compton scattering	Nuclear resonance scattering	Not observed

---

[28] For an X-ray tube operating at $120\,\text{kV}$ the average photon energy is about $70\,\text{keV}$ (Hsieh 2004).

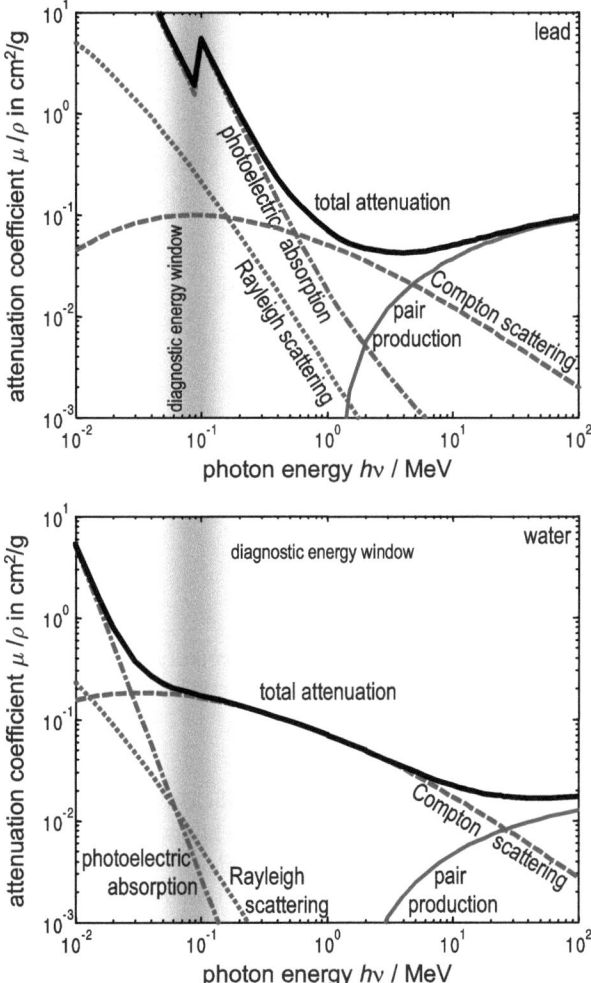

Fig. 2.17. Mass attenuation coefficient, μ/ρ, versus incident photon energy for lead (*top*) and water (*bottom*). For the diagnostic energy window of CT, $E = [50\,\mathrm{keV}\text{–}140\,\mathrm{keV}]$, photoelectric absorption is dominant for lead and Compton scattering is dominant for water. Pair production is possible for quantum energies of $E > 1\,\mathrm{MeV}$ (compiled with data from the web database XCOM [Berger et al. 2004])

Radiation by Compton scattering is a significant contribution to the patient's total dose, as well as to the radiologist, in interventional imaging procedures. Therefore, the radiologist must wear a shielding lead apron. In Fig. 2.18a, it is schematically shown that the patient becomes a source of radiation himself. For that reason, it is important to know the spatial iso-dose lines of a CT system to be obtained by a standard scattering phantom. In Fig. 2.18b,c, so-called scatter diagrams of a Philips TOMOSCAN CT system are shown in top and side view.

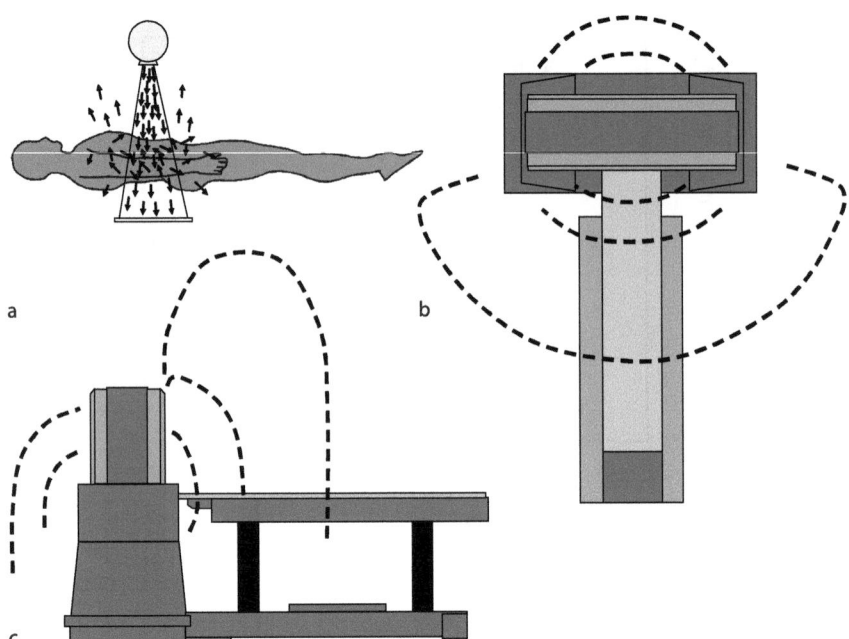

Fig. 2.18. a Due to Compton scattering the patient becomes a source of radiation and organs outside the region of examination are subjected to X-rays as well. Radiation protection must be designed on the basis of scatter diagrams, **b** and **c**, for a particular CT system

The discussion of the mass attenuation coefficient with respect to issues of radiation protection shows how important it is to know the physical mechanisms of X-ray photon attenuation. The physical principles of attenuation are different for diverse biomaterials. In Figs. 2.19–2.21 the functional behavior of the attenuation coefficients of some important biomaterials are given, again versus the quantum energy of the photons. The data are obtained from the web database XCOM (Berger et al. 2004).

For reference, and without a detailed discussion, the coefficients are presented for the elements iron (Fe, $Z = 26$), titanium (Ti, $Z = 22$), platinum (Pt, $Z = 78$), and gold (Au, $Z = 79$), as well as for the mixtures and compounds amalgam[29] (12% Sn, $Z = 50$; 51% Hg, $Z = 80$; 37% Ag, $Z = 47$) and compact bone (6% H, $Z = 1$; 28% C, $Z = 6$; 3% N, $Z = 7$; 41% O, $Z = 8$; 7% P, $Z = 15$; 15% Ca, $Z = 20$). The element mixtures of soft tissue and blood (after ICRP[30]) behave very similarly to water, as shown in Fig. 2.17 (bottom).

As a consequence of the variability of the attenuation as a function of the material characteristics presented in Figs. 2.19–2.21, the mathematical problem of CT is not as simple as expressed in (2.25), because in an anatomical slice, the attenuation

---

[29] Main composition $8Ag_3Sn + 33Hg \rightarrow 8Ag_3Hg_4 + Sn_8Hg$.
[30] ICRP – International Commission on Radiological Protection.

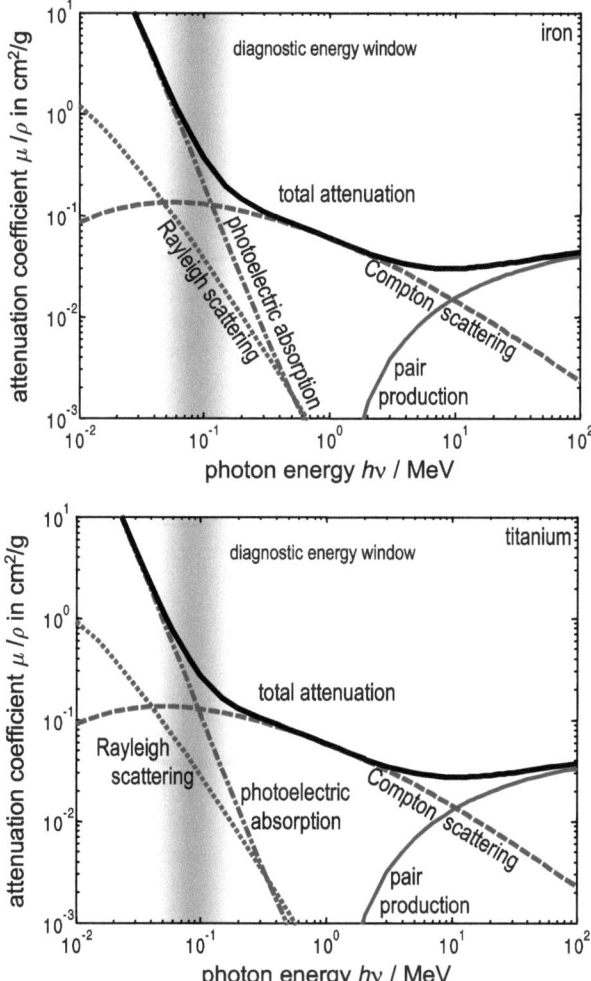

Fig. 2.19. Mass attenuation coefficient, μ/ρ, versus incident photon energy for iron (*top*) and titanium (*bottom*). For the diagnostic energy window of CT, $E = [50\,\text{keV}–140\,\text{keV}]$, photoelectric absorption is dominant in the low-energy part and Compton scattering is dominant in the high-energy part of the diagnostic window (compiled with data from the web database XCOM [Berger et al. 2004])

coefficient is a function of the spatial coordinates and photon energy. However, on the other hand, the spatial distribution of the attenuation coefficient μ is the main focus of diagnostic interest. The gray-value differences in CT images represent distinct material composites and densities of the organs. The normal CT image shall be seen as standard, as radiologists learn, and any variation of that standard is potentially pathologic.

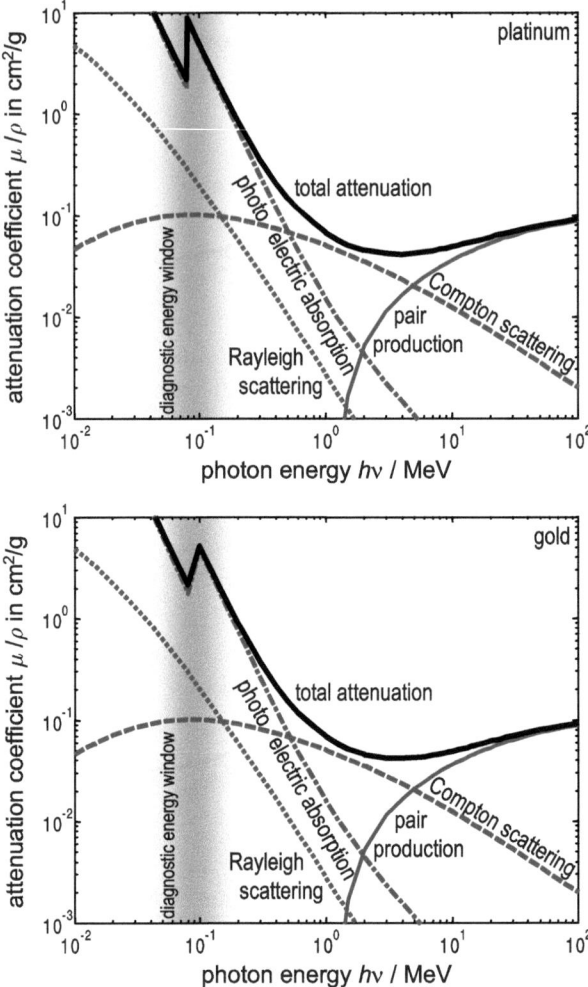

Fig. 2.20. Mass attenuation coefficient, μ/ρ, versus incident photon energy for platinum (*top*) and gold (*bottom*). For the diagnostic energy window of CT, $E = [50\,\mathrm{keV}\text{–}140\,\mathrm{keV}]$, photo-electric absorption is dominant (compiled with data from the web database XCOM [Berger et al. 2004])

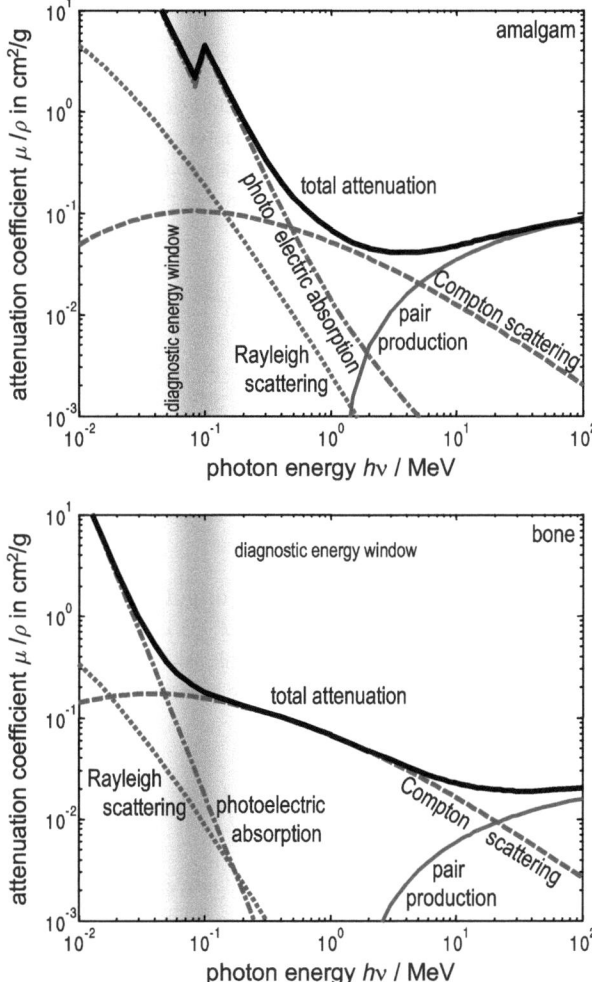

Fig. 2.21. Mass attenuation coefficient, μ/ρ, versus incident photon energy for amalgam (*top*: 12% Sn, $Z = 50$; 51% Hg, $Z = 80$; 37% Ag, $Z = 47$) and for compact bone (*bottom*: 6% H, $Z = 1$; 28% C, $Z = 6$; 3% N, $Z = 7$; 41% O, $Z = 8$; 7% P, $Z = 15$; 15% Ca, $Z = 20$). For the diagnostic energy window of CT, $E = [50\,\text{keV}–140\,\text{keV}]$, photoelectric absorption is dominant for amalgam and Compton scattering is dominant for bone (compiled with data from the web database XCOM [Berger et al. 2004])

2.4
Problems with Lambert–Beer's Law

From a mathematical point of view, the complicated dependency of the attenuation coefficient on the property of the penetrated material means that the differential equation (2.20) cannot be fully integrated as shown in (2.22). In the case of spatially varying attenuation, $\mu(\eta)$, the solution for the intensity after a running length, s, is given by

$$I(s) = I(0)\, e^{-\int_0^s \mu(\eta)\,d\eta}\,, \tag{2.42}$$

summing up the unknown coefficients along the X-ray beam within the exponent. A second point to mention is the energy dependence of the attenuation values, which is not modeled by (2.25). Hence, (2.42) must be extended to

$$I(s) = \int_0^{E_{\max}} I_0(E)\, e^{-\int_0^s \mu(E,\eta)\,d\eta}\,dE\,. \tag{2.43}$$

The difference between (2.42) and (2.43) is the origin of what is called the beam-hardening artifact. However, in practice, only (2.42) is used. Therefore, this is the basic model that will be used throughout this book. As one is especially interested in the spatial variation of the attenuation, i.e., in the inversion of the integral operation in (2.42), the projection integral

$$p(s) = -\ln\left(\frac{I(s)}{I(0)}\right) = \int_0^s \mu(\eta)\,d\eta \tag{2.44}$$

is defined for convenience. It is essentially the negative logarithm of the ratio of the incoming and outgoing number of photons.

In the technical realization of CT, one is restricted to a certain discretization of the reconstructed images. When an X-ray beam runs through the continuous matter, such a partitioning of the penetrated material cannot be physically measured. Nevertheless, one can model the matter to be discrete and calculate the total attenuation for this case. If the attenuation coefficient varies discontinuously, as schematically indicated in Fig. 2.22, the intensity reduction while running through the tissue can be estimated without solving a differential equation.

In this model, the attenuation

$$I_{i+1} = I_i - \mu_{i+1}I_i\Delta\eta = I_i(1 - \mu_{i+1}\Delta\eta) \tag{2.45}$$

of the intensity I_i resulting in the intensity I_{i+1} when running through the tissue, $i + 1$, of thickness $\Delta\eta$, and attenuation coefficient, μ_{i+1}, can be obtained. Consequently, the total attenuation is a series of products

$$I_n = I_0(1 - \mu_1\Delta\eta)(1 - \mu_2\Delta\eta)(1 - \mu_3\Delta\eta)\cdots(1 - \mu_i\Delta\eta)\cdots(1 - \mu_n\Delta\eta)\,. \tag{2.46}$$

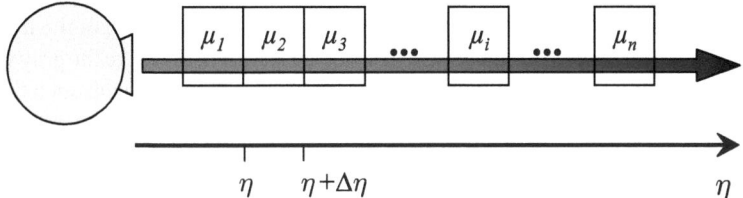

Fig. 2.22. Attenuation of an X-ray beam while running through a tissue that is modeled by a sequence of discontinuous partitions

If the discretization of the tissue is very fine, i.e., $\Delta \eta$ is chosen to be small, the factor terms in parentheses can be interpreted as the Taylor expansion of the exponential function. With

$$e^{-\delta} \approx 1 - \delta \qquad (2.47)$$

for small δ, the approximation

$$I_n \approx I_0 \, e^{-\mu_1 \Delta \eta} \, e^{-\mu_2 \Delta \eta} \, e^{-\mu_3 \Delta \eta} \cdots e^{-\mu_i \Delta \eta} \cdots e^{-\mu_n \Delta \eta} \qquad (2.48)$$

can be obtained. By taking the limit $\delta \to 0$ the exact equation

$$I_n = I_0 \, e^{-\sum\limits_{i=1}^{n} \mu_i \Delta \eta} \to I_0 \, e^{-\int \mu(\eta)\, d\eta} \qquad (2.49)$$

can be obtained. This view on the modeling of X-ray attenuation is closely related to a statistical view of photon number reduction by photon–matter interaction that will be introduced at the end of this chapter. To summarize the facts above, it can be said that the attenuation of X-ray through matter is well understood and thus,

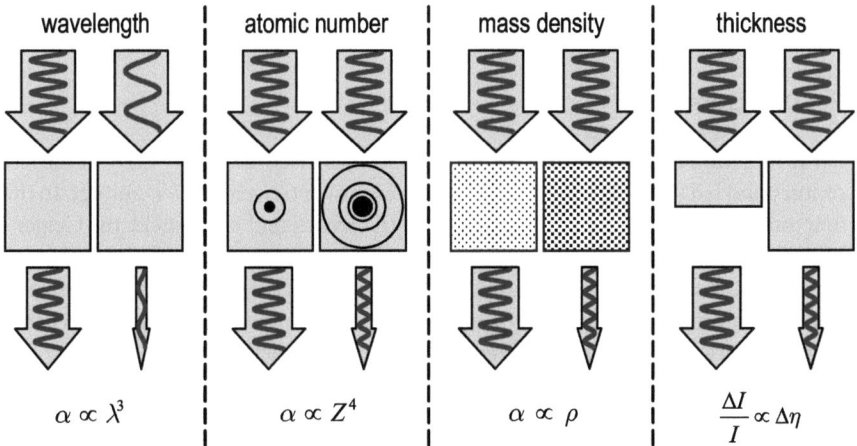

Fig. 2.23. Causes of attenuation: Wavelength of the incident beam, atomic number, mass density, and thickness of the medium (after Laubenberger and Laubenberger [1999])

the gray values of the CT images are a direct physical representation of the material properties. This is clearly an advantage in comparison to MRI, where the gray values can be tuned by various parameters of complicated scanning protocols such that the direct relation to physical properties is lost. In CT, high values of the attenuation coefficient, μ, are due to a high density or high atomic number of the medium. In Fig. 2.23, the relationship between the attenuation and the wavelength of the incident beam, as well as the atomic number, density, and thickness of the material, is schematically summarized.

In ▸ Chaps. 5–7 the reconstruction of non-superimposed object slices will be derived and discussed in detail. Within the calculation of the spatial distribution of the attenuation coefficient, $\mu(x, y)$, the projection modeled by (2.42) and (2.49) must be reversed in some sense.

2.5
X-ray Detection

Since the main interaction principles between X-ray photons and matter have been explained in the section above, no new effects have to be introduced to explain the interaction between X-ray radiation and the detector material. Consequently, X-ray quanta are not measured directly, but are detected via their interaction products (for example, emitted photoelectrons). The overall detection efficiency is primarily determined by the geometric efficiency (also called the *fill factor*) and the quantum efficiency (also called the *capture efficiency*) (Cunningham 2000). The geometric efficiency refers to the X-ray sensitive area of the detector as a fraction of the total exposed area, and the quantum efficiency refers to the fraction of incident quanta that are absorbed and contribute to the signal. The overall detection efficiency is the product of the geometric and the quantum efficiency.

2.5.1
Gas Detectors

X-ray radiation is able to ionize gases. This fact was discovered very early in the last century and led to the development of the well-known *Geiger–Müller* counter. In the first tomographic experiments carried out by Cormack and Hounsfield, the Geiger–Müller counter was used as a detector in pencil-beam geometry. In the early days of clinical CT, gas-based detector arrays were also manufactured for what are known as third-generation scanners in fan-beam geometry. Even today, some scanners using high-pressure xenon are in use. The photoelectric interaction,

$$h\nu + Xe \rightarrow Xe^+ + e^-, \tag{2.50}$$

describes the first part of the detection process chain. Xenon ions and electrons are attracted by high voltage to a cathode and an anode respectively. A series of alternating cathode and anode pairs forms the detector array. Figure 2.24 schematically shows the reaction chain starting with the photoelectric ionization to the recombination of the free charged particles at the electrode surfaces. The current produced by recombination is a measure of the X-ray intensity entering the detector.

A weak quantum efficiency (a low probability for the photoelectric absorption) can be compensated for by a high pressure and by tall ionization chambers. Another advantage of tall chambers is improved directional selectivity of the detector element. Since the ionization probability is proportional to the travel length of X-ray quanta inside an element (one cathode–anode block), detection of X-ray quanta with an oblique entrance will be suppressed. In this way, a tall detector element has a built-in collimation. However, septa between detector elements are insensitive regions that will decrease the geometric efficiency of the detector.

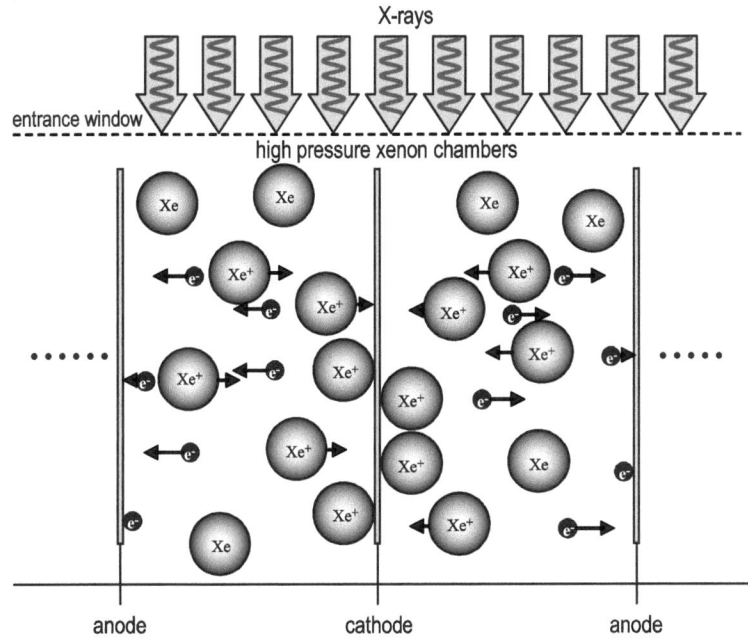

Fig. 2.24. Schematic cut-out of two adjacent ionizing chambers of a xenon high-pressure detector array. Since the chambers are communicating, all detector elements have the same Xe pressure and, therefore, the same sensitivity

2.5.2
Solid-State Scintillator Detectors

Today, almost all modern CT systems are equipped with scintillator detectors. Such a detector essentially consists of two main components: A scintillator medium and a photon detector. In a first step, the short-wave X-ray radiation entering the detector is converted into long-wave radiation (light) inside the scintillation media. Typical scintillator materials used are cesium iodide (CsI), bismuth germanate (BGO) or cadmium tungstate (CdWO$_4$). The choice of material is critical and depends on the desired quantum efficiency for the conversion from X-ray to light and on the time constant for the conversion process, which determines the *afterglow* of the detector.

For a very fast fluorescence decay, i.e., a very small time constant, as required by modern sub-second scanners, ceramic materials made of rare earth oxides like gadolinium oxysulphide (Gd$_2$O$_2$S) are used (Kalender 2000). To assess the quality of a detector material with respect to the desired behavior of high quantum efficiency, the mass attenuation coefficient must be known. In Fig. 2.25 it can be seen

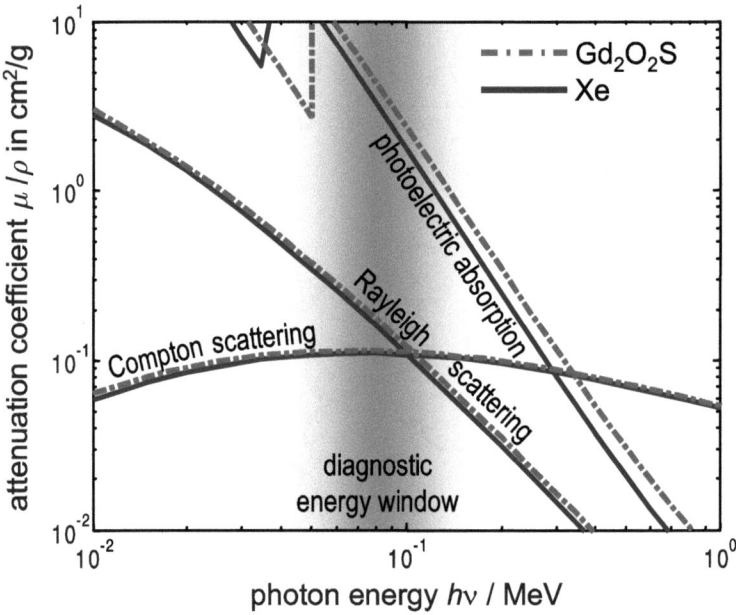

Fig. 2.25. Mass attenuation coefficient for the main photon–matter interaction principles in detector materials. Photoelectric absorption is more than one magnitude higher than scattering processes in xenon and the Gd$_2$O$_2$S-ceramic. Due to the fact that, inside the relevant diagnostic energy window, the mass attenuation coefficient due to photoelectric absorption for Gd$_2$O$_2$S is higher than for xenon, the quantum efficiency of the ceramic detector material is superior. This holds especially true, since the density of the solid-state detector material is obviously higher than the density of xenon

that the Gd_2O_2S ceramic has a higher probability of energy conversion via photo-electric absorption than xenon. Due to the fact that the mass density ρ of the gas xenon is three magnitudes smaller than the mass density of the solid-state detector material, the effective absorption of quanta inside the solid-state detector, and therefore its quantum efficiency, is significantly higher. This can be only partially compensated by long Xe chambers and a high gas pressure.

In Fig. 2.26 the components of a scintillator detector unit are shown schematically. On the right-hand side of Fig. 2.26 a detector unit of the Philips Tomoscan EG is shown. One unit consists of 16 detector elements. X-rays that have been scattered may undergo deflection through a small angle and finally reach the detector. This is an undesired detection because it reduces the image contrast. To suppress the measurement of scattered X-ray quanta, collimator lamella are attached to each element. Such an anti-scatter collimator grid is directed toward the X-ray focus to filter out photons not traveling in the line of sight between X-ray source and detector. Without an anti-scatter grid, the image quality would be significantly reduced.

However, there is an obvious disadvantage of the anti-scatter grid. To block an oblique entrance of scattered X-ray photons effectively, a minimum lamella thickness of 0.1 mm is required. In practice, the detector elements have a total geometric efficiency of about 50–80% (Kalender 2003). This decreased fill factor leads to an undesired reduction in spatial detector resolution.

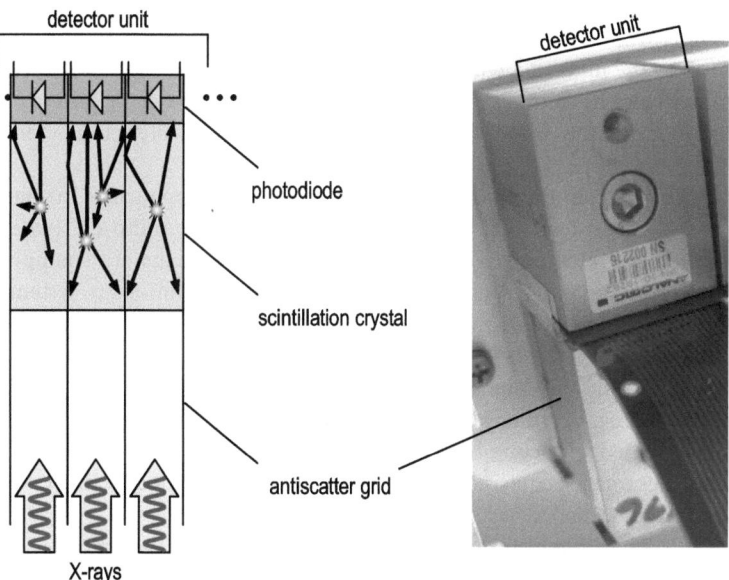

Fig. 2.26. Schematic drawing and photograph of a detector unit. Single detection channels are separated by thin anti-scatter lamella. The scintillator medium converts the X-ray quanta to light, which subsequently is detected by a photodiode mounted on the crystal

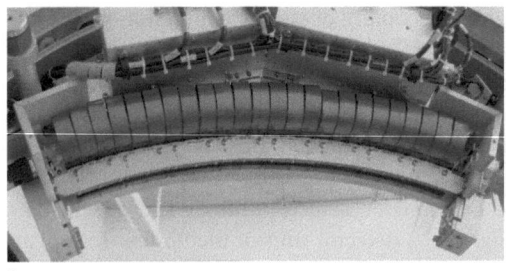

a   b

Fig. 2.27a–b. Uncovered detector array of the compact Philips Tomoscan EG scanner. **a** Ribbon cable connections of 24 detector units, each consisting of 16 channels, above the antiscatter grid. **b** Arrangement of the detector units in a circle segment

On the rotating sampling unit, the series of detector elements is arranged in a circle segment with the X-ray source at its center. In Fig. 2.27 a detector array with 24 units, each with 16 channels, is shown.

2.5.3
Solid-State Flat-Panel Detectors

Crystal- and ceramic-based solid-state detectors described in the previous section can be extended to multi-row or multi-slice detector systems. The key feature of such multi-row arrays is a maximum effective X-ray-sensitive area. This feature is quantified by the fill factor, which is explained below. Xenon high-pressure gas detector array systems cannot be easily extended to flat area detector systems. Therefore, all modern multi-slice CT systems are based on solid-state scintillation detectors.

In Fig. 2.28, a cone-beam detector system is drawn schematically. In contrast to detector systems in technical CT systems, for example, in micro-CT (where flat-panel detectors are employed) almost all clinical CT systems are equipped with cylindrical detector units. As illustrated in Fig. 2.28 the multi-array system forms a cylinder barrel with the X-ray source as its center. If the number of rows of such multi-slice detector systems is chosen to be very high, the orientation of the respective X-ray fan inside the cone beam, relative to the axial slice, becomes significant, and the requirements of image reconstruction increase. However, modern CT systems are equipped with cone-beam sampling units due to improved spatial resolution and faster image acquisition. Image reconstruction in cone-beam systems will be discussed in ▸ Chap. 8.

In Fig. 2.29, different technical configurations of multi-slice detector systems used in Siemens (left) and General Electric (right) multi-slice CT (MSCT) systems are shown. The Siemens detector is a UFC^{TM} adaptive array detector. This detector unit is partitioned into either 16 slices measuring 0.75 mm thick or 16 slices measuring 1.5 mm thick. The flexibility is achieved by the interconnection of pairs of

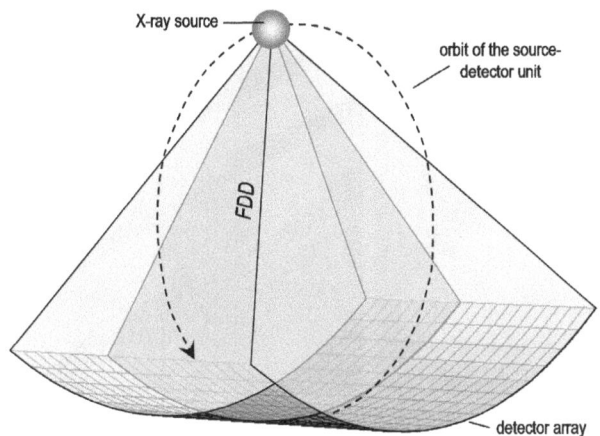

Fig. 2.28. Cone-beam illumination of a cylindrical detector system. The multi-array system forms a cylinder barrel with the X-ray focus as its center. The source-detector sampling unit, with diameter *FDD* (*Focus-Detector-Distance*), is rotating on a circle (*dashed line*) with the iso-center inside the field of measurement

detector segments. The General Electric detector type, shown here, has spatially equidistant rows of detector elements. However, an electronic combination of the rows results in flexible slice thickness as well.

Without reconstruction mathematics that are adapted to the cone-beam situation, image artifacts arise for the higher X-ray fan angulations, even in a 16-row detector system. This leads to clinically unacceptable image quality. In any case, the reconstruction algorithms used for the cone-beam geometry need to be revised for practical implementation. Therefore, it is a consequent step to leave the multi-row detector arrays and move in the direction of real flat-panel detector systems. These systems have recently become commercially available. However, they were not originally designed for use in CT systems, but rather to compete with established radiography systems using film cassettes, computed radiography cassettes, and image intensifiers.

In Fig. 2.30, the typical composition of such a digital flat-panel X-ray detector is shown. In Fig. 2.30a, the key aspect of the construction is illustrated. Each sensor element consists of a photodiode and a thin-film transistor (TFT). Both are made of amorphous silicon on a single glass substrate. The pixel[31] matrix is coated with an X-ray-sensitive layer. Multi-chip modules are used as read-out electronics at the edge of the detector field. The X-ray-sensitive coating is a cesium iodide layer used, for example, in General Electric CT systems. The basis is the single glass substrate with a silicon matrix of 2,048 × 2,048 sensors, each 200 μm in size. The monolithic structuring is done with thin-film technology such that a composition of a set of medium-sized sub-panels, which have undesired dead zones at the in-

---

[31] *Pixel* is an acronym composed of *picture* and *element*, using the abbreviation '*pix*' for picture.

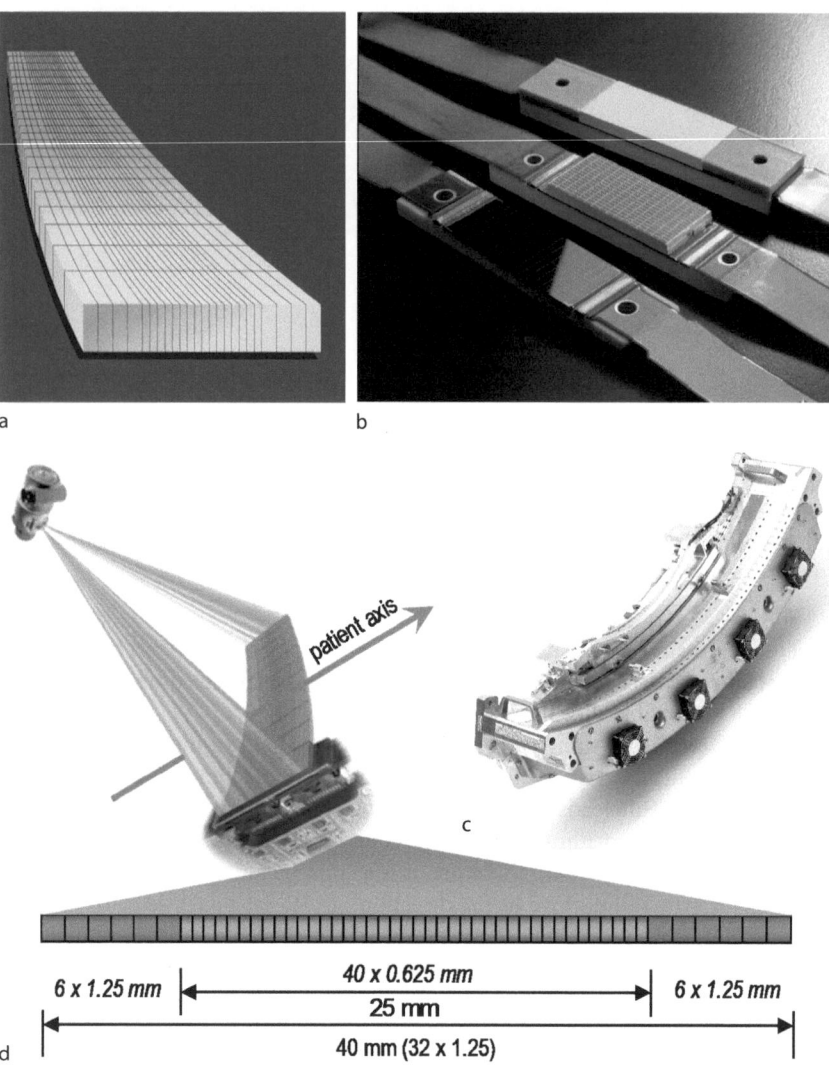

a   b

c

6 x 1.25 mm   40 x 0.625 mm   6 x 1.25 mm
25 mm

d   40 mm (32 x 1.25)

Fig. 2.29a–d. Detector parts used for multi-slice detector technology – courtesy of Siemens Medical Solutions (**a, c**), General Electric Medical Systems (**b**), and Philips Medical Systems (**d**)

terfaces, potentially producing imaging artifacts, is not required. The final coating with cesium iodide (CsI) is the required scintillator layer of the detector. The CsI layer is applied directly onto the pixel matrix by a physical deposition process.

The production technique is known from semiconductor production. Physical and chemical processing steps, i.e., the combination of photolithography and further etching phases, are applied to produce these finely structured detector

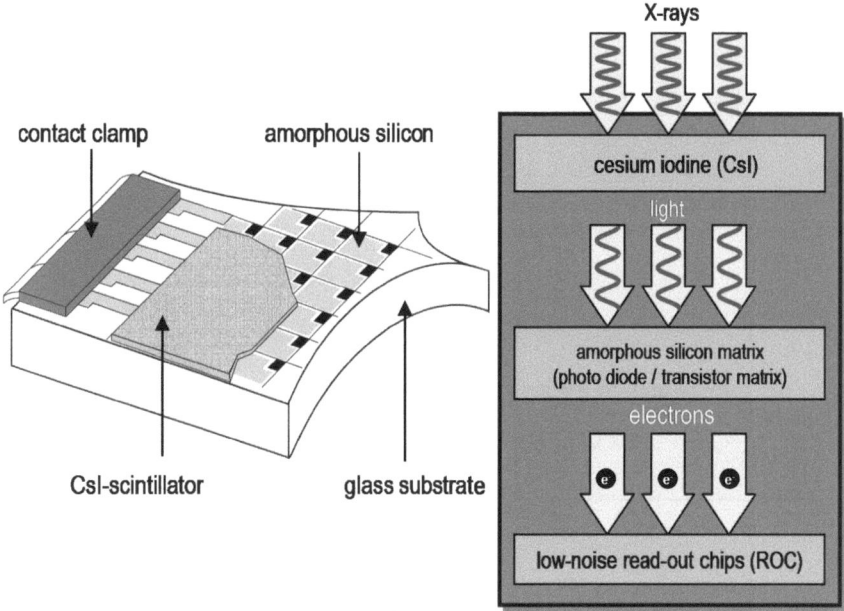

Fig. 2.30. *Left*: Composition of a digital flat-panel X-ray detector. *Right*: Schematic illustration of the signal conversion chain inside the detector (courtesy of General Electric Medical Systems: Brunst [2002])

elements. This leads to an X-ray-sensitive detector field with a desired high fill factor,

$$f = \frac{X\text{-}ray \ sensitive \ area \ of \ the \ detector}{total \ area \ of \ the \ detector} \ , \tag{2.51}$$

that cannot be produced by the simple combination of the detector types described in ▸ Sects. 2.5.1 and 2.5.2.

In Fig. 2.30, the signal processing flow inside the X-ray detector is illustrated schematically. X-ray quanta entering the detector are converted into visible light in the upper CsI scintillator layer. The light photons are guided to the photodiodes of the next processing layer. The photons are absorbed, thereby producing an electric charge in the photodiodes that is proportional to the intensity of the X-ray radiation. During detector exposure, the electric charge is integrated and stored in the detection element, which acts as capacitor. The actual read-out process is initialized by the thin-film transistor, which switches the charge to the read-out electronics via the data link. There, an amplification and the analog-to-digital conversion is performed on the same chip, resulting in fast operation with low noise. Figure 2.31 illustrates that the flat-panel detector system has a very high fill factor. In Fig. 2.32, a pre-assembly, raw detector with ribbon cable data links as well as a post-assembly, digital flat-panel detector that is ready for application, are shown. Its 4,194,304 pixels are integrated onto an active $41 \times 41 \, cm^2$ area.

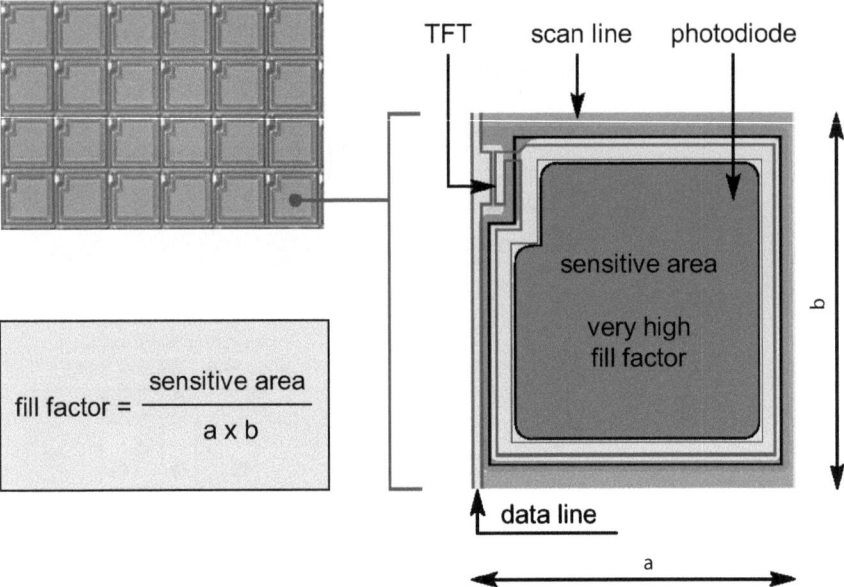

Fig. 2.31. Schematic illustration of the flat detector field and a single detector element of the field. The quality of the detector is influenced by the size of the X-ray-sensitive detector area. Insensitive areas such as splices of space for electronics must be minimized. The measure of this geometric detector quality is the fill factor, i.e., the ratio between the X-ray-sensitive area and the total area of the sensor (courtesy of General Electric Medical Systems: Brunst [2002])

A highly desired system property of the digital scintillation detector system is the linear dynamics over a wide range of illumination. In this way, high- and low-dose applications have the same contrast information, i.e., excellent contrast resolution. Figure 2.33 shows the characteristic response diagrams of the signal dynamics of an analog film system and a digital scintillator system respectively[32]. Film systems are capable of imaging objects with high contrast within a very narrow exposure range only. If the object is over- or underexposed, the contrast of the image can easily be too low. Due to the linear characteristic response curve of the digital system, it is robust against over- and underexposure. However, within the range of low-contrast imaging, flat-panel systems do not achieve the quality of dedicated CT detector systems described in the sections above (Kalender 2003).

In addition to the linear characteristic diagrams of the detector dynamics, another important advantage of the flat-panel detector is its excellent spatial resolution. To optimize the spatial resolution, cesium iodide is evaporated onto the matrix so that direct contact between the scintillation material and the carrying photodiode matrix is established. This manufacturing step is designed so that cesium io-

---

[32] Flat-panel detector systems were developed for use in simple radiography systems to replace film systems or computed radiography cassettes.

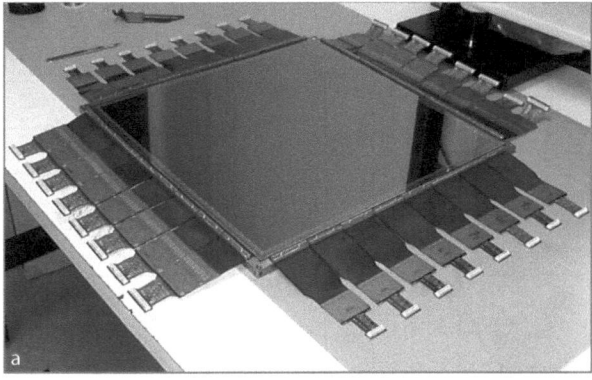

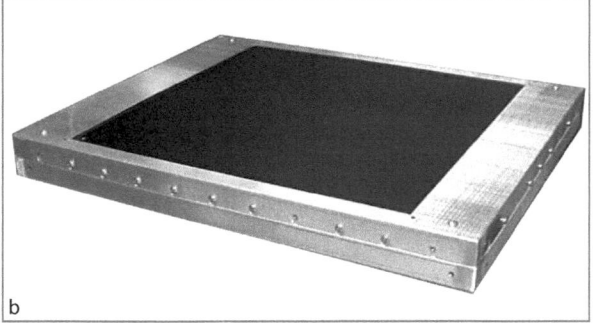

Fig. 2.32. Realization of a 41 cm × 41 cm flat-panel detector integrating 2,048 × 2,048 pixels. **a** Pre-assembly panel with ribbon cable for read-out electronics. **b** Complete flat-panel system ready for use (courtesy of General Electric Medical Systems: Brunst [2002])

dide grows anisotropically, forming needles on the matrix. In Fig. 2.34, the needle structure of CsI is demonstrated in an electron microscope picture. If X-ray quanta are converted into visible light inside the CsI structure[33], the emerging photons are, in all likelihood, traveling along the needles, because they act as a fiber-optical cable. In this way photons are guided directly onto the photodiode or in the opposite direction. The photons that are traveling in the opposite direction face a mirror on the top side of the CsI layer that ensures that eventually almost all photons find their way to the photodiode. This light guidance effect of the CsI fiber structure is the reason for the high quantum efficiency of the digital flat-panel detectors. The X-ray-sensitive CsI coating can be made very thick to obtain a high quantum efficiency and to also suppress broad photon scattering, which would reduce the spatial resolution. The scintillation light is bundled by the CsI fibers onto a small point on the photodiode matrix. However, an isotropic CsI layer must always find a compromise between high quantum efficiency and high spatial resolution.

---

[33] Typically, a cloud of about 1,000 optical photons, each with energy of approximately 2 eV, is generated by an X-ray quantum of several tens of keV.

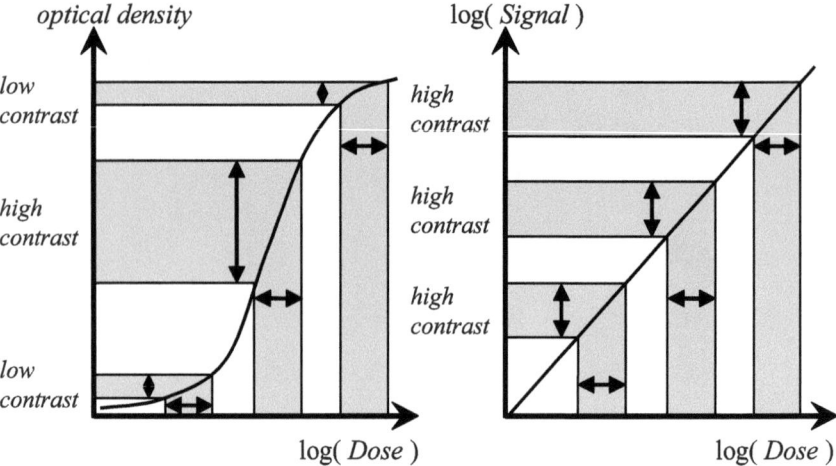

Fig. 2.33. Comparison of dynamic characteristics between a film detector system (*left*) and a digital scintillation detector (*right*). Besides a desired linearity of the digital detector, the dynamic range of the scintillation detector of 300–10,000:1 (compared with 25–100:1 of the film detector) is also superior (Brunst 2002)

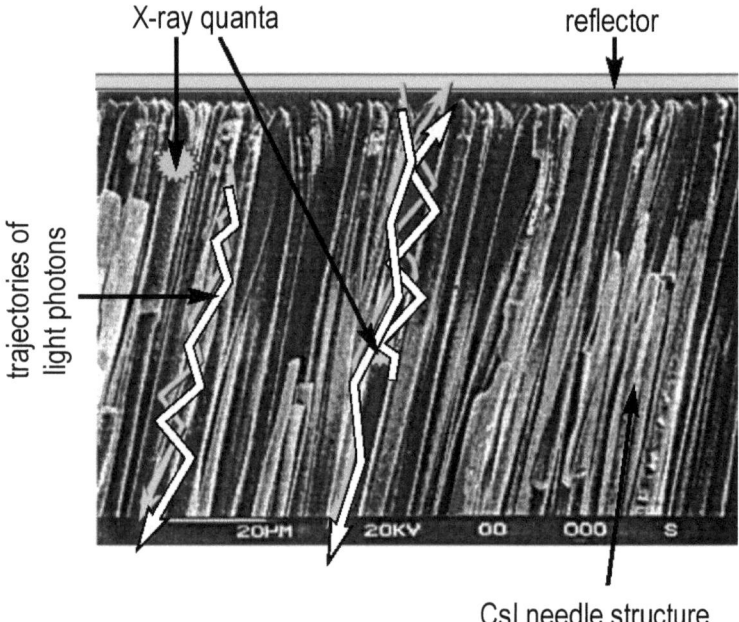

Fig. 2.34. The cesium iodide scintillation layer of a flat-panel detector element (picture taken with an electron microscope, courtesy of General Electric Medical Systems: Brunst [2002])

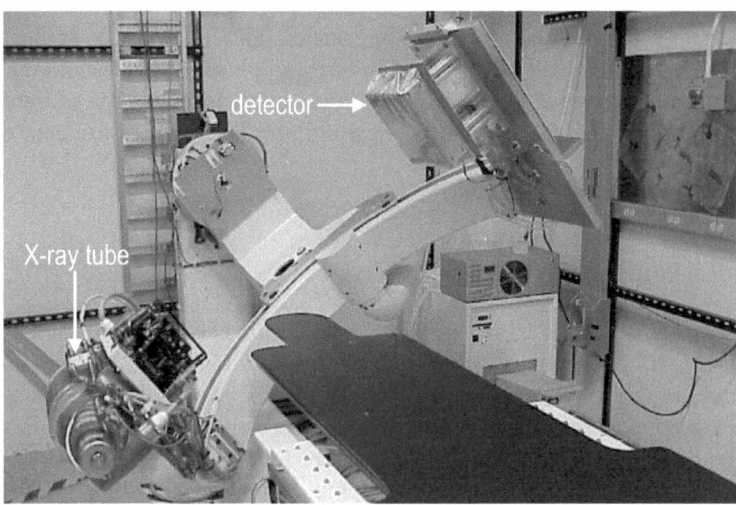

Fig. 2.35. A digital flat-panel detector mounted on an experimental rotating C-arm device for testing cone-beam reconstruction algorithms (courtesy of General Electric Medical Systems)

In Fig. 2.35 a digital flat-panel detector mounted on an experimental C-arm device is shown. The device can be rotated and, therefore, used to demonstrate the feasibility of cone-beam tomographic imaging. C-arm-based tomographic devices are relatively small and inexpensive compared with complete CT systems. On the other hand, they exhibit mechanical instabilities with reduced spatial resolution. However, there is a demand for interventional three-dimensional imaging modalities. This is because surgeons would like to add intra-operative three-dimensional CT information to endoscopic or ultrasound imaging in minimally invasive operations, but typically there is not enough space for a normal CT system in an operating room (Härer et al. 1999). Compared with clinical CT systems, a disadvantage of the flat-panel C-arm-based apparatus is the small diameter (30 cm or less) of the imaging field (Härer 1999).

2.6
X-ray Photon Statistics

X-ray radiation consists of X-ray quanta with specific statistical properties that must be considered in many fields of X-ray signal processing applications. It will be shown later in this book that particular reconstruction methods can be applied in CT that are based on the statistical nature of X-ray quanta. Furthermore, the evaluation of image quality is based upon the physically correct model of noise statistics.

The next sections will focus on the statistical properties of the X-ray source and on those of the X-ray detector. As with many stochastic processes, the statistics of X-ray generation and interaction with matter can be understood in the limit of processes with binomial probability densities. Additionally, Lambert–Beer's law of photon absorption will be discussed in terms of photon statistics and the most important moments of Poisson-distributed random variables will be addressed. In the final sections the central-limit theorem and non-Poisson measurement will be discussed.

2.6.1
Statistical Properties of the X-ray Source

Statistical reconstruction techniques like the maximum likelihood expectation maximization (MLEM) approach, which will be introduced in ▸ Sect. 6.5, are based on the exact knowledge of quantum number statistics. The probability density function is explicitly incorporated in the method. Therefore, a mathematical model of the physical nature of X-ray quanta must first be found.

To derive the statistical photon model, the section begins with a discussion of the X-ray quanta generation statistics inside the focus area on the X-ray tube anode. As mentioned at the beginning of this chapter, the lattice atoms of the anode material are bombarded with fast electrons that are accelerated during their transit from the cathode to the anode. In Fig. 2.36, this situation is illustrated. Let N electrons arrive at the active focus volume in the time window $[0, T]$ and, let each of

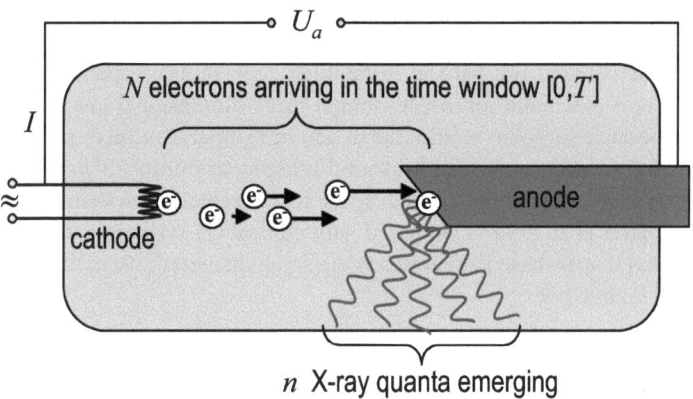

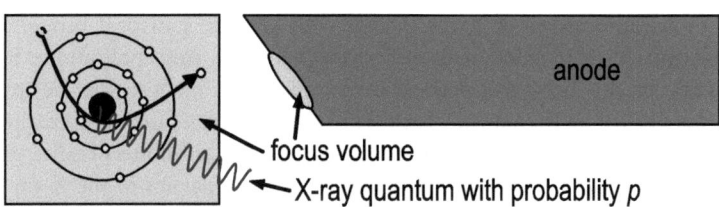

Fig. 2.36. X-ray quanta generation as a statistical counting process

the N electrons have a probability, p (with $0 \leq p \leq 1$), of interacting in the same time interval with one of the target atoms in such a way that an X-ray quantum emerges.

It is also assumed that each collision process between fast electrons and target atoms in the lattice is statistically independent of all other collisions inside the focus volume. Then, the probability, P, that the random variable, $\mathcal{N}$, i.e., the number of emerging X-ray quanta, is assigned exactly to the number n is

$$P(\mathcal{N} = n) = \binom{N}{n} p^n (1 - p)^{N-n}. \tag{2.52}$$

This means that the number of X-ray quanta, $\mathcal{N}$, is a binomially distributed random variable with $\mathcal{N}$ out of $\{0, \ldots N\}$. (2.52) is the well-known Bernoulli distribution, which measures the probability that n successful outcomes occur in a sequence of N independent trials. The term

$$\binom{N}{n} = \frac{N!}{n!(N-n)!} \tag{2.53}$$

is the binomial coefficient and can be visualized by *Pascal*'s triangle[34], illustrated in Fig. 2.37. The term is sometimes also read as "N chooses n".

It is well known that the statistics of independent coin-throwing experiments can be modeled by the Bernoulli distribution. The probability of heads and tails is $p = 0.5$. The Bernoulli distribution is the simplest discrete distribution, and it is a building block for other, more complicated discrete distributions.

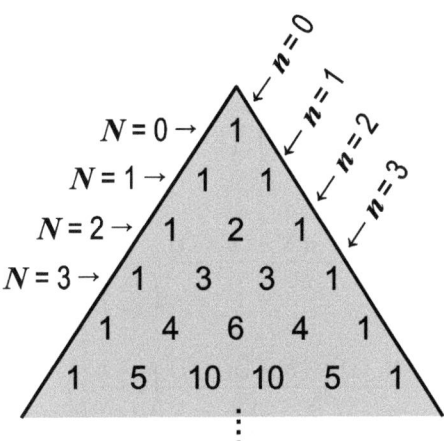

Fig. 2.37. Pascal's triangle

---

[34] Actually, 500 years before *Pascal* studied the number triangle, the Chinese mathematician *Yang Hui* described the triangle first. Therefore, in China it is known as *Yang Hui*'s triangle.

Obviously, the generation of X-ray quanta can be statistically interpreted as a counting process. (2.52) is a normalized discrete probability distribution, where

$$\sum_{n=0}^{N} P(\mathcal{N} = n) = \sum_{n=0}^{N} \binom{N}{n} p^n (1-p)^{N-n} = (p + (1-p))^N = 1. \tag{2.54}$$

However, a realistic magnitude of the number of fast electrons arriving at the focus volume in the time interval, $[0, T]$, can be approximated via the anode current. Let $I = 160\,\mathrm{mA}$, a typical current used for medical CT. Then, the number of electrons arriving in 1 ms is

$$N = T\frac{I}{e} = 10^{-3}\,\mathrm{s} \cdot \frac{0.16\,\mathrm{A}}{1.602 \cdot 10^{-19}\,\mathrm{As}} \approx 10^{15}. \tag{2.55}$$

On the other hand, the probability, p, of a successful collision is very small. It has been mentioned above that this small probability, or efficiency, of the X-ray quanta creation process is the reason for the considerable heat problem of X-ray tubes.

In practice, the number of arriving electrons and the probability of successful collision can be expressed in the limits $N \to \infty$ and $p \to 0$ respectively. However, since an X-ray tube produces X-ray quanta, there must be a positive number of photons in the limit of $Np \to n^*$, where $E[\mathcal{N}] = n^* > 0$ is the *expectation value* of the number of X-ray quanta. This leads to a certain limit of the Bernoulli distribution that will be derived below.

Assume that

$$Np = n^* \tag{2.56}$$

and substitute (2.56) into (2.52), resulting in

$$
\begin{aligned}
P(\mathcal{N} = n) &= \binom{N}{n} p^n (1-p)^{N-n} = \binom{N}{n}\left(\frac{n^*}{N}\right)^n \left(1 - \frac{n^*}{N}\right)^{N-n} \\
&= \frac{N!}{n!(N-n)!}\left(\frac{n^*}{N}\right)^n \left(1 - \frac{n^*}{N}\right)^{N-n} \\
&= \frac{N(N-1)\cdots(N-(n-1))}{n!}\; \frac{1}{\left(\frac{N}{n^*}\right)^n \left(1 - \frac{n^*}{N}\right)^n}\left(1 - \frac{n^*}{N}\right)^N \\
&= \frac{N(N-1)\cdots(N-(n-1))}{n!\left(\frac{N}{n^*} - 1\right)^n}\left(1 - \frac{n^*}{N}\right)^N \\
&= \frac{1}{n!}\left(\frac{N^n\left(1 - \frac{1}{N}\right)\cdots\left(1 - \frac{(n-1)}{N}\right)}{N^n\left(\frac{1}{n^*} - \frac{1}{N}\right)^n}\right)\left(1 - \frac{n^*}{N}\right)^N.
\end{aligned}
\tag{2.57}
$$

In the limit $N \to \infty$, the second term in the last row of (2.57) is $(n^*)^n$. The last term is given by

$$\lim_{N \to \infty} \left(1 - \frac{n^*}{N}\right)^N = e^{-n^*}, \qquad (2.58)$$

which can easily be shown, because

$$\lim_{N \to \infty} \left(1 - \frac{n^*}{N}\right)^N = \lim_{N \to \infty} \left(\exp\left(\ln\left(1 - \frac{n^*}{N}\right)^N\right)\right)$$

$$= \lim_{N \to \infty} \left(\exp\left(N \ln\left(1 - \frac{n^*}{N}\right)\right)\right). \qquad (2.59)$$

With the expansion of the logarithm,

$$\ln(1 \pm x) = -\sum_{i=1}^{\infty} \frac{(\mp 1)^i x^i}{i} \quad \text{for } -1 < x < +1, \qquad (2.60)$$

one obtains

$$\lim_{N \to \infty} \left(1 - \frac{n^*}{N}\right)^N = \lim_{N \to \infty} \left(\exp\left(N\left\{-\frac{n^*}{N} - \frac{1}{2}\left(\frac{n^*}{N}\right)^2 - \frac{1}{3}\left(\frac{n^*}{N}\right)^3 \cdots\right\}\right)\right). \qquad (2.61)$$

Factoring out $\left(\frac{n^*}{N}\right)$ leads, finally, to

$$\lim_{N \to \infty} \left(1 - \frac{n^*}{N}\right)^N = \lim_{N \to \infty} \left(\exp\left(-n^* - \frac{1}{2}n^*\left(\frac{n^*}{N}\right) - \frac{1}{3}n^*\left(\frac{n^*}{N}\right)^2 \cdots\right)\right) = e^{-n^*}. \qquad (2.62)$$

By substituting (2.62) into (2.57), one obtains

$$\lim_{N \to \infty} \binom{N}{n} p^n (1-p)^{N-n} = \frac{1}{n!}(n^*)^n e^{-n^*}. \qquad (2.63)$$

For a large number of fast electrons, the probability, P, that the random variable, $\mathcal{N}$, is exactly assigned to n, can be calculated via the Poisson distribution density function

$$P(\mathcal{N} = n) = \frac{(n^*)^n}{n!} e^{-n^*}. \qquad (2.64)$$

The parameter n^* of the Poisson distribution, i.e., the expectation value $E[\mathcal{N}]$ of the number of X-ray quanta, is a measure of the radiation intensity. Thus, X-ray generation is a Poisson process. Obviously, the Poisson distribution arises when the number of fast electrons interacting with the target medium is much larger than the expected number of X-ray quanta that emerge. In Fig. 2.38, a comparison between the binomial and the Poisson distribution is shown. In the left figure, the density functions for $N = 10$ coin-throwing experiments, each with the probability $p = 0.5$ of resulting in heads, are drawn. Obviously, the symmetric binomial distribution (illustrated with vertical lines) gives the correct model. If N increases and

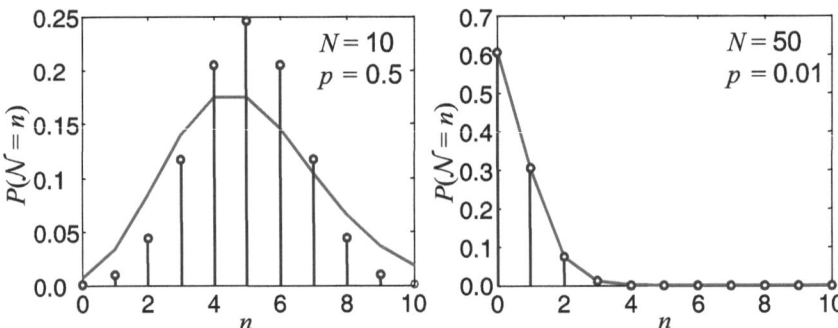

Fig. 2.38. Comparison between binomial and Poisson distribution density functions. The binomial density function is given by the *vertical lines*; the Poisson density function is given by the *solid line*, although the Poisson process is also discrete. *Left*: Ten coin flipping experiments with $p = 0.5$. *Right*: Fifty fast electrons interacting with the target medium creating an X-ray photon with a probability of $p = 0.01$

p decreases, for example, let $N = 50$ be the number of fast electrons and $p = 0.01$ be the efficiency of successful interaction in the anode material, then the binomial and Poisson model result in approximately the same distribution density function as illustrated in the right hand figure.

As shown above for the binomial density function, the probability values for the Poisson process must also in total be 1. With the Taylor expansion for the exponential function, it can be verified that

$$\sum_{n=0}^{\infty} P\left(\mathcal{N} = n\right) = \sum_{n=0}^{\infty} \frac{\left(n^*\right)^n}{n!} e^{-n^*} = e^{-n^*} \sum_{n=0}^{\infty} \frac{\left(n^*\right)^n}{n!} = e^{-n^*} e^{+n^*} = 1. \quad (2.65)$$

2.6.2
Statistical Properties of the X-ray Detector

Reviewing the X-ray detector types explained in ▶ Sect. 2.5 in terms of their statistics, it becomes clear that the X-ray photon detections via the photoelectric effect can also be seen as statistically independent processes. In statistics, such detectors are called Bernoulli detectors. Let p, with $0 \le p \le 1$, be the probability that entering X-ray quanta ionizes a xenon atom of the detector. It has been derived in the previous section that X-ray quanta leaving the X-ray tube are Poisson-distributed. Therefore, the detection process must be modeled for a Bernoulli detector, which receives Poisson-distributed quanta.

The probability that n quanta are detected, if N quanta enter the detector, is given by the conditional probability

$$P_D(n|N) = \begin{cases} \binom{N}{n} p^n (1-p)^{N-n} & \text{for } n = 0, 1, \dots, N \\ 0 & \text{for } n > N \end{cases} . \quad (2.66)$$

Since the source is modeled by the Poisson distribution,

$$P_S(\mathcal{N} = N) = \frac{(n^*)^N}{N!} e^{-n^*}, \tag{2.67}$$

this term must be multiplied with (2.66). In this way, the probability of a Bernoulli detection of n quanta of a Poisson source can be expressed as

$$
\begin{aligned}
P_{SD}(\mathcal{N} = n) &= \sum_{N=n}^{\infty} P_S(N) P_D(n|N) = \sum_{N=n}^{\infty} \frac{(n^*)^N}{N!} e^{-n^*} \binom{N}{n} p^n (1-p)^{N-n} \\
&= e^{-n^*} \sum_{N=n}^{\infty} \frac{(n^*)^N}{N!} \frac{N(N-1)\cdots(N-(n-1))}{n!} p^n (1-p)^{N-n} \\
&= e^{-n^*} \sum_{N=n}^{\infty} (n^*)^n (n^*)^{N-n} \frac{p^n (1-p)^{N-n}}{(N-n)! n!} \\
&= e^{-n^*} \sum_{N=n}^{\infty} \frac{(pn^*)^n ((1-p)n^*)^{N-n}}{(N-n)! n!}.
\end{aligned}
\tag{2.68}
$$

In the last part of (2.68), terms that do not include the number of entering photons, N, are factored out of the sum. Thus, one obtains

$$P_{SD}(\mathcal{N} = n) = e^{-n^*} \frac{(pn^*)^n}{n!} \sum_{N=n}^{\infty} \frac{((1-p)n^*)^{N-n}}{(N-n)!}. \tag{2.69}$$

Shifting the sum index gives

$$P_{SD}(\mathcal{N} = n) = e^{-n^*} \frac{(pn^*)^n}{n!} \sum_{N=0}^{\infty} \frac{((1-p)n^*)^N}{N!}. \tag{2.70}$$

The sum in (2.70) is the Taylor expansion of the exponential function $e^{(1-p)n^*}$. Ultimately, one obtains

$$
\begin{aligned}
P_{SD}(\mathcal{N} = n) &= e^{-n^*} \frac{(pn^*)^n}{n!} e^{(1-p)n^*} \\
&= \frac{(pn^*)^n}{n!} e^{-n^*} e^{+n^*} e^{-pn^*} \\
&= \frac{(pn^*)^n}{n!} e^{-pn^*}.
\end{aligned}
\tag{2.71}
$$

Equation (2.71) reveals that the number of Bernoulli-detected X-ray quanta of a Poisson source is again a Poisson-distributed random variable. The final result is merely scaled by the detection probability, p, i.e., the efficiency of the detector.

Equation (2.71) is exceedingly important because it explains why X-ray quanta show Poisson statistics after traveling through an absorbing object. Obviously, the attenuation processes inside the object are guided by binomial statistics, since the mechanisms for absorption are the same as those inside the detector. The statistical chain from generation of the quanta inside the X-ray tube, via the attenuation inside the object of interest, to the measurement by the X-ray detector, is called a cascaded Poisson process.

2.6.3
Statistical Law of Attenuation

Similar to the situation in ▸ Sect. 2.4, the traveling length, s, of X-ray quanta through an object of interest is subdivided into m slices. In Fig. 2.39, the geometric situation for a statistical model of the attenuation process is given. The subdivisions are non-overlapping, and the attenuation processes are assumed to be statistically independent.

An X-ray quantum arriving at position η has the probability

$$p_t = 1 - \mu(\eta)\Delta\eta ,\qquad(2.72)$$

of passing the object interval 2 and, in turn, the probability

$$p_a = \mu(\eta)\Delta\eta ,\qquad(2.73)$$

of being absorbed in that interval. Generally, the transmission probability of the X-ray quantum running through interval i is given by

$$p_{t,i} \approx 1 - \mu(i\Delta\eta)\Delta\eta ,\qquad(2.74)$$

with

$$\Delta\eta = \frac{s}{m} \quad\text{and}\quad \eta = i\frac{s}{m} .\qquad(2.75)$$

In the limit $m \to \infty$, (2.74) becomes exact. Since the intervals are disjoint and the attenuation processes are assumed to be statistically independent, the transmission probability for the entire traveling length, s, is given by

$$p_{t,s} \approx \prod_{i=1}^{m} p_{t,i} .\qquad(2.76)$$

The logarithm of (2.76) gives

$$\ln(p_{t,s}) = \lim_{m\to\infty} \sum_{i=1}^{m} \ln\left(1 - \mu\left(i\frac{s}{m}\right)\frac{s}{m}\right) .\qquad(2.77)$$

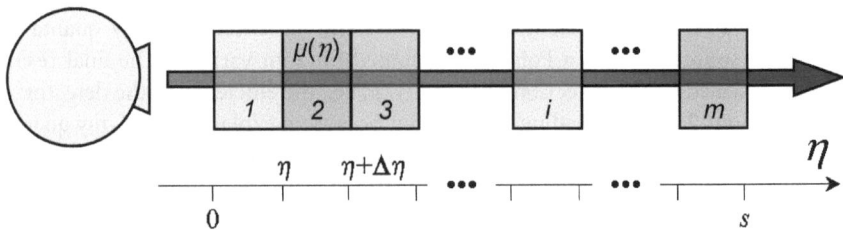

Fig. 2.39. Subdivision of the X-ray quanta traveling length into the m disjoint, statistically independent intervals of thickness $\Delta\eta$

Due to the limit in (2.77), the Taylor expansion for the logarithm in (2.60) can be truncated after the linear term, such that

$$\ln(p_{t,s}) = - \lim_{m \to \infty} \sum_{i=1}^{m} \mu\left(i\frac{s}{m}\right)\frac{s}{m} \tag{2.78}$$

is obtained. However, due to the limit, the right-hand side of (2.78) is essentially the integral

$$\ln(p_{t,s}) = - \int_0^s \mu(\eta)\,d\eta . \tag{2.79}$$

This means, the probabilities of X-ray quantum transmission and absorption during traveling through the entire object are given by

$$p_{t,s} = e^{-\int_0^s \mu(\eta)\,d\eta} \quad \text{and} \quad p_{a,s} = 1 - e^{-\int_0^s \mu(\eta)\,d\eta} \tag{2.80}$$

respectively. The probability that $n < N$ quanta are transmitted, if N quanta are leaving the X-ray source, is again given by the Bernoulli distribution function,

$$P(\mathcal{N} = n) = \binom{N}{n} p^n (1-p)^{N-n} , \tag{2.81}$$

with an expectation value,

$$n^* = E[\mathcal{N}] = \sum_{n=0}^{N} \mathcal{N}P(\mathcal{N} = n) , \tag{2.82}$$

that can be calculated via the definition of the binomial series, i.e.,

$$\sum_{n=0}^{N} P(n) = \sum_{n=0}^{N} \binom{N}{n} p^n q^{N-n} = (p+q)^N , \tag{2.83}$$

with $p + q = 1$. Applying the operator $p \cdot \partial/\partial p$ to both sides of (2.83) yields

$$
\begin{aligned}
p\frac{\partial}{\partial p}\left(\sum_{n=0}^{N} P(n)\right) &= \sum_{n=0}^{N} \binom{N}{n} pn p^{n-1} q^{N-n} \\
&= \sum_{n=0}^{N} n\binom{N}{n} p^n q^{N-n} \\
&= pN(p+q)^{N-1} .
\end{aligned}
\tag{2.84}
$$

Since the middle row of (2.84) is the explicit calculation of the expectation value in (2.82) and $p + q = 1$, it has been shown that

$$
\begin{aligned}
E[\mathcal{N}] = n^* &= \sum_{n=0}^{N} \mathcal{N}P(\mathcal{N} = n) \\
&= \sum_{n=0}^{N} n\binom{N}{n} p^n (1-p)^{N-n} = Np .
\end{aligned}
\tag{2.85}
$$

This means that one has an ideal detector if $p = 1$, and the detector is switched off if $p = 0$ (Epstein 2003). For the expectation value of the X-ray quanta number, one obtains

$$n^* = N p_{t,s} = N e^{-\int_0^s \mu(\eta) \, d\eta} \qquad (2.86)$$

when (2.80) is substituted into (2.85). It has already been mentioned that the number of X-ray quanta is proportional to the radiation intensity. With $n^* \propto I(s)$ and $N \propto I(0)$, (2.86) finally reads

$$I(s) = I(0) e^{-\int_0^s \mu(\eta) \, d\eta} . \qquad (2.42)$$

The statistical discussion of the attenuation process also leads us to *Lambert–Beer's law*. It should be noted that the same result is obtained if the Poisson distribution function (2.64) is used instead of the binomial distribution function in (2.81).

2.6.4
Moments of the Poisson Distribution

To evaluate the image quality of the CT imaging process quantitatively, one has to discuss the moments of the Poisson distribution function. The first moment of the Poisson distribution is given by

$$\mu = E[\mathcal{N}] = \sum_n \mathcal{N} P(\mathcal{N} = n) , \qquad (2.87)$$

i.e., the expectation value, μ, of the number of X-ray quanta assigned with the random variable $\mathcal{N}$. Further, the second central moment is given by

$$\sigma^2 = E\left[(\mathcal{N} - \mu)^2\right] = \sum_n (\mathcal{N} - \mu)^2 P(\mathcal{N} = n) , \qquad (2.88)$$

i.e., the variance, σ^2, of the distribution. The square root of the variance is the standard deviation, σ.

An important value for quantitative signal or image quality evaluation is the *signal-to-noise ratio*, defined as the ratio,

$$SNR(\mathcal{N}) = \frac{\mu}{\sigma} , \qquad (2.89)$$

between signal mean[35] and standard deviation.

The expectation value of a binomially distributed random variable has already been used in the previous section. The result of (2.85) was $E[\mathcal{N}] = Np$. To calculate

---

[35] The mean is an estimate of the expectation value only. However, in this book the symbol is the same.

the expectation value of a Poisson-distributed random variable, it is started with the same term,

$$E[\mathcal{N}] = \sum_{n=0}^{\infty} \mathcal{N} P(\mathcal{N} = n), \qquad (2.90)$$

and utilized the definition of the Poisson distribution function (2.64) such that

$$
\begin{aligned}
E[\mathcal{N}] &= \sum_{n=0}^{\infty} n \frac{(n^*)^n}{n!} e^{-n^*} = \sum_{n=1}^{\infty} n \frac{(n^*)^n}{n!} e^{-n^*} \\
&= e^{-n^*} \sum_{n=1}^{\infty} n \frac{(n^*)^n}{n!} = e^{-n^*} \sum_{n=1}^{\infty} n^* \frac{(n^*)^{n-1}}{(n-1)!} \qquad (2.91) \\
&= n^* e^{-n^*} \sum_{n=0}^{\infty} \frac{(n^*)^n}{(n)!} = n^* e^{-n^*} e^{+n^*} = n^*.
\end{aligned}
$$

Obviously, the parameter

$$E[\mathcal{N}] = n^*$$

is the expectation value. This is consistent with the result of (2.85), because $n^* = Np$ and, within the limits $N \to \infty$ and $p \to 0$, the Bernoulli distribution transforms to the Poisson distribution function.

The variance of a Poisson-distributed random variable can also be calculated explicitly. Starting again with

$$E\left[(\mathcal{N} - \mu)^2\right] = \sum_{n=0}^{\infty} (\mathcal{N} - \mu)^2 P(\mathcal{N} = n) \qquad (2.92)$$

and substituting the definition of the Poisson distribution function and the result of (2.91), one derives

$$
\begin{aligned}
&E\left[(\mathcal{N} - \mu)^2\right] \\
&= \sum_{n=0}^{\infty} (n - n^*)^2 \frac{(n^*)^n}{n!} e^{-n^*} \\
&= e^{-n^*} \left\{ \sum_{n=1}^{\infty} n^2 \frac{(n^*)^n}{n!} - 2n^* \sum_{n=1}^{\infty} n \frac{(n^*)^n}{n!} + (n^*)^2 \sum_{n=0}^{\infty} \frac{(n^*)^n}{n!} \right\} \\
&= e^{-n^*} \left\{ n^* \sum_{n=1}^{\infty} n \frac{(n^*)^{n-1}}{(n-1)!} - 2(n^*)^2 e^{+n^*} + (n^*)^2 e^{+n^*} \right\} \\
&= e^{-n^*} \left\{ n^* \sum_{n=0}^{\infty} (n+1) \frac{(n^*)^n}{(n)!} - 2(n^*)^2 e^{+n^*} + (n^*)^2 e^{+n^*} \right\} \qquad (2.93) \\
&= e^{-n^*} \left\{ n^* \sum_{n=0}^{\infty} n \frac{(n^*)^n}{(n)!} + n^* \sum_{n=0}^{\infty} \frac{(n^*)^n}{(n)!} - 2(n^*)^2 e^{+n^*} + (n^*)^2 e^{+n^*} \right\} \\
&= e^{-n^*} \left\{ (n^*)^2 e^{+n^*} + n^* e^{+n^*} - 2(n^*)^2 e^{+n^*} + (n^*)^2 e^{+n^*} \right\} \\
&= e^{-n^*} e^{+n^*} \left\{ (n^*)^2 + n^* - 2(n^*)^2 + (n^*)^2 \right\} = n^*.
\end{aligned}
$$

The results of (2.91) and (2.93) characterize the Poisson distribution. Obviously, it holds that

$$\mu = \sigma^2 , \tag{2.94}$$

which means that the expectation value equals the variance of the random variable

$$E[\mathcal{N}] = E\left[(\mathcal{N} - \mu)^2\right] . \tag{2.95}$$

For the signal-to-noise ratio of Poisson-distributed signals, like X-ray radiation, one obtains

$$SNR(\mathcal{N}) = \frac{n^*}{\sqrt{n^*}} = \sqrt{n^*} . \tag{2.96}$$

The signal-to-noise ratio increases with the square root of the number of quanta. Considering (2.55) and (2.56), the signal-to-noise ratio is proportional to the square root of the X-ray tube current.

2.6.5
Distribution for a High Number of X-ray Quanta

For a high number of quanta, the factorial in (2.64) can be approximated by the Stirling equation,

$$n! \approx n^n e^{-n} \sqrt{2\pi n}. \tag{2.97}$$

In this case one obtains

$$
\begin{aligned}
P(\mathcal{N} = n) &= \frac{(n^*)^n}{n!} e^{-n^*} \approx \frac{(n^*)^n}{n^n e^{-n} \sqrt{2\pi n}} e^{-n^*} \\
&= \frac{1}{\sqrt{2\pi n}} e^{-(n^* - n)} \left(\frac{n^*}{n}\right)^n \\
&= \frac{1}{\sqrt{2\pi n^*}} e^{-(n^* - n)} \left(\frac{n^*}{n}\right)^{n + \frac{1}{2}}
\end{aligned}
\tag{2.98}
$$

as an approximation of the Poisson distribution density function. Analogous to the derivative of (2.58), $\exp(\ln(\cdot))$ is applied onto the bracketed term in the last row of equation (2.98), such that

$$
\begin{aligned}
P(\mathcal{N} = n) &= \frac{1}{\sqrt{2\pi n^*}} e^{-(n^* - n)} e^{\ln\left(\frac{n^*}{n}\right)^{n + \frac{1}{2}}} \\
&= \frac{1}{\sqrt{2\pi n^*}} e^{-(n^* - n)} e^{(n + \frac{1}{2}) \ln\left(\frac{n^*}{n}\right)} .
\end{aligned}
\tag{2.99}
$$

With the Taylor expansion of the logarithm (2.60), one obtains

$$
\begin{aligned}
P(\mathcal{N} = n) &= \frac{1}{\sqrt{2\pi n^*}} e^{-(n^* - n)} e^{(n + \frac{1}{2}) \ln\left(1 + \frac{n^* - n}{n}\right)} \\
&\approx \frac{1}{\sqrt{2\pi n^*}} e^{-(n^* - n)} e^{(n + \frac{1}{2})\left(\frac{n^* - n}{n} - \frac{1}{2}\left(\frac{n^* - n}{n}\right)^2\right)} ,
\end{aligned}
\tag{2.100}
$$

which can be truncated after the quadratic term. As one is interested in very large numbers of X-ray quanta, n, the term $(n + 1/2)$ can be substituted with n, such that

$$P\left(\mathcal{N} = n\right) \approx \frac{1}{\sqrt{2\pi n^*}} e^{-(n^*-n)} e^{n\left(\frac{n^*-n}{n} - \frac{1}{2}\left(\frac{n^*-n}{n}\right)^2\right)}$$

$$= \frac{1}{\sqrt{2\pi n^*}} e^{-(n^*-n)} e^{n^*-n} e^{-\frac{1}{2}n\left(\frac{n^*-n}{n}\right)^2} \qquad (2.101)$$

$$= \frac{1}{\sqrt{2\pi n^*}} e^{-\frac{(n^*-n)^2}{2n}}.$$

For highly probable numbers of events, i.e., $n \approx n^*$, or $(n - n^*)/n \ll 1$, (2.101) finally transforms into

$$P\left(\mathcal{N} = n\right) \approx \frac{1}{\sqrt{2\pi n^*}} e^{-\frac{(n^*-n)^2}{2n^*}}, \qquad (2.102)$$

which is the probability density function of Gaussian distributed random variables.

(2.102) is a consequence of what is called the *Central Limit Theorem*. The Central Limit Theorem ensures that, regardless of the underlying distribution of mutu-

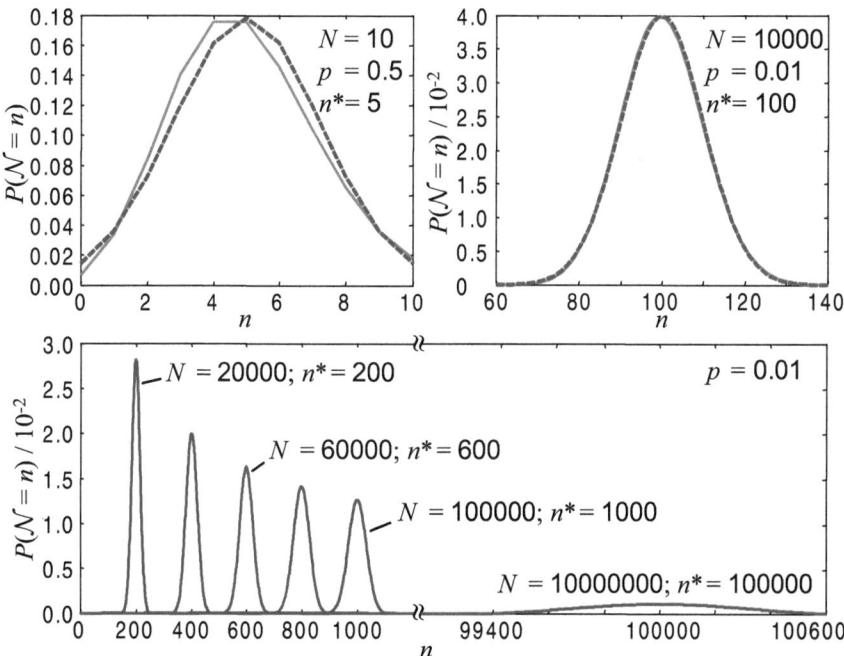

Fig. 2.40. *Upper curves*: Comparison between Poisson and normal distribution density functions. The Gaussian density functions are shown by *dashed lines*; the Poisson density functions are shown by *solid lines*. *Left*: Ten coin flipping experiments. *Right*: 10^4 fast electrons interacting with the X-ray tube anode material. *Lower curves*: Gaussian distribution density functions for increasing mean and variance respectively ($n^* = \mu = \sigma^2$)

ally independent random variables, $\mathcal{N}_i$, as long as one is summing enough random variables, $S = \sum_{i=1} \mathcal{N}_i$, the resulting density function of S is Gaussian. In Fig. 2.40, a comparison between a Poisson and a Gaussian density function is shown. The upper figures illustrate the deviations of the density functions for expectation values of 5 and 100 respectively. The Poisson density function is plotted as a solid line and the Gaussian density function as a dashed line.

Although (2.102) guarantees that the distribution density function of the random X-ray quanta generation process has a Gaussian appearance, this does not mean that the model has two independent parameters,

$$P(\mathcal{N} = n) \approx \frac{1}{\sqrt{2\pi}\sigma} e^{-\frac{(\mu-n)^2}{2\sigma^2}}, \qquad (2.103)$$

i.e., mean, μ, and standard deviation, σ. As shown in Fig. 2.40, the Gaussian density function behaves like the Poisson density function. Since $\mu = \sigma^2$, the density function will be flattened as the mean increases. For the counting processes considered here, the Gaussian density function is therefore limited to a one-parameter distribution.

2.6.6
Non-Poisson Statistics

A scintillator-based X-ray detector can be considered, in principle, as an ideal Bernoulli detector in the sense that an incident X-ray photon is converted into light that is consecutively detected by a photodiode or photomultiplier, in which an electron or a burst of electrons, respectively, is produced. The number of electrons, or electron bursts, counted in a certain time interval obeys the Poisson statistics if the statistics of the incident X-ray photons are Poisson-distributed as well. It has been shown in the previous sections that this is indeed the case. However, at high-energy flux rates it is not possible to detect individual X-ray photons. In such a situation, the detector works as an integrating detector rather than as a counting detector, and the average current of the photodiode is taken as a measure of X-ray intensity. If this is the case, the brightness of the scintillation light becomes important. The brightness must not be considered in counting situations, where the number of 'light on' events in a time interval is the quantity of interest. However, here the brightness of the incident radiation is a random variable as well.

In a situation where a photomultiplier is used to detect the scintillator photons, it can easily be shown that the resulting signal after the electron avalanche is non-Poisson. Let m be the electron amplification factor of each dynode cascade step, i.e., $\mathcal{N}_{out} = m \cdot \mathcal{N}_{in}$. Since $\mathcal{N}_{in}$ is a Poisson-distributed random variable, it holds that $\sigma_{in}^2 = E[\mathcal{N}_{in}]$. Further, it is true that $\sigma_{out}^2 = (m \cdot \sigma_{in})^2 = m^2 \cdot E[\mathcal{N}_{in}]$. However, the latter equality reveals that $\mathcal{N}_{out}$ cannot be a Poisson-distributed variable because $m^2 \cdot E[\mathcal{N}_{in}] \neq E[\mathcal{N}_{out}]$.

Generally, raw intensity data acquired from the data acquisition system (DAS) of CT is processed through several correcting steps to develop *corrected data*. This preprocessing chain is mandatory in the correction of a number of physical effects. Steps of the chain include corrections for offsets, non-linearity of detectors, temperature variations, detector afterglow, etc. In daily routine, an air or blank scan is carried out to compensate for sensitivity variations of the detector elements and to cope with structured form filters as the tube-side bowtie filters (cf. ▶ Sect. 11.4, Fig. 11.10) respectively. This is done in order to obtain a calibration function that allows the assumption of a homogeneous illumination of the detector array within the image reconstruction steps. Additionally, a water beam hardening correction is performed. In a subsequent step, the logarithm of this corrected data is taken to achieve the projection or sinogram data. These sinogram values generally do not obey Poisson statistics and it can be shown that reconstruction algorithms that expect Poisson-distributed data sometimes lead to suboptimal results. However, a remedy has been proposed by Nuyts et al. (2001). A noise-equivalent count (NEC) scaling or shifting can be applied in order to transform arbitrary sinogram noise into noise that is approximately Poisson-distributed. The NEC scaling factor, α, can be obtained via

$$E\left[(\alpha \mathcal{N} - E\left[\alpha \mathcal{N}\right])^2\right] = E\left[(\alpha \mathcal{N} - \alpha \mu)^2\right] = \alpha^2 E\left[(\mathcal{N} - \mu)^2\right] \tag{2.104}$$

and

$$E\left[\alpha \mathcal{N}\right] = \alpha E\left[\mathcal{N}\right] , \tag{2.105}$$

leading to

$$\alpha = \frac{E\left[\mathcal{N}\right]}{E\left[(\mathcal{N} - \mu)^2\right]} . \tag{2.106}$$

The NEC shifting constant, γ, can be obtained similarly via

$$E\left[((\mathcal{N} + \gamma) - E\left[\mathcal{N} + \gamma\right])^2\right] = E\left[((\mathcal{N} + \gamma) - E\left[\mathcal{N}\right] - \gamma)^2\right] = E\left[(\mathcal{N} - E\left[\mathcal{N}\right])^2\right] \tag{2.107}$$

and

$$E\left[\mathcal{N} + \gamma\right] = E\left[\mathcal{N}\right] + \gamma . \tag{2.108}$$

With the characteristic of Poisson-distributed data,

$$E\left[((\mathcal{N} + \gamma) - E\left[\mathcal{N} + \gamma\right])^2\right] = E\left[\mathcal{N} + \gamma\right] , \tag{2.109}$$

the shifting constant is given by

$$\gamma = E\left[(\mathcal{N} - E\left[\mathcal{N}\right])^2\right] - E\left[\mathcal{N}\right] . \tag{2.110}$$

The non-Poisson sinogram values may be subjected either to NEC scaling or to NEC shifting.

3 Milestones of Computed Tomography

Contents

3.1	Introduction	75
3.2	Tomosynthesis	76
3.3	Rotation–Translation of a Pencil Beam (First Generation)	79
3.4	Rotation–Translation of a Narrow Fan Beam (Second Generation)................	83
3.5	Rotation of a Wide Aperture Fan Beam (Third Generation)	84
3.6	Rotation–Fix with Closed Detector Ring (Fourth Generation)	87
3.7	Electron Beam Computerized Tomography.................................	89
3.8	Rotation in Spiral Path	90
3.9	Rotation in Cone-Beam Geometry	91
3.10	Micro-CT	93
3.11	PET-CT Combined Scanners	96
3.12	Optical Reconstruction Techniques	98

3.1
Introduction

Conventional X-ray imaging suffers from the severe drawback that it only produces two-dimensional projections of a three-dimensional object. This results in a reduction in spatial information (although an experienced radiologist might be able to compensate for this). In any case, a projection represents an averaging. The result of the averaging can be imagined if one were to overlay several radiographic sections at the light box for diagnosis. It would be difficult for even an expert to interpret the results, as averaging comes along with a considerable reduction in contrast, compared with the contrast present in one slice.

Figure 3.1 shows conventional images of the cranium (left) and of the knee (right). These images show the high attenuation of X-rays within, for example, the cranial bone and most notably around the dental fillings. The small differences in attenuation that characterize soft tissue, however, are not visible at all. The mor-

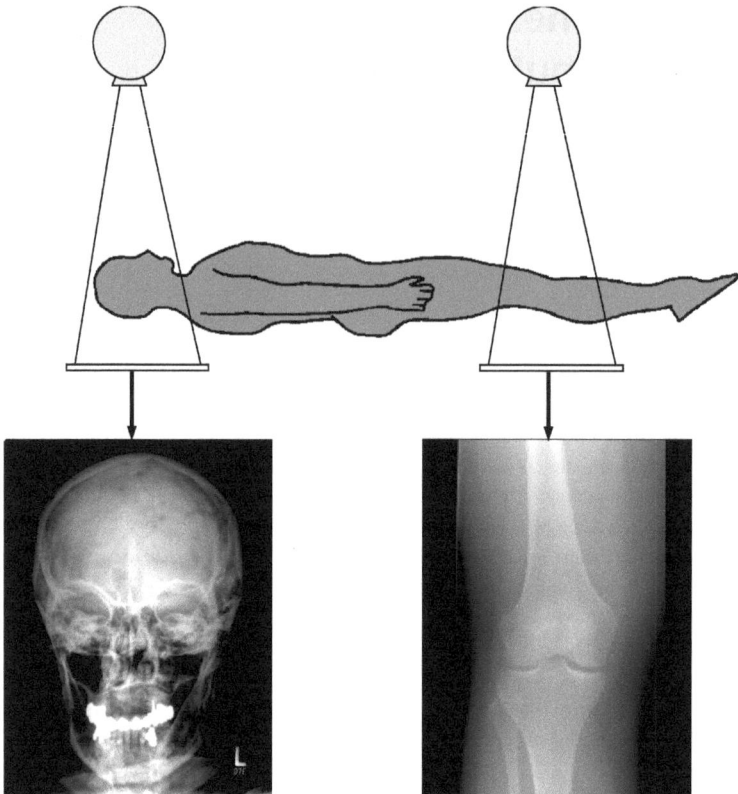

Fig. 3.1. Conventional planar X-ray leads to low-contrast images that do not allow clinical diagnostics involving soft tissue. The figure shows the cranial bone (*left*), in which details of spatial structures in the brain cannot be recognized. On the image of the knee (*right*), even bone structures have low contrast. This low contrast is caused by the averaging process during the X-ray's passage through the body

phology of the brain, in particular, is completely lost in the averaging process. In the knee, even the bone structures are imaged with poor contrast due to the superimposition.

3.2
Tomosynthesis

In the 1920s, the desire to undo the averaging process that characterizes conventional X-ray radiography led to the first tomographic concept. The word *tomography* itself is composed of the two Greek words *tomos* (slice) and *graphein* (draw). The word tomography was considerably influenced by the Berlin physician Grossmann, whose *Grossmann* tomograph was able to image one single slice of the body (Grossmann 1934). Figure 3.2 shows a historic picture of the *Grossmann* tomograph (left) (Lossau 1995), as well as a modern tomosynthesis scanner (right) (Härer et al. 2003).

The principle of the conventional or analog geometric tomography method is illustrated in Fig. 3.3. During image acquisition, the X-ray tube is linearly moved in one direction, while the X-ray film is synchronously moved in the opposite direction. For this reason, only points in the plane of the rotation center are imaged sharply. All points above and below this region are blurred, more so at greater distances from the center of rotation. Hence, the method can be interpreted as "blurring tomography." It is called "tomosynthesis" if there is digital post-processing of the projection images.

The blurred information above and below the center of rotation does not disappear, but is superimposed on the sharp image as a kind of gray veil or haze. Therefore, a substantial reduction in contrast is noticeable. However, the gain in quality compared with a simple radiograph is clearly visible in the example of a tomosynthetically acquired slice sequence of the knee in Fig. 3.3 (Härer et al. 1999, 2003). Table 3.1 gives a historic overview of the milestones of analog tomography methods.

Due to the increased availability of electronic X-ray detectors, tomosynthesis systems are currently regaining scientific attention (Stevens 2000). Figure 3.2 shows the Siregraph T.O.P. from Siemens Medical Solutions, a modern tomosynthesis scanner (right). In this system, projection images are measured during movement and stored digitally using an image intensifier system. This allows subsequent image reconstruction that is superior to the analog blurring technique. Figure 3.3

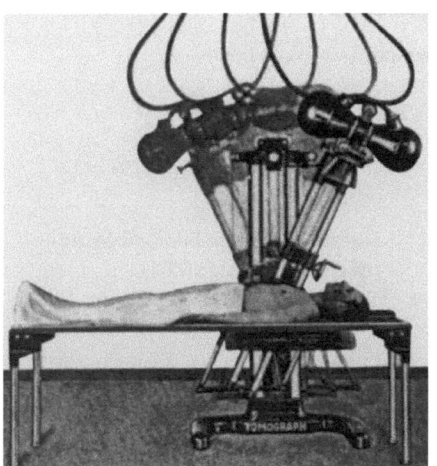
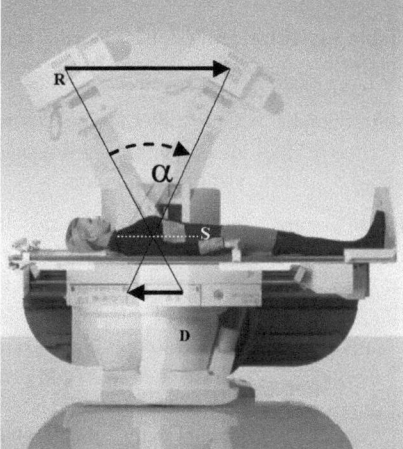

Fig. 3.2. Digital tomosynthesis is superior to conventional or analog geometric tomography. *Left*: The historic *Grossmann* tomograph (Lossau 1995; Grossmann 1934). *Right*: The Siregraph T.O.P. from Siemens, equipped with an image intensifier system (D), which is synchronously moved with the X-ray tube (R), and therefore defines a slice (S). Courtesy of Siemens Medical Solutions, from (Härer et al. 1999, 2003)

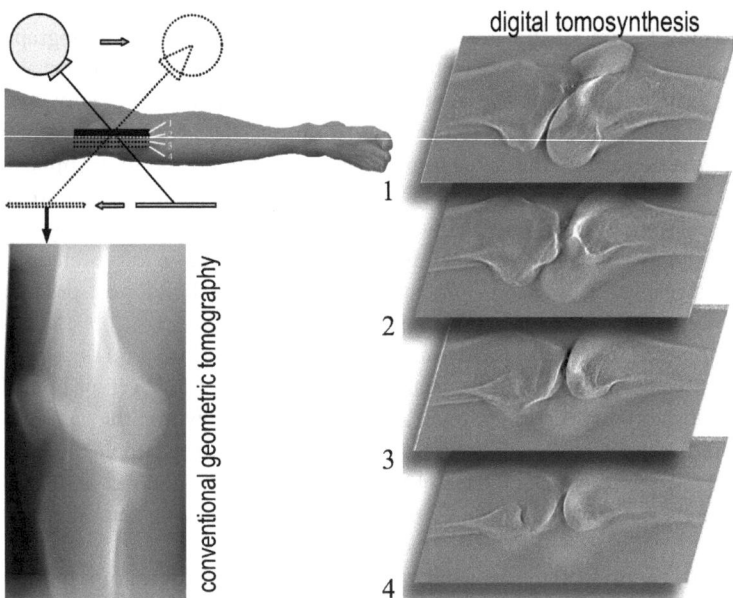

Fig. 3.3. The first attempts to create radiographic slices of the human body were carried out using conventional or analog geometric tomography, also referred to as tomosynthesis if the acquired X-ray images are digitally post-processed. In this process, the X-ray tube and the detector are synchronously moved in opposite directions. For this reason, structures situated above and below the plane of the center of rotation are blurred (tomosynthesis images of a knee phantom: Courtesy of Siemens Medical Solutions (Härer et al. 1999, 2003))

Table 3.1. Historic milestones of analog tomography (see Webb 1990)

Class	Mechanical arrangement	Motion	Inventor
Stratigraphy	X-ray tube and film cassette are perpendicular to a rigid pivoting pendulum	Linear, circular or spiral	Alessandro Vallebona (1930)
Planigraphy	X-ray tube and film cassette move in parallel equidistant planes with reciprocal motions	Linear, circular or spiral	André Edmond Marie Bocage (1921), Ziedses des Plantes (1921), Ernst Pohl (1932)
Tomography	X-ray tube and film cassette are attached to a rigid pendulum, but the detector is always parallel to the tomographic plane	Linear	Gustav Grossmann (1935)

enables a direct comparison between the analog image of the knee given by conventional geometric tomography (bottom left) and single slices acquired with a spacing of 8 mm using tomosynthesis. The blurring angle, α, determines the extent to which the contribution from structures above and below the slice (S) are suppressed in the blurring. Besides the class of trajectory (see Table 3.1) of the X-ray tube (R) and detector (D), the blurring angle, α, determines the quality of the image at S. The larger the angle, α, the smaller the extrinsic slice artifacts, i.e., the amplitude of the signal from structures located exterior to the slice being reconstructed (Härer et al. 2003).

A related method, called "orthopan tomography," is now widespread in dental radiology. In this method, a panoramic view of rows of teeth is produced on an imaging plane that is curved to follow the jaw. However, sophisticated trajectories allow a reconstruction of slices by geometric tomography as well.

Nowadays, the term "tomography", despite competitive modalities such as MRI (magnetic resonance imaging) or PET (positron emission tomography), is still most commonly associated with computed tomography (CT) or, more precisely, with X-ray CT. CT is also referred to as CAT (computerized axial tomography)[1].

Computed tomography avoids the superimposition of blurred planes and produces such a large contrast that even soft tissue can be imaged well. The resulting leap in the quality of diagnostic imaging led to the enormous success of CT.

Historically, four distinct generations of CT have emerged. Their classification relates to both the way that X-ray tubes and detectors are constructed, and the way that they move around the patient. Figures 3.4, 3.8, 3.9, and 3.12 illustrate the different generations schematically.

3.3
Rotation–Translation of a Pencil Beam (First Generation)

The first generation of CT involves an X-ray tube that emits a single needle-like X-ray beam, which is selected from the X-ray cone by means of an appropriate pinhole collimator. This geometry is referred to as "pencil beam". A single detector is situated on the opposite side of the measuring field and the X-ray tube. The detector is moved synchronously along with the X-ray tube. This displacement is linear and is repeated for different projection angles, γ (cf. Fig. 3.4). Depending on the specific attenuation properties of the tissue, the intensity of the X-ray is attenuated on its path through the body.

The amount of X-ray attenuation is measured by the detector and subsequently digitally recorded. For each angle, γ, this step yields a simple one-dimensional radiograph. However, from this initial radiograph, it is still not possible to determine the spatial distribution of the tissue attenuation coefficients. It is clear that in order to determine the location of two consecutive objects on one projection line, the situation needs to be viewed sideways as well. This is the approach taken by CT. CT views an object from all sides, since the projection angle is varied from $0°$ to $180°$.

---

[1] In the early years terms like *Computerized Transaxial Tomography* (CTAT) or *Digital Axial Tomography* (DAT) were used (Bushong 2000).

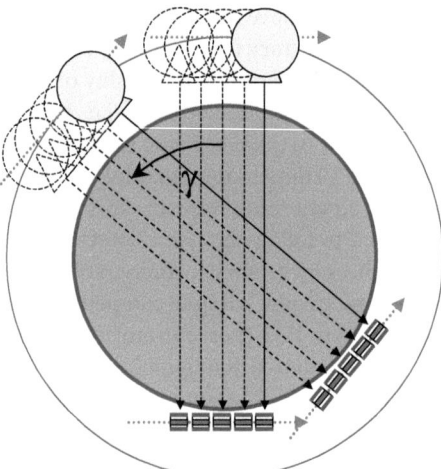

Fig. 3.4. The first generation of CT is equipped with a pencil beam and a single detector. These are moved linearly, and the configuration is rotated through different projection angles, γ. Each point inside the field of view needs to be X-rayed from all "sides," so the X-ray source and detector are rotated through 180°. The X-ray pencil beam is extracted from the beam characteristic of the source, by using an appropriate pin-hole collimator

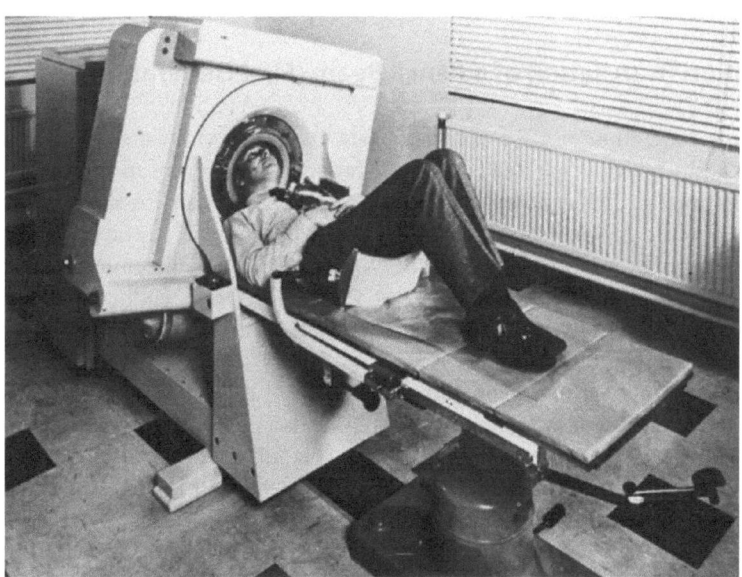

Fig. 3.5. The first head scanner was built in the EMI Central Research Laboratories in London. Courtesy of General Electric Medical Systems

The first CT scanner, built by the company EMI (*Electric and Musical Industries Ltd.*), was based on this principle. In 1972 Godfrey N. Hounsfield realized the scanner in the EMI Central Research Laboratories (Hounsfield 1973). For his invention, he jointly won the Nobel Prize for medicine together with Allen M. Cormack in 1979. Figure 3.5 shows a picture of the EMI head scanner. The first experiments by

Fig. 3.6. Hounsfield's historic experiment, which he established in London at the EMI Central Research Laboratories in the 1960s (*top*) and the first image produced (*bottom*). Courtesy of General Electric Medical Systems

Hounsfield were performed in 1969 with a radioactive Americium source ($^{241}_{95}$Am) being linearly displaced along with the detector. Hounsfield collected data in steps of 1°, through the 180° necessary for image reconstruction. The first reconstruction of a two-dimensional slice took 9 days – a clinically unacceptable processing time. However, for the first time, a two-dimensional slice image could be achieved that did not originate from averaging or blurring information, as in conventional geometric tomography mentioned in the section above. Figure 3.6 shows the historical layout of the experiment (top) as well as the first image it produced (bottom).

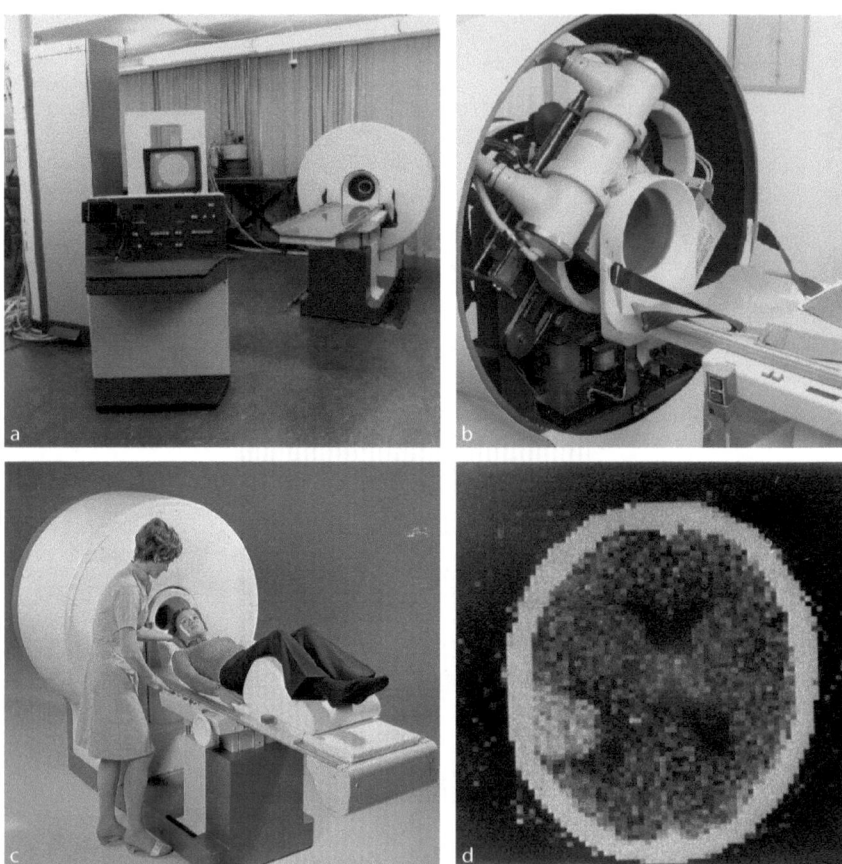

Fig. 3.7. One of the first commercially available, first-generation CT scanners was the Siemens Siretom in 1974, shown here **a** closed and **b** with covers removed. With this scanner it was possible to produce **c** axial images of the head with **d** an image matrix of 80 × 80 pixels. Courtesy of Siemens Medical Solutions

The actual reconstruction of a two-dimensional slice is only possible if the projection values are digitized. The image section to be visualized is determined by the measurement principle, since the pencil beam virtually cuts the object at that slice. However, the calculation of the spatial distribution of the attenuation values at that very slice is mathematically complex and will be described in ▸ Chaps. 5–7. Due to the complexity of the mathematical methods of Johann Radon's reconstruction technique, published in 1917, it remained only theory for half a century. A practical solution to the reconstruction problem was not possible until the development of computer technology[2]. It is from this that CT gets its name.

The first commercial scanners from EMI had a narrow focused X-ray beam and a single sodium iodide (NaI) scintillation detector. This pencil-beam principle, which is not practiced any more, is of fundamental importance as it is here that the mathematical methods of reconstruction can be understood most easily. Indeed, the mathematical methods of more modern geometries can be obtained from pencil-beam geometry using suitable coordinate transformations. Figure 3.7 shows one of the first tomographs of the first generation produced by Siemens in 1974. The subsequent rapid development of CT has been, and still is, driven by three essential goals: Reduction of acquisition time, reduction of X-ray exposure, and, last but not least, reduction of cost. Throughout the course of optimizing these factors there are several historical stages, which are briefly described in the following sections.

3.4
Rotation–Translation of a Narrow Fan Beam (Second Generation)

The computed tomograph of the second generation has an X-ray source with a narrow fan beam and a short detector array consisting of approximately 30 elements (cf. Fig. 3.8). However, since the aperture angle of the fan beam is small, the X-ray tube and detector array still need to be translated linearly before the projection angle is adjusted for another projection. In the earliest of the second-generation scanners, the angle of the fan beam was about $10°$.

Despite the need for linear displacement, the acquisition time was reduced to a few minutes per slice, as the detector array could measure several intensities simultaneously. However, the measuring field was still small. For this reason, and due to their long acquisition times, the first- and second-generation scanners were mainly restricted to use in imaging the cranium. Figure 3.7c gives an idea of the scale of the measuring field. The cranium can be fixed in the scanner and shows no large intrinsic movement during acquisition time (relative to the spatial resolution that could be achieved at the time).

---

[2] In ▸ Sect. 3.12, it is shown that this is not entirely historically correct. In 1977, Edholm (1977) proposed an analog, optical procedure for tomographic image reconstruction.

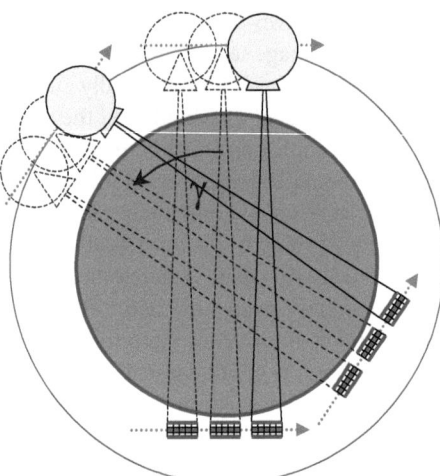

Fig. 3.8. The CT scanners of the second generation have one X-ray source with fan-beam geometry, as well as a short detector array. The X-ray fan is created from a conical X-ray source by means of a slit-shaped collimator. The fan angle is about 10° and the detector is made up of 30 elements. In this case, the detector and tube also need to be moved linearly before the projection angle is adjusted, as the entire field of view is not X-rayed by the fan

This is certainly not the case when imaging the area of the thorax or the abdomen, as the intrinsic movements of the heart and lung, as well as the movement of the diaphragm and soft organs of the abdomen, produce artifacts in the reconstructed images. The mathematical methods of reconstruction demand that all points of one slice are X-rayed from all angles from 0° through 180°. One object moving out of the imaging plane during rotation of the X-ray tube, due to patient movement, will inevitably result in errors in image reconstruction. Such image errors, referred to as motion artifacts, are described in detail in ▶ Chap. 9.

3.5
Rotation of a Wide Aperture Fan Beam (Third Generation)

The main goal of developments in the 1970s was to reduce acquisition time down to less than 20 s. This was intended to give enough time to acquire an image of the abdomen with minimal error, while the patients held their breath. A major step toward achieving this goal was an extension of the second generation's fan-beam concept, i.e., the introduction of a substantially larger angle of the X-ray fan and a correspondingly longer detector array. Figure 3.9 gives a schematic illustration of this principle of third-generation scanners.

Nowadays, the angle of an X-ray fan beam is typically between 40° and 60° and the detector array is usually constructed as an arc with between 400 and 1,000 elements. In this way it is now possible to simultaneously X-ray the entire measuring field, which is currently wide enough to cover the torso, for each projection angle, γ. As a result, the third generation of CT systems can completely abandon linear displacement of the X-ray tube. The acquisition time for third-generation systems is reduced considerably, since a continuous rotation can take place without interruption for linear displacement. The majority of CT scanners currently in use are fan-beam systems of the third generation. Figure 3.10 shows an experimental set-up of a third-generation scanner with a mechanical magnification unit in the Philips research labs and Fig. 3.11 shows a typical Siemens CT scanner of 1981.

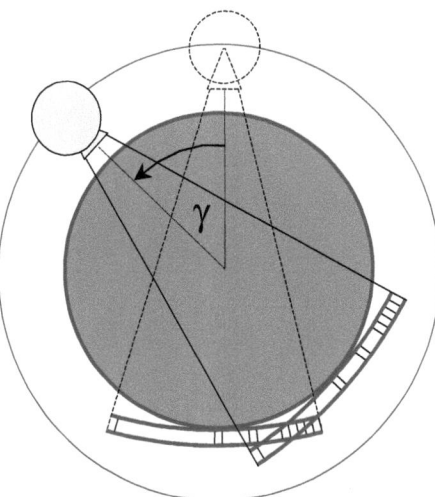

Fig. 3.9. The third generation of CT scanners has a substantially larger angle of the X-ray fan and a longer detector array, such that the entire measuring field can be X-rayed simultaneously for one single projection angle, γ. Hence, the need for linear displacement of the source-detector system is removed. The third generation represents an extension of the principle behind the second generation. The fan angle is usually around 40° to 60° and the detector array is made of up to 1,000 detector elements. As a consequence of the larger aperture, the X-ray is exploited to a greater extent. Also, the scanner can rotate continuously without stopping for linear displacement. This results in a major reduction of acquisition time

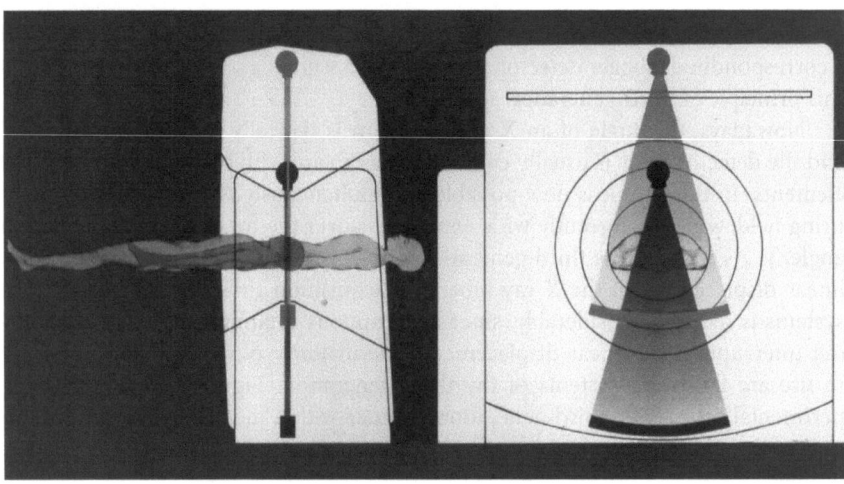

Fig. 3.10. a Principle of geometric enlargement in 1977. Two possible positions of the tube/detector construction as to optimize the scanner for both head and body scan, while having a limited detector size. **b** Detector module of 192 channels. **c** Lab model including geometric enlargement construction. Courtesy of Leon de Vries, Philips Medical Systems

Typical image errors of third-generation scanners are called "ring artifacts." A single defective, or even insufficiently calibrated, detector in the array results in characteristic ring-shaped errors in the reconstructed tomogram. These kinds of

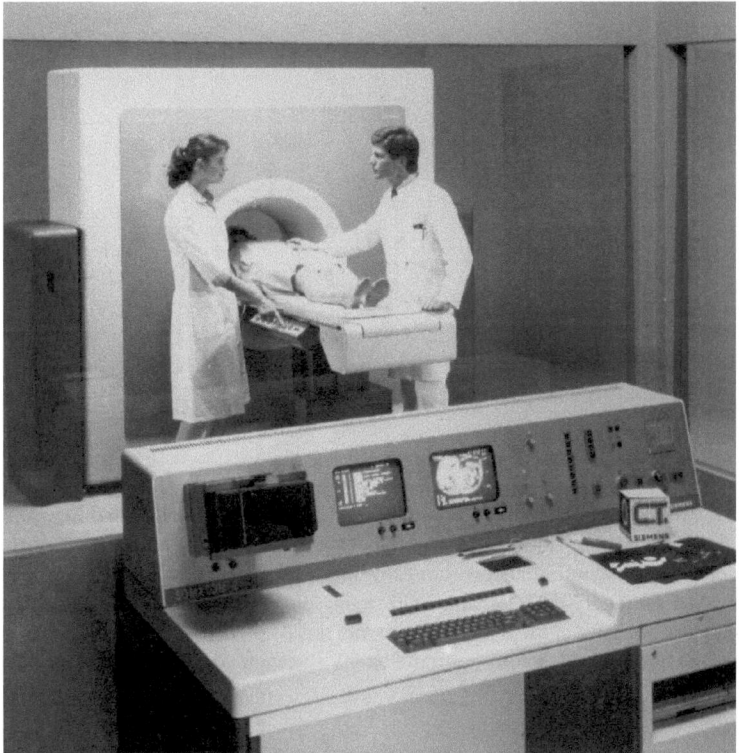

Fig. 3.11. An example of a whole body scanner from 1981: The Siemens Somatom family DR. This third generation of computed tomography scanners has a measuring field of approximately 53 cm in diameter. With 512 solid state detectors, acquisition times of about 3 s were possible for a 360° rotation (courtesy of Siemens Medical Solutions)

artifacts are discussed in detail in ▸ Sect. 9.6.5. While they are seen in images from third-generation scanners, they cannot appear in images from fourth-generation tomographs.

3.6
Rotation–Fix with Closed Detector Ring (Fourth Generation)

The fourth generation of CT scanners does not differ from the third-generation tomographs with respect to the X-ray tube. The fan-beam source also rotates continuously around the measuring field without any linear displacement. The difference is in the closed, stationary detector ring with up to 5,000 single elements. The X-ray tube either rotates outside (Fig. 3.12, left) or inside (Fig. 3.12, right) the detector ring.

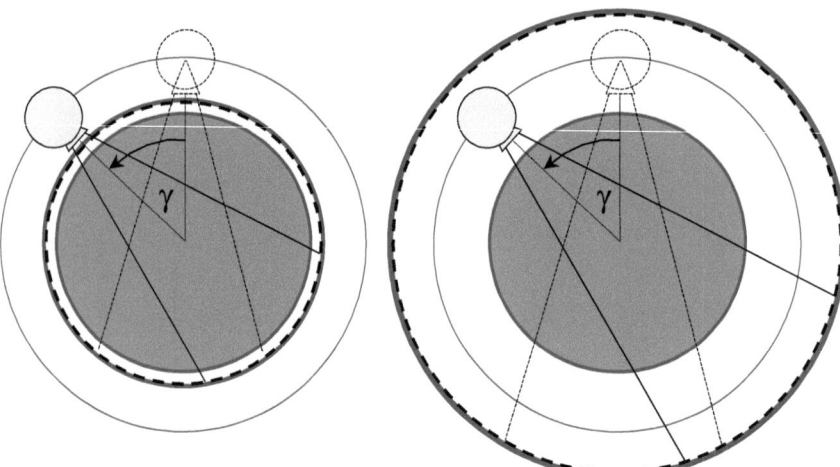

Fig. 3.12. Computed tomography scanners of the fourth generation do not differ from those of the third generation with respect to the X-ray source. As with the third generation, the X-ray source rotates continuously without any linear displacement. However, the fourth generation has a stationary detector ring. Within the fourth generation systems there are two distinct types: The path of the X-ray source is either outside (*left*) or inside (*right*) the detector ring. The number of detectors is around 5,000. The angle of the fan beam depends on the measuring field

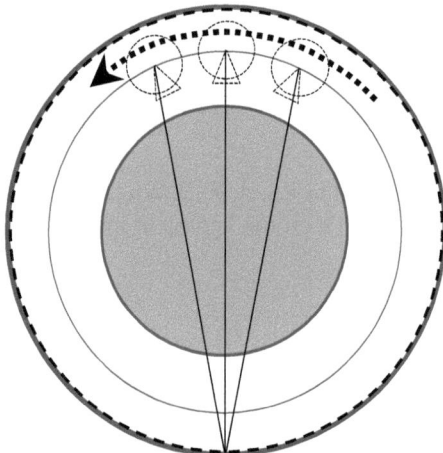

Fig. 3.13. The fourth generation of CT scanners uses an "inverse fan," whose center is situated at a single detector. Unlike the third-generation scanners, it is in principle possible to measure an arbitrarily dense inverse fan, as long as a single detector element can be read at a high enough frequency while the X-ray source moves continuously

If the X-ray tube is outside the detector ring, it is necessary to prevent the X-rays from radiating through the detectors from behind. Therefore, the detector ring is dynamically tilted away from the path of the tube. In this way, the line of sight between the tube and the appropriate section of the detector ring only passes through the patient (and patient table) and not through the electronics behind the detectors.

The fourth-generation tomographs establish "inverse fans," which are centered on detectors rather than on the X-ray focus. Figure 3.13 shows an inverse fan schematically. An inverse fan is also referred to as a "detector fan."

An inverse fan can be very dense, limited only by the sampling rate at which individual detectors can be read out. As a result, unlike third-generation tomographs, this scanner is not limited to the spatial resolution of a single fan beam.

3.7
Electron Beam Computerized Tomography

If there is a need for even shorter acquisition times, the concept of moving sampling systems must be left behind entirely. One approach to achieving this is using electron beam computerized tomography (EBCT). This type of CT was developed for cardiac imaging.

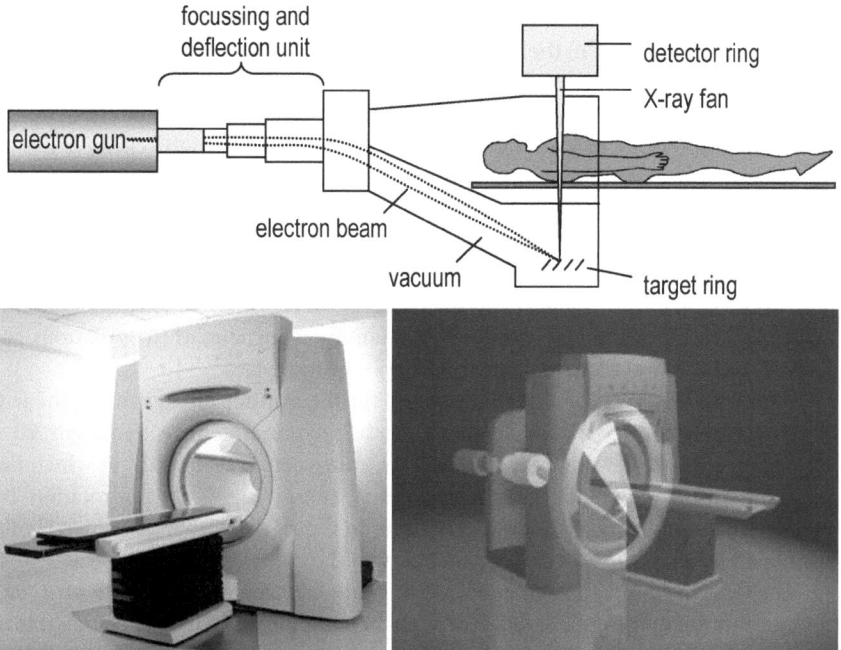

Fig. 3.14. Realization of an electron beam computerized tomography (EBCT) scanner by Imatron, for fast cardiac imaging. Courtesy of General Electric Medical Systems

In EBCT, there is no longer a localized X-ray tube rotating around the patient as in conventional CT technology. Instead the patient is situated, in a manner of speaking, *inside* the X-ray tube. An electron beam is focused onto wolfram target rings, which are arranged in a half circle around the patient, and generates the desired X-ray fan beam upon impact with the wolfram target. The X-ray irradiation is measured with a stationary detector ring.

Such systems have mainly been sold to cardiologists by the company Imatron. The electron beam technique is able to acquire an image slice in 50 ms. Figure 3.14 shows a diagram of an EBCT system as well as an illustration of a modern Imatron system. Further technical details can be found, for example, in Weisser (2000).

3.8
Rotation in Spiral Path

Another development, which was a great leap forward in capability from that of the third generation, led to what is identified by Bushberg et al. (2002) as the sixth generation of tomography scanners. This refers to the introduction of slip-ring technology, which enables spiral or helical sampling respectively.

As the X-ray tube must be continuously supplied with energy, the rate of circular movement was previously limited by the attachment of an electric cable, which was mounted on a spool. In this process the cable was unwound in one direction and carefully wound up in the other. This represented a huge obstacle to the reduction of acquisition time. The X-ray sampling unit had to stop and start again after a certain angle of rotation. Although data could be collected throughout both clockwise and counter-clockwise rotations, limits were placed on high velocities due to the increasing torsional moment. This problem was solved by the introduction of slip-ring technology. In this technology, the energy is provided via sliding contacts, situated between the outside of the sampling unit, in what is called the *gantry*, and the rotating sampling unit. This enables the sampling unit carrying the X-ray source and, in the case of third-generation scanners, the detector array, to rotate continuously. As a result of slip-ring technology, rotation frequencies of two rotations per second, "sub-second scanners", are nowadays commonplace.

Figure 3.15 shows different technologies for the transfer of energy. On the left, slip-ring technology, as discussed above, can be seen on a Siemens Somatom AR.T CT scanner. However, there are also smaller, more compact devices, in which the sampling unit is independent from an external energy supply during rotation, by means of accumulators. An example is given on the right of Fig. 3.15, which shows a mobile Philips Tomoscan M CT scanner with the covers removed.

The slip-ring innovation enabled a new acquisition technique. Along with a continuous motion of the patient table through the sampling unit, it became possible to measure data in the shape of a spiral[3]. The spiral-CT technique was demonstrated

---

[3] Strictly speaking, the X-ray tube trajectory is a helix.

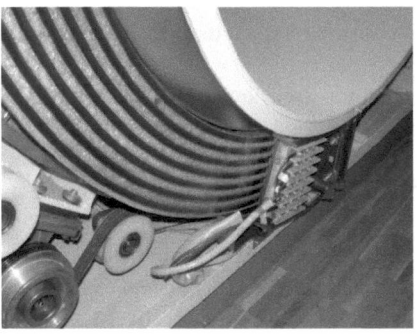 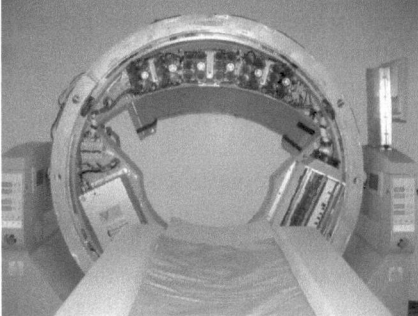

Fig. 3.15. Different methods of energy transfer to the continuously rotating sampling unit. In the Siemens Somatom AR.T (*left*), slip-ring technology is used. In the compact mobile CT scanner of the Philips Tomoscan M (*right*) an accumulator array is used

Fig. 3.16. Willi Kalender in the mid-1980s at the Laboratories of Siemens (by courtesy of W. Kalender)

successfully on a prototype by Kalender in 1989 (Kalender et al. 1989). Figure 3.16 shows Willi Kalender at the laboratories of Siemens Medical Solutions in the mid-1980s.

3.9
Rotation in Cone-Beam Geometry

As already mentioned, there has so far not been a common definition of the generations of development of CT. In Bushberg et al. (2002), scanners equipped with a cone-shaped X-ray beam and a plane detector are referred to as the seventh generation.

However, even within the cone-beam scanner itself, one needs to distinguish between systems that use only a small cone opening, as in the case of a multi-slice (multi-line) detector system or indeed a symmetric X-ray cone. In particular, the necessary reconstruction methods differ extensively between these systems, as demonstrated in ▸ Chap. 8.

To understand the motivation behind the development of cone-beam CT systems, recall that the step from the pencil-beam to the fan-beam concept came along with the advantages that the X-ray source was exploited more effectively, and that there was a reduction in acquisition time. In ▸ Sect. 2.2.3 it has been discussed that the efficiency of the energy transformation throughout the generation of X-ray radiation is just about 1%.

As the heat produced inside the X-ray tube essentially defines the physical capacitance, and therefore limits the measuring time, the next straightforward step in the development of CT scanners involved the use of a cone-shaped X-ray beam, which is already produced in the X-ray tube. Both the pencil-beam and the fan-beam geometry are created by means of appropriate pin-hole or slit collimators, which re-shape the original X-ray source intensity profile, reducing efficiency.

Technologically, there are three important problems that had to be solved before the successful application of cone-beam geometry to CT imaging.

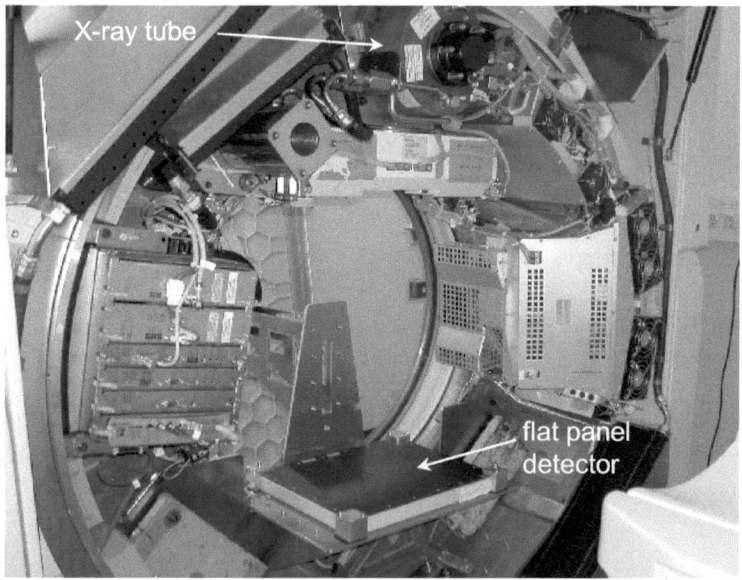

Fig. 3.17. Prototype of a cone-beam CT scanner, equipped with a flat detector. Courtesy of Siemens Medical Solutions

First of all, a flat-panel detector, which did not exist at the time, had to be introduced to replace the line or multi-line detector arrays. Second, the huge amount of raw data that quickly emerge on sub-second scanners in particular had to be transferred from the rotating sampling unit to the image reconstruction computer. The required bandwidth for the data transfer poses a challenge even today. Third, there is the problem of reconstruction, whose mathematics is slightly more sophisticated compared with the two-dimensional methods. This will be discussed in detail in ▸ Chap. 8. Figure 3.17 shows a prototype of a cone-beam CT scanner at the Siemens laboratories.

3.10
Micro-CT

Recently, micro-CTs, which essentially comprise a miniaturized design of the cone-beam CTs mentioned in the previous section, and which are typically used for non-destructive, three-dimensional microscopy, have become commercially available. The X-rayed measuring field, usually as small as $2\,cm^3$ in volume, is so small that medical applications might seem to be ruled out. Indeed, these scanners are more commonly used for material testing and analysis, but medical applications are on their way to taking center stage. An example in human medicine is the analysis of trabecular structures in bones. Micro-CTs are also ideal scanners for radiological examinations of small animals (De Clerck et al. 2003).

Figure 3.18 shows a commercial micro-CT system, manufactured by the Belgian company SkyScan. It is produced as a desktop device and has a measurement chamber that is entirely shielded against scattered X-ray beams by lead walls, so that no further means of protection are necessary. The object to be examined is placed on a rotating specimen disk, which is controlled by a step motor.

Figure 3.18a shows the preparation of an examination of a Neanderthal man's tooth. This specimen is about 200,000 years old. The curator of the Department of Preservation of Archeological Monuments in Koblenz is currently interested in the structure of the tooth's interior. Figure 3.18b shows a cone-beam X-ray of the tooth. As well as the nerve channel, the fissure lines are imaged very well.

The two most crucial components of micro-CTs are the X-ray tube and the two-dimensional detector array. In particular, it is the size of the focus and the size of the detector elements that, apart from the mechanical accuracy of the rotary motion, determine spatial resolution. In Fig. 2.7 it has already been illustrated how the size of the focus affects the representation of detail. It follows that the micro-CT system needs a micro-focus tube. In ▸ Sect. 9.2 it will be shown that it is the modulation transfer function,

$$MTF_{\text{imaging system}}(q) = MTF_{\text{X-ray source}}(q)MTF_{\text{detector}}(q) = \left|\frac{\sin(\pi b_F q)}{\pi b_F q}\right|\left|\frac{\sin(\pi b_D q)}{\pi b_D q}\right|$$

$$(3.1)$$

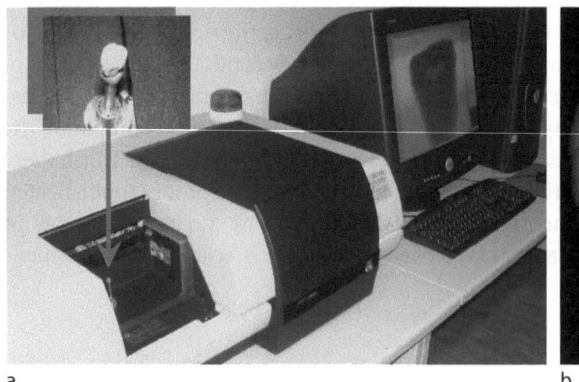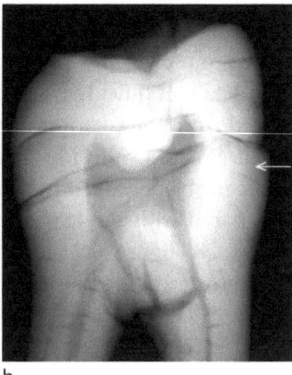

a   b

Fig. 3.18. a Micro-CT desk device of SkyScan Inc. at the Institute of Medical Engineering, University of Lübeck, Germany. The object to be examined is placed on a rotating specimen holder inside the measurement chamber. The measurement field is about $2\,cm^3$ in volume. **b** A simple X-ray image of a tooth from an archeological dig illustrates an impressive variety of structures. As well as the nerve channel, fine fissure lines that have been formed inside this Neanderthal man's tooth over the millennia are recognizable. The *small arrow* indicates the location of a reconstructed slice, which will be examined in more detail. The planar $1,024 \times 1,024$ pixel charge-coupled device (CCD) array is cooled down and has a detector size of $b_D < 10\,\mu m$

that quantifies the resolution of objects with a spatial frequency q (given in line pairs per millimeter). In the case of micro-CTs, an X-ray focus size of $b_F < 10\,\mu m$ is desirable. Using such a small target area for the electrons, the anode current cannot be very large. The current is typically less than $100\,\mu A$. Since the current controls the intensity of the X-ray spectrum, there are certain constraints with respect to the materials being examined. A 12-bit X-ray charge-coupled device (CCD) chip with a pixel matrix of $1,024 \times 1,024$ is used as a detector, which is connected to the scintillation crystal by fiber optics. The size of the picture elements has an order of magnitude of around $b_D < 10\,\mu m$. SkyScan specifies a resolution of about $10\,\mu m$ (Sassov 1999, 2002a, 2002b). As micro-CTs are cone-beam X-ray systems, three-dimensional reconstruction methods are required to calculate the images (Wang et al. 1998).

Figure 3.19 shows the difference in spatial resolution between a micro-CT and a clinical CT system, using the example of the tooth from a Neanderthal man. In a reconstructed slice, fine fissures are identifiable in the image from the micro-CT (Fig. 3.19a). The reconstruction using a clinical CT (Fig. 3.20b) is not able to resolve these details. The location of the section shown in Fig. 3.19a,b is marked with an arrow in Fig. 3.18b. When comparing the three-dimensional reconstructions and

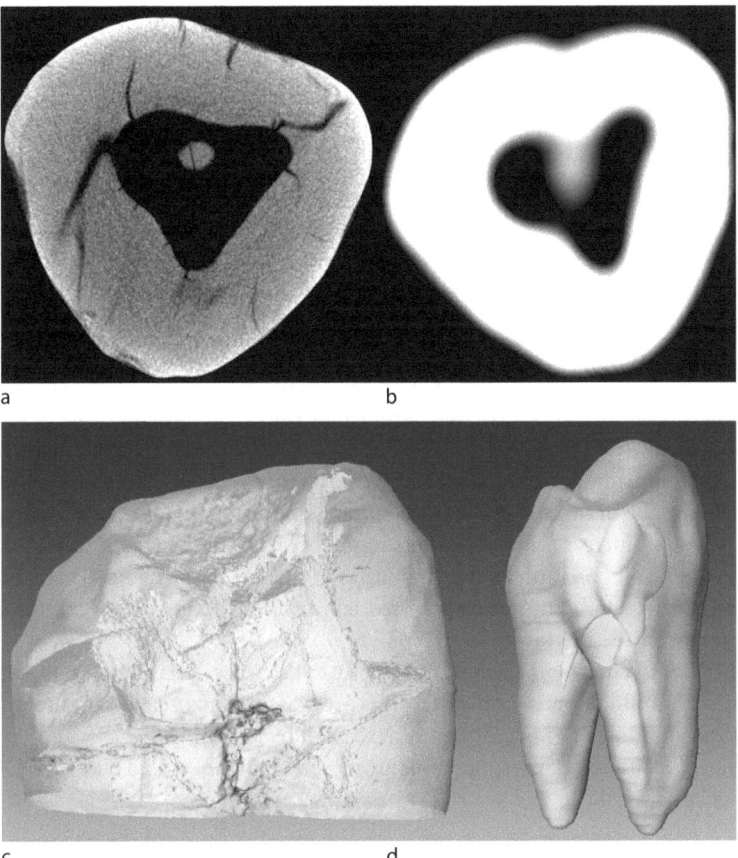

Fig. 3.19. Comparative inspection of image resolutions, using the example of the tooth from an archeological dig. An approximately 200,000-year-old Neanderthal man's tooth was examined using **a, c** micro-CT and **b, d** ordinary clinical CT. The position of the reconstructed section is marked in Fig. 3.18b. The difference is immediately obvious. The fine fissures that have formed inside the tooth over the millennia cannot be imaged using clinical CT. **c** In the three-dimensional reconstruction of the micro-CT image the fissures can be identified as fissure planes. For comparison, the entire tooth imaged by the ordinary CT scanner has been reconstructed in **d**. Given the dimensions of the tooth, even this reconstruction is impressive, but it does not reveal the micro-structures evident in the micro-CT image

visualizations in Fig. 3.19c,d, there is a distinct difference as well. Fissure planes passing through the entire tooth are only visible on the micro-CT image.

3.11
PET-CT Combined Scanners

With the exception of contrast-enhanced angiography and perfusion techniques, CT, on its own, is only able to provide morphological information, i.e., information on the shape of objects. On the other hand, positron emission tomography (PET) provides information on the metabolism, i.e., the biomedical function of an anatomical region. As already described in ▶ Chap. 2, CT is based on the attenuation of X-ray radiation. Different organs, having different absorption properties, are therefore only imaged according to their shape, and the patient is considered to be passive during imaging.

In PET, the patient is injected with a radioactively marked *tracer*, which is metabolized inside the body. One very important tracer is ^{18}F-FDG (2–(*Fluorine-18-Fluoro-2-Deoxy-D-Glucose*), which can be used to trace glucose metabolism. This is especially important for oncology studies, since, due to their faster metabolism, the ^{18}F-FDG uptake of tumors exceeds the uptake of non-malignant tissue. Compared with ordinary glucose, ^{18}F-FDG differs in the presence of the tracer atom. As such, it behaves like glucose only at the beginning of the metabolic chain. Thereafter, the molecule is detected, but not catabolized any further, leading to an accumulation of the tracer inside the tumor.

^{18}F-FDG is a positron emitter[4], so wherever the tracer accumulates, the process of positron annihilation is intensified. This process was shown as a daughter process of pair production in Fig. 2.15, and is described by (2.40). In the case of ^{18}F-FDG, however, a proton, p, decays into a neutron, n, a neutrino, ν, and a positron, e^+, as a parent process

$$p \rightarrow n + e^+ + \nu . \tag{3.2}$$

Upon collision with an electron close to where this decay occurs, the positron entirely de-materializes, becoming two gamma ray photons that fly away in opposite directions. The gamma ray photons are then measured by two detectors, located opposite each other, in what is called a coincidence measurement. By means of filtered backprojection, the location of the de-materialization can be reconstructed, and tumors are represented as *hot spots* inside the image. The mathematical fundamentals of the image reconstruction of PET techniques will be described later. For further details of the technique, and of applications of PET, the reader is referred, for example, to Ruhlmann et al. (1998).

An interesting approach to imaging diagnostics is the combination of both morphological and functional imaging methods. The goal of displaying function along with morphology in a single image has been realized for some time using methods of image registration[5]. Registration is an image processing step that must over-

---

[4] Positrons, e^+, are the antiparticles of electrons, e^-.

[5] Registration is an image processing step, aligning the coordinate systems of two modalities. A detailed illustration of these techniques can be found, e.g., in Hajnal et al. (2001).

come problems caused by the different positioning of the patient in two different scanners, and changes resulting from the time that passes between the two acquisitions.

In the case of combined techniques, the patient is successively scanned using the different imaging modalities. Using a combined PET-CT scanner, such as, for instance, the *biograph* from Siemens (Fig. 3.20, top), the images are acquired almost simultaneously, with the patient effectively in the same position, so that the location of a tumor relative to the surrounding anatomy can be displayed immediately. The bottom of Fig. 3.20 shows an intestinal tumor being imaged as a *hot spot*.

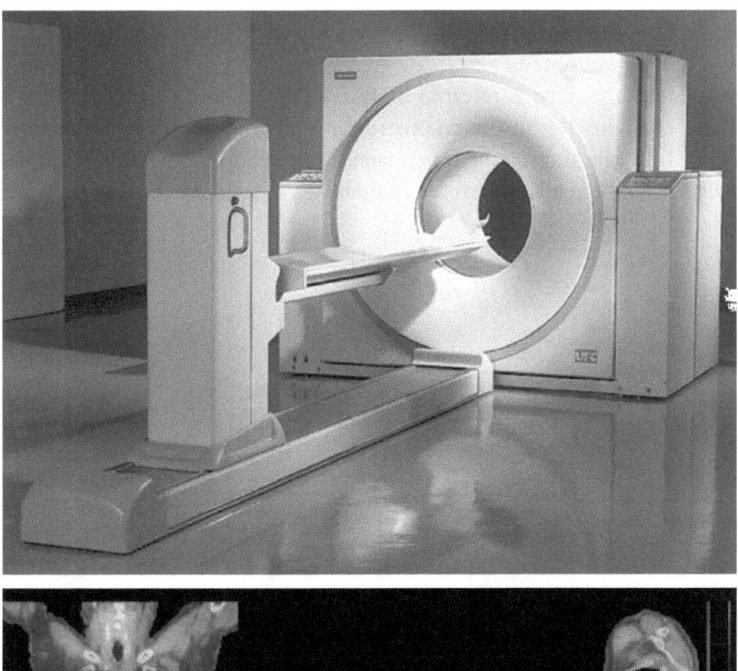

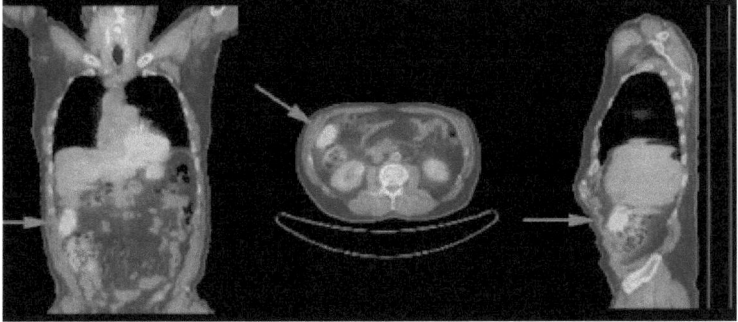

Fig. 3.20. The *biograph* scanner from Siemens is an integrated PET-CT concept that can image metabolic and anatomic information at the same time. Courtesy of Siemens Medical Solutions

3.12
Optical Reconstruction Techniques

The term "computed tomography" is historically connected to the development of computers, which have been successfully applied in the field of image reconstruction since the 1960s. It was mentioned in ▸ Sect. 3.3, that a practically applicable solution to the problem of image reconstruction was not possible before the introduction of computer technology. In fact, there is also the possibility of achieving reconstruction in an optico-photographic fashion. In 1977 Edholm introduced a tricky optical configuration, which is illustrated in Fig. 3.21.

In order to acquire an image, an X-ray fan is produced using slit collimation, which radiates through the object and whose projection profile is recorded on a film as a single line. The film is moved linearly and synchronously with the rotation of the object being examined.

The resulting pattern is referred to as a sinogram, and can be used for reconstruction. The first step in the reconstruction chain is the one-dimensional illumination of the sinogram, which is imaged by a cylinder lens onto a second film. The cylinder lens spreads the profile spatially. By displacing the sinogram film and

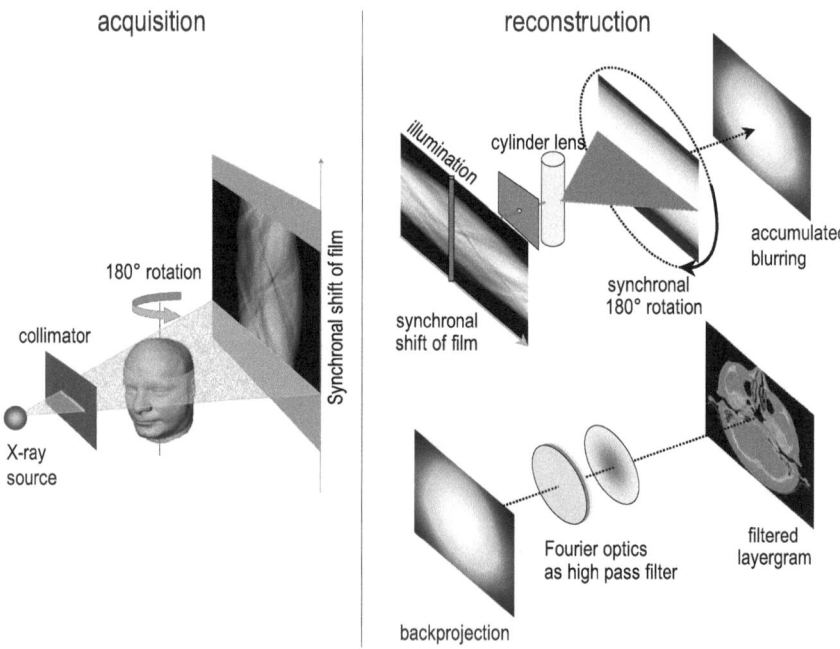

Fig. 3.21. Principle of the optical reconstruction method for analog CT based on the ideas of Edholm (1977)

simultaneously rotating the cylindrical lens, a simple backprojection is achieved since all projection profiles are superimposed. In a final step the resulting back-projection is high-pass filtered using Fourier optics to produce the image reconstruction. Nowadays, computer technology is so advanced, that results are superior to those of optical reconstruction. Therefore, optical techniques are no longer pursued.

4 Fundamentals of Signal Processing

Contents

4.1	Introduction	102
4.2	Signals	102
4.3	Fundamental Signals	102
4.4	Systems	104
4.5	Signal Transmission	106
4.6	Dirac's Delta Distribution	109
4.7	Dirac Comb	112
4.8	Impulse Response	115
4.9	Transfer Function	116
4.10	Fourier Transform	118
4.11	Convolution Theorem	124
4.12	Rayleigh's Theorem	125
4.13	Power Theorem	125
4.14	Filtering in the Frequency Domain	126
4.15	Hankel Transform	128
4.16	Abel Transform	132
4.17	Hilbert Transform	133
4.18	Sampling Theorem and Nyquist Criterion	135
4.19	Wiener–Khintchine Theorem	141
4.20	Fourier Transform of Discrete Signals	144
4.21	Finite Discrete Fourier Transform	145
4.22	z-Transform	147
4.23	Chirp z-Transform	148

4.1
Introduction

The entire theory of the continuously developing area of signal processing is beyond the scope of this chapter. In this chapter, only the principal methods of signal processing that are important for computed tomography (CT) will be described. In contrast to most essays on signal processing, descriptions of the signals in the spatial domain will be used – since it is in this form that signals are present in X-ray and CT imaging modalities.

4.2
Signals

The signals being considered here are one-dimensional spatial signals: $s(x)$ – for instance from a detector array of a CT scanner, or two-dimensional images: $f(x, y)$ – e.g., one slice of a 3D CT image. The signals are either continuous or, after sampling, discrete functions of one or more variables.

4.3
Fundamental Signals

In signal processing, some fundamental signals are extraordinary in the sense that the deformation of a signal throughout its transmission enables conclusions to be drawn about the system itself. These fundamental signals are given in (4.1) to (4.7), as well as in Fig. 4.1.

Heaviside step function

$$\text{step}(x) = \begin{cases} 1 & x \geq 0 \\ 0 & \text{otherwise} \end{cases} \qquad (4.1)$$

Rectangular function

$$\text{rect}(x) = \begin{cases} 1 & |x| \leq 1/2 \\ 0 & \text{otherwise} \end{cases} \qquad (4.2)$$

Triangle function

$$\text{tri}(x) = \begin{cases} 1 - |x| & |x| \leq 1 \\ 0 & \text{otherwise} \end{cases} \qquad (4.3)$$

Sinc function

$$\text{si}(x) \equiv \text{sinc}(x) = \frac{\sin(\pi x)}{\pi x} \qquad (4.4)$$

Gaussian function

$$\text{gauss}(x) = \frac{1}{\sqrt{\pi}} e^{-x^2} \qquad (4.5)$$

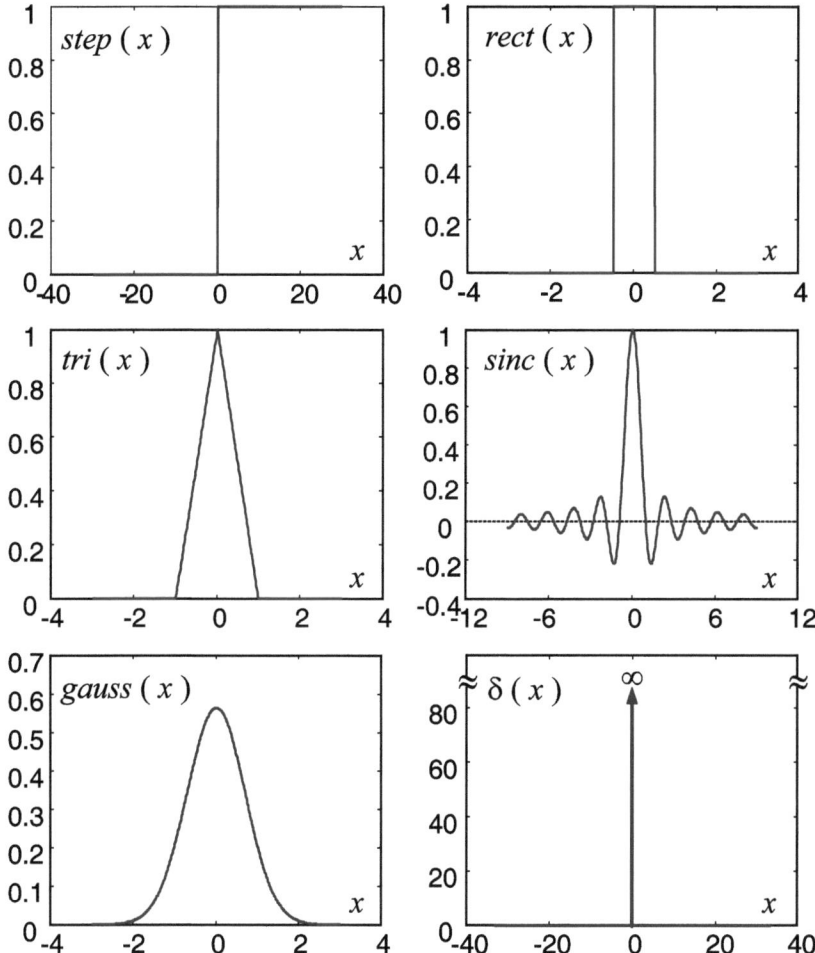

Fig. 4.1. Illustration of some of the fundamental functions of signal processing. By analyzing the changes that these signals experience within linear and complex systems, conclusions can be drawn about the systems themselves. Dirac's *delta* "function" is of particular importance. It is connected to the "impulse response" of the system, which will be used frequently in later chapters. Due to the importance of the *delta* impulse, which is actually a distribution, this "function" is described in detail in ▸ Sects. 4.6–4.7

Delta distribution

$$\int\limits_{-\infty}^{+\infty} f(x)\delta(x - x_0)\,\mathrm{d}x = f(x_0) \qquad (4.6)$$

or formally,

$$\delta(x - x_0) = \begin{cases} 0 & x \neq x_0 \\ \infty & x = x_0 \end{cases} \quad \text{and} \quad \int\limits_{-\infty}^{+\infty} \delta(x - x_0)\,\mathrm{d}x = 1 . \quad (4.7)$$

4.4
Systems

Systems process signals. In particular, one refers to a "transmission system," and describes it by an output signal, $g(x)$, corresponding to a specific input signal, $s(x)$. The transformation equation can be formally written as

$$g(x) = \mathcal{L}\{s(x)\} \qquad (4.8)$$
$$g(x, y) = \mathcal{L}\{f(x, y)\} . \qquad (4.9)$$

A transformation can also be represented schematically by a block diagram, as shown in Fig. 4.2.

In general, transmission systems are described in terms of their attributes *linearity*, *spatial invariance*, *shift-invariance*, *causality*, and *stability*. These properties are briefly introduced in the following sections.

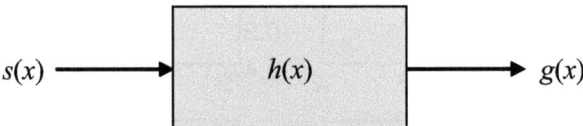

Fig. 4.2. Schematic block diagram illustrating the signal transmission. $h(x)$ is called the impulse response

4.4.1
Linearity

A system is linear if, for all $a_i \in \mathbb{C}$,

$$\mathcal{L}\left\{\sum_i a_i s_i(x)\right\} = \sum_i a_i g_i(x) , \qquad (4.10)$$

or, in the two-dimensional case,

$$\mathcal{L}\left\{\sum_i a_i f_i(x, y)\right\} = \sum_i a_i g_i(x, y) . \qquad (4.11)$$

An example of a linear transmission system in image processing is an edge filter for horizontal edges, which, in its simplest form, can be described by

$$\mathcal{L}\{f_i(x, y)\} \equiv f_i(x, y) - f_i(x, y + 1) = g_i(x, y) . \qquad (4.12)$$

The edge filter detects gray-value changes in images, like those provided in Fig. 4.3. The two images on the left of Fig. 4.3 contain horizontal gray-value edges with a step

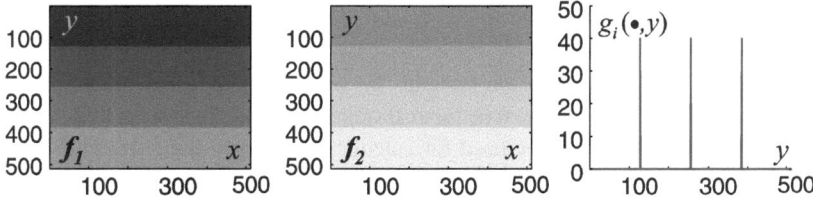

Fig. 4.3. Due to the linearity of the edge filter, two images that differ in their absolute gray-scale values, but which have horizontal edges of the same step size, yield the same response. The response, $g_i(\bullet, y)$, is illustrated on the right as a vertical profile

size of 40 units of gray-scale intensity at each step. The gray-value level in the middle image is 100 units higher than in the image on the left. It is important that the edge filter gives the same response to the gray-value change in the left image

$$f_1(x, y) = 40 \rightarrow f_1(x, y + 1) = 80 \tag{4.13}$$

as it does to the gray-value change in the central image

$$f_2(x, y) = 140 \rightarrow f_2(x, y + 1) = 180 . \tag{4.14}$$

The response of the edge filter is displayed as a profile in Fig. 4.3 (right). The horizontal edges yield to the same response for both images.

4.4.2
Position or Translation Invariance

A transmission system is said to be position invariant, or alternatively shift invariant if, for arbitrary x_0, y_0

$$\mathcal{L}\{s(x - x_0)\} = g(x - x_0) \tag{4.15}$$

or, in the two-dimensional case

$$\mathcal{L}\{f(x - x_0, y - y_0)\} = g(x - x_0, y - y_0) . \tag{4.16}$$

Although (4.15) and (4.16) look like self-evident properties, one cannot assume that real systems are position or translation invariant. For example, image distortion can violate these properties. A common acronym for linear translation invariant systems is "LTI systems."

4.4.3
Isotropy and Rotation Invariance

During the transmission of generalized images with two or more dimensions, their shape does not change in the presence of isotropy or rotation invariance. The example of an edge filter provided in Fig. 4.3 is only an isotropic system if the result of the filter is independent of the orientation of the edge and always yields the same response.

4.4.4
Causality

In the case of causal transmission systems, the output signal is not known prior to the input signal. This is always true for real online systems. In system theory, non-causal systems are likely to be used for calculations, due to their straightforward mathematical treatment (Lüke 1999). Strictly speaking, the term "causality" is only applicable in the case of time signals. Generally, in image processing applications, an image is present in its entirety and signal processing takes place with respect to its spatial coordinates (Werner 2000).

4.4.5
Stability

A system is said to be "amplitude-stable" if it responds to an amplitude-bounded input signal with an amplitude-bounded output signal. Those systems are also called "BIBO" systems (*bounded input–bounded output*). In engineering, unstable systems can be problematic due to, for instance, the overflow of a number format that can cause feedback processes in a software module to be controlled by absurd results.

4.5
Signal Transmission

To analyze transmission systems, the fundamental signals (4.1) to (4.7) are used because their modulation allows conclusions to be made quite easily about the system. The concept of the utilization of fundamental signals will be explained in the following paragraphs. One of these signals is the normalized rectangular function

$$s_0(x) = \frac{1}{X_0} \, \text{rect}\left(\frac{x}{X_0}\right), \tag{4.17}$$

with width X_0 and height $1/X_0$. This function is based on the fundamental rectangular signal (4.2).

Figure 4.4 shows how a signal consisting of two consecutive normalized rectangular signals is deformed by a linear transmission system. The deformation is

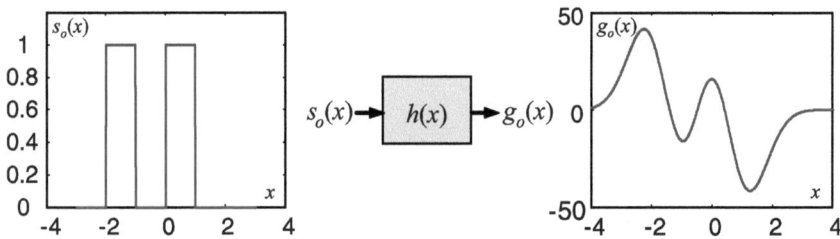

Fig. 4.4. Transformation via transmission of the input signal, $s_0(x)$, consisting of two consecutive rectangular signals. The linear transmission system is described by $h(x)$, which is called the "impulse response" of the system

given by the system's specific property, which is described by what is called the "impulse response," $h(x)$. The impulse response of a system will be described in detail in ▶ Sect. 4.8.

To aid the understanding of sampling, and therefore digitization of an analog signal, the step function based on the rectangular function is used as an approximation of an arbitrary signal, $s(x)$. Therefor, the so-called gate property of the rectangular function is used. Figure 4.5 shows this behavior where a segment of a waveform of finite length is selected. The analog signal, $s(x)$, is approximated at the position nX_0 by the rectangular function

$$re(x) = s(nX_0) \operatorname{rect}\left(\frac{x - nX_0}{X_0}\right), \tag{4.18}$$

where $s(nX_0)$ denotes the height of the signal, $s(x)$, at the position nX_0 and, the *rect* function is responsible for the gating.

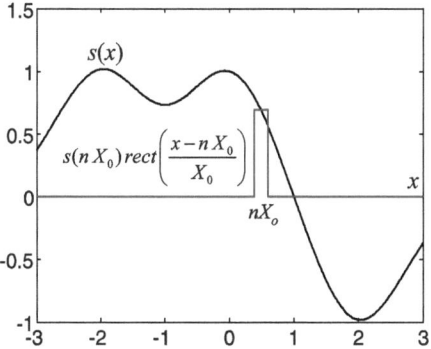

Fig. 4.5. Gate property of the rectangular function. The analog signal, $s(x)$, is approximated by the rectangular function having width X_0 and height $s(nX_0)$ at the position nX_0

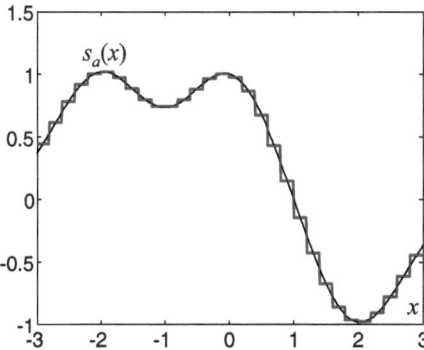

Fig. 4.6. Approximation of a signal by a step function. The smaller the width of consecutive rectangles, the more accurately the step function approximates the original analog signal

For the approximation of the entire signal, $s(x)$, rectangular functions are combined resulting in the "step function," as follows

$$s(x) \approx s_a(x) = \sum_{n=-\infty}^{\infty} s(nX_0) \operatorname{rect}\left(\frac{x - nX_0}{X_0}\right). \tag{4.19}$$

Figure 4.6 illustrates how an analog signal is approximated discretely by a sequence of rectangles, which are normalized accordingly.

The smaller the width of each rectangle, the more accurately the step function approximates the analog signal. Using the definition of the normalized rectangular function (4.17) gives

$$s_a(x) = \sum_{n=-\infty}^{\infty} s(nX_0)s_0(x - nX_0)X_0. \tag{4.20}$$

If the transmission system is linear and position- or translation-invariant, the transmitted signal satisfies

$$g(x) \approx g_a(x) = \sum_{n=-\infty}^{\infty} s(nX_0)g_0(x - nX_0)X_0. \tag{4.21}$$

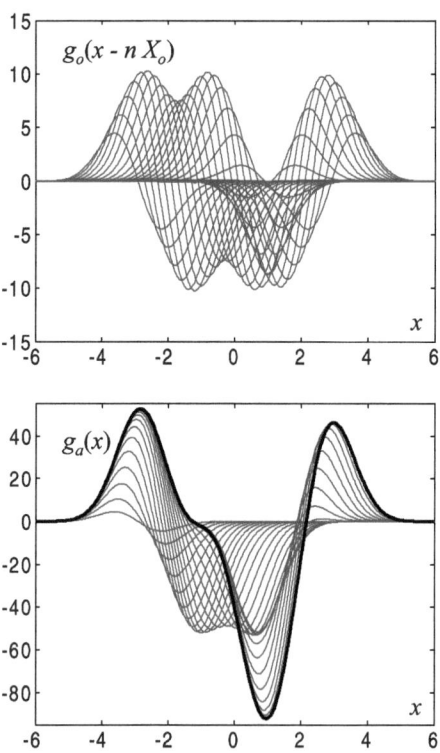

Fig. 4.7. Approximation of the output signal by superposition of individual system responses. The *bottom image* shows the result of the summing, as well as the convergence of the sum of the respective system responses of the particular fundamental rectangular signals – shown in the *top image* – toward the overall system response (cf. Fig. 4.11)

In this way, the approximated output signal results from the superposition of the system response to the differently weighted rectangular impulses. Figure 4.7 shows the individual system responses to the rectangular impulses and the result of the summation. At the top of Fig. 4.7, the individual responses to all rectangles of the step function from Fig. 4.6 are shown.

The description of an entire transmission by a summation of individual responses, as given by (4.21), is a typical characteristic of linear systems. This is also to be seen in the theory of linear differential equations, where the sum of weighted, individual results represents the general result. At the bottom of Fig. 4.7, it can be seen how the sum of the individually transmitted fundamental signals converges with the overall result (cf. also Fig. 4.11).

4.6
Dirac's Delta Distribution

The signal $s(x)$, as introduced in the previous paragraph, is more accurately approximated by using a normalized rectangular impulse of smaller width X_0 in (4.17), while the area of the impulse remains unity in its normalized representation. Hence, the narrower the impulse, the larger the amplitude, $1/X_0$.

The limit

$$\delta(x) = \lim_{X_0 \to 0} \frac{1}{X_0} \operatorname{rect}\left(\frac{x}{X_0}\right) \tag{4.22}$$

defines Dirac's delta impulse, or more precisely the δ-distribution, which was introduced above as an elementary "function" in (4.6) and (4.7). Figure 4.8 shows schematically how (4.22) tends toward the limit. The limit gives a "spike" impulse of infinite height and vanishing width.

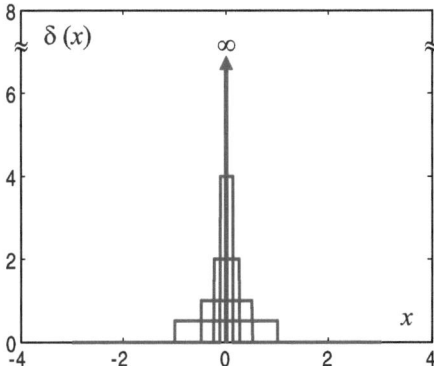

Fig. 4.8. Tendency toward the limit from a normalized rectangular impulse to a spike impulse of infinite height. The resulting δ-distribution is of great importance in the theory of sampling of analog signals

It should be noted that alternative limit representations of the δ-distribution exist, such as

$$\delta(x) = \lim_{X_0 \to 0} \frac{1}{2X_0} e^{\left(\frac{-|x|}{X_0}\right)} \tag{4.23}$$

or

$$\delta(x) = \lim_{\varepsilon \to \infty} \frac{\sin(\pi\varepsilon x)}{\pi x} . \tag{4.24}$$

These will be used later on (cf. (4.179)). Further representations can be found in Barrett and Swindell (1981).

The application of the limit in (4.20) means that the location of the gating must be replaced by

$$nX_0 \to \xi \tag{4.25}$$

and the width of the gating by

$$X_o \to d\xi . \tag{4.26}$$

In this way one obtains the important property

$$s(x) = \int_{-\infty}^{+\infty} s(\xi)\delta(x - \xi)\,d\xi$$

$$= s(x) * \delta(x) , \tag{4.27}$$

or for the two-dimensional case of an image

$$f(x, y) = \int_{-\infty}^{+\infty} \int_{-\infty}^{+\infty} f(\xi, \eta)\delta(x - \xi, y - \eta)\,d\xi\,d\eta$$

$$= f(x, y) * \delta(x, y) , \tag{4.28}$$

which is a so-called convolution. (4.27) is symmetric with respect to the permutation of the integration variables, such that

$$s(\xi) = \int_{-\infty}^{+\infty} s(x)\delta(x - \xi)\,dx \tag{4.29}$$

is satisfied. By simple substitution, it holds that

$$s(-\xi) = \int_{-\infty}^{+\infty} s(y - \xi)\delta(y)\,dy . \tag{4.30}$$

(4.29) is called the "sifting property" of the δ-distribution.

Two further properties are mentioned explicitly here, as they will be needed in the following chapters. First of all, the scaling property for $\delta(ax)$, with $a \neq 0$, should be analyzed. The familiar interpretation as a stretching of the function is not applicable in the case of an impulse of width zero. However, from the definition in (4.22), it follows that the normalization of the integral is given as

$$\int_{-\infty}^{+\infty} \delta(x)\,dx = 1 ,$$ (4.31)

which raises the question once again, of what meaning $\delta(ax)$ has as an integrand. To address this question, consider the simple substitution $y = ax$, resulting in

$$\int_{-\infty}^{+\infty} \delta(ax)\,dx = \frac{1}{a}\int_{-\infty}^{+\infty} \delta(y)\,dy = \frac{1}{a}\int_{-\infty}^{+\infty} \delta(x)\,dx .$$ (4.32)

Since the δ-distribution is symmetric, $\delta(-x) = \delta(x)$, and therefore in general,

$$\delta(ax) = \frac{1}{|a|}\delta(x) .$$ (4.33)

Occasionally, the argument of the δ-distribution itself is a complicated function, $g(x)$, of the spatial variable x. This further property, which will be required in ▸ Sect. 5.6 for the calculation of the properties of the simple backprojection, is briefly described here.

If $g(x)$ has a single root at $x = x_0$, it is evident that $\delta(g(x))$ vanishes everywhere except for the infinitesimal neighborhood of $x = x_0$. Taking the Taylor expansion of $g(x)$ in the region of the root x_0 gives

$$\delta(g(x)) = \delta\left(g(x_0) + (x - x_0) \left.\frac{dg}{dx}\right|_{x_0} \right) .$$ (4.34)

Higher order terms of the Taylor expansion are not required, as the neighborhood around x_0 is arbitrarily small. Since $g(x_0) = 0$ is satisfied, it immediately follows that

$$\delta(g(x)) = \frac{\delta(x - x_0)}{\left| \left.\dfrac{dg}{dx}\right|_{x_0}\right|} .$$ (4.35)

In this way, one can exploit the scaling property, (4.33), of the δ-distribution. If the function $g(x)$ has more than one root, (4.35) can be generalized by

$$\delta(g(x)) = \sum_i \frac{\delta(x - x_i)}{\left| \left.\dfrac{dg}{dx}\right|_{x_i}\right|} .$$ (4.47)

The fundamental properties of the δ-distribution are given in Table 4.1. Additional properties can be found, for example, in Messiah (1981), Klingen (2001) or in Bracewell (1965).

(4.36) to (4.46) imply that one is allowed to replace the left side with the expression on the right side, if it occurs as an integrand of an integral over x.

Table 4.1. Fundamental properties of the δ-distribution according to Messiah (1981), Klingel (2001), and Bracewell (1965)

$$\delta(x) = \delta(-x) \tag{4.36}$$

$$a\delta(x) + b\delta(x) = (a + b)\delta(x) \tag{4.37}$$

$$\delta(g(x)) = \sum_i |g'(x_i)|^{-1}\,\delta(x - x_i) \tag{4.38}$$

with x_i being single roots of $g(x)$. Especially for $g(x) = ax$ it holds that

$$\delta(ax) = \frac{1}{|a|}\delta(x)\,, \text{ if } a \neq 0 \tag{4.39}$$

$$x\delta(x) = 0 \tag{4.40}$$

$$f(x)\delta(x - a) = f(a)\delta(x - a) \tag{4.41}$$

$$\int \delta(x - y)\delta(y - a)\,dy = \delta(x - a) \tag{4.42}$$

$$\delta(x - a) * \delta(x - b) = \delta(x - a - b) \tag{4.43}$$

$$\frac{d}{dx}\,\text{step}(x) = \delta(x) \tag{4.44}$$

$$\delta(x - a)\delta(x - b) = 0 \qquad \text{if } a \neq b^1 \tag{4.45}$$

$$\delta(x) = \frac{1}{2\pi}\int_{-\infty}^{+\infty} e^{ikx}\,dk \tag{4.46}$$

4.7
Dirac Comb

The sift property (4.29) of the δ-distribution is used for sampling an analog signal. This is to be done using the so-called Dirac comb

$$s_a(x) = \sum_{n=-\infty}^{+\infty} s(nX_0)\delta(x - nX_0) = s(x)\sum_{n=-\infty}^{+\infty} \delta(x - nX_0)\,. \tag{4.48}$$

Figure 4.9 shows how a function, $s(x)$, is sampled by the Dirac comb. Unlike the step function (4.19), which approximates the analog signal, the periodic, needle-shaped δ-distribution now measures the analog signal precisely at discrete points.

---

[1] The expression $\delta(x)\delta(x)$ is not defined.

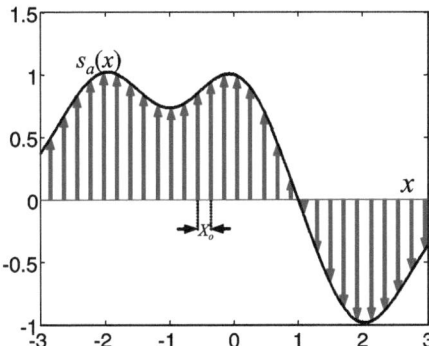

Fig. 4.9. The process of sampling an analog signal using a Dirac comb. Contrary to the concept of the step function in Fig. 4.6, here the signal $s(x)$ is sampled at equidistant discrete points with intervals of X_0. The comb is represented by *arrows* for visualization purposes only. Mathematically, the height of Dirac's delta impulse does not vary; rather, the sampled values are the weights of Dirac's delta impulse train

Due to the appearance of the impulse train, Bracewell (1965) denoted the Dirac comb by the Cyrillic letter Ш (pronounced "*shah*"), which is therefore shown by

$$s_a(x) = s(x)\,\text{Ш}(x) \tag{4.49}$$

where

$$\text{Ш}(x) = \sum_{n=-\infty}^{+\infty} \delta(x - nX_0)\,. \tag{4.50}$$

For CT the sift property has to be expressed two-dimensionally. In two dimensions, it is possible to construct δ-lines from spike impulses, $\delta(x, y)$, which can be considered as a contiguous "lining-up" of δ-spike impulses. Figure 4.10 shows how an anatomical object (in this case a section through the abdomen) is sampled along parallel lines.

Integration of the attenuation values, which are given by the spatial distribution $f(x, y)$, can be written as a one-dimensional integral in a Cartesian coordinate system, (ξ, η), which is rotated by angle γ with respect to the (x, y) patient frame

$$p_\gamma(\xi) = \int_{-\infty}^{+\infty} f(\xi\cos(\gamma) - \eta\sin(\gamma),\, \xi\sin(\gamma) + \eta\cos(\gamma))\,\mathrm{d}\eta\,. \tag{4.51}$$

The η-axis is aligned with the direction of a given projection, i.e., the lines along which attenuation information is summed up by the integration. In the particular case of CT, this is the direction of the X-rays.

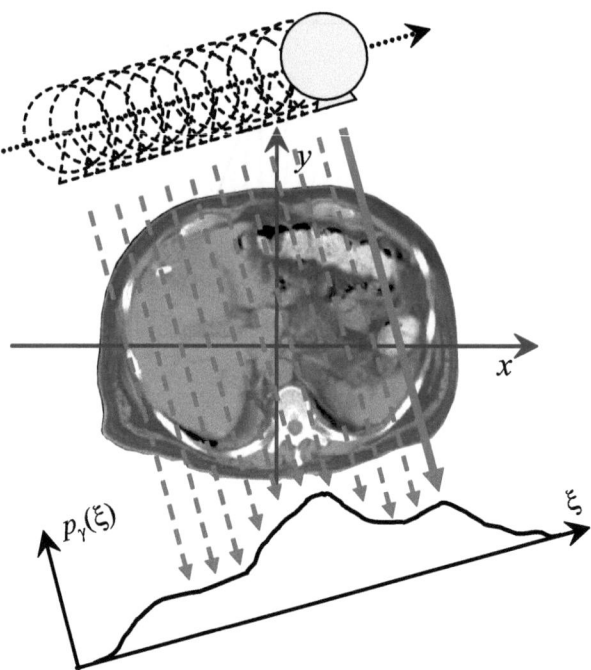

Fig. 4.10. Two-dimensional spatial sampling of an object along parallel lines in computed tomography. The response, $p_y(\xi)$, measures the sum of all attenuation values along the X-ray lines of the current projection angle, y. The signal of interest is the spatial distribution of the attenuation values, $f(x, y)$

(4.51) can also be described as a convolution with a δ-line, where the sift property of the δ-distribution is employed, giving

$$p_y(\xi) = f(x, y) * \delta(\mathbf{L}) = \iint\limits_{(x,y) \in \mathbb{R}^2} f(x, y)\delta((x, y) - \mathbf{L})\,dx\,dy \qquad (4.52)$$

or alternatively

$$f * \delta(\mathbf{L}) = \int f(\mathbf{r})\delta(\mathbf{r} - \mathbf{L})\,d\mathbf{r} = \int\limits_{\mathbf{r} \in \mathbf{L}} f(\mathbf{r})\,d\mathbf{r}. \qquad (4.53)$$

In this way, one obtains the projection of all values on the path of the X-ray along the line $\mathbf{L}$ through the object onto the ξ-axis.

4.8
Impulse Response

The way in which physical objects and systems are examined always comes down to the same principle. One tries to excite the system and waits for the response. If one has a detailed knowledge of the excitation, many properties of the system can be determined well. An exceptionally simple excitation of a system can be achieved by Dirac's delta impulse.

The response of a transmission system to the delta distribution, or δ-impulse, is given by

$$s(\xi) = \int_{-\infty}^{+\infty} s(x)\delta(\xi - x)\,dx, \tag{4.54}$$

which is essentially exploiting the sift property. In the case of a linear transmission system

$$g(x) = \mathcal{L}\{s(x)\}, \tag{4.55}$$

it reads

$$g(x) = s(x) * h(x) = \int_{-\infty}^{+\infty} s(\xi)h(x - \xi)\,d\xi, \tag{4.56}$$

where $h(x)$ denotes the already mentioned impulse response.

Figure 4.11 gives an example of the transmission of a signal $s(x)$ via the impulse response

$$h(x) = -\frac{2x}{\sqrt{\pi}}e^{-x^2}. \tag{4.57}$$

Note that the arbitrarily chosen impulse response function (4.57) is the same as the one used in ▸ Sect. 4.5.

The response given by (4.56) can be extended for two-dimensional systems by

$$g(x, y) = \int_{-\infty}^{+\infty}\int_{-\infty}^{+\infty} f(\xi, \eta)h(x - \xi, y - \eta)\,d\xi\,d\eta. \tag{4.58}$$

(4.58) describes the fundamental equation of system theory for imaging systems. The response of the system is called the *point-spread function*. It follows that a linear

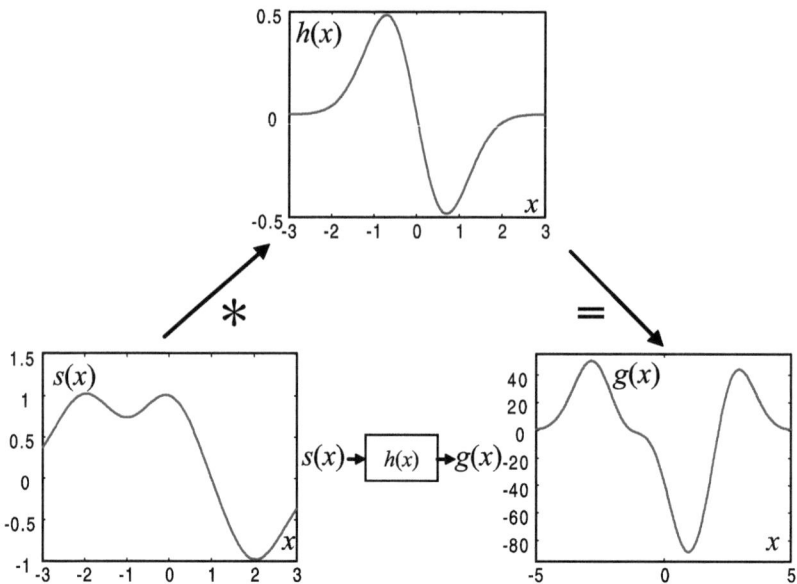

Fig. 4.11. Transmission of a signal. The transmitted signal is given by the convolution of the signal $s(x)$ with the system's impulse response $h(x)$

transmission system with a point-spread function, $h(x, y)$, responds to an input image, $f(x, y)$, with an output image

$$g(x, y) = f(x, y) * h(x, y),\qquad(4.59)$$

where (4.59) is the standard notation used to abbreviate (4.58).

4.9
Transfer Function

The question of how to determine the impulse response of a system still remains. For this purpose, specific functions are applied to the system of interest. Consider the following eigenvalue problem

$$\mathcal{L}\sigma = H\sigma,\qquad(4.60)$$

where $\mathcal{L}$ is again a linear operator, σ is a so-called eigenfunction of the system and H denotes a constant, i.e., the system's eigenvalue. Let the linear operation $\mathcal{L}$ be given by

$$\mathcal{L}\sigma \equiv \sigma * h(x).\qquad(4.61)$$

The eigenfunctions can be described by

$$\sigma(x) = e^{i2\pi ux} = \cos(2\pi ux) + i\sin(2\pi ux),\qquad(4.62)$$

where u is the spatial frequency of the system. The application of the linear operator to the eigenfunction gives

$$g(x) = \sigma(x) * h(x)$$

$$= \int_{-\infty}^{+\infty} h(\xi) e^{i2\pi u(x-\xi)} d\xi \qquad (4.63)$$

$$= e^{i2\pi ux} \int_{-\infty}^{+\infty} h(\xi) e^{-i2\pi u\xi} d\xi .$$

Hence, the eigenvalues are given by

$$H(u) = \int_{-\infty}^{+\infty} h(x) e^{-i2\pi ux} dx . \qquad (4.64)$$

The eigenvalues $H(u)$ provide the amplitudes and phases of the system that depend on the spatial frequency.

In the case of an image, (4.63) can again be extended so that in the two-dimensional domain,

$$H(u,v) = \int_{-\infty}^{+\infty} \int_{-\infty}^{+\infty} h(x,y) e^{-i2\pi(ux+vy)} dx \, dy \qquad (4.65)$$

is valid, where u denotes the spatial frequency in the x direction and v the spatial frequency in the y direction. If one defines

$$u = \frac{1}{\lambda_x}, \quad v = \frac{1}{\lambda_y} \quad \text{and} \quad k_x = \frac{2\pi}{\lambda_x}, \quad k_y = \frac{2\pi}{\lambda_y}, \qquad (4.66)$$

the k-space representation of the Fourier transform

$$H(k_x, k_y) = \int_{-\infty}^{+\infty} \int_{-\infty}^{+\infty} h(x,y) e^{-i(k_x x + k_y y)} dx \, dy \qquad (4.67)$$

can be obtained. This equation describes the Fourier transform, whose formal notation in one dimension is often given by

$$H = \mathcal{F}\{h\} \quad \text{or} \quad h \circ\!\!-\!\!\bullet H .$$

As one is interested in h, the Fourier transform must be inverted, resulting in

$$h = \mathcal{F}^{-1}\{H\} \qquad (4.68)$$

or

$$h(x) = \int_{-\infty}^{+\infty} H(u) e^{i2\pi ux} du , \qquad (4.69)$$

or alternatively for the *point-spread function* in the two-dimensional case

$$h(x,y) = \int_{-\infty}^{+\infty} \int_{-\infty}^{+\infty} H(u,v) e^{i2\pi(ux+vy)} du \, dv . \qquad (4.70)$$

This means that the transfer function of a system is the Fourier transform of the corresponding impulse response (Lüke 1999). Please note, if the k-space notation is used for the inverse Fourier transform – analog to (4.67) – a normalization factor of $(2\pi)^{-1}$ must be introduced for each dimension.

4.10
Fourier Transform

Due to the importance of the Fourier transform not only in the field of signal processing for CT, it is briefly introduced here. In the upcoming chapters, in which the reconstruction mathematics of CT will be considered, the following definition of a function's Fourier transform is used.

If $f(x)$ is a real or complex-valued function of the variable x, then its Fourier transform, if it exists, is the function

$$F(u) = \left(\frac{\alpha}{2\pi}\right)^{\frac{1}{2}} \int_{-\infty}^{+\infty} f(x)\,e^{-i\alpha ux}\,dx \equiv \mathcal{F}\{f(x)\}\,, \tag{4.71}$$

where α is a constant whose origin is different in the field of signal processing from that in other applications. In quantum mechanics, for example, one often chooses the reciprocal of Planck's constant (normalized by $1/2\pi$), which is $\alpha = 1/\hbar$. Throughout later chapters, the definition $\alpha = 2\pi$ is used.

$f(x)$ results from $F(u)$ by inversion of the Fourier transform

$$f(x) = \left(\frac{\alpha}{2\pi}\right)^{\frac{1}{2}} \int_{-\infty}^{+\infty} F(u)\,e^{i\alpha ux}\,du \equiv \mathcal{F}^{-1}\{F(u)\}\,. \tag{4.72}$$

Using this symmetric definition allows us to avoid confusion regarding which normalization term must be written in front of the integral of the transform and its inverse. In more general terms, with $f(x_1, x_2, x_3, \ldots, x_n)$ being a function of the n variables $x_1, x_2, x_3, \ldots, x_n$, the Fourier transform is defined by

$$F(u_1, \ldots, u_n) = \left(\frac{\alpha}{2\pi}\right)^{\frac{1}{2}n} \int_{-\infty}^{+\infty} \cdots \int_{-\infty}^{+\infty} f(x_1, \ldots, x_n)\,e^{-i\alpha(u_1 x_1 + \ldots + u_n x_n)}\,dx_1 \ldots dx_n \tag{4.73}$$

and its inverse[2] is given by

$$f(x_1, \ldots, x_n) = \left(\frac{\alpha}{2\pi}\right)^{\frac{1}{2}n} \int_{-\infty}^{+\infty} \cdots \int_{-\infty}^{+\infty} F(u_1, \ldots, u_n)\,e^{i\alpha(u_1 x_1 + \ldots + u_n x_n)}\,du_1 \ldots du_n\,. \tag{4.74}$$

Important properties of the Fourier transform are summarized in Table 4.2.

---

[2] It very often happens that the forward transform and the inverse transform are of unequal difficulty.

Table 4.2. Important properties of the Fourier transform according to Messiah (1981), Klingel (2001), and Bracewell (1965)

$f(x) =$	$F(u) =$	
$\left(\dfrac{\alpha}{2\pi}\right)^{\frac{1}{2}} \displaystyle\int_{-\infty}^{+\infty} F(u)\,e^{i\alpha ux}\,du$	$\left(\dfrac{\alpha}{2\pi}\right)^{\frac{1}{2}} \displaystyle\int_{-\infty}^{+\infty} f(x)\,e^{-i\alpha ux}\,dx$	(4.75)

Linearity

$af(x) + bg(x)$	$aF(u) + bG(u)$	(4.76)

Argument scaling

$f\left(\dfrac{x}{c}\right)$	$|c|\,F(cu)$	(4.77)
$|c|\,f(cx)$	$F\left(\dfrac{u}{c}\right)$	(4.78)
$f(-x)$	$F(-u)$	(4.79)

Transform of the complex conjugate

$f^*(x)$	$F^*(-u)$	(4.80)
$F^*(x)$	$f^*(u)$	(4.81)

Transform of the transform

$F(x)$	$f(-u)$	(4.82)

Derivative of the transform

$xf(x)$	$\dfrac{i}{\alpha}F'(u)$	(4.83)

Derivative of the original function

$f'(x)$	$i\alpha u F(u)$	(4.84)

Argument shifting

$f(x - x_0)$	$e^{-i\alpha ux_0}F(u)$	(4.85)
$e^{i\alpha u_0 x}f(x)$	$F(u - u_0)$	(4.86)

Transform of special functions

$\delta(x)$	$\left(\dfrac{\alpha}{2\pi}\right)^{\frac{1}{2}}$	(4.87)
$\delta(x - x_0)$	$\left(\dfrac{\alpha}{2\pi}\right)^{\frac{1}{2}} e^{-i\alpha ux_0}$	(4.88)
$\text{step}(x)$	$\dfrac{1}{\sqrt{2\pi\alpha}}\left(\pi\delta(u) - \dfrac{i}{u}\right)$	(4.89)
$\left(\dfrac{\chi}{\sqrt{\pi}}\right)^{\frac{1}{2}} e^{-\frac{1}{2}\chi^2 x^2}$	$\left(\dfrac{\alpha}{\chi\sqrt{\pi}}\right)^{\frac{1}{2}} e^{-\frac{1}{2}\frac{\alpha^2 u^2}{\chi^2}}$	(4.90)

Provided that a function, f, satisfies the mean value condition,

$$\mathcal{F}\{f(x)\} = (F(u^-) + F(u^+))/2 \,, \tag{4.91}$$

at discontinuity points, i.e., the mean of the unequal limits of $F(u)$ where u^- and u^+ denote the leftward and the dexter limit, the following equations hold

$$\mathcal{F}^{-1}\{\mathcal{F}\{f(x)\}\} = f(x) \,, \tag{4.92}$$

or alternatively

$$\mathcal{F}\{\mathcal{F}^{-1}\{F(u)\}\} = F(u) \,, \tag{4.93}$$

as equations of identical functions. This can be seen if one explicitly writes down the transformation, giving

$$f(x) = \mathcal{F}^{-1}\left\{\mathcal{F}\{f(x)\}\right\} = \mathcal{F}^{-1}\left\{\int\limits_{-\infty}^{+\infty} f(x) e^{-i2\pi ux}\, dx\right\} \,. \tag{4.94}$$

The application of the inverse transform requires renaming of the spatial variables, as otherwise the non-quadratic elements would not be considered in the following double integral. Therefore, one writes

$$f(x) = \int\limits_{-\infty}^{+\infty}\left\{\int\limits_{-\infty}^{+\infty} f(\xi) e^{-i2\pi u\xi}\, d\xi\right\} e^{i2\pi ux}\, du \,. \tag{4.95}$$

As the integration order inside the double integral can be permuted, it holds that

$$f(x) = \int\limits_{-\infty}^{+\infty}\int\limits_{-\infty}^{+\infty} f(\xi) e^{-i2\pi u(\xi-x)}\, d\xi du$$

$$= \int\limits_{-\infty}^{+\infty} f(\xi)\left\{\int\limits_{-\infty}^{+\infty} e^{-i2\pi u(\xi-x)}\, du\right\} d\xi \tag{4.96}$$

is true and with the definition of the δ-distribution (4.46), it finally follows that

$$f(x) = \int\limits_{-\infty}^{+\infty} f(\xi)\delta(\xi - x)\, d\xi \tag{4.97}$$

is also true, which comes from the sift property introduced above.

In this section, the Fourier transform of two specific functions, which will become more important in later chapters, will be calculated explicitly.

First of all, the Fourier transform of the two-dimensional rectangular function

$$\mathrm{rect}_\varepsilon(x, y) = \begin{cases} 1 & |x| \le \varepsilon/2 \text{ and } |y| \le \varepsilon/2 \\ 0 & \text{otherwise} \end{cases} \tag{4.98}$$

is determined, where ε is the width of the rectangular window. This gives the Fourier transform

$$F(u, v) = \int_{-\infty}^{+\infty} \int_{-\infty}^{+\infty} \mathrm{rect}_\varepsilon(x, y) e^{-i2\pi(ux+vy)} \, dx \, dy \equiv \mathcal{F}\{f(x, y)\} . \tag{4.99}$$

By inserting the relevant integration limits into (4.99), the integral is reduced to

$$\begin{aligned} F(u, v) &= \int_{-\varepsilon/2}^{+\varepsilon/2} \int_{-\varepsilon/2}^{+\varepsilon/2} e^{-i2\pi(ux+vy)} \, dx \, dy \\ &= \int_{-\varepsilon/2}^{+\varepsilon/2} e^{-i2\pi ux} \, dx \int_{-\varepsilon/2}^{+\varepsilon/2} e^{-i2\pi vy} \, dy \\ &= \left[-\frac{1}{i2\pi u} e^{-i2\pi ux} \right]_{-\varepsilon/2}^{+\varepsilon/2} \left[-\frac{1}{i2\pi v} e^{-i2\pi vy} \right]_{-\varepsilon/2}^{+\varepsilon/2} . \end{aligned} \tag{4.100}$$

In this way, the integral (4.99) can be solved, and by substituting the limits, one is able to recognize the complex notation of the *sine* function,

$$\begin{aligned} F(u, v) &= \frac{e^{+i\pi\varepsilon u} - e^{-i\pi\varepsilon u}}{i2\pi u} \frac{e^{+i\pi\varepsilon v} - e^{-i\pi\varepsilon v}}{i2\pi v} \\ &= \frac{\sin(\pi\varepsilon u)}{\pi u} \frac{\sin(\pi\varepsilon v)}{\pi v} , \end{aligned} \tag{4.101}$$

and the *sinc* function,

$$F(u, v) = \varepsilon^2 \, \mathrm{sinc}(\varepsilon u) \, \mathrm{sinc}(\varepsilon v) \tag{4.102}$$

respectively.

Figure 4.12 shows the two-dimensional *sinc* function, resulting from the Fourier transform of a square rectangular window.

For specific functions, the direct application of the Fourier transform might be more difficult than for the rectangular function. For the sign function, also referred to as the *signum* function,

$$\mathrm{sign}(x) = \begin{cases} 1 & \text{for } x > 0 \\ 0 & \text{for } x = 0 \\ -1 & \text{for } x < 0 \end{cases} , \tag{4.103}$$

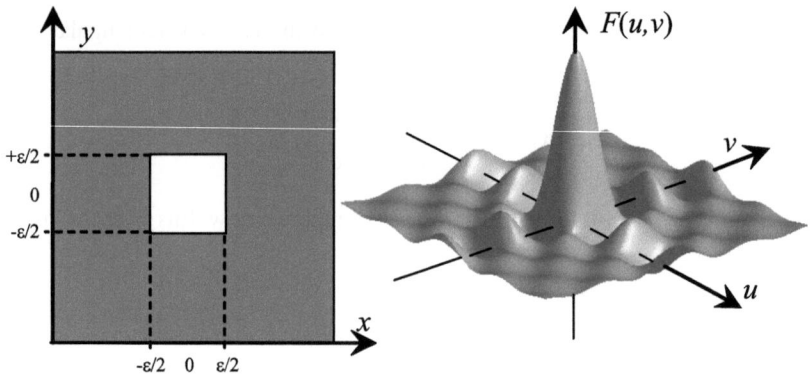

Fig. 4.12. Two-dimensional $[-\varepsilon/2, \varepsilon/2]$-rectangular window (*left*) and the corresponding Fourier transform (*right*): The two-dimensional *sinc* function (4.102). If the function in the spatial domain is axially symmetric, the Fourier transform is real, i.e., the imaginary part vanishes

which will be of importance in ▸Sect. 5.10, the convergence of the Fourier integral is not immediately obvious.

In this case, convergence of the integral is obtained by what is called a regular sequence, $g_\beta(x)$ (Fichtenholz 1982, Bracewell 1965). Here, the following scheme will be used.

A function $f(x)$, for which the improper integral

$$I = \int_0^\infty f(x)\,dx \tag{4.104}$$

does not exist, is assigned the function $g_\beta(x)f(x)$ with $g_\beta(x) = e^{-\beta x}$, such that the integral

$$\int_0^\infty g_\beta(x)f(x)\,dx \tag{4.105}$$

converges for $\beta > 0$. The integral (4.105) has a finite limit

$$I = \lim_{\beta \to 0} \int_0^\infty e^{-\beta x} f(x)\,dx\,, \tag{4.106}$$

which represents the generalized value[3] of the divergent integral (4.104).

In the case of the signum function, this formula leads to

$$\text{sign}(x) = \lim_{\beta \to 0} \begin{cases} e^{-\beta x} & \text{for } x > 0 \\ 0 & \text{for } x = 0 \\ -e^{+\beta x} & \text{for } x < 0 \,. \end{cases} \tag{4.107}$$

---

[3] Please note that the integral of (4.106) is the one-sided Laplace transform.

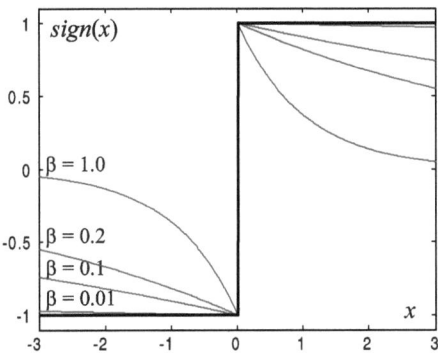

Fig. 4.13. The signum function approximated by a regular sequence of exponential functions. This approximation is necessary to ascertain the convergence of the integrals for explicit application of the Fourier transform

Figure 4.13 shows the convergent behavior of the signum function rewritten with the convergence-generating regular sequence. For the decreasing parameter, β, the surrogate function (4.107) tends rapidly toward the original function (4.103).

When calculating the Fourier transform, the transform is first applied to the surrogate function, before applying the limit to the convergence-generating regular sequence.

$$
\begin{aligned}
F(u) &= \left(\frac{\alpha}{2\pi}\right)^{\frac{1}{2}} \int_{-\infty}^{+\infty} \mathrm{sign}(x)\, e^{-i\alpha u x}\, dx \\
&= \lim_{\beta \to 0} \left(\frac{\alpha}{2\pi}\right)^{\frac{1}{2}} \left\{ -\int_{-\infty}^{0} e^{\beta x}\, e^{-i\alpha u x}\, dx + \int_{0}^{\infty} e^{-\beta x}\, e^{-i\alpha u x}\, dx \right\}.
\end{aligned}
\tag{4.108}
$$

Both of the partial integrals in (4.108) can be solved easily.

$$
\begin{aligned}
F(u) &= \lim_{\beta \to 0} \left(\frac{\alpha}{2\pi}\right)^{\frac{1}{2}} \left\{ -\int_{-\infty}^{0} e^{\beta x - i\alpha u x}\, dx + \int_{0}^{\infty} e^{-(\beta x + i\alpha u x)}\, dx \right\} \\
&= \lim_{\beta \to 0} \left(\frac{\alpha}{2\pi}\right)^{\frac{1}{2}} \left\{ -\int_{-\infty}^{0} e^{(\beta - i\alpha u)x}\, dx + \int_{0}^{\infty} e^{-(\beta + i\alpha u)x}\, dx \right\} \\
&= \lim_{\beta \to 0} \left(\frac{\alpha}{2\pi}\right)^{\frac{1}{2}} \left\{ \left[-\frac{e^{(\beta - i\alpha u)x}}{\beta - i\alpha u} \right]_{-\infty}^{0} + \left[-\frac{e^{-(\beta + i\alpha u)x}}{\beta + i\alpha u} \right]_{0}^{\infty} \right\}.
\end{aligned}
\tag{4.109}
$$

The substitution of the interval borders of the integrals and the evaluation of the limit with respect to β finally results in

$$
\begin{aligned}
F(u) &= \lim_{\beta \to 0} \left(\frac{\alpha}{2\pi}\right)^{\frac{1}{2}} \left\{ -\frac{1}{\beta - i\alpha u} + \frac{1}{\beta + i\alpha u} \right\} \\
&= \left(\frac{\alpha}{2\pi}\right)^{\frac{1}{2}} \frac{2}{i\alpha u}.
\end{aligned}
\tag{4.110}
$$

Since $\alpha = 2\pi$ is used in the later chapters, the result of the transformation is given as

$$\text{sign}(x) \; \circ\!\!-\!\!-\!\!\bullet \; \frac{1}{i\pi u} \; . \tag{4.111}$$

For more details on regular sequences, particularly well-behaved functions and generalized functions, the reader is referred to Fichtenholz (1982) or Bracewell (1965).

4.11
Convolution Theorem

An important property of the Fourier transform is given by the convolution theorem. The convolution of the two functions $s(x)$ and $h(x)$,

$$g(x) = s(x) * h(x) = \int_{-\infty}^{+\infty} s(\xi)h(x-\xi)\,d\xi \,, \tag{4.112}$$

is considered in the spatial domain. To do so, the Fourier transform of (4.112) has to be carried out

$$G(u) = \mathcal{F}\{s(x) * h(x)\} \,. \tag{4.113}$$

By applying the Fourier transform and permuting the integration order one obtains

$$
\begin{aligned}
G(u) &= \int_{-\infty}^{+\infty} \int_{-\infty}^{+\infty} s(\xi)h(x-\xi)\,d\xi e^{-i2\pi ux}\,dx \\
&= \int_{-\infty}^{+\infty} \int_{-\infty}^{+\infty} s(\xi)h(x-\xi)e^{-i2\pi ux}\,d\xi dx \,.
\end{aligned}
\tag{4.114}
$$

Substituting $y = x - \xi$ and $z = \xi$ into (4.114) gives

$$
\begin{aligned}
G(u) &= \int_{-\infty}^{+\infty} \int_{-\infty}^{+\infty} s(z)h(y)e^{-i2\pi u(y+z)}\,dzdy \\
&= \int_{-\infty}^{+\infty} \int_{-\infty}^{+\infty} s(z)h(y)e^{-i2\pi uy}e^{-i2\pi uz}\,dzdy \,.
\end{aligned}
\tag{4.115}
$$

Both integrals in (4.115) can therefore be separated into two factors, giving

$$
\begin{aligned}
G(u) &= \int_{-\infty}^{+\infty} h(y)e^{-i2\pi uy}\,dy \int_{-\infty}^{+\infty} s(z)e^{-i2\pi uz}\,dz \\
&= H(u)S(u) \,,
\end{aligned}
\tag{4.116}
$$

that is

$$h(x) * s(x) \; \circ\!\!-\!\!-\!\!\bullet \; H(u)S(u) \,, \tag{4.117}$$

and conversely

$$H(u) * S(u) \; \bullet\!\!-\!\!-\!\!\circ \; h(x)s(x) \,, \tag{4.118}$$

The proof of (4.118) can be found, for example, in Klingen (2001). The interested reader is referred to Bracewell (2003) for an extensive discussion of the two-dimensional convolution theorem.

4.12
Rayleigh's Theorem

The theorem of Rayleigh[4] proves that the integral over the squared modulus of
a function is equal to the integral of the squared modulus of its Fourier transform.
In the two-dimensional case this can be seen via

$$
\int_{-\infty}^{+\infty}\int_{-\infty}^{+\infty} f(x,y)f^*(x,y)\,dx\,dy = \int_{-\infty}^{+\infty}\int_{-\infty}^{+\infty} f(x,y)f^*(x,y)\,e^{-i2\pi(xu'+yv')}\,dx\,dy ,
$$

$$(4.119)$$

if u' and v' are assumed to be zero. The asterisks denote the complex conjugates of
the spatial signals and spectra respectively. Using the convolution theorem stated
in the section above, the right-hand side of (4.119) is the convolution of the Fourier
transforms, i.e., $F(u',v') * F^*(-u',-v')$. Therefore, one obtains

$$
\int_{-\infty}^{+\infty}\int_{-\infty}^{+\infty} f(x,y)f^*(x,y)\,dx\,dy = \int_{-\infty}^{+\infty}\int_{-\infty}^{+\infty} F(u,v)F^*(u-u',v-v')\,du\,dv . \quad (4.120)
$$

Since u' and v' are zero, one finally finds that

$$
\int_{-\infty}^{+\infty}\int_{-\infty}^{+\infty} f(x,y)f^*(x,y)\,dx\,dy = \int_{-\infty}^{+\infty}\int_{-\infty}^{+\infty} F(u,v)F^*(u,v)\,du\,dv , \quad (4.121)
$$

or, more simply,

$$
\int_{-\infty}^{+\infty}\int_{-\infty}^{+\infty} |f(x,y)|^2\,dx\,dy = \int_{-\infty}^{+\infty}\int_{-\infty}^{+\infty} |F(u,v)|^2\,du\,dv . \quad (4.122)
$$

The integrals represent the amount of energy in the system.

4.13
Power Theorem

The power theorem represents a generalized version of Rayleigh's theorem. By substituting f and f^* with f_1 and f_2^* respectively, one obtains

$$
\int_{-\infty}^{+\infty}\int_{-\infty}^{+\infty} f_1(x,y)f_2^*(x,y)\,dx\,dy = \int_{-\infty}^{+\infty}\int_{-\infty}^{+\infty} F_1(u,v)F_2^*(u,v)\,du\,dv . \quad (4.123)
$$

---

[4] Rayleigh's theorem corresponds to Parseval's theorem for Fourier series (see Bracewell
1965).

The derivation is the same as for Rayleigh's theorem. This expression can be interpreted as the power or energy of the signal. The left-hand side of (4.123) is evaluated as the product of a pair of canonically conjugate variables, such as an electric and magnetic field integrated over space. On the right-hand side, the frequency components are multiplied and integrated over the whole spectrum.

4.14
Filtering in the Frequency Domain

The convolution theorem provides an elegant method of applying filtering in the frequency domain. This will be discussed exemplarily on the basis of a high-pass and a low-pass filtering of an abdominal CT image. First of all, Fig. 4.14 provides a description of a circular aperture used for filtering, and the corresponding function in the frequency domain. The result of the high-pass and the low-pass filtering of the image is shown in Fig. 4.15.

In this filtering process, good results still do not come for free. As it has been shown in ▸ Sect. 4.10, (4.102), the Fourier transform of a rectangular function is a *sinc* function. This is also reflected by the result of filtering here. Strictly speaking, by applying a circular-rectangular filter, the sine in the result of the *sinc* function must be replaced by the Bessel function, and so

$$F(q) \propto \frac{J_1(2\pi\varepsilon q)}{q}, \tag{4.124}$$

where J_1 is the first-order Bessel function, which frequently appears in the solution of problems that have cylindrical symmetry, and $q = (u^2 + v^2)^{1/2}$ is the radial frequency variable.

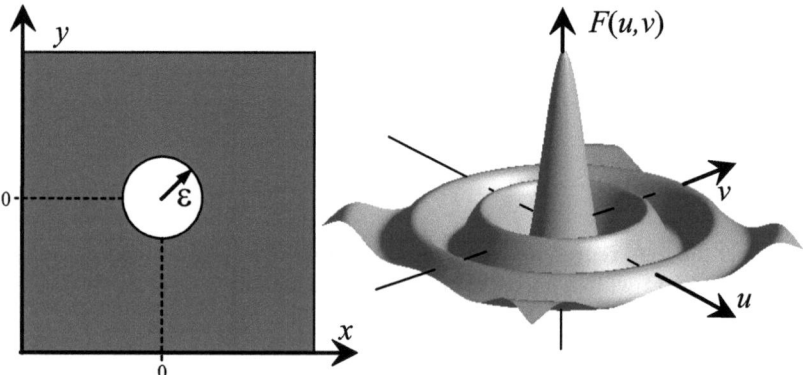

Fig. 4.14. Circular object (*left*) and the corresponding Fourier transform (*right*), which is given by the two-dimensional first-order Bessel function of the first kind in (4.124)

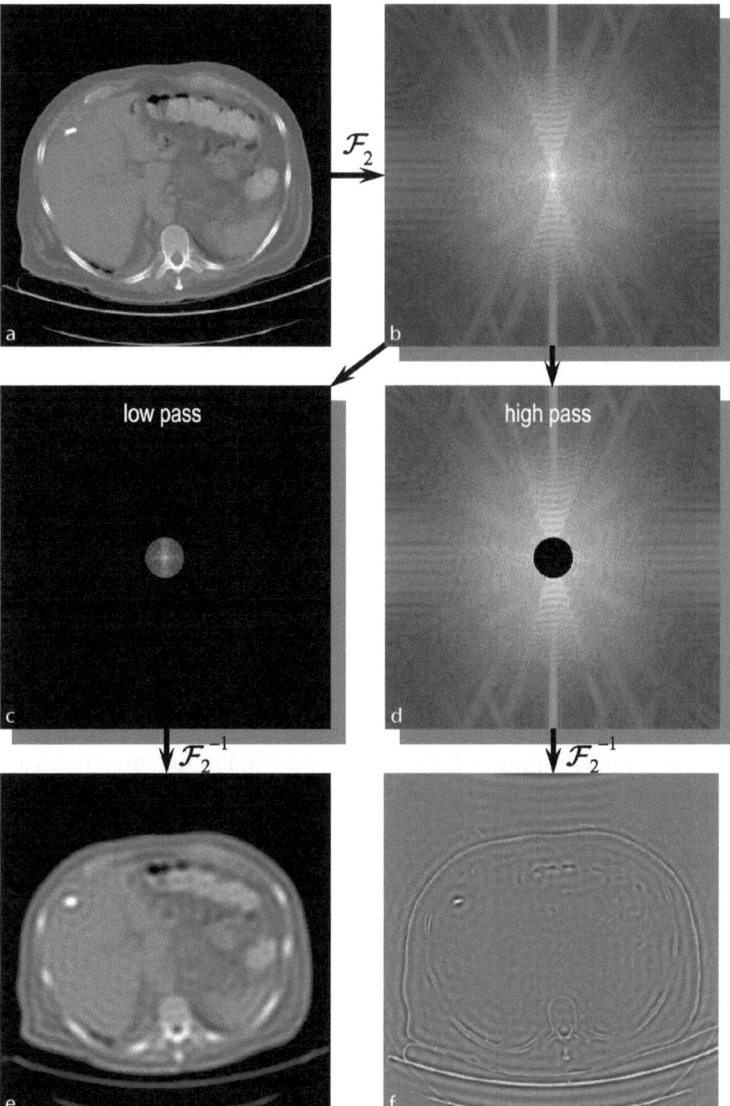

Fig. 4.15. Low-pass (**c**) and high-pass filtering (**d**) of an abdominal image (**a**) using a rectangular filtering function in the frequency domain (**b**). The figure only provides the magnitude of the complex spectrum in each case. With a circular aperture, only the low (**c**) or high frequencies (**d**) respectively remain in the spectrum. Therefore, each complementary spatial frequency is forced to be zero. The inverse transformation back to the spatial domain emphasizes the object's edges (for the high-pass, **f**) or results in a slightly blurred image (for the low-pass, **e**). In addition, one can see in both cases the characteristic waves that seem to emanate from the edges. These waves are, in fact, sidelobes, which are typically produced by sharp-edged filters in the frequency domain

The Bessel function of the first kind is defined in frequency space by the following integral

$$J_n(q) = \frac{1}{2\pi i^n} \int\limits_0^{2\pi} e^{-iq\cos(\theta)} e^{in\theta} \, d\theta \,, \tag{4.125}$$

where n is the order of the Bessel function.

In Fig. 4.14, similar to Fig. 4.12, the Fourier transform of a circular object can be seen. In analogy to the definition of the *sinc* function – cf. (4.3) – $F(q)$ can alternatively be expressed in terms of the *jinc* function (Bracewell 1965), defined as

$$\text{jinc}(q) = \frac{J_1(\pi q)}{2q} \,. \tag{4.126}$$

However, this will not be pursued in detail here. In many books about optics one can find derivations and examples of Bessel functions relating to diffraction at a circular aperture.

As a consequence of the filtering, one is able to recognize subsidiary waves of the *sinc* function, called the *sidelobes*, in areas of homogeneous gray values. For this reason, window functions are used to counteract these sidelobes. Specific window functions have been examined in detail in the literature (Harris 1978; Parzen 1961).

There are numerous proposals for appropriate window functions (such as Hamming, Hanning, Blackman or Kaiser [Azizi 1987]) that demonstrate less ripple in the result than the rectangular window does. In ▸ Sect. 7.2, which is concerned with the technical realization of image reconstruction for CT, the optimal window for filtered backprojection will be considered in detail.

4.15
Hankel Transform

The Hankel transform is a specific type of the two-dimensional Fourier transform with a radially symmetric transform kernel, $f(r)$, where $r = (x^2 + y^2)^{1/2}$. This type of transform is therefore also referred to as a Fourier–Bessel transform. Starting with the definition given in (4.73) and setting $\alpha = 2\pi$ leads to

$$F(u, v) = \int\limits_{-\infty}^{+\infty} \int\limits_{-\infty}^{+\infty} f(r) e^{-2\pi i(ux+vy)} \, dx \, dy \equiv \mathcal{F}\{f(r)\} \,. \tag{4.127}$$

With the problem having radial symmetry, the mathematical description is essentially simplified by introducing polar coordinates. Thus, by substituting

$$x + iy = r e^{i\theta} \tag{4.128}$$

and

$$u + iv = q e^{i\phi} , \tag{4.129}$$

so that the equations

$$x = r \cos(\theta) \tag{4.130}$$

$$y = r \sin(\theta) \tag{4.131}$$

$$r = \sqrt{x^2 + y^2} \tag{4.132}$$

and

$$u = q \cos(\phi) \tag{4.133}$$

$$v = q \sin(\phi) \tag{4.134}$$

$$q = \sqrt{u^2 + v^2} \tag{4.135}$$

describe the relationship between Cartesian and polar coordinates, (4.127) can be re-written as

$$F(q) = \int\limits_{-\infty}^{+\infty} \int\limits_{-\infty}^{+\infty} f(r) e^{-2\pi i r q (\cos(\phi) \cos(\theta) + \sin(\phi) \sin(\theta))} \, dx \, dy \equiv \mathcal{F}\{f(r)\} . \tag{4.136}$$

To carry out the integration, the area element $dx \, dy$ must now be replaced by $J \, dr \, d\theta$, where J is called the *Jacobian*, i.e.,

$$J \equiv \det\left(\frac{\partial(x, y)}{\partial(r, \theta)}\right) = \begin{vmatrix} \dfrac{\partial x}{\partial r} & \dfrac{\partial y}{\partial r} \\ \dfrac{\partial x}{\partial \theta} & \dfrac{\partial y}{\partial \theta} \end{vmatrix} = \begin{vmatrix} \cos(\theta) & \sin(\theta) \\ -r \sin(\theta) & r \cos(\theta) \end{vmatrix} \tag{4.137}$$

$$= r \left(\cos^2(\theta) + \sin^2(\theta)\right) = r .$$

In Fig. 4.16 the calculation of the infinitesimal area element can be seen graphically.

This leads to

$$F(q) = \int\limits_{0}^{+\infty} \int\limits_{0}^{2\pi} f(r)\, e^{-2\pi i r q (\cos(\phi)\cos(\theta) + \sin(\phi)\sin(\theta))} r \, d\theta \, dr \equiv \mathcal{F}\{f(r)\}. \qquad (4.138)$$

Using the addition law of the cosine

$$\cos(\alpha \pm \beta) = \cos(\alpha)\cos(\beta) \mp \sin(\alpha)\sin(\beta), \qquad (4.139)$$

it follows that

$$F(q) = \int\limits_{0}^{+\infty} \int\limits_{0}^{2\pi} f(r)\, e^{-2\pi i r q (\cos(\theta - \phi))} r \, d\theta \, dr. \qquad (4.140)$$

The constant phase shift about the angle ϕ in the argument of the cosine, can be removed by shifting the integration limits, giving

$$F(q) = \int\limits_{0}^{+\infty} \int\limits_{-\phi}^{2\pi - \phi} f(r)\, e^{-2\pi i r q (\cos(\theta))} r \, d\theta \, dr. \qquad (4.141)$$

Since the integrand of (4.141) is radially symmetric, the phase shift can be neglected[5] so that, after resorting the integrands, one obtains

$$F(q) = \int\limits_{0}^{+\infty} f(r) \left(\int\limits_{0}^{2\pi} e^{-2\pi i r q (\cos(\theta))} \, d\theta \right) r \, dr. \qquad (4.142)$$

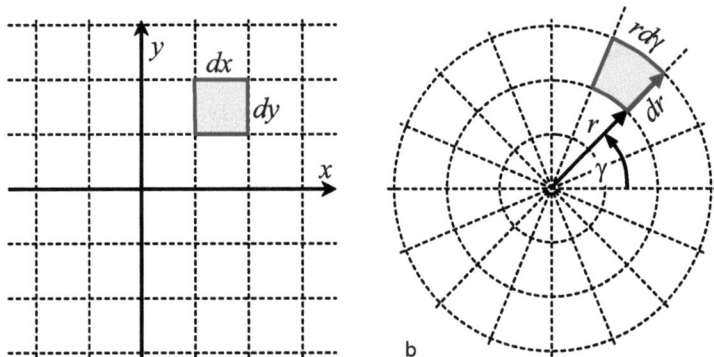

Fig. 4.16. The formation of an infinitesimal area element after transformation from **a** Cartesian to **b** polar coordinates

---

[5] It simply represents an initial angle in an integration over one full rotation.

The term in the bracket in (4.142) is just the 0th order Bessel function of the first kind – defined in general in (4.125) –, i.e., in the frequency space

$$J_0(q) = \frac{1}{2\pi} \int_0^{2\pi} e^{-iq\cos(\theta)}\, d\theta \tag{4.143}$$

thus, one may write

$$F(q) = 2\pi \int_0^{+\infty} f(r)J_0(2\pi qr)r\, dr \ . \tag{4.144}$$

(4.144) is known as the *Hankel* transform. It is a one-dimensional transform since f and F are functions of one variable. However, as functions of one variable, they may also represent two-dimensional functions.

To complete the description of the 0th order Hankel transform, the inverse transform given as

$$f(r) = 2\pi \int_0^{+\infty} F(q)J_0(2\pi qr)q\, dq \tag{4.145}$$

is needed. The pair of Hankel transform equations are conventionally denoted by

$$F(q) = \mathcal{H}_0\{f(r)\} \tag{4.146}$$

and

$$f(r) = \mathcal{H}_0\{F(q)\} \ . \tag{4.147}$$

In fact, Fig. 4.14 from the section above is an example of the Hankel transform, i.e., the Hankel transform of a disk

$$f(r) = \text{rect}\left(\frac{r}{2\varepsilon}\right) \tag{4.148}$$

is given by

$$F(q) = \frac{\varepsilon J_1(2\pi\varepsilon q)}{q} \ . \tag{4.149}$$

The rectangular function, $\text{rect}(r)$, has the Hankel transform $\text{jinc}(q)$, shown in integral formulation

$$\text{jinc}(q) = 2\pi \int_0^{+\infty} \text{rect}(r)J_0(2\pi qr)r\, dr \tag{4.150}$$

and

$$\text{rect}(r) = 2\pi \int_0^{+\infty} \text{jinc}(q)J_0(2\pi qr)q\, dq \tag{4.151}$$

respectively.

4.16
Abel Transform

Computed tomography of rotationally symmetric objects results in a mathematically interesting transform, which is closely connected to both the Hankel transform
and the Fourier transform. In the case of the projections of a radially symmetric

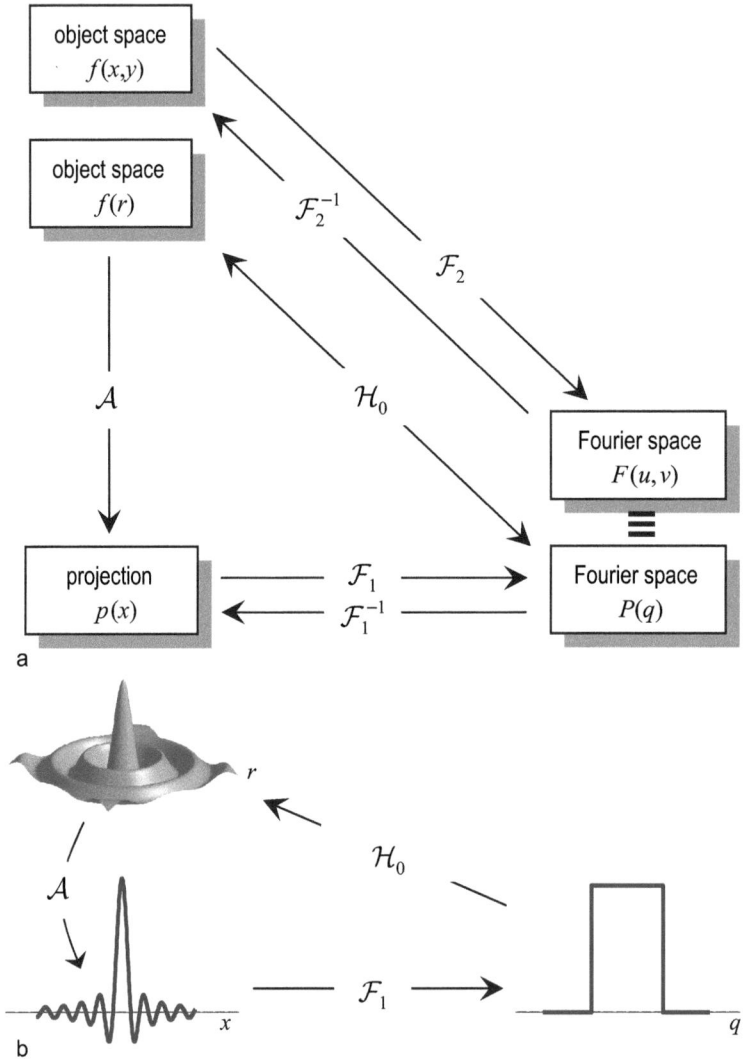

Fig. 4.17. a Relationship between the Fourier transform, the Hankel transform, and the Abel
transform. The Hankel transform can be replaced by the Abel transform and a subsequent
one-dimensional Fourier transform. **b** Example of the Fourier–Abel–Hankel cycle using
jinc(r)

object, (4.51) can be simplified, since the projection onto the x-axis is given by

$$p(x) = \int_{-\infty}^{+\infty} f(r)\, dy \,, \tag{4.152}$$

where the radius r is given by (4.132). Substituting

$$y = \sqrt{r^2 - x^2} \,, \tag{4.153}$$

gives

$$p(x) = 2 \int_{x}^{+\infty} \frac{f(r)r}{\sqrt{r^2 - x^2}}\, dr \,. \tag{4.154}$$

From (4.153) it is clear that the minimum radius is x. The factor 2 in (4.154) is due to contributions from values above and below the x-axis.

(4.154) is referred to as the *Abel transform* of the object $f(r)$. The Abel transform is conventionally denoted by

$$p(x) = \mathcal{A}\{f(r)\} \,. \tag{4.155}$$

For a radially symmetric object function, $f(x, y) = f(r)$, the relationship in Fig. 4.17 holds.

It follows that

$$f(r) = \mathcal{H}_0 \mathcal{F}_1 \mathcal{A}\{f(r)\} \tag{4.156}$$

also holds. (4.156) is called the Fourier–Abel–Hankel cycle. Since the Fourier–Abel–Hankel operation is cyclic, it holds that

$$\mathcal{H}_0 \mathcal{F}_1 \mathcal{A} = \mathcal{A} \mathcal{H}_0 \mathcal{F}_1 = \mathcal{F}_1 \mathcal{A} \mathcal{H}_0 = \mathcal{I} \tag{4.157}$$

where $\mathcal{I}$ is the identity operator.

4.17
Hilbert Transform

The Hilbert transform of a function, $f(x)$, is defined[6] by

$$\mathcal{H}\{f(x)\} = \frac{1}{\pi} \int_{-\infty}^{+\infty} \frac{f(x')}{x - x'}\, dx' \,. \tag{4.158}$$

---

[6] Due to the divergence at $x = x'$ the integral (4.158) needs to be interpreted in the sense of *Cauchy's* fundamental theorem (Bronstein and Semendjajew 1979), i.e., $\lim_{\varepsilon \to 0} \left\{ \int_{-\infty}^{x'-\varepsilon} + \int_{x'+\varepsilon}^{\infty} \right\}$.

It should be noted that the definition in (4.158) describes a convolution of the function $f(x)$ with the function $(\pi x)^{-1}$, i.e.,

$$\mathcal{H}\{f(x)\} = \frac{1}{\pi x} * f(x) \,. \tag{4.159}$$

This convolution can be applied in the frequency domain. Therefore, one needs to recall the derivation of the transform (4.111) and apply it to the *signum* function in the frequency domain. This gives

$$\frac{1}{\pi x} \;\circ\!\!\!-\!\!\!\bullet\; -\mathrm{i}\,\mathrm{sign}(u) = \begin{cases} -\mathrm{i} & \text{for positive } u \\ 0 & \text{for } u = 0 \\ +\mathrm{i} & \text{for negative } u \end{cases} \tag{4.160}$$

In this way, the Hilbert transform has the properties of not altering the amplitude, but of shifting the phase by $+\pi/2$ or $-\pi/2$ respectively (according to the sign of u). Instead of carrying out the integral in (4.158), the Hilbert transform is obtained via the convolution theorem. Let

$$F(u) = \mathcal{F}\{f(x)\} \,, \tag{4.161}$$

then

$$\mathcal{F}\left\{\frac{1}{\pi x} * f(x)\right\} = \mathcal{F}\{\mathcal{H}\{f(x)\}\} = \mathcal{H}\{\mathcal{F}\{f(x)\}\} = \mathcal{H}\{F(u)\} = -\mathrm{i}\cdot\mathrm{sign}(u)\cdot F(u) \,. \tag{4.162}$$

The Hilbert transform is thus obtained by

$$\mathcal{H}\{f(x)\} = \mathcal{F}^{-1}\{-\mathrm{i}\cdot\mathrm{sign}(u)\cdot F(u)\} \,. \tag{4.163}$$

By applying the Hilbert transform twice in succession, the phases become inverted, with the final result being the negative of the original function, thus

$$\mathcal{H}\{\mathcal{H}\{f(x)\}\} = -f(x) \,. \tag{4.164}$$

This can be derived in the frequency domain as well.

$$\begin{aligned} \mathcal{F}\{\mathcal{H}\{\mathcal{H}\{f(x)\}\}\} &= \mathcal{H}\{\mathcal{H}\{F(u)\}\} \\ &= \mathcal{H}\{-\mathrm{i}\cdot\mathrm{sign}(u)\cdot F(u)\} \\ &= -\mathrm{i}\cdot\mathrm{sign}(u)\cdot\mathcal{H}\{F(u)\} \\ &= -\mathrm{i}\cdot\mathrm{sign}(u)\cdot-\mathrm{i}\cdot\mathrm{sign}(u)\cdot F(u) \\ &= -F(u) \,. \end{aligned} \tag{4.165}$$

Therefore, the inverse transform is obtained by

$$f(x) = -\left(\frac{1}{\pi x}\right) * \mathcal{H}\{f(x)\} \,, \tag{4.166}$$

or alternatively

$$f(x) = -\frac{1}{\pi} \int_{-\infty}^{+\infty} \frac{\mathcal{H}\{f(x')\}}{x - x'} \,\mathrm{d}x' \,. \tag{4.167}$$

As a consequence, four successive Hilbert transforms result in the identity operation, i.e.,

$$\mathcal{H}\mathcal{H}\mathcal{H}\mathcal{H} = \mathcal{I} \,. \tag{4.168}$$

4.18
Sampling Theorem and Nyquist Criterion

Due to sampling, a continuous signal is transformed into a digital[7] one. The resulting, spatially discrete signal only coincides with the continuous signal, for instance, the continuously distributed attenuation coefficients at the sampled points. The signal values situated between the sample points are not captured. In this section, it will be briefly considered how accurately the continuous spatial signal, $s(\xi)$, needs to be sampled, in order to measure a discrete signal, s_i, that sufficiently represents the original one.

Figure 4.18 shows that it is always possible to construct different continuous functions passing all sample points as long as the sampling interval is larger than zero. For this reason, quite differently shaped functions might look identical after sampling, i.e., the sampling is not unique (Stearns and Hush 1999).

The question of which changes in information content the continuous signal experiences due to sampling can be answered by means of the Nyquist–Shannon sampling theorem.

According to the Nyquist–Shannon sampling theorem, the sampled signal bears the same amount of information as the continuous signal if the sampling frequency is larger than twice the highest frequency (or alternatively the bandwidth of band-limited signals) contained in the spectrum of the continuous signal. In other words, a signal, $s(\xi)$, must be sampled at least twice during the cycle with its highest frequency.

In this way, if the signal $s(\xi)$ satisfies

$$S(u) = 0 \ \text{ for } \ u \geq \frac{u_N}{2} , \tag{4.169}$$

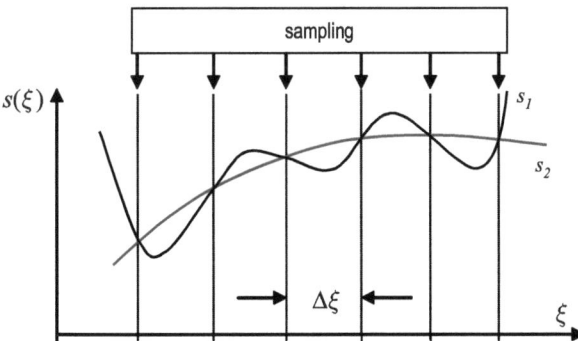

Fig. 4.18. Different functions with identical sampling points. Both functions share the same sampled values, although the shapes of the two functions are quite different

---

[7] However, a digital signal is not achieved alone from sampling, since it requires quantization as well.

then the sampling rate must be larger than u_N, which is called the Nyquist frequency. This criterion protects against errors due to band overlap, the "aliasing", where a high-frequency signal digitized at too low a sampling rate will be indistinguishable from a low frequency signal. In two dimensions this phenomenon is referred to as the Moiré effect. With a predetermined sampling period, the signal needs to be filtered by a corresponding low-pass filter (Azizi 1987). For practical applications, one has to keep in mind that the sampling theorem is a theoretical statement that does not take into account the violation of certain assumptions, including quantization errors or spatial irregularities of the arrangement of the sampling points (*jitter*) (Kiencke 1998).

The sampling theorem can be justified in the frequency domain by considering a signal that has been sampled using the comb function,

$$s_a(\xi) = \sum_{n=-\infty}^{+\infty} s(n\Delta\xi)\delta(\xi - n\Delta\xi) = s(\xi) \sum_{n=-\infty}^{+\infty} \delta(\xi - n\Delta\xi) = s(\xi)\,\text{III}(\xi)\,, \quad (4.170)$$

in the spatial domain. In (4.170), the spatial coordinate, ξ, is continuous and the discrete sequence values, $s(n\Delta\xi)$, are the strengths of the impulses located at each sample point.

Here, one initially calculates the Fourier transform of the comb function (Lüke 1999; Klingen 2001), which gives another comb function. This result can be understood by means of the relations introduced earlier. First, according to (4.88), the Fourier transform of the comb function gives

$$\mathcal{F}\left(\sum_{n=-\infty}^{+\infty} \delta(\xi - n\Delta\xi)\right) = \sum_{n=-\infty}^{+\infty} e^{-2\pi i u n \Delta\xi}\,. \quad (4.171)$$

Writing the right hand side of (4.171) as a limit

$$\sum_{n=-\infty}^{+\infty} e^{-2\pi i u n \Delta\xi} = \lim_{N\to\infty} \sum_{n=-N}^{N} e^{-2\pi i u n \Delta\xi} \quad (4.172)$$

and substituting $\psi = \exp(-2\pi i u \Delta\xi)$ such that,

$$\sum_{n=-\infty}^{+\infty} e^{-2\pi i u n \Delta\xi} = \lim_{N\to\infty} \sum_{n=-N}^{N} \psi^n\,, \quad (4.173)$$

the right-hand side of (4.173) can be represented as the limit of a geometric series (Papula 2000), namely

$$\lim_{N\to\infty} \sum_{n=-N}^{N} \psi^n = \lim_{N\to\infty} \psi^{-N} \frac{1 - \psi^{2N+1}}{1 - \psi}$$

$$= \lim_{N\to\infty} e^{2\pi i u \Delta\xi N} \frac{1 - e^{-2\pi i u \Delta\xi(2N+1)}}{1 - e^{-2\pi i u \Delta\xi}} \quad (4.174)$$

$$= \lim_{N\to\infty} e^{2\pi i u \Delta\xi N} \frac{e^{-\pi i u \Delta\xi(2N+1)}\left(e^{+\pi i u \Delta\xi(2N+1)} - e^{-\pi i u \Delta\xi(2N+1)}\right)}{e^{-\pi i u \Delta\xi}\left(e^{+\pi i u \Delta\xi} - e^{-\pi i u \Delta\xi}\right)}\,.$$

With $K = 2N + 1$, one obtains the expression

$$\mathcal{F}\left(\sum_{n=-\infty}^{+\infty} \delta(\xi - n\Delta\xi)\right) = \lim_{K \to \infty} \frac{\left(e^{+K\pi iu\Delta\xi} - e^{-K\pi iu\Delta\xi}\right)}{\left(e^{+\pi iu\Delta\xi} - e^{-\pi iu\Delta\xi}\right)} \tag{4.175}$$

and further

$$\mathcal{F}\left(\sum_{n=-\infty}^{+\infty} \delta(\xi - n\Delta\xi)\right) = \lim_{K \to \infty} \left(\frac{\sin(K\pi u\Delta\xi)}{\sin(\pi u\Delta\xi)}\right). \tag{4.176}$$

Since K is an integer, the result on the right-hand side of (4.176) does not change if the quantity $(u\Delta\xi)$ is replaced by $(u\Delta\xi - n)$, where n is also an integer, as sine is a periodic function and the zeros of the numerator and the denominator still appear at integer numbers, $(u\Delta\xi)$. Thus, it holds that

$$\lim_{K \to \infty} f_K(u) = \lim_{K \to \infty} \left(\frac{\sin(K\pi u\Delta\xi)}{\sin(\pi u\Delta\xi)}\right) = \lim_{K \to \infty} \left(\frac{\sin(K\pi(u\Delta\xi - n))}{\sin(\pi(u\Delta\xi - n))}\right). \tag{4.177}$$

(4.177) also implies that, due to translation invariance, there is a fundamental interval, $(-1/2\Delta\xi, +1/2\Delta\xi)$, which repeats itself periodically. This means that the function (4.177) is indeed identical within the intervals $(\{n - 1/2\}/\Delta\xi, \{n + 1/2\}/\Delta\xi)$ for all integer numbers, $n \in \mathbb{Z}$. Figure 4.19 shows the shape of $f_K(u)$ in (4.177) for an increasing parameter $K = \{3, 11, 51, 999\}$.

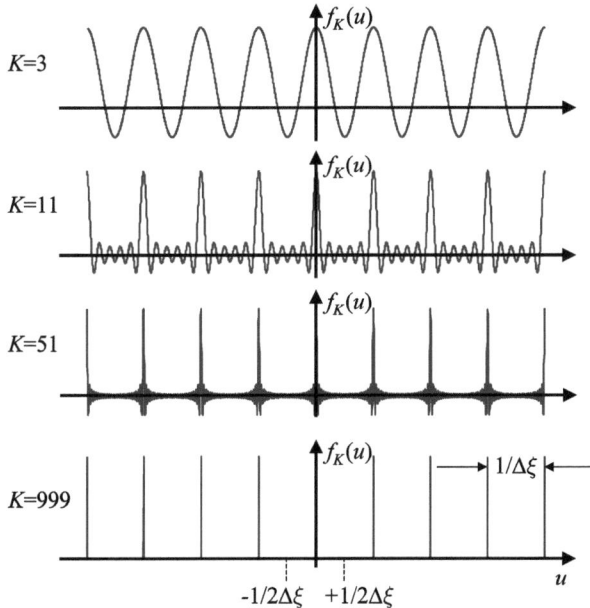

Fig. 4.19. The function $f_K(u)$ tends toward the comb function for increasing parameter K. In this way, it is shown that the Fourier transformation of the comb function is once again a comb function

For each $n \in \mathbb{Z}$, in the neighborhood of $(u\Delta\xi - n) = 0$, one can approximate the denominator in (4.177) with the linear term of its Taylor expansion, i.e.,

$$\sin(\pi(u\Delta\xi - n)) \approx \pi(u\Delta\xi - n) , \tag{4.178}$$

so that, by means of definitions (4.24) and (4.46) of the δ-distribution,

$$\delta(u) = \int_{-\infty}^{\infty} e^{-i2\pi ux}\, dx = \lim_{\varepsilon \to \infty} \int_{-\varepsilon/2}^{\varepsilon/2} e^{-i2\pi ux}\, dx = \lim_{\varepsilon \to \infty} \frac{\sin(\pi\varepsilon u)}{\pi u} \tag{4.179}$$

it follows that

$$\lim_{K \to \infty} \left(\frac{\sin(K\pi(u\Delta\xi - n))}{\pi(u\Delta\xi - n)} \right) = \delta(u\Delta\xi - n), \tag{4.180}$$

for $(\{n - 1/2\}/\Delta\xi, \{n + 1/2\}/\Delta\xi]$ and all $n \in \mathbb{Z}$. Since (4.180), as already discussed above, is periodic – repeating the fundamental interval $(-1/2\Delta\xi, +1/2\Delta\xi]$ – and effective for arbitrary integer, n, (cf. Fig. 4.19), the expression on the right-hand side must be interpreted as a sequence of δ-impulses, allowing us to write

$$\mathcal{F}\left(\sum_{n=-\infty}^{+\infty} \delta(\xi - n\Delta\xi) \right) = \sum_{n=-\infty}^{+\infty} \delta(u\Delta\xi - n) . \tag{4.181}$$

This, along with the scaling property (4.39), gives

$$\mathcal{F}\left(\sum_{n=-\infty}^{+\infty} \delta(\xi - n\Delta\xi) \right) = \frac{1}{|\Delta\xi|} \sum_{n=-\infty}^{+\infty} \delta\left(u - \frac{n}{\Delta\xi}\right). \tag{4.182}$$

Taking advantage of the convolution theorem (4.118) and by means of (4.182) for sampling (4.170)

$$s_a(\xi) = s(\xi) \sum_{n=-\infty}^{+\infty} \delta(\xi - n\Delta\xi) \tag{4.183}$$

in the frequency domain, one obtains the expression

$$S_a(u) = S(u) * \frac{1}{\Delta\xi} \sum_{n=-\infty}^{+\infty} \delta\left(u - \frac{n}{\Delta\xi}\right). \tag{4.184}$$

The convolution can be written explicitly as

$$S_a(u) = \int_{-\infty}^{\infty} S(v) \frac{1}{\Delta\xi} \sum_{n=-\infty}^{+\infty} \delta\left(u - \frac{n}{\Delta\xi} - v\right) dv , \tag{4.185}$$

so that, by permuting the summation and the integration in (4.185),

$$S_a(u) = \frac{1}{\Delta\xi} \sum_{n=-\infty}^{+\infty} \int_{-\infty}^{\infty} S(v)\delta\left(u - \frac{n}{\Delta\xi} - v\right) dv \tag{4.186}$$

can be obtained.

Once again, the final result is obtained by means of the sift property of the δ-distribution, thus

$$S_a(u) = \frac{1}{\Delta\xi} \sum_{n=-\infty}^{+\infty} S\left(u - \frac{n}{\Delta\xi}\right). \qquad (4.187)$$

The Fourier transform of the sampled signal arises from the convolution of the signal spectrum with the comb function in the frequency domain. Thus, the spectrum of the continuous signal is repeated periodically with a period of $1/\Delta\xi$. Figure 4.20 illustrates this behavior schematically.

In the case of sub-sampling, i.e., when the original signal's maximal frequency is *not* below half the sampling frequency, then the sampled signal's periodically repeated spectra overlap. This results in aliasing errors, which are physically based on the beat effect.

(4.170) to (4.184) can be interpreted in a very general manner as

$$s(x) \cdot \text{III}(x) \; \circ\!\!-\!\!\bullet \; S(u) * \text{III}(u) \qquad (4.188)$$

and

$$s(x) * \text{III}(x) \; \circ\!\!-\!\!\bullet \; S(u) \cdot \text{III}(u), \qquad (4.189)$$

where $\text{III}(x)$ and $\text{III}(u)$ denote the comb function in the spatial and in the frequency domain respectively. The basic meaning of (4.188) and (4.189) is

1. A sampled signal has a periodic spectrum,
2. A periodic signal has a line spectrum[8].

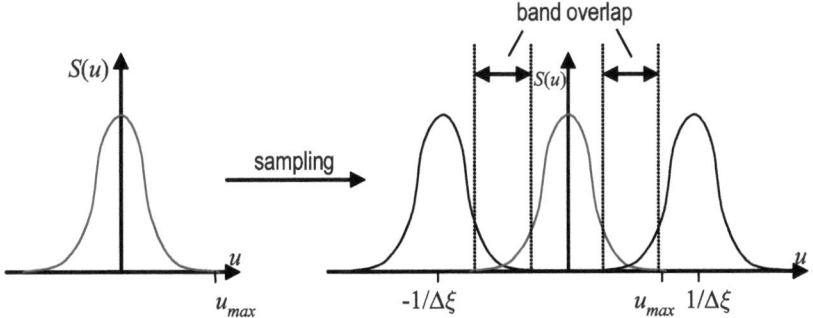

Fig. 4.20. Sampling results in a periodic recurrence of the continuous signal in the frequency domain. In the presence of under-sampling, the spectra are so close to each other that they overlap. This overlapping in the frequency domain appears as beating in the spatial domain. This effect is called aliasing

---

[8] The lines represent the coefficients of the Fourier series.

If the frequency or spatial functions, originating from the convolution with the comb function, are free from any overlaps, the sampled function related by the Fourier transform can be reconstructed faultlessly.

At this point, an important question remaining is how to regain an analog signal from a sampled signal. For the reconstruction of the signal provided in Fig. 4.9, low-pass filtering is necessary. An ideal low-pass filter

$$H_{\text{lp}}(u) = \Delta\xi \, \text{rect}\left(\frac{u}{1/\Delta\xi}\right) \tag{4.190}$$

is a rectangular function of the height $\Delta\xi$ inside the interval $[-1/(2\Delta\xi), +1/(2\Delta\xi)]$, i.e., a multiplicative rectangular filter

$$S(u) = S_{\text{a}}(u)\Delta\xi \, \text{rect}\left(\frac{u}{1/\Delta\xi}\right) \tag{4.191}$$

in the frequency domain. Referring to the convolution theorem (4.117), this expression can be represented in the spatial domain by a convolution

$$s(\xi) = s_{\text{a}}(\xi) * \text{sinc}\left(\frac{\xi}{\Delta\xi}\right). \tag{4.192}$$

This means that the signal recovery is a convolution of the sampling points with the *sinc* function. Figure 4.21 illustrates this principle, which is called the *Shannon–Whittaker* interpolation.

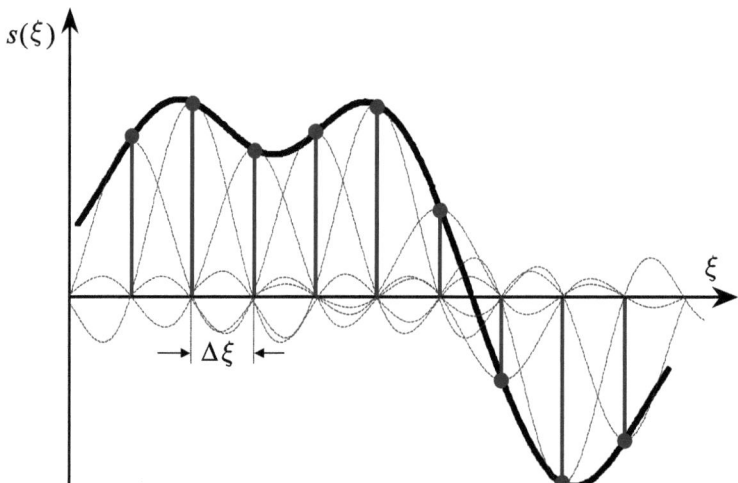

Fig. 4.21. The sampled values of a spatially varying function are reconstructed by low-pass filtering the signal. The use of an ideal low-pass filter, which is represented by a rectangular filter in the frequency domain, corresponds to the convolution of the sampling points with the *sinc* function

For a deeper understanding, as well as more on the fundamental importance of the sampling theorem, the reader is referred to Lüke (1999). For further details on the Fourier transform of comb functions and the generalized boundary value for the solution of (4.182), the reader is referred to Klingen (2001) or Barrett and Swindell (1981).

4.19
Wiener–Khintchine Theorem

The Wiener–Khintchine theorem establishes a connection between the spatial variation of the signal, the amplitude spectrum, the autocorrelation function, and the energy density spectrum[9]. Here, the Wiener–Khintchine theorem will be presented mathematically in two dimensions. This represents a good exercise for understanding its relation to the Fourier transform and the δ-distribution. Starting with the definition of the autocorrelation function of a general image $f(x, y)$

$$C(x, y) = \int\limits_{-\infty}^{+\infty} \int\limits_{-\infty}^{+\infty} f^*(\xi, \eta) f(x + \xi, y + \eta) \, d\xi \, d\eta \,, \qquad (4.193)$$

it can be shown via the inverse Fourier transform of the complex conjugate image, $f^*(x, y)$, and the displaced image, $f(x + \xi, y + \eta)$ respectively that the autocorrelation function is given by the inverse Fourier transform of the energy density spectrum. In Fig. 4.22, the intermediate steps necessary for the derivation of the autocorrelation function via the Fourier transform are shown for the example of an abdominal image. First, the definition of the Fourier transform is incorporated as follows:

$$C(x, y) = \int\limits_{-\infty}^{+\infty} \int\limits_{-\infty}^{+\infty} \left\{ \int\limits_{-\infty}^{+\infty} \int\limits_{-\infty}^{-\infty} F^*(u, v) \, e^{-i2\pi(u\xi+v\eta)} \, du \, dv \right\}$$
$$\times \left\{ \int\limits_{-\infty}^{+\infty} \int\limits_{-\infty}^{-\infty} F(u', v') \, e^{i2\pi(u'(\xi+x)+v'(\eta+y))} \, du' \, dv' \right\} d\xi \, d\eta \,. \qquad (4.194)$$

The permutation of the integration order results in

$$C(x, y) = \int\limits_{-\infty}^{+\infty} \int\limits_{-\infty}^{+\infty} \left\{ \int\limits_{-\infty}^{+\infty} \int\limits_{-\infty}^{-\infty} \int\limits_{-\infty}^{+\infty} \int\limits_{-\infty}^{-\infty} F^*(u, v) F(u', v') \right.$$
$$\left. \times e^{-i2\pi(u\xi+v\eta)} \, e^{i2\pi(u'(\xi+x)+v'(\eta+y))} \, du' \, dv' \, du \, dv \right\} d\xi \, d\eta \,. \qquad (4.195)$$

---

[9] To be precise, the following relation is the Wiener–Khintchine theorem only if f is a homogeneous and ergodic field (Jan 2006).

Re-sorting the exponents

$$
C(x, y) = \int\limits_{-\infty}^{+\infty} \int\limits_{-\infty}^{+\infty} \left\{ \int\limits_{-\infty}^{+\infty} \int\limits_{-\infty}^{-\infty} \int\limits_{-\infty}^{+\infty} \int\limits_{-\infty}^{-\infty} F^*(u, v) F(u', v') \right.
$$

$$
\left. \times\, e^{i2\pi(u'x+v'y)}\, e^{i2\pi(\xi(u'-u)+\eta(v'-v))}\, du'\, dv'\, du\, dv \right\} d\xi\, d\eta ,
$$

(4.196)

and once again re-sorting the integration order, yields

$$
C(x, y) = \int\limits_{-\infty}^{+\infty} \int\limits_{-\infty}^{-\infty} \int\limits_{-\infty}^{+\infty} \int\limits_{-\infty}^{-\infty} F^*(u, v) F(u', v') e^{i2\pi(u'x+v'y)}
$$

$$
\times \left\{ \int\limits_{-\infty}^{+\infty} \int\limits_{-\infty}^{+\infty} e^{i2\pi(\xi(u'-u)+\eta(v'-v))}\, d\xi\, d\eta \right\} du'\, dv'\, du\, dv .
$$

(4.197)

Due to property (4.46) the bracketed integral in the above equation is equivalent to a δ-distribution, giving

$$
C(x, y) = \int\limits_{-\infty}^{+\infty} \int\limits_{-\infty}^{-\infty} \int\limits_{-\infty}^{+\infty} \int\limits_{-\infty}^{-\infty} F^*(u, v) F(u', v') e^{i2\pi(u'x+v'y)} \delta(u'-u, v'-v)\, du'\, dv'\, du\, dv .
$$

(4.198)

Using the sifting property (4.29) of the δ-distribution it follows from (4.198), that

$$
C(x, y) = \int\limits_{-\infty}^{+\infty} \int\limits_{-\infty}^{-\infty} F^*(u, v) F(u, v) e^{i2\pi(ux+vy)}\, du\, dv
$$

(4.199)

and therefore

$$
C(x, y) = \int\limits_{-\infty}^{+\infty} \int\limits_{-\infty}^{-\infty} |F(u, v)|^2\, e^{i2\pi(ux+vy)}\, du\, dv
$$

(4.200)

or

$$
C(x, y) = \mathcal{F}_2^{-1}\{|F(u, v)|^2\} = \mathcal{F}_2^{-1}\{S(u, v)\}
$$

(4.201)

where $S(u, v)$ is the energy density spectrum of the image $f(x, y)$.

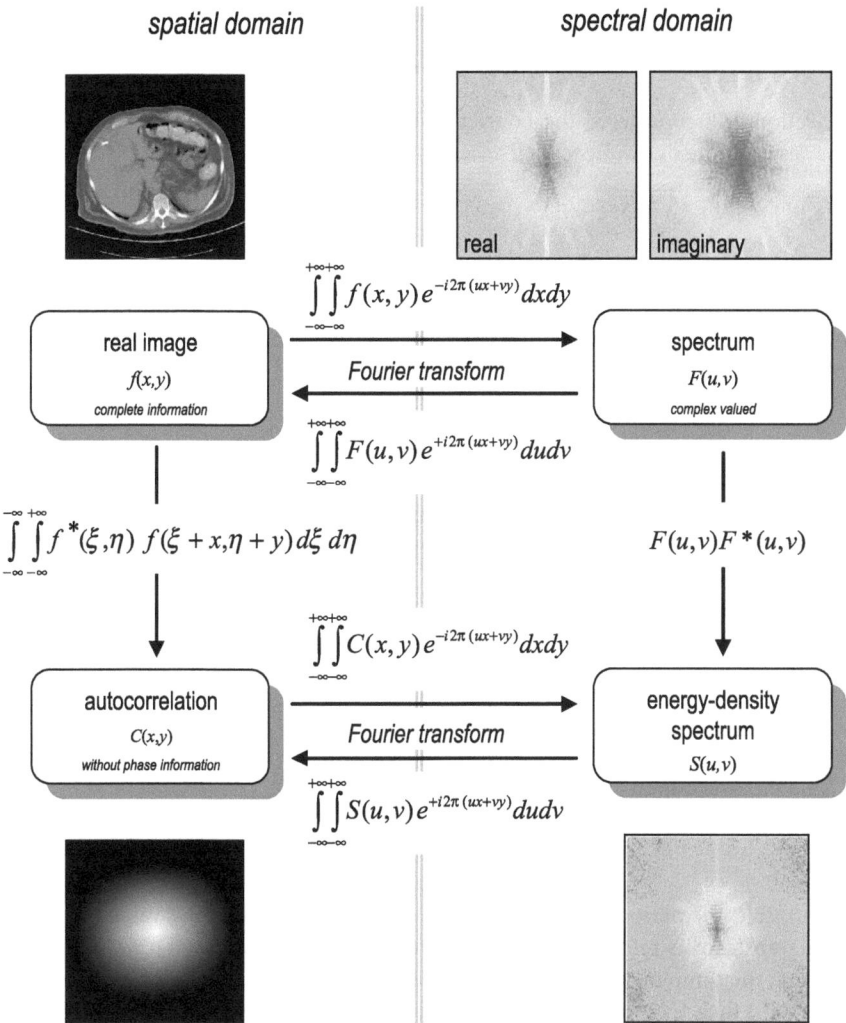

Fig. 4.22. The Wiener–Khintchine theorem for image information. The corresponding transforms are given for an abdominal tomogram

The Wiener–Khintchine theorem is a special case of the general cross correlation function of two different images $f(x, y)$ and $g(x, y)$, for which

$$C_{\text{fg}}(x, y) = \mathcal{F}_2^{-1}\{F^*(u, v)G(u, v)\} \tag{4.202}$$

holds. Result (4.202) is similar to the convolution theorem (4.117).

4.20
Fourier Transform of Discrete Signals

▸ Section 4.18 considered the Fourier transform of sampled signals. Considering the result in the one dimensional case,

$$s_a(\xi) = s(\xi) \sum_{n=-\infty}^{+\infty} \delta(\xi - n\Delta\xi) \quad \circ\!\!-\!\!-\!\!-\!\!\bullet \quad S_a(u) = \frac{1}{|\Delta\xi|} \sum_{n=-\infty}^{+\infty} S\left(u - \frac{n}{\Delta\xi}\right) \quad (4.203)$$

has been obtained.

In this section the spectrum should be calculated directly by a Fourier transform of the sampled signal. The application of the Fourier transform to the left-hand side of (4.203) gives

$$S_a(u) = \int_{-\infty}^{+\infty} s(\xi) \sum_{n=-\infty}^{+\infty} \delta(\xi - n\Delta\xi) \, e^{-i2\pi u\xi} d\xi . \quad (4.204)$$

In (4.204) the function to be sampled can be moved inside the summation. Also, the order of the integration and the summation can be permuted, giving

$$S_a(u) = \int_{-\infty}^{+\infty} \sum_{n=-\infty}^{+\infty} s(\xi)\delta(\xi - n\Delta\xi) \, e^{-i2\pi u\xi} d\xi$$

$$= \sum_{n=-\infty}^{+\infty} \int_{-\infty}^{+\infty} s(\xi)\delta(\xi - n\Delta\xi) e^{-i2\pi u\xi} d\xi . \quad (4.205)$$

Using the sift property of the δ-distribution leads to

$$S_a(u) = \sum_{n=-\infty}^{+\infty} s(n\Delta\xi) e^{-i2\pi un\Delta\xi} . \quad (4.206)$$

As a result, the periodic spectrum only depends on the sampling values, $s(n\Delta\xi)$. This transform is referred to as the discrete Fourier transform (DFT). In the literature, the sampling interval is sometimes set to $\Delta\xi = 1$ (the unit of the sampling interval is therefore "one detector" or "one pixel"). In this case, the spectrum of the discrete signal $s(n)$ is given by

$$S_a(u) = \sum_{n=-\infty}^{+\infty} s(n) e^{-i2\pi nu} . \quad (4.207)$$

(4.206) is the complex notation of the Fourier series, so one may write

$$S_a(u) = \sum_{n=-\infty}^{+\infty} s_n e^{-i2\pi un\Delta\xi} , \quad (4.208)$$

where s_n denotes the nth Fourier coefficient, which can be calculated by

$$s_n = \Delta\xi \int_{0}^{\frac{1}{\Delta\xi}} S_a(u) e^{i2\pi un\Delta\xi} du . \quad (4.209)$$

4.21
Finite Discrete Fourier Transform

In numerical calculations of the Fourier transform of discrete signals, only a finite number of frequencies must be calculated. Using the discrete Fourier transform, a discrete frequency spectrum is assigned to a discrete signal.

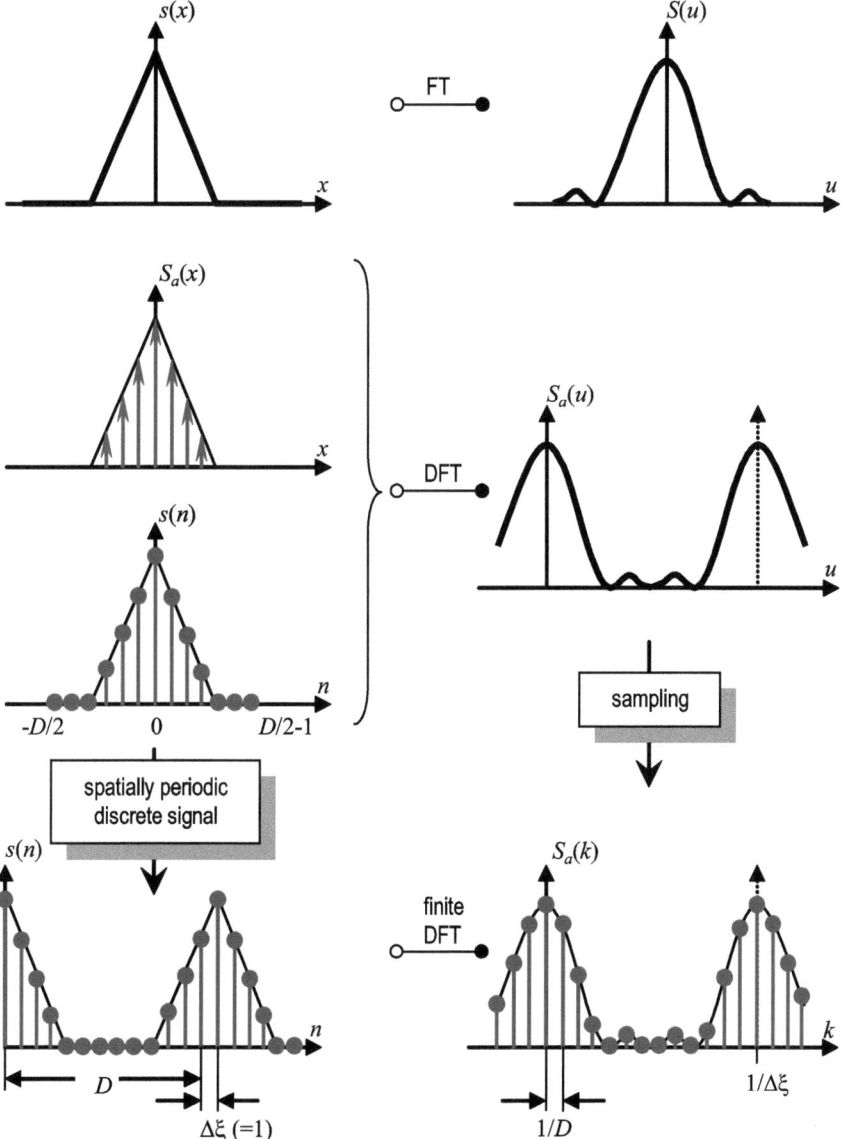

Fig. 4.23. Relationship among the continuous signal, the sampled signal, and spatially limited and spatially discrete signals, as well as their respective spectra (adapted from Lüke [1999])

If, as in the concrete case of a detector array in CT, the number of values of the discrete signal, $\{s(0), s(1), \ldots, s(D-1)\}$, is limited to D, then, according to the sampling theorem, a sampling interval equal to the reciprocal length, $1/D$, of the detector array length is adequate (as mentioned above, the width of one detector element is set to $\Delta\xi = 1$). The discrete sampled signal of the detector array, $s_a(n)$, which is periodic with period D, is assigned to the sampled spectrum, $S_a(k)$. As shown above, the spectrum is also periodic. It follows that a calculation of D spectral values of one period is sufficient. This is called the finite discrete Fourier transform (finite DFT). The discrete frequency points are

$$u_k = \frac{k}{D} \quad \text{for } k = \{0, \ldots, D-1\} \tag{4.210}$$

and the discrete frequency periodic Fourier transform of a spatially discrete, periodic signal within one period is

$$S_a(k) = \sum_{n=0}^{D-1} s_a(n) e^{-i2\pi n \frac{k}{D}}, \tag{4.211}$$

Table 4.3. Classification of the Fourier transform. For each particular type, the table indicates how the domain is being changed by the transform (the sampling length $\Delta\xi$ is set to one)

	Continuous space	**Discrete space**
	Fourier transform	Discrete Fourier transform
	Type: $\mathbb{R} \to \mathbb{R}$	Type: $\mathbb{Z} \to [0,1]$
Continuous frequency	$S(u) = \int_{-\infty}^{+\infty} s(x) e^{-i2\pi ux}\, dx$	$S_a(u) = \sum_{n=-\infty}^{+\infty} s(n) e^{-i2\pi nu}$
	Type: $\mathbb{R} \to \mathbb{R}$	Type: $[0,1] \to \mathbb{Z}$
	$s(x) = \int_{-\infty}^{+\infty} S(u) e^{i2\pi ux}\, du$	$s(n) = \int_{0}^{1} S_a(u) e^{i2\pi nu}\, du$
	Fourier series	Discrete Fourier transform
	Type: $[0,D] \to \mathbb{Z}$	Type: $\mathbb{Z}_D \to \mathbb{Z}_D$
Discrete frequency	$S(k) = \frac{1}{D} \int_{0}^{D} s(x) e^{-ik\frac{2\pi}{D}x}\, dx$	$S_a(k) = \sum_{n=0}^{D-1} s_a(n) e^{-i2\pi n \frac{k}{D}}$
	Type: $\mathbb{Z} \to [0,D]$	Type: $\mathbb{Z}_D \to \mathbb{Z}_D$
	$s(x) = \sum_{k=-\infty}^{+\infty} S(k) e^{ik\frac{2\pi}{D}x}$	$s_a(n) = \frac{1}{D} \sum_{k=0}^{D-1} S_a(k) e^{i2\pi n \frac{k}{D}}$

and the inverse transform is

$$s_a(n) = \frac{1}{D} \sum_{k=0}^{D-1} S_a(k) e^{i2\pi n \frac{k}{D}}. \tag{4.212}$$

Figure 4.23 shows this relationship schematically. In summary, the transforms can be classified as shown in Table 4.3.

4.22
z-Transform

The "z-transform" represents the generalization of the Fourier transform for spatially discrete signals. For a brief description of the z-transform and, subsequently, the "chirp z-transform", the notation described here will follow that in Lüke (1999) and Stearns and Hush (1999).

The z-transform is used in cases where the discrete Fourier transform (4.207) does not converge for all series. This is the case if the sequence of the signal values, $s(n)$, cannot be summed absolutely; therefore, the summation

$$S = \sum_{n=-\infty}^{+\infty} |s(n)| \tag{4.213}$$

is not finite[10]. In such cases, convergence can be achieved using a similar trick to the one that was used in (4.105). Convergence is assured by an exponential weighting, so that

$$\sum_{n=-\infty}^{+\infty} |s(n) e^{-\sigma n}| < \infty. \tag{4.214}$$

Then, the discrete Fourier transform according to (4.207) exists, so

$$s(n) e^{-\sigma n} \circ\!\!-\!\!\bullet \sum_{n=-\infty}^{+\infty} s(n) e^{-\sigma n} e^{-i2\pi n u} = \sum_{n=-\infty}^{+\infty} s(n) e^{-(\sigma+i2\pi u)n}. \tag{4.215}$$

Thus, by substituting $z = e^{(\sigma+i2\pi u)}$, the right-hand side of (4.215) becomes

$$S(z) = \sum_{n=-\infty}^{+\infty} s(n) z^{-n}. \tag{4.216}$$

(4.216) is referred to as the z-transform. Decomposing the complex frequency variable into its absolute value and phase,

$$z = |z| e^{i\varphi} = e^{\sigma} e^{i2\pi u} \tag{4.217}$$

---

[10] Typically, delta impulses appear in the spectrum $S_a(u)$.

allows the interpretation in the complex z-plane. In this plane, at each circle of radius

$$|z| = e^{\sigma} , \qquad (4.218)$$

one period of the Fourier spectrum $S_a(u)$ of the corresponding spatially discrete signal, $s(n)e^{-\sigma n}$, is located. This generalization of the Fourier transform includes a larger class of signals. The circular domain, in which the Fourier spectra converge, i.e., (4.214) is fulfilled, represents the region of convergence for the z-transform. For spatially limited signals the region of convergence covers the entire z-plane – with exception of the origin. If the region of convergence contains $|z| = 1$, then the Fourier and the z-transform are related to each other by the substitution $z = e^{i2\pi u}$ (where $\sigma = 0$). This means that the z-transform calculated over the unit circle is identical to the discrete Fourier transform (DFT). If the region of convergence does not contain the unit circle, the DFT is not defined for this series (Stearns and Hush 1999).

4.23
Chirp z-Transform

The chirp z-transform is an algorithm for the calculation of the z-transform of a series of finite length with sampling values in the frequency domain that are set at equal intervals along a generalized path in the z-plane (Stearns and Hush 1999). To preserve the reference to the Fourier transform, only paths representing arc sections of the unit circle are taken into account (cf. Fig. 4.24). The chirp z-transform will be important when considering the "linogram sampling" of the Radon space in ▸ Sect. 5.5.

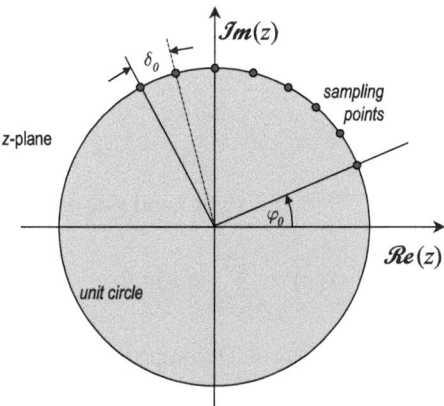

Fig. 4.24. The chirp z-transform can be used to calculate M sampling values of the z-transform on the unit circle

Let $\{s(n)\}_{n=0,\ldots,N-1}$ be a series of N sampled values in the spatial domain, whose spectrum is of interest. According to (4.216), the z-transform is then given by

$$S(z) = \sum_{n=0}^{N-1} s(n)z^{-n} . \tag{4.219}$$

Since in this case the z-transform is calculated over the unit circle by the substitution $z = e^{i2\pi u}$, it is identical to the discrete Fourier transform.

Consider the Fourier transform of M sampling values on the unit circle, starting from the angle φ_0 and sampling at intervals of δ_0. The sampling along an arc can be described by

$$z_k = AB^{-k} \text{ with } k = 0, \ldots, M - 1 , \tag{4.220}$$

where

$$A = e^{i\varphi_0} \text{ and } B = e^{-i\delta_0} . \tag{4.221}$$

Substituting into (4.219) gives

$$S(z) = \sum_{n=0}^{N-1} s(n)\left(AB^{-k}\right)^{-n} = \sum_{n=0}^{N-1} s(n)A^{-n}B^{nk} \text{ where } k = 0, \ldots, M - 1. \tag{4.222}$$

The fact that $(k-n)^2 = k^2 - 2kn + n^2$, results in the *Bluestein* identity (Bluestein 1970)

$$kn = \frac{1}{2}\left[k^2 + n^2 - (k - n)^2\right] , \tag{4.223}$$

which, inserted into (4.222), gives the expression

$$S(z_k) = \sum_{n=0}^{N-1} s(n)A^{-n}B^{\frac{n^2}{2}}B^{\frac{k^2}{2}}B^{-\frac{(k-n)^2}{2}} = B^{\frac{k^2}{2}}\sum_{n=0}^{N-1}\left(s(n)A^{-n}B^{\frac{n^2}{2}}\right)B^{-\frac{(k-n)^2}{2}} . \tag{4.224}$$

Furthermore, by replacing

$$g_n = s(n)A^{-n}B^{\frac{n^2}{2}} \text{ and } h_n = B^{-\frac{n^2}{2}} , \tag{4.225}$$

the sum of products from (4.222) becomes a linear convolution

$$S(z_k) = B^{\frac{k^2}{2}}\sum_{n=0}^{N-1} g_n h_{k-n} . \tag{4.226}$$

With (4.226) the efforts pay back, because the convolution can be performed very efficiently in the Fourier domain by means of the fast Fourier transform (FFT).

The name "chirp" relates to the form of

$$h_n = e^{i\delta_0 \frac{n^2}{2}} = e^{in\left(\delta_0 \frac{n}{2}\right)} , \tag{4.227}$$

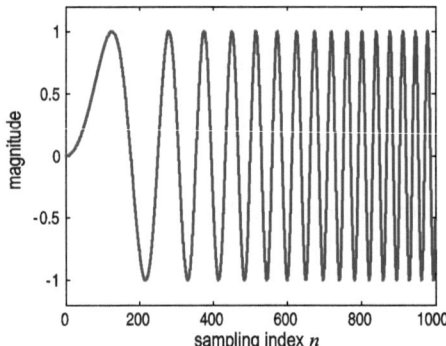

Fig. 4.25. Example of a chirp signal: The sinusoidal behavior of the signal's imaginary part (4.227) with linearly increasing frequency

whose imaginary part is illustrated in Fig. 4.25. h_n is a complex oscillation with a linearly increasing frequency.

5 Two-Dimensional Fourier-Based Reconstruction Methods

Contents

5.1	Introduction	151
5.2	Radon Transformation	153
5.3	Inverse Radon Transformation and Fourier Slice Theorem	163
5.4	Implementation of the Direct Inverse Radon Transform	167
5.5	Linogram Method	170
5.6	Simple Backprojection	175
5.7	Filtered Backprojection	179
5.8	Comparison Between Backprojection and Filtered Backprojection	183
5.9	Filtered Layergram: Deconvolution of the Simple Backprojection	187
5.10	Filtered Backprojection and Radon's Solution	191
5.11	Cormack Transform	194

5.1
Introduction

On 30 April 1917 the Austrian mathematician Johann Radon presented his work *On the Determination of Functions from their Integrals along Certain Manifolds*[1] at the annual meeting of the *Royal Saxonian Society of Physical and Mathematical Sciences*[2]. The most important theorems of this work have been extracted and translated[3] here using Radon's own terminology to introduce the variables used throughout the subsequent chapters.

> *If one integrates a function of two variables (x, y) – a function $f(x, y)$ in the plane – that satisfies suitable regularity conditions, along an arbitrary straight line L, then the values $p(L)$ of these integrals define a line function.*

---

[1] Original German title: *Über die Bestimmung von Funktionen durch ihre Integralwerte längs gewisser Mannigfaltigkeiten.*

[2] Physikalisch-Mathematische Königlich Sächsische Gesellschaft der Wissenschaften.

[3] Translation by R. Lohner, School of Mathematics, Georgia Institute of Technology, Atlanta, GA, USA, first printed in Deans (1983).

The problem that is solved in this paper is the inversion of this linear functional transformation, i.e., answers to the following questions are given: Is every line function that satisfies suitable regularity conditions obtainable by this process? If this is the case, is the point function f then uniquely determined by p, and how can it be found?

Let $f(x, y)$ be a real function defined for all real points $\mathbf{r} = [x, y]$ that satisfies the following regularity conditions:

a) $f(x, y)$ is continuous.
b) The following double integral, which is to be taken over the whole plane, is convergent:

$$\iint \frac{|f(x, y)|}{\sqrt{x^2 + y^2}} \, dx \, dy . \tag{5.1}$$

c) For an arbitrary point $\mathbf{r} = [x, y]$ and any $R \geq 0$, let

$$\overline{f}_{\mathbf{r}}(R) = \frac{1}{2\pi} \int_0^{2\pi} f(x + R \cos(\gamma), y + R \sin(\gamma)) \, d\gamma . \tag{5.2}$$

Then for every point $\mathbf{r}$,

$$\lim_{R \to \infty} \overline{f}_{\mathbf{r}}(R) = 0 . \tag{5.3}$$

Thus the following theorems hold.

Theorem I: *The integral of f along the straight line L with the equation*

$$x \cos(\gamma) + y \sin(\gamma) = \xi , \tag{5.4}$$

given by

$$p(\xi, \gamma) = p(-\xi, \gamma + \pi) = \int_{-\infty}^{+\infty} f(\xi \cos(\gamma) - \eta \sin(\gamma), \xi \sin(\gamma) + \eta \cos(\gamma)) \, d\eta \tag{5.5}$$

is "in general" well defined. This means that on any circle those points that have tangent lines (see Fig. 5.1 – not given in Radon's original paper) for which p does not exist form a set of linear measure zero.

Theorem II: *If the mean value of $p(\xi, \gamma)$ is formed for the tangent lines of the circle with center $\mathbf{r} = (x, y)$ and radius R:*

$$\overline{P}_{\mathbf{r}}(R) = \frac{1}{2\pi} \int_0^{2\pi} p(x \cos(\gamma) + y \sin(\gamma) + R, \gamma) \, d\gamma , \tag{5.6}$$

then this integral is absolutely convergent for all $\mathbf{r} = (x, y)$ and R.

Theorem III: *The value of $f(x, y)$ is completely determined by p and can be computed as follows:*

$$f(\mathbf{r}) = -\frac{1}{\pi} \int_0^{\infty} \frac{d\overline{p}_{\mathbf{r}}(R)}{R} . \tag{5.7}$$

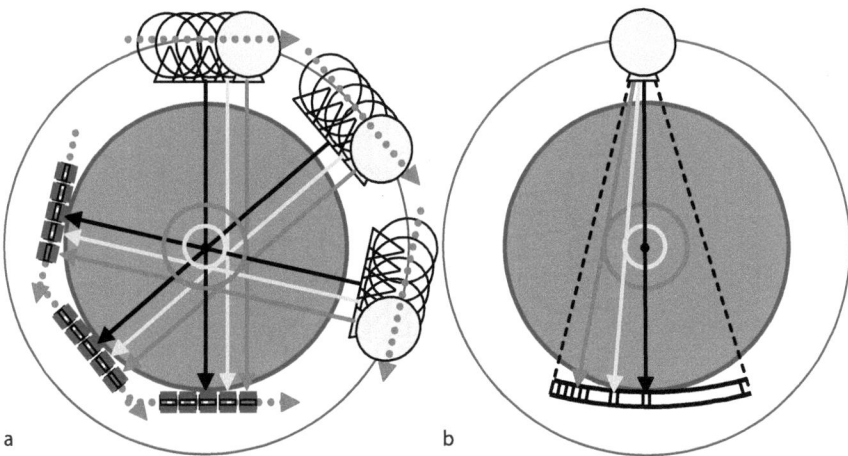

Fig. 5.1. a Due to the circular orbit of the X-ray source, the projection paths – visualized by the straight lines **L** on which the X-ray photons are running through the tissue – are organized as tangents of circles with an iso-center at point $\mathbf{r} = (0,0)$ inside the measurement field. **b** This circular sampling structure can be found in fan-beam geometry as well. In ▸ Sect. 7.7 the coordinate transformation between pencil- and fan-beam geometry will be discussed in detail

The following sections of this chapter will explain how Radon's results can be understood in terms of modern computed tomography (CT). In particular, they demonstrate how the integral inversion is technically realized and in ▸ Chap. 7 the practical aspects of the implementation are discussed in detail.

5.2
Radon Transformation

Today, CT systems use fan-beam geometry[4]. However, it is more instructive to introduce CT reconstruction by describing those methods that use pencil geometry, in which, for every direction of projection, the X-ray beam is collimated to a pencil shape and moved linearly in the direction parallel to a linear X-ray detector array. This concept has already been mentioned in ▸ Sect. 3.3 with regard to the first generation of CT systems. Sequentially, the X-ray sampling unit is rotated by an angle γ, and, subsequently, moved linearly. This process is repeated until any point of the object to be reconstructed is illuminated by at least 180°. Figure 5.2 shows this sampling process schematically.

---

[4] The three-dimensional extension of the fan-beam concept to cone-beam geometry will be discussed in ▸ Chap. 8.

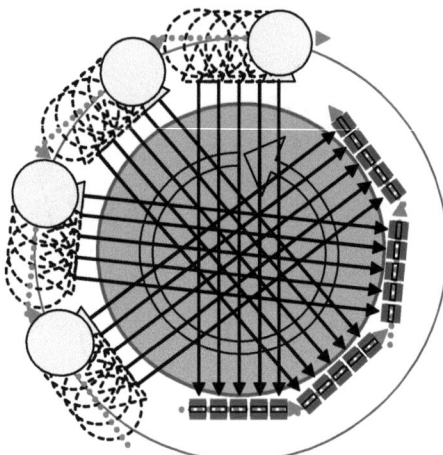

Fig. 5.2. Schematic drawing of pencil- or needle-beam geometry in first-generation CT systems. The X-ray beam, confined to a pencil or needle shape by collimation, produces a parallel projection via a parallel shift of the X-ray tube in the direction parallel to the linear detector array for any projection angle. The sectional plane defined by this projection geometry is the tomographic slice that is to be reconstructed

For a fixed projection angle γ and a particular linear shift position of the X-ray source, one is interested in the projection integral

$$p(s) = \int_0^s \mu(\eta)\,\mathrm{d}\eta \,. \tag{5.8}$$

If the attenuation coefficient changes in discrete steps only, as Fig. 5.3 suggests, one obtains the projection sum

$$p(s) = \sum_{k=1}^s \mu_k \Delta\eta \tag{5.9}$$

instead of the projection integral of (5.8).

Note that, since the X-ray source is moved parallel to the X-ray detector array, the integration length s used in (5.8) and (5.9) is constant.

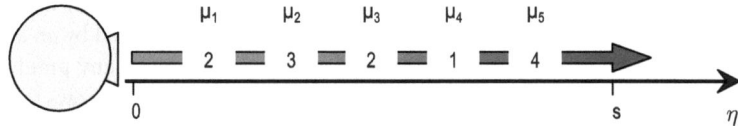

Fig. 5.3. Attenuation of the radiation intensity when passing through inhomogeneous tissue. Due to the inhomogeneity of the attenuation coefficients along the integration length, the values cannot be reconstructed by a simple inversion. However, the effort required by the reconstruction is paid back, since exactly this spatial distribution of attenuation coefficients forms the image of clinical interest

On the other hand, there are a number of parallel paths through the object, which define a sectional plane or CT slice. Therefore, it is sensible to write the projection integral as a function of the linear position of the source – or the location of the corresponding detector element, ξ – and the projection angle γ, and no longer as a function of the constant path s between X-ray tube and detector element. For this purpose, a coordinate system (ξ, η) is defined, which rotates on the gantry together with the X-ray source and the detector. This coordinate system will be referred to as the rotating sampling system – in contrast to the fixed patient coordinate system (x, y). Figure 5.4 visualizes the correspondence between the two coordinate systems.

By a repetition of the alternating rotation and shifting sequence of the rigidly connected source-detector system – often referred to as rotating *disk* on the *gantry* – a set of parallel projections $\{p_{\gamma_1}(\xi), p_{\gamma_2}(\xi), p_{\gamma_3}(\xi), \ldots\}$ is measured, from which the spatial distribution of the object points or, more precisely, the spatial distribution of the attenuation coefficients within a selected slice of the patient, has to be reconstructed. In Fig. 5.4 an axial cranial tomography is used as an example to demonstrate the correspondence of the coordinate systems that will be used extensively throughout this chapter.

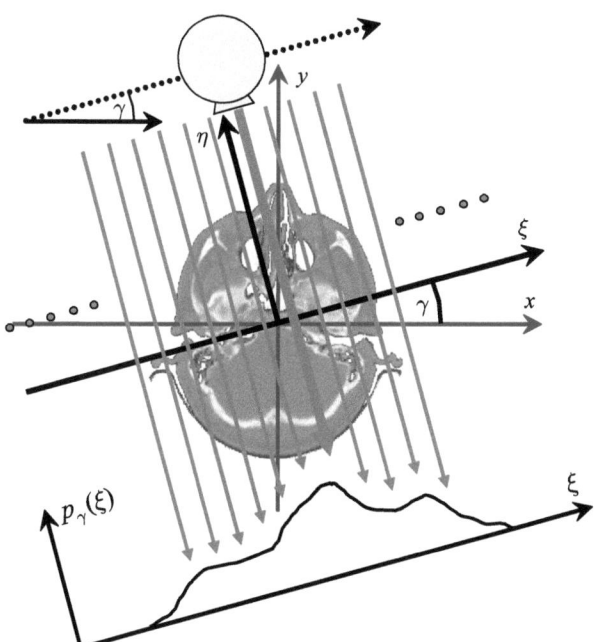

Fig. 5.4. The sampling disk rotating on the gantry of the tomograph carries the X-ray tube and the detector. When both elements are shifted synchronously and parallel to each other for any angle γ, the X-rays parallel to each other in the rotating sampling system (ξ, η) define the slice that is to be reconstructed. Here, as an example, this slice is axial cranial tomography

In this special case of constant integration length, s, the projection integral under projection angle γ defined in the rotating (ξ, η) system, is written as

$$p_\gamma(\xi) = \int\limits_0^s \mu(\xi, \eta)\,\mathrm{d}\eta\,. \tag{5.10}$$

Figure 5.5 visualizes the projection integral for the passage of X-ray photons through the tissue. In practice, the spatial distribution of the attenuation coefficients that are to be reconstructed is a discrete set of values only. This is obviously due to the limited resolution of the real sampling, displaying, and storing system. However, in this chapter signals are assumed to be continuous. This again allows a more instructive introduction to the theory of image reconstruction. The reconstruction procedures being implemented in the field are actually based on discrete signals and will be discussed in detail in ▸ Chap. 7.

(5.10) represents an integration along the path of the traveling photons given by the position ξ of the X-ray source and the X-ray detector respectively under a certain projection angle γ. In practice, one is not interested in expressing the attenuation values, μ, as a function of the rotating sampling system (ξ, η). In fact, the image of the spatial distribution of attenuation values must be given as a function of the fixed

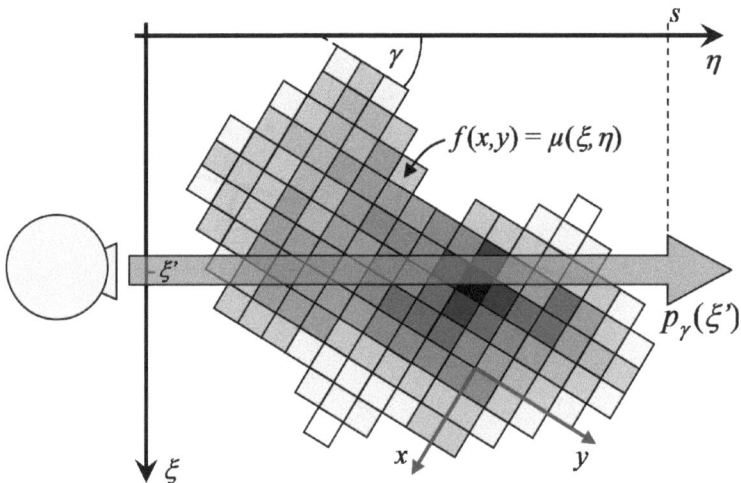

Fig. 5.5. The object is seen to rotate if viewed from a point traveling with the X-ray tube in the (ξ, η) frame. The spatial distribution of the reconstructed attenuation coefficients is a discrete set of values in the actual measuring situation

patient coordinate system (x, y). To describe the relation between the coordinate systems (ξ, η) and (x, y), in a first step the unit vectors

$$\mathbf{n}_\xi = \begin{pmatrix} \cos(\gamma) \\ \sin(\gamma) \end{pmatrix} \tag{5.11}$$

and

$$\mathbf{n}_\eta = \begin{pmatrix} -\sin(\gamma) \\ \cos(\gamma) \end{pmatrix} \tag{5.12}$$

are defined. These vectors span the rotating (ξ, η) frame. Figure 5.6 visualizes the rotating unit vectors that are attached to the rotating sampling disk on the gantry and the patient at rest in the fixed (x, y) system.

To describe the projection path – and along with that the projection integral (5.10) – in the fixed patient coordinate system the geometric relations

$$\xi = \left(\mathbf{r}^{\mathrm{T}} \cdot \mathbf{n}_\xi \right) = x \cos(\gamma) + y \sin(\gamma) \tag{5.13}$$

and

$$\eta = \left(\mathbf{r}^{\mathrm{T}} \cdot \mathbf{n}_\eta \right) = -x \sin(\gamma) + y \cos(\gamma) \tag{5.14}$$

are given between the two coordinate systems for a particular point within the object to be imaged, $\mathbf{r} = (x, y)^{\mathrm{T}}$. With the help of (5.13) and (5.14), the variables of the attenuation coefficient in (5.10) can be substituted such that

$$\begin{aligned} f(x, y) &= \mu \left(\xi(x, y), \eta(x, y) \right) \\ &= \mu \left(\left(\mathbf{r}^{\mathrm{T}} \cdot \mathbf{n}_\xi \right), \left(\mathbf{r}^{\mathrm{T}} \cdot \mathbf{n}_\eta \right) \right) \end{aligned} \tag{5.15}$$

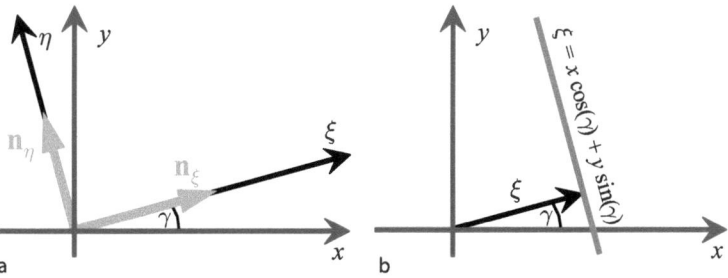

Fig. 5.6. a Definition of the rotating sampling system (ξ, η) by unit vectors. The fixed (x, y) coordinate system represents the patient coordinate system at rest. **b** A projection line through the object can be given in terms of the Hessian normal form

expresses the attenuation values f in the fixed (x, y) system. From a physical point of view, f and μ are identical attenuation coefficients for the tissue at the point $\mathbf{r} = (x, y)^T$ in the fixed (x, y) system and at the point $\boldsymbol{\rho} = (\xi, \eta)^T$ in the rotating (ξ, η) system. In Fig. 5.7 this relation is schematically shown.

The relation of the attenuation coefficients in the rotating sampling and the fixed patient system is summarized in the functional chart in Fig. 5.8.

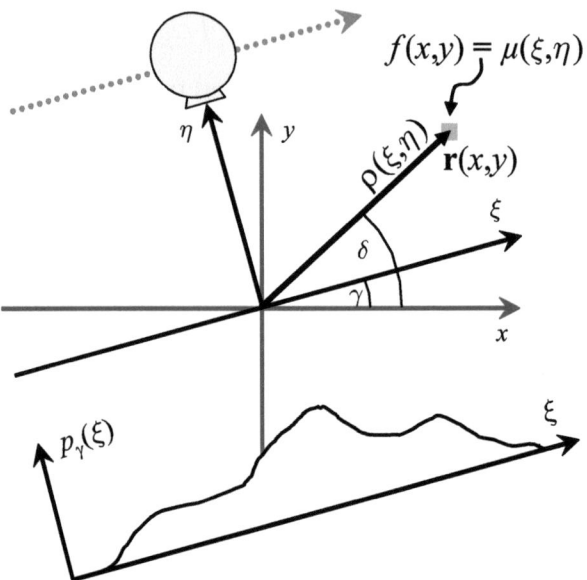

Fig. 5.7. The attenuation coefficients f at the point $\mathbf{r}$ in the fixed (x, y) patient system and μ at the point $\boldsymbol{\rho}$ in the rotating (ξ, η) sampling system

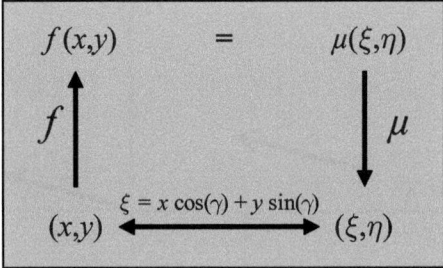

Fig. 5.8. Functional correspondence of the attenuation coefficients defined in the fixed (x, y) and the rotating (ξ, η) coordinate system

From the view point of signal processing, the stepwise shift of the X-ray source represents a sampling process of a continuous projection signal. Let $\mathbf{L}$ be a δ-line in the sectional plane, the sampling can then be described as follows

$$f * \delta(\mathbf{L}) = \int_{\mathbb{R}^2} f(\mathbf{r})\delta(\mathbf{r} - \mathbf{L})\,d\mathbf{r} \qquad (5.16)$$

or, equivalent by

$$f * \delta(\mathbf{L}) = \int_{\mathbf{r} \in \mathbf{L}} f(\mathbf{r})\,d\mathbf{r} \,. \qquad (5.17)$$

Figure 5.9 shows the sampling process in a plane, again schematically. The line $\mathbf{L}$ is the set of all points in the tissue that is to be examined. Since $\mathbf{L}$ is the path of X-ray photons for a certain position (ξ, γ) of the X-ray tube, the meaning of (5.16) or (5.17) is that all attenuation values in the tissue along the line $\mathbf{L}$ are integrated. In fact, this reflects the physical process of X-ray attenuation by matter.

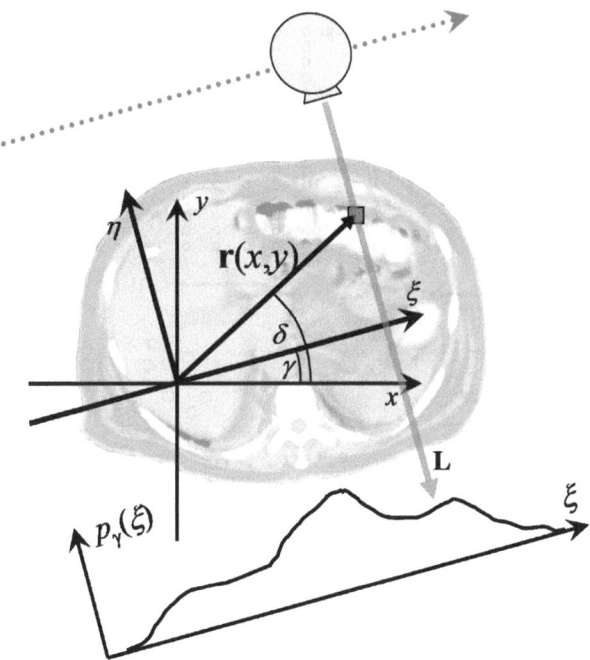

Fig. 5.9. Line sampling of the object. All points of the tissue lying on the line $\mathbf{L}$ are passed through by X-ray photons. The resulting exponential decay of the X-ray intensity is due to the attenuation values lying on the line $\mathbf{L}$

Due to the sifting property of the δ-distribution (5.16) results in the sum of all points $\mathbf{r}$ of the sectional plane that lie on line $\mathbf{L}$. Since the physical interpretation of the line $\mathbf{L}$ is the X-ray photon path, $\mathbf{L}$ may be substituted by $(\mathbf{r}^T\mathbf{n}_\xi) = \xi$, because ξ is the detector position for counting the remaining photons. The projection integral (5.16) can thus be written as follows

$$f * \delta(\mathbf{L}) = \int f(\mathbf{r})\delta((\mathbf{r}^T \cdot \mathbf{n}_\xi) - \xi)\,d\mathbf{r}$$

$$= \int\limits_{-\infty}^{\infty} \int\limits_{-\infty}^{\infty} f(x, y)\delta(x\cos(\gamma) + y\sin(\gamma) - \xi)\,dx\,dy \qquad (5.18)$$

$$= p(\xi)\,.$$

As mentioned above the projections vary with the projection angle γ. Therefore, the parameter γ must be included in a complete description of the projection process:

$$p = p(\xi, \gamma) = p_\gamma(\xi)\,. \qquad (5.19)$$

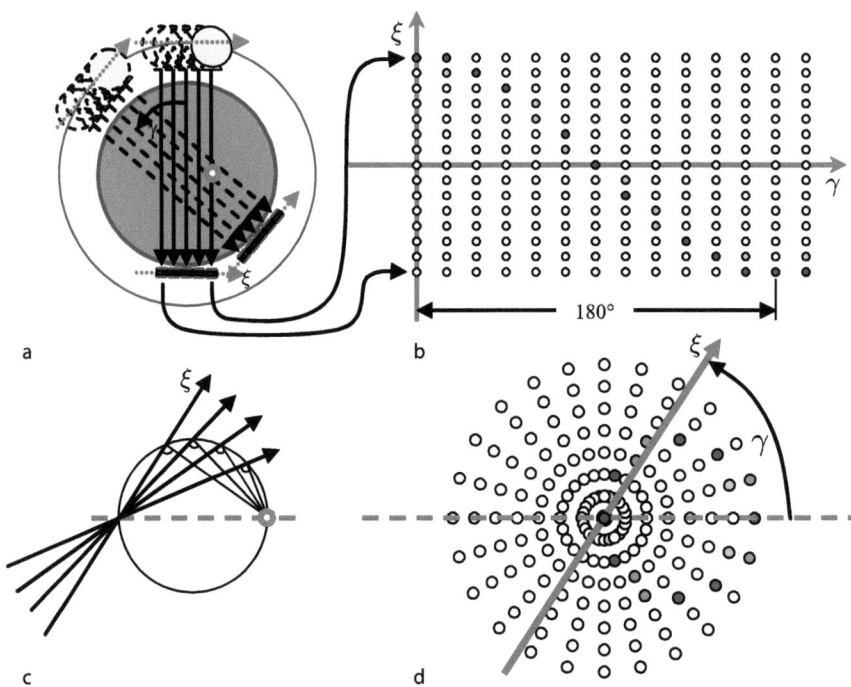

a   b   c   d

Fig. 5.10. a Arrangement of the measured projection values $p_\gamma(\xi)$. Projection data are usually displayed in a Cartesian (ξ, γ) grid. Such a Radon space diagram is often called a *sinogram*. Note that the sinogram is related to the data space of a *Hough* transformation (see Lehmann et al. [1997]). Occasionally, one also uses a polar coordinate grid (**d**) to visualize Radon space data (**b**). The *circles* in the polar arrangement of the projection values stem from the theorem of Thales (**c**)

$p_\gamma(\xi)$ is called the two-dimensional Radon transform of the object. One may write

$$f(x, y) \; \circ\!\!\xrightarrow{\mathcal{R}_2}\!\!\bullet \; f * \delta(\mathbf{L}) = p_\gamma(\xi) \tag{5.20}$$

or

$$p_\gamma(\xi) = \mathcal{R}_2\{f(x, y)\} \tag{5.21}$$

respectively. A detailed discussion of the properties of the Radon transform can be found in Helgason (1999).

Figure 5.10 schematically shows a diagram of Radon-transformed data. Typically, the projection values $p_\gamma(\xi)$ of the Radon space are arranged in a Cartesian (ξ, γ) diagram. In such a diagram, the projection values of objects that lie outside the rotation center of the tomograph produce a sinusoidal trace. Therefore, this graphical representation of projection values is often called a *sinogram*. Figure 5.10 shows the formation of the sinusoidal curve for a single object point.

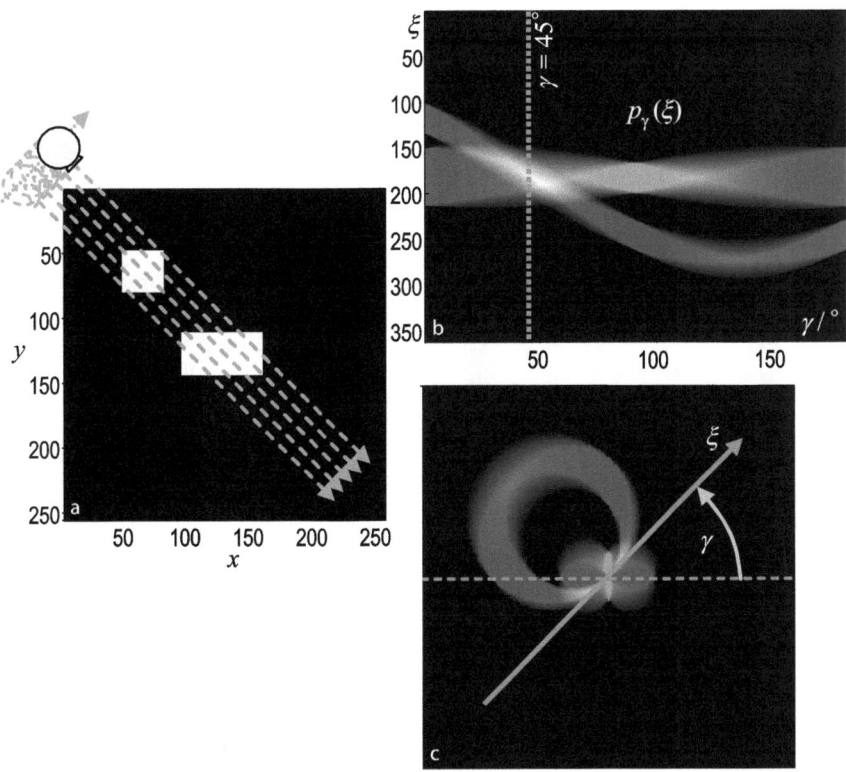

Fig. 5.11. a Synthetic 256 × 256 pixel image. The homogeneous attenuation values of two objects are simulated by gray values set to 1. **b** Cartesian Radon space of the synthetic image provided in **a**. **c** Radon space of the synthetic image in polar coordinates

Projection data are acquired from 0 to 180° only, since, according to the physical symmetry, the X-ray path back through the object that is to be examined provides no additional information to the forward path. Figure 5.11a shows a synthetic tomogram with an homogeneous square shifted away from the rotation center and an homogeneous rectangle at the rotation center. Figures 5.11b and c display the Radon space corresponding to Fig. 5.11a. In the diagram the projection values $p_\gamma(\xi)$ are displayed in gray values. High integral attenuation is represented by white and low attenuation by black. Figure 5.11a shows the projection values in the Cartesian (ξ, y) plane. The sinusoidal trace of the projection values in the Radon space for the square in the object space can be easily recognized. The further an object is apart from the rotation center the larger is the modulation amplitude of the corresponding sine trace in the Radon space. The rectangle therefore does not show this global sinusoidal path. Nevertheless, the rectangular shape can be identified due to the characteristic path of the values in the Radon space as well. Under a view of $\gamma = 45°$ the rectangle and the square lie on the same line according to the projection geometry. This situation, indicated by the dashed line in Fig. 5.11b, is reflected by an exceedingly high value at $p_{45°}(181)$.

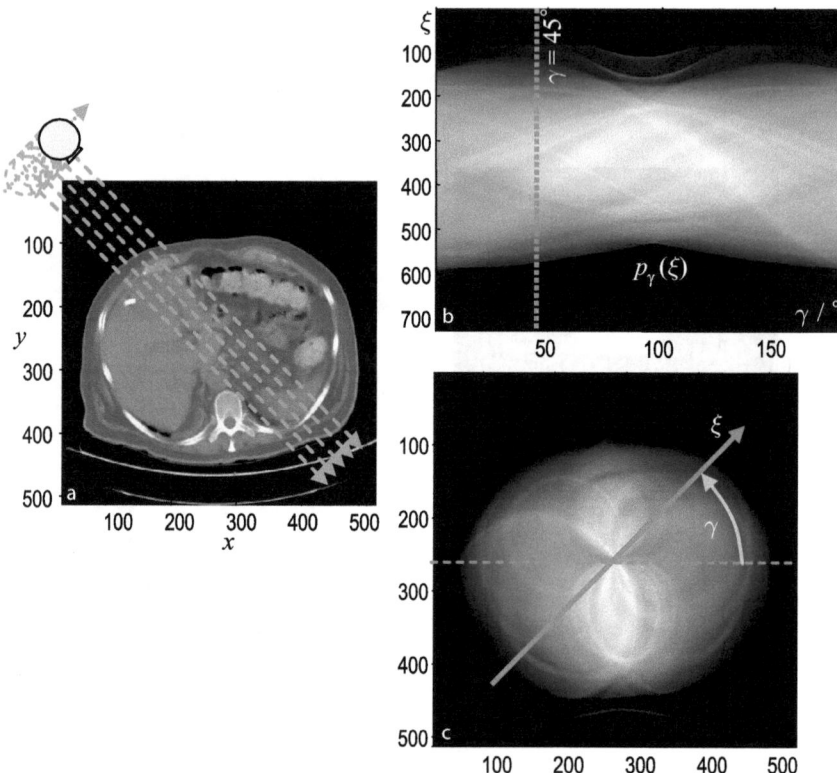

Fig. 5.12. a Computed tomographic slice image of the abdomen. **b** Cartesian Radon space of the abdomen provided in **a**. **c** Radon space of the abdomen provided in **a** in polar coordinates

Figure 5.11c shows the Radon space in polar coordinates. Both object representations are well separated from each other. The further an object is apart from the rotation center the larger the corresponding circle in the polar Radon space. For the rectangle in the center of Fig. 5.11a the mirror symmetrical nature in the object space leads to a mirror symmetric signature in the Radon space. The pattern of the polar Radon space representation is a result of the theorem of Thales (Papula 2000).

As an example, Fig. 5.12a shows a tomographic slice of an abdomen. Figure 5.12b and c again represent the corresponding Radon space. The interpretation of these diagrams is more complicated, as in the case of the synthetic tomogram, since each point of the object produces its own sinusoidal trace of different amplitude, phase, and gray value. However, this data set represents the basis for the following reconstruction methods.

5.3
Inverse Radon Transformation and Fourier Slice Theorem

(5.21) presents a first important result. However, in practice one faces the inverse problem already formulated by Johann Radon in 1917. One is interested in the spatial distribution of attenuation values $f(x, y) = \mu(\xi, \eta)$, having measured only the projection data $p_\gamma(\xi)$. The Fourier slice theorem that is summarized in Scheme 5.1 does indeed solve this problem. It is divided into three main steps.

While steps 1 and 3 can be directly carried out by the simple application of the Fourier and the inverse Fourier transforms respectively, step 2 requires knowledge of a certain identity of the Fourier transforms of $f(x, y)$ and $p_\gamma(\xi)$. To understand this identity, in a first step the Radon space data $p(\xi, \gamma) = p_\gamma(\xi)$ are seen as one-dimensional functions of the detector coordinate ξ parameterized by the projection angle γ. This means that for every projection angle γ a one-dimensional spectrum

$$P(q, \gamma) = P_\gamma(q) = \int_{-\infty}^{\infty} p_\gamma(\xi) e^{-2\pi i q \xi} d\xi \qquad (5.22)$$

Scheme 5.1 Fourier slice theorem

1. Calculation of the Fourier transform of $p_\gamma(\xi)$

$$p_\gamma(\xi) \circ\!\!-\!\!\!-\!\!\!-\!\!\bullet P_\gamma(q)$$

2. Construction of the Fourier transform of f from P

$$P_\gamma(q) \bullet\!\!-\!\!\!-\!\!\!-\!\!\bullet F(u, v)$$

3. Calculation of the inverse Fourier transform of F provides the desired function f

$$F(u, v) \bullet\!\!-\!\!\!-\!\!\!-\!\!\circ f(x, y)$$

is obtained. This transformation completes step 1 of Scheme 5.1. To carry out step 3 one has to keep in mind that F is a function of the Cartesian spectral coordinates u and v, because the spatial distribution of attenuation coefficients $f(x, y)$ is defined in Cartesian coordinates as well. The transformation into a spectral representation does not change the nature of the coordinate system, which means

$$F(u, v) \quad \bullet\!\!-\!\!\!-\!\!\circ \quad f(x, y) . \tag{5.23}$$

Using an inverse two-dimensional Fourier transform, this step can also be carried out easily.

If, however, the Fourier transform does not change the nature of the coordinate system, it is obvious that a change in coordinates has to be carried out in step 2, because the Radon transform $p_\gamma(\xi)$, along with its Fourier transform $P_\gamma(q)$, are given in polar coordinates (ξ, γ) and (q, γ) respectively. Figure 5.13 shows this relation schematically. The transformations in the vertical direction of Fig. 5.13 are given by the Fourier transform in each case. However, from left to right a change in coordinates from polar to Cartesian coordinates has to be carried out.

As shown in Fig. 5.13, the spatial vector $\boldsymbol{\xi} = \boldsymbol{\xi}(\xi, \gamma)$ points in the same direction as the spectral vector $\mathbf{q} = \mathbf{q}(q, \gamma)$. This is due to the rotational variance of the Fourier transform.

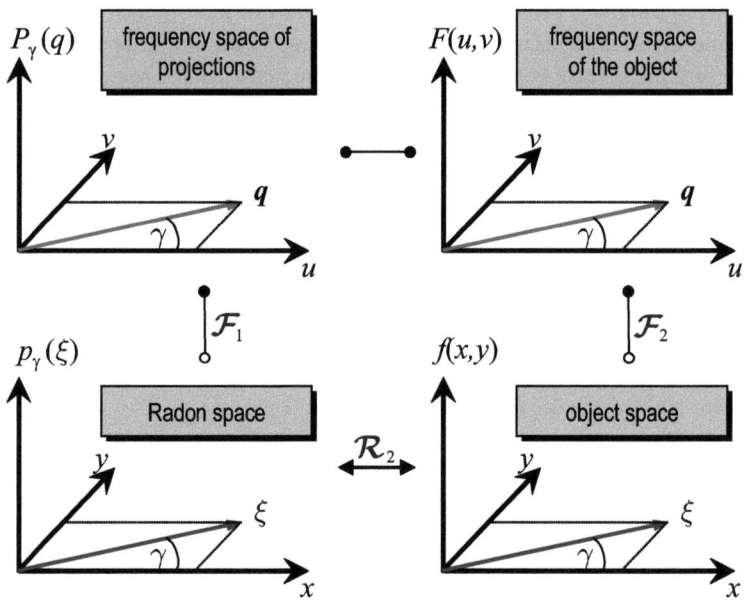

Fig. 5.13. Schematic representation of the Fourier slice theorem. Since the Fourier transform does not change the nature of coordinate systems, a change from polar to Cartesian coordinates has to be carried out to find the correspondence between $F(u, v)$ and $P_\gamma(q)$. The polar nature of the Radon space is inherently produced by the circular sampling of the Cartesian object space

To change the coordinate systems it is briefly repeated here that a point with the polar coordinates (q, γ) has the Cartesian coordinates

$$u = q \cos(\gamma)$$
$$v = q \sin(\gamma) \, . \tag{5.24}$$

When replacing the Cartesian coordinates with polar coordinates of the function $F(u, v)$ on the left-hand side of transformation (5.23), one obtains

$$F(u(q, \gamma), v(q, \gamma)) = F(q \cos(\gamma), q \sin(\gamma))$$
$$= F_{\text{polar}}(q, \gamma) \, . \tag{5.25}$$

The Fourier slice theorem ensures that $F_{\text{polar}}(q, \gamma)$ is identical to $P_\gamma(q)$, thus

$$F(q \cos(\gamma), q \sin(\gamma)) = P(q, \gamma) = P_\gamma(q) \, . \tag{5.26}$$

This means that a linear radial intersection of the two-dimensional Fourier spectrum $F(u, v)$ of the spatial distribution of attenuation values $f(x, y)$ at angle γ equals the one-dimensional Fourier transform $P_\gamma(q)$ of the measured Radon values $p_\gamma(\xi)$ that result from the projection of f under the angle γ. Figure 5.14 shows this important relation schematically in a functional diagram.

The functional relation shown in Fig. 5.14 can be derived directly from the co-ordinate transformation rules. Starting with

$$P_\gamma(q) = \int\limits_{-\infty}^{\infty} p_\gamma(\xi) \, e^{-2\pi i q \xi} \, d\xi \, , \tag{5.27}$$

it must be recalled that the projection values $p_\gamma(\xi)$ in (5.27) result from the projection integral (5.10), such that

$$P_\gamma(q) = \int\limits_{-\infty}^{\infty} \left\{ \int\limits_{0}^{s} \mu(\xi, \eta) \, d\eta \right\} e^{-2\pi i q \xi} \, d\xi \, . \tag{5.28}$$

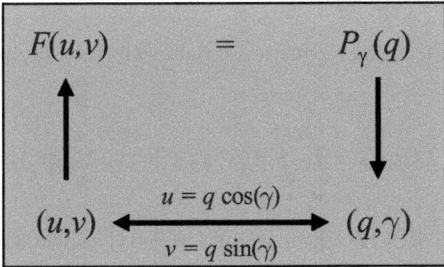

Fig. 5.14. Schematic functional presentation of the Fourier slice theorem identity. In the spectral representation $F(u, v)$ and $P_\gamma(q)$ can be identified with each other if a change in coordinates from Cartesian to polar coordinates is carried out

The integration in the direction of the X-ray path through the object, i.e., the inner term of (5.28), can be extended to infinity, if one assumes that outside the object to be examined no attenuation occurs. Hence,

$$P_\gamma(q) = \int\limits_{-\infty}^{\infty} \int\limits_{-\infty}^{\infty} \mu(\xi,\eta)\, e^{-2\pi i q\xi}\, d\xi\, d\eta \,. \tag{5.29}$$

As a next step, the coordinate system has to be changed from the rotating sampling system (ξ,η) to the patient system (x,y) at rest. Both systems are of Cartesian nature. Thus, the integration element, $d\xi\, d\eta$, can be directly replaced with $dx\, dy$, such that

$$P_\gamma(q) = \int\limits_{-\infty}^{\infty} \int\limits_{-\infty}^{\infty} \mu(\xi(x,y),\eta(x,y))\, e^{-2\pi i q(\mathbf{r}^T\cdot\mathbf{n}_\xi)}\, dx\, dy \,. \tag{5.30}$$

The inner product $(\mathbf{r}^T\mathbf{n}_\xi) = \xi$, introduced in (5.13), in the exponent of (5.30) also includes the transition to the fixed (x,y) system. With the correspondence represented by (5.15) it follows that

$$P_\gamma(q) = \int\limits_{-\infty}^{\infty} \int\limits_{-\infty}^{\infty} f(x,y)\, e^{-2\pi i q(\mathbf{r}^T\cdot\mathbf{n}_\xi)}\, dx\, dy \,. \tag{5.31}$$

To prove that (5.31) is identical to the two-dimensional Fourier transform of the object, one starts with the Cartesian formulation of the Fourier transform of the tomographic section $f(x,y)$ in the fixed patient coordinate system

$$F(u,v) = \int\limits_{-\infty}^{\infty} \int\limits_{-\infty}^{\infty} f(x,y)\, e^{-2\pi i(xu+yv)}\, dx\, dy \,. \tag{5.32}$$

Inserting the polar relations from (5.24) in the exponent of (5.32) one obtains

$$F(u,v) = \int\limits_{-\infty}^{\infty} \int\limits_{-\infty}^{\infty} f(x,y)\, e^{-2\pi i(xq\cos(\gamma)+yq\sin(\gamma))}\, dx\, dy \,. \tag{5.33}$$

After factoring out the radial frequency variable q in the exponent, (5.33) simplifies to

$$F(u,v) = \int\limits_{-\infty}^{\infty} \int\limits_{-\infty}^{\infty} f(x,y)\, e^{-2\pi i q(\mathbf{r}^T\cdot\mathbf{n}_\xi)}\, dx\, dy \tag{5.34}$$

through substitution of (5.13). The integrand in (5.34) is identical to the one in (5.31), whereby it is shown that

$$F(u,v)\big|_{\substack{u=q\cos(\gamma)\\ v=q\sin(\gamma)}} = P_\gamma(q) \,. \tag{5.35}$$

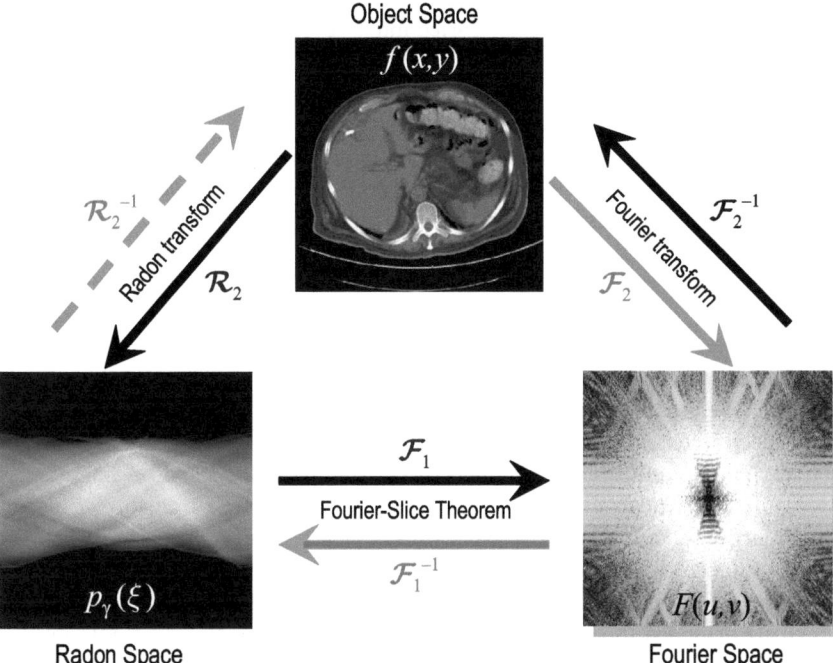

Fig. 5.15. Schematic summary of the relations among the spatial object domain (shown as an axial abdomen tomogram), the Radon space (given over an interval of 180° from the object), and the Fourier space (only absolute values are shown). The Fourier domain results directly from the spatial domain by a two-dimensional Fourier transform of the object, but can also be obtained by the Fourier slice theorem using a set of one-dimensional Fourier transforms of the projection profiles in the Radon space

This is a core result of the Fourier slice theorem and the most important result for all Fourier-based reconstruction methods. (5.35) can also be used in the reverse direction to give a Radon transform when the image $f(x, y)$ is known. Figure 5.15 summarizes the relation among the spatial domain of the object, the Radon space, and the Fourier space of both. The Fourier slice theorem can be expressed in one sentence. It states that the one-dimensional Fourier transform of the projection profile can be identified with a radial line in the Cartesian Fourier space of the object drawn at the angle of the corresponding measurement.

5.4
Implementation of the Direct Inverse Radon Transform

Using the main result of the Fourier slice theorem stated by (5.35), a simple direct algorithmic scheme for tomographic object reconstruction $f(x, y)$ from the measured projection data $p_\gamma(\xi)$ can be given. In such a scheme the polar spectral data

$P_y(q)$ of the projections have to be sorted into a spectral Cartesian coordinate system (u, v). In this way, the function $F(u, v)$ is assembled. An inverse Fourier transform of $F(u, v)$ back into the spatial domain directly leads to the desired tomogram $f(x, y)$. Figure 5.16 shows the direct reconstruction scheme as a flow chart.

Unfortunately, there is a practical difficulty one is faced with the simple direct inverse Radon transform. Due to data handling problems and dose considerations, a CT scanner is obviously limited with regard to the number of projections $\{p_{y_1}(\xi), p_{y_2}(\xi), \dots, p_{y_N}(\xi)\}$ that can be measured within one revolution of the sampling disk. Therefore, the values of the Fourier transform $F(u, v)$ of the attenuation coefficients $f(x, y)$ are only known for a limited number of measured radial lines in the (u, v) space.

The sampling discretization of the projection angle y itself, however, is not the problem. The problem lies in the position, or more precisely, the distribution of the polar spectral data in the Cartesian coordinate system. The spectral representation of the tomographic spatial domain is only known on radial, "star-like" lines.

Figure 5.17a shows the radial configuration of the Fourier transform $P_y(q)$ of the projection data measured at angle y in the Cartesian (u, v) space. Obviously, the spectral projection data lie on circles in the (u, v) space. Since the fast Fourier transform (FFT) needs data lying on a regular rectangular grid, the data $P_y(q)$, which are arranged in circles, must be defined on a Cartesian grid before inverse transformation. This process is called regridding.

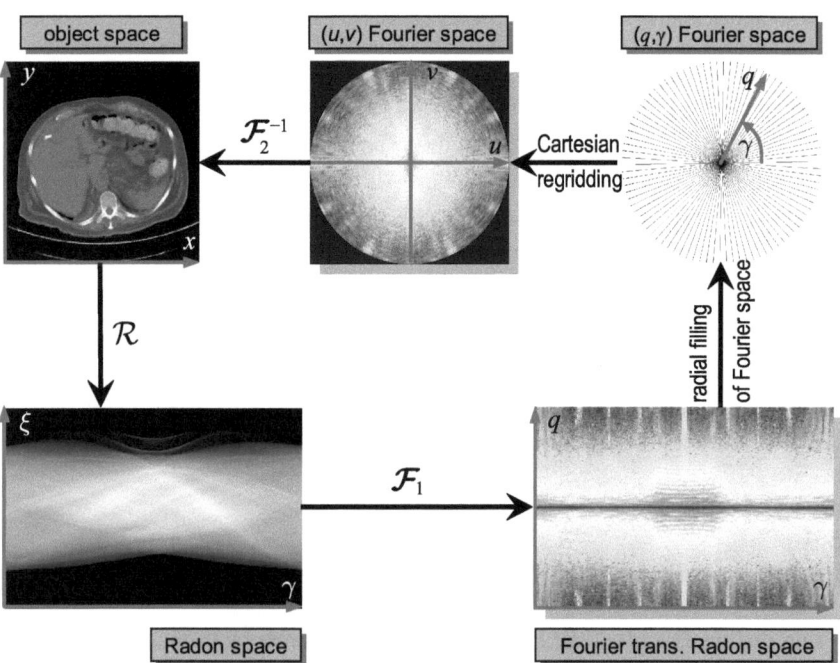

Fig. 5.16. Flow chart of the direct reconstruction scheme based on the Fourier slice theorem

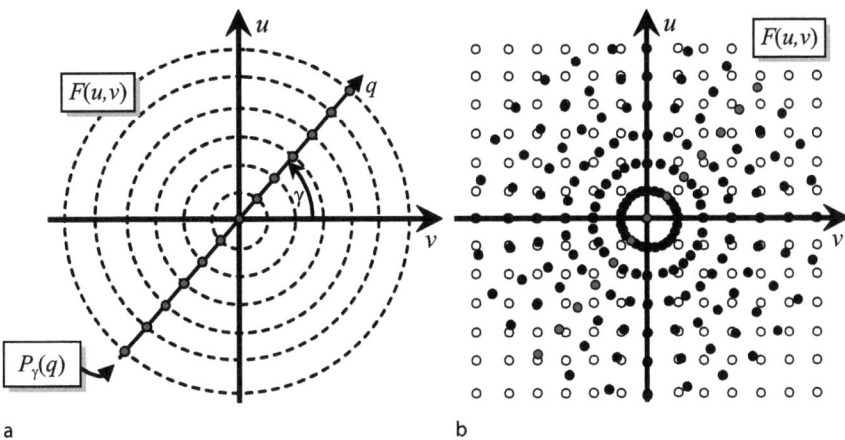

a   b

Fig. 5.17. a Radially arranged values $P_\gamma(q)$ obtained from the one-dimensional Fourier transform of projection measurements. **b** Regridding of the spectral projection data, initially being arranged in a polar coordinate frame, onto a Cartesian grid. The radial and the Cartesian sample points of the spectral data of $P_\gamma(q)$ and $F(u,v)$ only lie directly above each other on the u and v axes that span the Fourier space of the object

In order to perform the direct inverse Radon transformation using an FFT algorithm, the Cartesian grid has to be created from the polar configuration of spectral data via an appropriate interpolation (e.g., nearest-neighbor, bilinear, etc.). As a direct consequence of the radial data arrangement – as illustrated in Fig. 5.17b – the density of measured spectral data in the (u,v) space decreases for higher frequencies. Therefore, the interpolation error increases at high spatial frequencies. Unfortunately, this effect leads to a degradation of image quality because high spatial frequencies represent image areas that are rich in detail.

As a counter-measure, the geometrical sampling may be changed in order to obtain those projection values in the Radon space that lie exactly on a Cartesian grid in the Fourier space of the object. Due to the fact that the Radon space and Fourier space are connected via the Fourier transform, the complementary variables ξ and q behave reciprocally to each other, i.e., when the sampling points in the Radon space are arranged on concentric circles (see Fig. 5.18a), the sampling points in the corresponding Fourier space (see Fig. 5.18b) lie on circles as well, and vice versa. Obviously, this is due to the fact that the sampling points in the Radon space are equidistant in all radial directions.

If, on the other hand, the points in the Fourier space are expected to lie on a Cartesian grid, this has consequences for the required locations of the sampling points in the Radon space. Looking at Fig. 5.18d it can be seen that the Fourier points are located on the boundary of concentric squares. Since in the spectral domain the sampling points on the diagonals of these squares have greater separation than the sampling points on the u and v axes, the quadratic shape is not preserved by the inverse Fourier transform back into Radon space. Due to the reciprocal relation

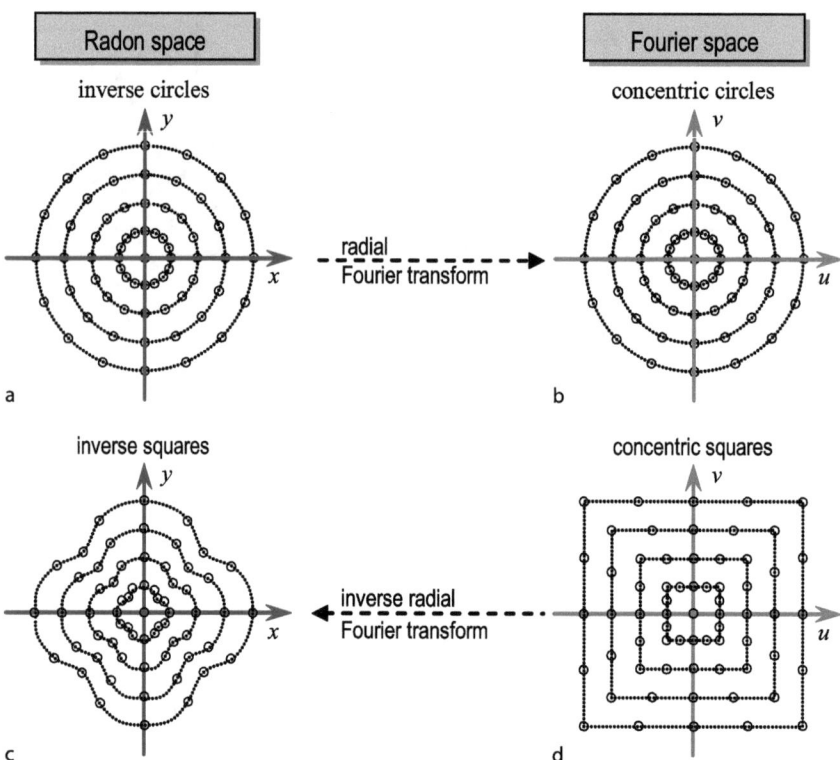

Fig. 5.18. a Radially arranged values of $p_{\gamma(\xi)}$ obtained from projection measurements are lying on concentric circles. **b** After the radial Fourier transform, the values of $P_\gamma(q)$ are lying on concentric circles as well. **c** Arrangement of sampling points in the Radon space on inverse concentric squares consisting of four semi-circles. **d** Regular arrangement of points on concentric quadrants in the Fourier space

between Fourier and Radon space variables, the sampling points on the diagonals in the Radon space must lie closer to each other than the corresponding points on the x- and y-axes. One obtains what is called "inverse squares" as visualized in Fig. 5.18c. The formation of inverse squares during the measurement process by the linogram method will be discussed in the following section.

5.5
Linogram Method

One way to overcome the problem of two-dimensional interpolation when using the direct inverse Radon transform is called the "linogram method" (Jacobson 1996; Magnusson 1993). As motivated by the discussion in the previous section, spectral sampling points on concentric squares in the Fourier space are created instead of concentric circles by a sophisticated non-equidistant sampling procedure. The inverse chirp z-transform (Stearns and Hush 1999; cf. ▸ Sect. 4.23) is optimally adapted to this very situation such that no interpolation is necessary.

The linogram method can be seen as a direct Fourier method. However, instead of the sinogram sampling arrangement of the projection data introduced above, here a slightly different data structure, namely a linogram, has to be created or measured. For this linogram sampling arrangement, the overall sampling length b varies with the projection angle γ such that

$$b(\gamma) = \begin{cases} b(0°)\cos(\gamma) & \text{for } -45° \leq \gamma < 45° \\ b(0°)\sin(\gamma) & \text{for } 45° \leq \gamma < 135° . \end{cases} \tag{5.36}$$

$b(0°)$ represents the overall sampling length[5] for the projection angle $\gamma = 0°$. Figure 5.19 shows the corresponding sampling points, for an exemplary projection angle for each of the two intervals given by (5.36). The detector arrays that are fixed for this kind of sampling arrangement (drawn in bold in Fig. 5.20 for each case of (5.36)) coincide with the x- and y-axes respectively of the fixed patient coordinate system. For comparison, the rotating detector array of the sinogram sampling arrangement is also shown.

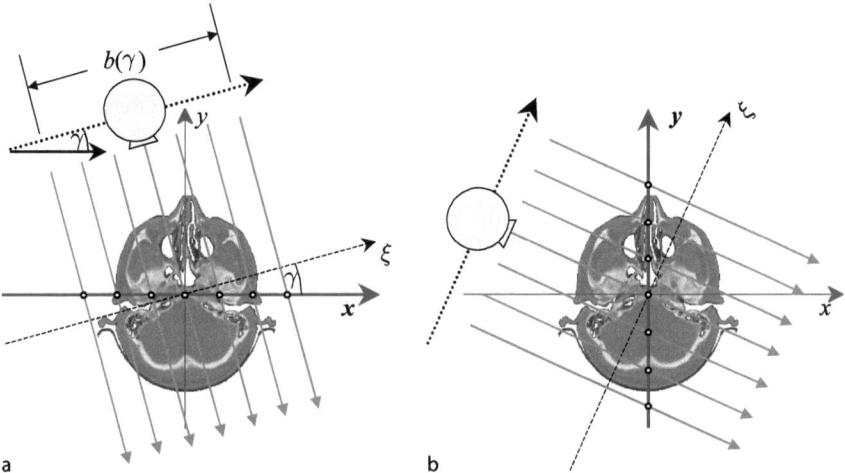

a   b

Fig. 5.19. Linogram sampling **a** in the angle interval $-45° \leq \gamma < 45°$ and **b** in the angle interval $45° \leq \gamma < 135°$. The sampling points are indicated on virtual fixed detector lines in each case (see the bold x-axis in **a** and the bold y-axis in **b**). Obviously, the sampled overall length $b(\gamma)$ has its minimum at $45°$

---

[5] $b(0°) = D\Delta\xi$ similar to the sinogram sampling arrangement, where D is the number of sampling points.

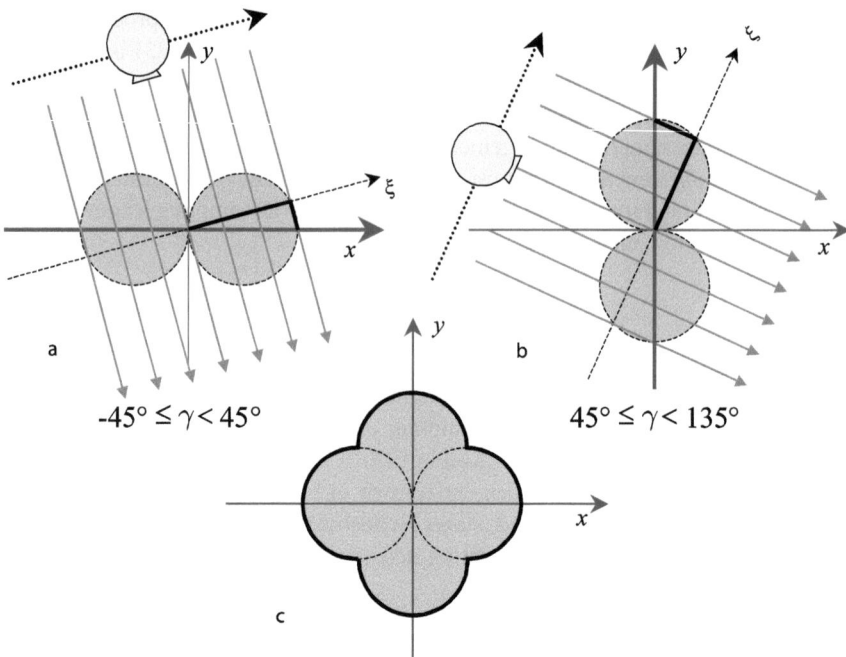

Fig. 5.20. For linogram sampling, two circles are the underlying shapes that confine the length of the radial sampling line. The circles can be constructed with the theorem of Thales for each of the angle intervals **a** $-45° \leq \gamma < 45°$ and **b** $45° \leq \gamma < 135°$. **c** Overall, a Radon space sampling point confinement is established by four semi-circles

For a fixed distance between the sampling points on the linear detector array, it is obvious that the overall length $b(\gamma)$ that is actually sampled, has its minima at the projection angles $\gamma = 45°$ and $\gamma = 135°$, and its maxima at $\gamma = 0°$ and $\gamma = 90°$ respectively. When the projection angle γ is varied, the length $b(\gamma)$ draws a circle according to the theorem of Thales. In this way, two circles are the underlying shape of data arrangement in the Radon space domain for each of the two angle intervals, as presented in Fig. 5.20.

Overall, the allowed sampling area in the Radon space is defined by four semi-circles that are constructed with the theorem of Thales for each of the two angle intervals given by (5.36). The reciprocal relation between the Radon space and the Fourier space has already been mentioned above. If an overall length b has been sampled in the Radon space, a sampling point distance of b^{-1} is the consequence in the Fourier space. Since the number of sampling points D is not altered by the Fourier transform, the corresponding overall length of the radial Fourier transform B must change when the projection angle is varied. Obviously, the overall length in the Fourier space $B(\gamma)$ has its maxima at the projection angles $\gamma = 45°$ and $\gamma = 135°$, and its minima at $\gamma = 0°$ and $\gamma = 90°$ respectively.

The Fourier space length is given by

$$
B(\gamma) = \begin{cases} \dfrac{D}{b(0°)\cos(\gamma)} & \text{for } -45° \leq \gamma < 45° \\[3mm] \dfrac{D}{b(0°)\sin(\gamma)} & \text{for } 45° \leq \gamma < 135° \end{cases}
\tag{5.37}
$$

as a direct consequence of (5.36).

(5.37) leads to concentric sampling areas with a quadratic form in the Fourier space. In this way, the spectral sampling points in the (u, v) space are arranged on horizontal and vertical lines, as required. However, the distribution of sampling points on the boundaries of the Fourier space squares does not completely solve the interpolation problem. Along the horizontal and vertical lines of the sampling directions given by the Fourier space squares, equidistant sampling points can be achieved when choosing a non-constant projection angle increment, $\Delta\gamma$. In fact, equidistant sampling points in the spectral domain are obtained by choosing constant increments for

$$
\Delta\tan(\gamma) \quad \text{for } -45° \leq \gamma < 45°
\tag{5.38}
$$

and

$$
\Delta\cot(\gamma) \quad \text{for } 45° \leq \gamma < 135°
\tag{5.39}
$$

respectively. This results directly from the geometry of the squares in the Fourier space. Figure 5.21a and b summarize the linogram method schematically. According to the segmentation into the angle intervals $-45° \leq \gamma < 45°$ and $45° \leq \gamma < 135°$ during the linogram-based data acquisition in the Radon space, these areas can be treated separately and fitted together in a later step. After separation of the angle intervals, a radial Fourier transform is applied that leads to the desired equidistant vertical arrangement of points for the interval $-45° \leq \gamma < 45°$ and the equidistant horizontal arrangement for the interval $45° \leq \gamma < 135°$.

The sampling point density in the Fourier space is inhomogeneous, i.e., the density of spectral sampling points increases for decreasing spatial frequencies. This over-representation of slowly varying object information is compensated for by a linear weighting of the frequency space. Since the inverse chirp-z transform treats all points with the same weight, an uncompensated transformation would have the effect of a low-pass filtering.

Hereafter, the inverse chirp-z transform is applied in the vertical direction for the angle interval $-45° \leq \gamma < 45°$ and in the horizontal direction for the comple-

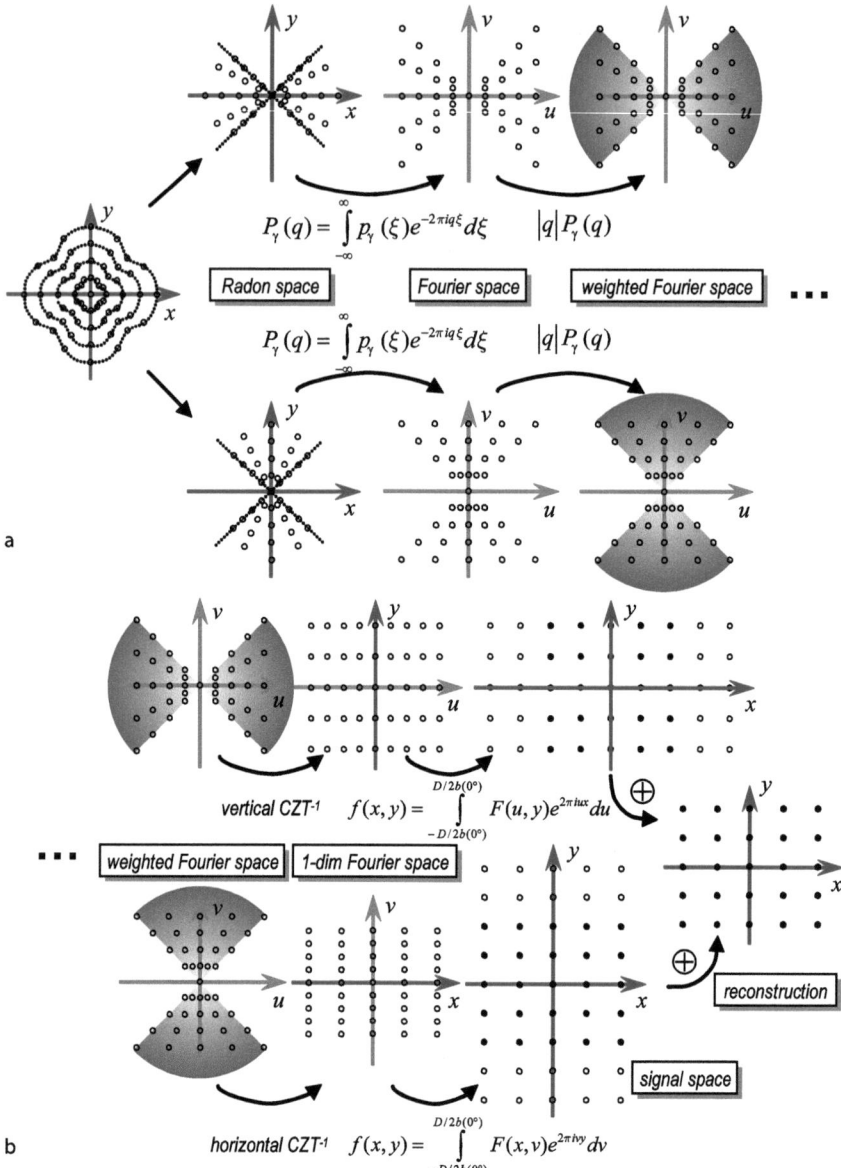

$$P_\gamma(q) = \int\limits_{-\infty}^{\infty} p_\gamma(\xi)e^{-2\pi i q\xi}d\xi \qquad |q|P_\gamma(q)$$

Radon space   Fourier space   weighted Fourier space   · · ·

$$P_\gamma(q) = \int\limits_{-\infty}^{\infty} p_\gamma(\xi)e^{-2\pi i q\xi}d\xi \qquad |q|P_\gamma(q)$$

a

vertical CZT⁻¹ $$f(x,y) = \int\limits_{-D/2b(0°)}^{D/2b(0°)} F(u,y)e^{2\pi i ux}du$$

· · ·   weighted Fourier space | 1-dim Fourier space

reconstruction

signal space

horizontal CZT⁻¹ $$f(x,y) = \int\limits_{-D/2b(0°)}^{D/2b(0°)} F(x,v)e^{2\pi i vy}dv$$

b

Fig. 5.21a,b. Schematic presentation of the linogram-based reconstruction method

ment angle interval $45° \leq \gamma < 135°$. This results in frequency representation along the relevant directions such that a heterogeneous space of mixed spatial and spectral coordinates is produced for each case.

Along the spectral direction of this space, another one-dimensional inverse Fourier transform is applied, i.e.,

$$f_1(x, y) = \int_{-\frac{D}{2b(0°)}}^{\frac{D}{2b(0°)}} F(u, y) e^{2\pi i u x} \, du \quad \text{for } -45° \leq \gamma < 45° \tag{5.40}$$

and

$$f_2(x, y) = \int_{-\frac{D}{2b(0°)}}^{\frac{D}{2b(0°)}} F(x, v) e^{2\pi i v y} \, dv \quad \text{for } -45° \leq \gamma < 135° . \tag{5.41}$$

Each of the separately treated angle intervals leads to a separate image in the signal space of the object, which consists of only half of all projections. Finally, the complete reconstruction is obtained by adding the partial reconstructions f_1 and f_2

$$f(x, y) = f_1(x, y) + f_2(x, y) . \tag{5.42}$$

Note that only mutually overlapping areas lead to the desired image. For a more detailed discussion of the linogram method, the reader is referred to C. Jacobson and M. Magnusson-Seger (Jacobson 1996; Magnusson 1993).

5.6
Simple Backprojection

As discussed above, the Fourier slice theorem induces a general direct reconstruction procedure. Having measured projections at an adequately high number of angles such that the spectral (u, v) space can be filled densely with data points, a simple inverse two-dimensional Fourier transform has to be applied to reconstruct the attenuation values, $f(x, y)$, of the object. However, due to the above-mentioned problems of Cartesian *regridding*, a different reconstruction strategy for sinogram data sampling is used in practice. In fact, today, all modern CT systems implement "filtered backprojection".

To understand what the term *filtered* means in this context, let us step back for a moment. At first glance, one might think that the image that is to be reconstructed could potentially be obtained by a simple backprojection of the measured projection profiles $p_\gamma(\xi)$, just as suggested in Fig. 1.2. Such a strategy would smear back the profile values in the direction from which the radiation came.

Mathematically, this process can be modeled by the following equation

$$g(x, y) = \int_0^\pi p_\gamma(\xi) \, d\gamma$$

$$= \int_0^\pi p_\gamma (x \cos(\gamma) + y \sin(\gamma)) \, d\gamma . \tag{5.43}$$

Although this kind of simple backprojection intuitively seems to reverse the process of projection, the procedure does not result in the desired distribution of attenuation values. In fact, the backprojection defined by (5.43) is not an adequate reconstruction method for the original morphology of the objects in the image $f(x, y)$, because each point in the image grid receives non-negative contributions from all other points of the original image. This problem becomes immediately clear when discussing the reconstruction of all points outside the support of the image, i.e., all points $f(x, y) = 0$. Due to the fact that the projection profile $p_y(\xi)$ is a non-negative function, the simple backprojection smears back non-negative values over the entire image. In this way, positive values are assigned to image pixels even outside the object. The backprojections from other directions cannot compensate for this incorrect pixel entry, because the entire set of projection profiles in the sinogram is a set of non-negative functions.

Looking at the backprojection process in detail by substituting (5.16) or (5.18) into (5.43), it follows that

$$g(x, y) = \int_0^\pi \iint_{\text{all } \mathbf{r} \in L} f(\mathbf{r}) \, d\mathbf{r} \, dy$$

$$= \int_0^\pi \iint_{\mathbf{r} \in \mathbb{R}^2} f(\mathbf{r}) \delta(\mathbf{r} - L) \, d\mathbf{r} \, dy \tag{5.44}$$

$$= \int_0^\pi \iint_{\mathbf{r} \in \mathbb{R}^2} f(\mathbf{r}) \delta\left((\mathbf{r}^T \cdot \mathbf{n}_\xi) - \xi\right) \, d\mathbf{r} \, dy \, .$$

The line L, on which the backprojection takes place, can again be substituted by the Hessian normal form $(\mathbf{r}^T \cdot \mathbf{n}_\xi) = \xi$ and by taking advantage of the δ-distribution symmetry

$$g(x, y) = \int_0^\pi \iint_{\mathbf{r}' \in \mathbb{R}^2} f(\mathbf{r}') \delta\left((\mathbf{r}^T \cdot \mathbf{n}_\xi) - (\mathbf{r}'^T \cdot \mathbf{n}_\xi)\right) \, d\mathbf{r}' \, dy \tag{5.45}$$

is obtained. Changing the order of integration in (5.45) results in

$$g(x, y) = \iint_{\mathbf{r}' \in \mathbb{R}^2} f(\mathbf{r}') \left(\int_0^\pi \delta\left((\mathbf{r} - \mathbf{r}')^T \cdot \mathbf{n}_\xi\right) dy \right) d\mathbf{r}' \, . \tag{5.46}$$

As indicated in Fig. 5.22, let φ be the angle between the difference vector $(\mathbf{r} - \mathbf{r}')$ and the x-axis.

In this way the inner product of $(\mathbf{r} - \mathbf{r}')$ and $\mathbf{n}_\xi$ in (5.46) can be replaced by

$$g(x, y) = \iint_{\mathbf{r}' \in \mathbb{R}^2} f(\mathbf{r}') \left(\int_0^\pi \delta\left(|\mathbf{r} - \mathbf{r}'| \cos(\varphi - y)\right) dy \right) d\mathbf{r}' \, . \tag{5.47}$$

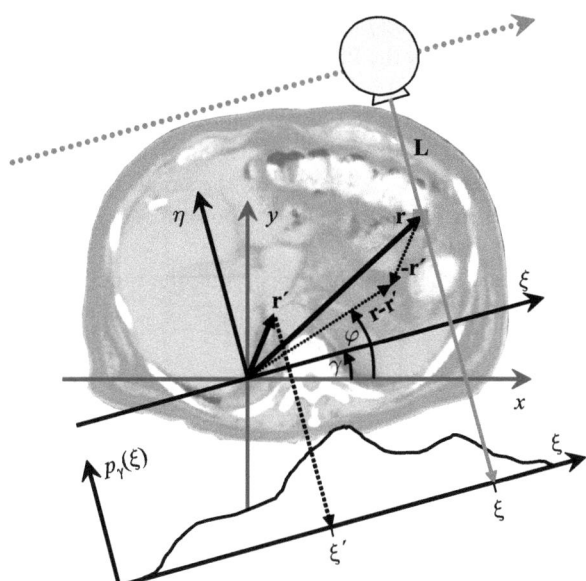

Fig. 5.22. Reconstruction geometry for simple backprojection

When applying the rule (4.38) for calculations using the δ-distribution of functions, one obtains

$$g(x, y) = \iint_{\mathbf{r}' \in \mathbb{R}^2} f(\mathbf{r}') \left(\int_0^\pi \frac{\delta(\gamma - \gamma_0)}{|\mathbf{r} - \mathbf{r}'| \left| \sin\left(\pm \frac{\pi}{2}\right) \right|} \, d\gamma \right) d\mathbf{r}' \tag{5.48}$$

for the inner integral, since the argument of the δ-distribution is the function $|\mathbf{r} - \mathbf{r}'| \cos(\varphi - \gamma)$ of which the only zero value, γ_0, lies in the interval $0 \leq \gamma < \pi$ at $\gamma_0 = (\varphi - \gamma) = \pi/2$ (or $-\pi/2$, as the sign depends on φ). The denominator of the inner integral in (5.48) is constant in terms of integration along the projection angle, γ. The integration of the δ-distribution in the numerator results in the value 1, so that

$$g(x, y) = \iint_{\mathbf{r}' \in \mathbb{R}^2} f(\mathbf{r}') \frac{1}{|\mathbf{r} - \mathbf{r}'|} \, d\mathbf{r}'$$

$$= \int_{-\infty}^{+\infty} \int_{-\infty}^{+\infty} f(x', y') \frac{1}{|(x - x', y - y')|} \, dx' \, dy' \tag{5.49}$$

is obtained.

Obviously, (5.49) represents a convolution of the original image $f(x, y)$ with the function

$$h(x, y) = \frac{1}{|(x, y)|} \tag{5.50}$$

and thus,

$$g(x, y) = f(x, y) * h(x, y). \tag{5.51}$$

In fact, $h(x, y)$ is the *point-spread function* (PSF) of the imaging system. Looking at Fig. 5.23a it becomes clear why the PSF is just $|\mathbf{r}|^{-1}$. If the image $f(x, y)$ consists of one single point only, i.e., $f(x, y) = \delta(x, y)$, this point appears as a δ-distribution in the projection profiles as well. The PSF answers the question of how the image of an ideal object point is blurred or spread by the imaging system due to the backsmearing process. It can be seen in Fig. 5.23b that for the simple backprojection process, the density of the lines in $g(x, y)$ around the point in $f(x, y)$ geometrically decreases with $|\mathbf{r}|^{-1}$.

This result is explicitly obtained by substituting the function $f(x, y)$ with the δ-distribution in (5.49)

$$
\begin{aligned}
g(x, y) &= \int_{-\infty}^{+\infty} \int_{-\infty}^{+\infty} f(x', y') \frac{1}{|(x - x', y - y')|} \, dx' \, dy' \\
&= \int_{-\infty}^{+\infty} \int_{-\infty}^{+\infty} \delta(x', y') \frac{1}{|(x - x', y - y')|} \, dx' \, dy' .
\end{aligned}
\tag{5.52}
$$

With the sifting property of the δ-distribution, (5.52) immediately reduces to

$$
g(x, y) = \frac{1}{|(x, y)|} .
\tag{5.53}
$$

In this case the reconstructed image $g(x, y)$ itself is the representation of the PSF. Due to the fact that (5.50) and (5.53) yield the same result, the function $h(x, y)$ is also called the *impulse response*.

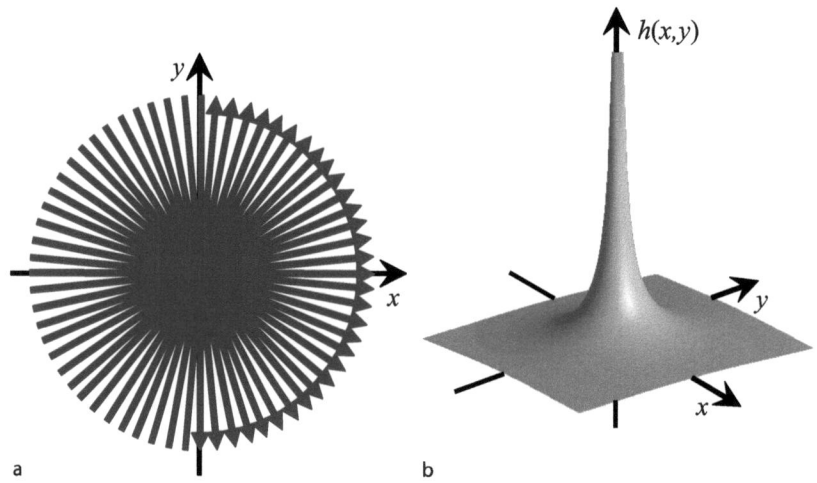

Fig. 5.23. Simple backprojection reconstruction of a single point. **a** The line density around the object that is to be reconstructed decreases with the distance from the point. **b** The *point-spread function* of a single point

As an aside, this result can also be found in electrodynamics, where the electric field around a very thin and infinitely long straight conductor decreases by $1/r$. In three-dimensional space the density of electric field lines around a point charge decreases by $1/r^2$. This decrease along with the distance is known from Coulomb's law.

One might think that (5.44) represents a sensible reconstruction method, because in Fig. 5.23 the original point of the object can be detected as the maximum value in the image $g(x, y)$. However, one has to keep in mind that any point of the object to be reconstructed is blurred by $h(x, y)$. It will be shown later that the convolution of the image with the PSF results in an overall unacceptable blurring of the image. In fact, adequate reconstruction strategies can be found that include a suitable deconvolution of the image, thus compensating for the blurring effects seen in naive backprojection.

5.7
Filtered Backprojection

What went wrong with the simple backprojection introduced in the previous section? To answer this question one has to focus on the Fourier slice theorem again. The identities of the Fourier transform of the desired image $F(u, v)$ and the Fourier transform of the measured projections $P_y(q)$ have been shown in ▸ Sect. 5.3. Since $P_y(q)$ is found as a radial line along the q-axis under the polar angle y in the Cartesian (u, v) space, it is reasonable to have a closer look at the change from Cartesian to polar coordinates.

In fact, an image reconstruction procedure – the filtered backprojection – can be derived as a clever result of the coordinate transformation. As a first step to obtaining the image $f(x, y)$ from projections, the inverse Fourier transform

$$f(x, y) = \int_{-\infty}^{\infty} \int_{-\infty}^{\infty} F(u, v) e^{2\pi i(xu+yv)} \, du \, dv \tag{5.54}$$

applied to $F(u, v)$ must be expressed in polar coordinates. To do this one substitutes

$$\begin{aligned} u &= q \cos(y) \\ v &= q \sin(y) \, . \end{aligned} \tag{5.55}$$

The infinitesimal area integration element $du \; dv$ in (5.54) is transformed to $J \, dq \, dy$, where J is the *Jacobian*, i.e.,

$$J \equiv \det\left(\frac{\partial(u, v)}{\partial(q, y)}\right) = \begin{vmatrix} \dfrac{\partial u}{\partial q} & \dfrac{\partial v}{\partial q} \\ \dfrac{\partial u}{\partial y} & \dfrac{\partial v}{\partial y} \end{vmatrix} = \begin{vmatrix} \cos(y) & \sin(y) \\ -q \sin(y) & q \cos(y) \end{vmatrix} \tag{5.56}$$

$$= q \left(\cos^2(y) + \sin^2(y)\right) = q \, .$$

In Fig. 5.24 the use of the *Jacobian* is demonstrated. The area elements $du\,dv$, which must cover the entire two-dimensional plane in (5.54), have the same square shape, independent of their actual location in the plane (compare Fig. 5.24a). However, it can be seen in Fig. 5.24b that in polar coordinates the "tiles" $q\,dq\,dy$ change their shape as a function of distance from the origin.

When using the new infinitesimal mosaic of integration elements $q\,dq\,dy$ for the entire two-dimensional plane being described in polar coordinates, the inverse Fourier transform is given as

$$f(x,y) = \int_0^{2\pi} \int_0^{+\infty} F\left(q\cos(y), q\sin(y)\right) e^{2\pi iq(x\cos(y)+y\sin(y))} q\,dq\,dy . \qquad (5.57)$$

The outer integral in (5.57) can be split into two parts when treating the projections for the angles $y = [0, \pi]$ and $y = [\pi, 2\pi]$ separately. In this way one obtains

$$f(x,y) = \int_0^{\pi} \int_0^{+\infty} F(q,y) e^{2\pi iq(x\cos(y)+y\sin(y))} q\,dq\,dy$$
$$+ \int_{\pi}^{2\pi} \int_0^{+\infty} F(q,y) e^{2\pi iq(x\cos(y)+y\sin(y))} q\,dq\,dy . \qquad (5.58)$$

The π-shift of the angle interval in the second integral term with respect to the first term can be expressed by a phase shift of the functional terms, such that

$$f(x,y) = \int_0^{\pi} \int_0^{+\infty} F(q,y) e^{2\pi iq(x\cos(y)+y\sin(y))} q\,dq\,dy$$
$$+ \int_0^{\pi} \int_0^{+\infty} F(q,y+\pi) e^{2\pi iq(x\cos(y+\pi)+y\sin(y+\pi))} q\,dq\,dy . \qquad (5.59)$$

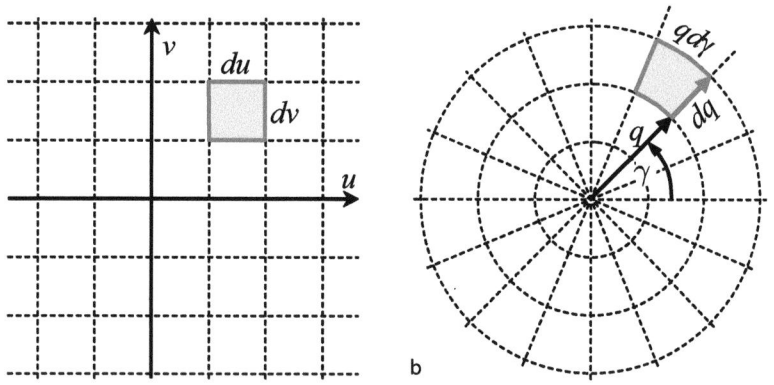

Fig. 5.24. Shape of the infinitesimal area integration elements in **a** Cartesian and **b** polar Fourier coordinates

According to the symmetry properties of the Fourier transform (see e.g., Bracewell [1965]; Klingen [2001]), i.e.,

$$\mathcal{Re}\{F(q,\gamma)\} \equiv \mathcal{Re}\{F(-q,\gamma+\pi)\} = \mathcal{Re}\{F(-q,\gamma)\} \equiv \mathcal{Re}\{F(q,\gamma+\pi)\} \quad (5.60)$$

and

$$\mathcal{Im}\{F(q,\gamma)\} \equiv \mathcal{Im}\{F(-q,\gamma+\pi)\} = -\mathcal{Im}\{F(-q,\gamma)\} \equiv -\mathcal{Im}\{F(q,\gamma+\pi)\}, \quad (5.61)$$

for real projection data, (5.59) can be written as

$$
\begin{aligned}
f(x,y) &= \int_0^\pi \int_0^{+\infty} \left(\mathcal{Re}\left(F(q,\gamma)\right) + i\mathcal{Im}\left(F(q,\gamma)\right)\right) e^{2\pi i q(x\cos(\gamma)+y\sin(\gamma))} q \, dq \, d\gamma \\
&\quad + \int_0^\pi \int_0^{+\infty} \left(\mathcal{Re}\left(F(q,\gamma)\right) - i\mathcal{Im}\left(F(q,\gamma)\right)\right) e^{-2\pi i q(x\cos(\gamma)+y\sin(\gamma))} q \, dq \, d\gamma \\
&= \int_0^\pi \int_0^{+\infty} \left(\mathcal{Re}\left(F(q,\gamma)\right) + i\mathcal{Im}\left(F(q,\gamma)\right)\right) e^{2\pi i q(x\cos(\gamma)+y\sin(\gamma))} q \, dq \, d\gamma \\
&\quad - \int_0^\pi \int_{-\infty}^0 \left(\mathcal{Re}\left(F(-q,\gamma)\right) - i\mathcal{Im}\left(F(-q,\gamma)\right)\right) e^{2\pi i q(x\cos(\gamma)+y\sin(\gamma))} q \, dq \, d\gamma .
\end{aligned}
$$
$$(5.62)$$

Using the symmetry of F again one obtains

$$
\begin{aligned}
f(x,y) &= \int_0^\pi \int_0^{+\infty} F(q,\gamma) e^{2\pi i q(x\cos(\gamma)+y\sin(\gamma))} q \, dq \, d\gamma \\
&\quad + \int_0^\pi \int_{-\infty}^0 F(q,\gamma) e^{2\pi i q(x\cos(\gamma)+y\sin(\gamma))} (-q) \, dq \, d\gamma ,
\end{aligned}
$$
$$(5.63)$$

which can finally be written as one term

$$f(x,y) = \int_0^\pi \int_{-\infty}^{+\infty} F(q,\gamma) e^{2\pi i q(x\cos(\gamma)+y\sin(\gamma))} |q| \, dq \, d\gamma . \quad (5.64)$$

In keeping with the Fourier slice theorem as stated by (5.35), it is furthermore true that

$$F(q\cos(\gamma), q\sin(\gamma)) = P_\gamma(q) . \quad (5.65)$$

According to this, the relation

$$f(x,y) = \int_0^\pi \int_{-\infty}^{+\infty} P_\gamma(q) e^{2\pi i q\xi} |q| \, dq \, d\gamma \quad (5.66)$$

can be derived by substituting Hessian's normal form (5.13) of a straight line into the exponent of (5.64). For the angular integration in (5.66), one writes

$$
f(x, y) = \int_0^\pi \left\{ \int_{-\infty}^{+\infty} P_\gamma(q)\, |q|\, e^{2\pi i q \xi}\, dq \right\} d\gamma
$$

(5.67)

$$
= \int_0^\pi h_\gamma(\xi)\, d\gamma .
$$

For an object point $\mathbf{r} = (x, y)^T$ and a certain projection angle γ, ξ is the projection coordinate of the point $\mathbf{r}$. In fact, ξ is the detector position of the sampling system. The second row of (5.67) is the backprojection of a new term

$$
h_\gamma(\xi) = \int_{-\infty}^{+\infty} P_\gamma(q)\, |q|\, e^{2\pi i q \xi}\, dq .
$$

(5.68)

Let us call $h_\gamma(\xi)$ the filtered projection, because $h_\gamma(\xi)$ is actually the high-pass filtered projection signal $p_\gamma(\xi)$. The high-pass nature of (5.68) can be understood by the convolution theorem. Here, the corresponding multiplication of the signal and the filter characteristic is given in frequency space. Without the term $|q|$, (5.68) would directly result in $p_\gamma(\xi)$, i.e., the inverse Fourier transform of $P_\gamma(q)$. Multiplying $P_\gamma(q)$ by $|q|$, however, results in high-pass filtering, as the linearly increasing frequency variable results in a linear weighting of the spatial spectrum of $p_\gamma(\xi)$. With (5.67), the filtered backprojection principle is defined. It can be structured into the three main steps as summarized in Scheme 5.2.

According to the convolution theorem, the necessary filtering in (5.68) can also be formulated in the spatial domain ξ, where the product $|q|P_\gamma(q)$ in the spectral domain becomes a convolution in the spatial domain, thus

$$
h_\gamma(\xi) = \int_{-\infty}^{+\infty} p_\gamma(z) g(\xi - z)\, dz .
$$

(5.69)

Function $g(\xi - z)$ in the spatial domain is the inverse Fourier transform of the weighting function $|q|$ in the spectral domain.

Scheme 5.2 Filtered backprojection

1. Calculation of the Fourier transform of $p_\gamma(\xi)$

$$
p_\gamma(\xi) \circ\!\!-\!\!-\!\!-\!\!-\!\!\bullet P_\gamma(q)
$$

2. Back transform of the high-pass filtered $P_\gamma(q)$

$$
|q|P_\gamma(q) \bullet\!\!-\!\!-\!\!-\!\!-\!\!\circ h_\gamma(\xi)
$$

3. Backprojection on the line $\xi = x \cos(\gamma) + y \sin(\gamma)$

$$
f(x, y) = \int_0^\pi h_\gamma(\xi)\, d\gamma
$$

(5.70)

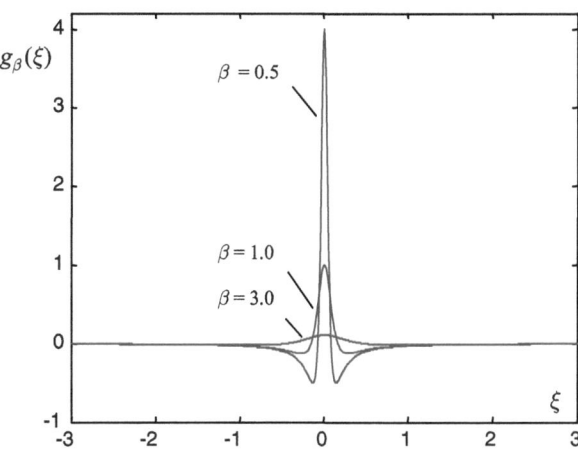

Fig. 5.25. Approximation of the impulse response of the ideal backprojection filter in the spatial domain

In terms of signal processing, the convolution kernel $g(\xi)$ is the impulse response of the high-pass filter. Unfortunately, $|q|$ is not a square integrable function. Therefore, the mathematical recipe using a convergence-generating regular sequence of functions must be applied to obtain $g(\xi)$ by inverse Fourier transform. This trick has been introduced in ▸ Sect. 4.10. To do so, the function

$$G_\beta(q) = |q|\, e^{-\beta|q|} \qquad (5.71)$$

is defined, carrying the convergence-generating function $\exp(-\beta|q|)$. In this way the necessary convergence is produced, leading to the inverse Fourier transform

$$g_\beta(\xi) = \frac{\beta^2 - (2\pi\xi)^2}{(\beta^2 + (2\pi\xi)^2)^2} \,. \qquad (5.72)$$

Finally, the limit $\beta \to 0$ of (5.72) has to be taken

$$g(\xi) = \lim_{\beta \to 0} g_\beta(\xi) = -\frac{1}{(2\pi\xi)^2} \,, \qquad (5.73)$$

which leads to the impulse response of the high pass in the spatial domain. In Fig. 5.25 the approximation of the impulse response is given for decreasing parameters β.

5.8
Comparison Between Backprojection and Filtered Backprojection

In this section the results of filtered backprojection and simple backprojection are directly compared. A very simple software phantom is used that consists of a single square with homogeneous attenuation coefficients located at the iso-center of the sampling system. Figure 5.26 shows the phantom in a 256×256 pixel image. In the same figure, the corresponding data in the Radon space are presented in the typical Cartesian sinogram arrangement. The Radon space values are quite easy to understand. The profile of the projection values $p_y(\xi)$ at $0°$ and $90°$ (as for $180°$ and $270°$) consists of a rectangle function with the width of the square defined by the software phantom. At angles $45°$ and $135°$ (as for $225°$ and $315°$) the projection profile of $p_y(\xi)$ is a triangle function. To aid visualization, the characteristic $p_y(\xi)$ profiles are schematically drawn on the Radon space figure. Overall, the Radon space data do not follow a global sinusoidal trace, since the square is placed at the iso-center.

The concepts for reconstruction of the tomographic image $f(x, y)$ from the projection values $p_y(\xi)$ that were derived in ▸ Sects. 5.6 and 5.7 are demonstrated by a successive backprojection. In Fig. 5.27 the intermediate results, while increasing the number N_p of backprojections $N_p = \{1, 3, 10, 180\}$, are presented row-wise.

The left and right columns show the simple backprojection and the filtered backprojection respectively. Already from the first backprojection the reconstruction recipe of including a high-pass filter becomes clear. Those areas in the image to be reconstructed that are blurred by the simple backprojection of $p_y(\xi)$, are considered right from the beginning, i.e., the PSF (5.50) is compensated for a priori by the negative values of $h_y(\xi)$. After the third backprojection ($N_p = 3$) it can be seen that the simple unfiltered backprojection reconstructs positive attenuation values in image areas that are located outside the square. These incorrect values cannot be compensated for by further backprojections.

In the center of the image, an object is reconstructed that becomes smooth in later backprojections. This blurring of the sharp original square defined in the

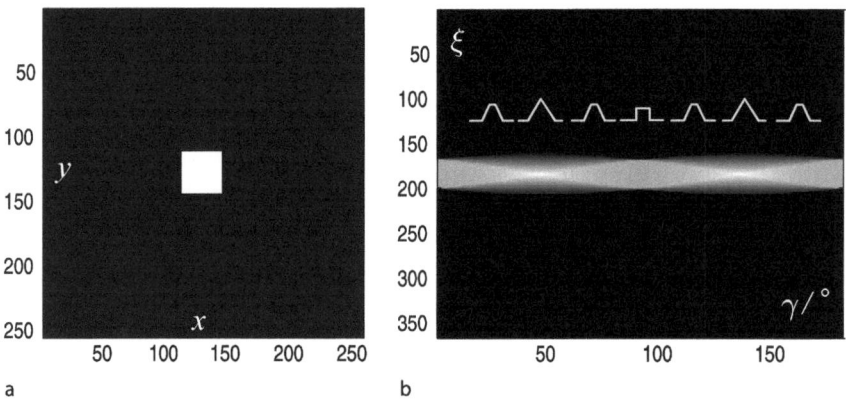

a   b

Fig. 5.26a,b. A simple software phantom consisting of a single square of homogeneous attenuation values $f(x, y) \equiv \mu$ at the iso-center. On the *left*, the corresponding Radon space is given. The projection values $p_y(\xi)$ are not globally sinusoidally modulated. However, the projection profiles oscillate between a rectangle and a triangle profile

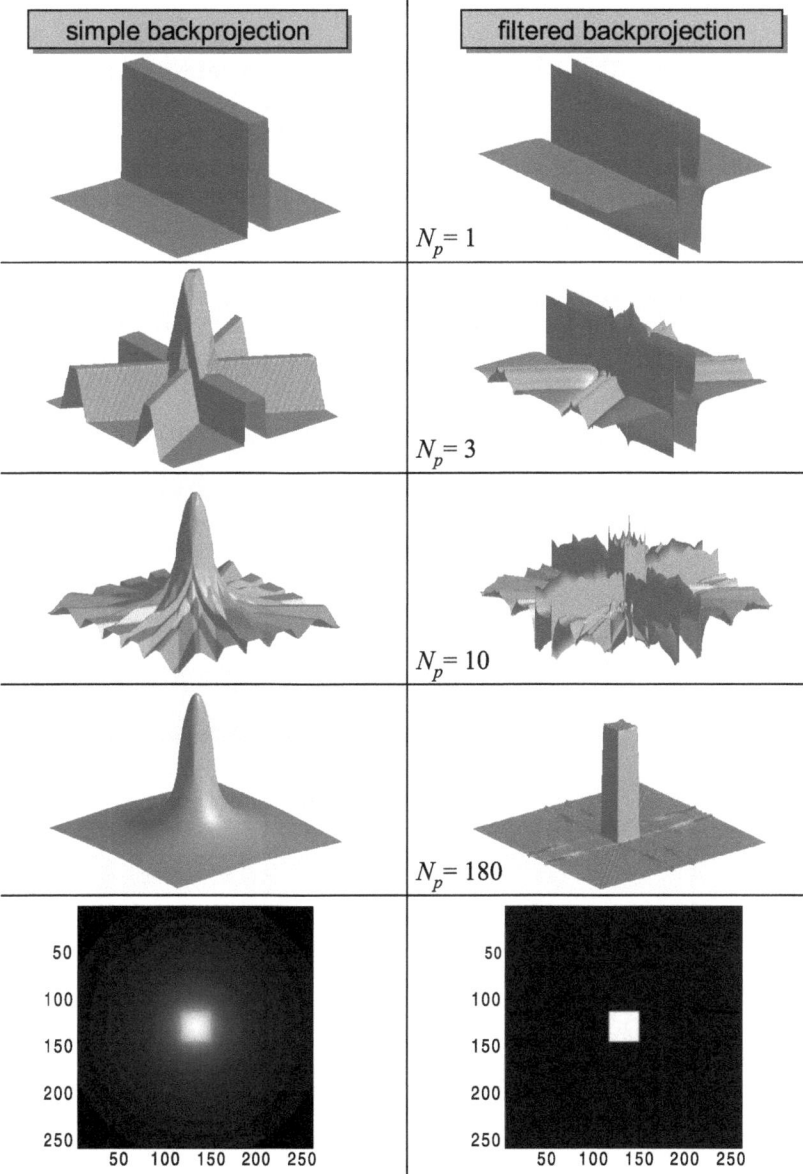

Fig. 5.27. Successive reconstruction of the tomographic software phantom image from projection data. In each reconstruction phase the results of the simple backprojection (*left column*) and the filtered backprojection (*right column*) are presented in direct correspondence. The number of projections N_P per row is increased such that, from top to bottom, intermediate results are presented for $N_P = \{1, 3, 10, 180\}$

software phantom is unacceptable, because it would make diagnostics impossible in clinical situations. For the filtered backprojection, the intermediate results are harder to interpret. This is due to the sensitive mutual compensation process of all filtered backprojections. However, even after 10 filtered backprojections, the original object is clearly recognizable with sharp edges. The last row in Fig. 5.27 shows the reconstructed intensity image after $N_p = 180$ backprojections. The quality difference between unfiltered and filtered backprojection is visually obvious. The stripes established tangential to the object for $N_p = 180$ are due to the *Gibbs* phenomenon (Epstein 2003).

Figure 5.28 shows the unfiltered and the filtered backprojection for a real abdomen tomogram. The location of the slice that is to be reconstructed is drawn into the planning overview[6] (see Fig. 5.28a). Figure 5.28b shows the correspond-

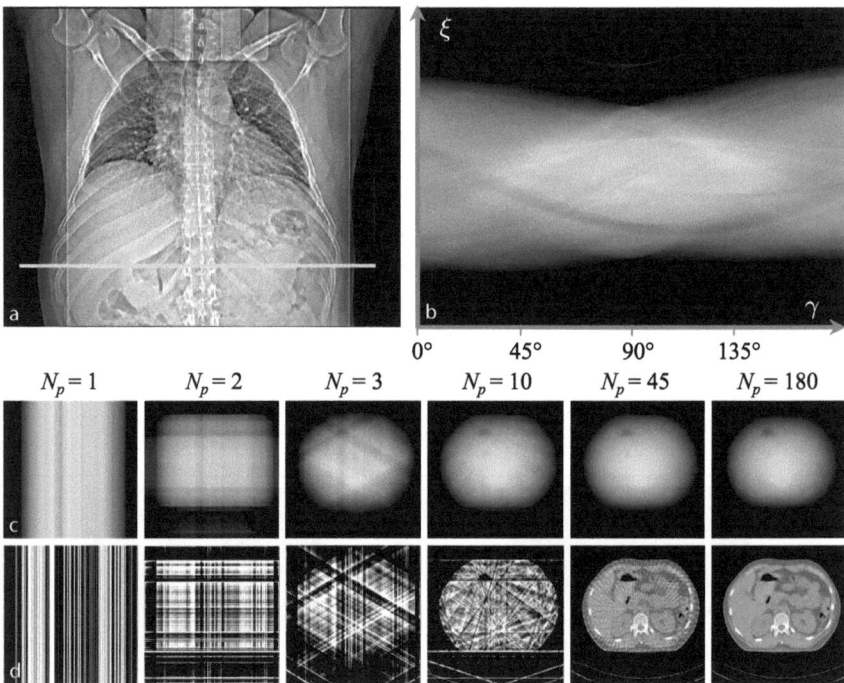

Fig. 5.28a–d. Successive reconstruction of a tomographic abdomen image from projection data. Image **a** shows the position of the axial abdomen section in a planning overview. In image **b** the corresponding sinogram, i.e., the complete Radon space data of the sectional plane, can be seen. Column-wise the simple backprojections (*row* **c**) and the corresponding filtered backprojections (*row* **d**) are presented for the following projection numbers: $N_p = \{1, 2, 3, 10, 45, 180\}$

---

[6] Called scanogram, topogram, scout view or pilot view depending on the manufacturer.

Fig. 5.29. Projection pattern for the filtered backprojection in the image space

ing Radon space, i.e., the projection values $p_y(\xi)$ in a Cartesian (ξ, y) diagram over an angle interval of 180°. Rows c and d of the same figure show the successive superpositions of the backprojections and the filtered backprojections respectively. By convolving the original image with the PSF derived in (5.50), which is included in the simple backprojection imaging process, the expected blurred image is obtained.

Compared with the direct reconstruction method introduced in ▸ Sect. 5.4 the sampling pattern of the filtered backprojection is more appropriate to the problem of the circular projection acquisition scheme. It has been shown that direct reconstruction leads to sparsely distributed points in the high-frequency domain of the Fourier space of the image to be reconstructed. The filtered backprojection avoids the problematic sampling pattern in the two-dimensional Fourier space; it rather distributes reconstruction points directly into the desired image space. Figure 5.29 shows the sampling pattern of the filtered backprojection for 12 projection angles.

5.9
Filtered Layergram: Deconvolution of the Simple Backprojection

Since the blurring of image $g(x, y)$ produced by a simple backprojection is due to an inherent convolution with the PSF (5.50), the image $g(x, y)$ can be further processed to obtain the correct distribution $f(x, y)$ of attenuation values. In this section the respective deconvolution is derived.

The simple backprojection image $g(x, y)$ is the result of the convolution

$$g(x, y) = f(x, y) * h(x, y) \tag{5.74}$$

where $f(x, y)$ is the desired image and $h(x, y)$ is the PSF

$$h(x, y) = \frac{1}{|(x, y)|} . \tag{5.75}$$

Thus,

$$g(x, y) = f(x, y) * \frac{1}{\sqrt{x^2 + y^2}} . \tag{5.76}$$

When using the two-dimensional extension of the convolution theorem

$$f(x, y) * h(x, y) \circ\!\!-\!\!\bullet F(u, v)H(u, v) , \tag{5.77}$$

(5.76) can be described as a multiplication in the spectral domain, which means

$$\mathcal{F}_2\{g(x, y)\} = \mathcal{F}_2\{f(x, y)\}\,\mathcal{F}_2\left\{\frac{1}{\sqrt{x^2 + y^2}}\right\} ; \tag{5.78}$$

hence,

$$G(u, v) = F(u, v)\mathcal{F}_2\left\{\frac{1}{\sqrt{x^2 + y^2}}\right\} . \tag{5.79}$$

$\mathcal{F}_2$ represents the two-dimensional Fourier transform. In the case of the PSF $h(x, y) = |\mathbf{r}|^{-1}$ the Fourier transform has to be calculated as follows:

$$H(u, v) = \mathcal{F}_2\left\{\frac{1}{\sqrt{x^2 + y^2}}\right\} \equiv \int_{-\infty}^{\infty} \int_{-\infty}^{\infty} \frac{1}{\sqrt{x^2 + y^2}} e^{-i2\pi(xu+yv)}\, dx\, dy . \tag{5.80}$$

Due to the radial symmetry of the PSF, polar coordinates are introduced so that – analogous to (5.55) and (5.56) –

$$H(u, v) = \mathcal{F}_2\left\{\frac{1}{r}\right\} \equiv \int_0^{2\pi} \int_0^{\infty} \frac{1}{r} e^{-i2\pi(ur\cos(\gamma)+vr\sin(\gamma))} r\, dr\, d\gamma \tag{5.81}$$

is obtained. When introducing polar coordinates for the spectral domain at the same time, it is further true that

$$H(q, \gamma) = \int_0^{2\pi} \int_0^{\infty} e^{-i2\pi(q\cos(\phi)r\cos(\gamma)+q\sin(\phi)r\sin(\gamma))}\, dr\, d\gamma . \tag{5.82}$$

This is the same expression as in ▸ Sect. 4.15, (4.138) and leads to the Hankel transform, so that

$$H(q) = \mathcal{H}_0\left\{\frac{1}{r}\right\} = \int\limits_0^{+\infty}\int\limits_0^{2\pi} e^{-i2\pi qr\cos(\gamma)}\,dy\,dr = 2\pi\int\limits_0^{+\infty} J_0(2\pi qr)\,dr \;. \tag{5.83}$$

With the substitution $z = 2\pi qr$ it follows that

$$H(q) = \mathcal{H}_0\left\{\frac{1}{r}\right\} = \frac{1}{q}\int\limits_0^{+\infty} J_0(z)\,dz \;. \tag{5.84}$$

After Abramowitz and Stegun (1970), the integral of the Bessel function of 0th order in (5.84) is

$$\int\limits_0^{+\infty} J_0(z)\,dz = 1 \tag{5.85}$$

such that

$$H(q) = \mathcal{H}_0\left\{\frac{1}{r}\right\} = \frac{1}{|\mathbf{q}|} = \frac{1}{\sqrt{u^2 + v^2}} \;. \tag{5.86}$$

Substituting this result into (5.79), it follows that

$$G(u, v) = F(u, v)\frac{1}{|\mathbf{q}|} \;. \tag{5.87}$$

In this way, the deconvolution of the blurred image $g(x, y)$ can be described as

$$f(x, y) = \mathcal{F}_2^{-1}\{|\mathbf{q}|\,G(u, v)\} = \mathcal{F}_2^{-1}\{|\mathbf{q}|\,\mathcal{F}_2\{g(x, y)\}\} \;. \tag{5.88}$$

Scheme 5.3 Filtered layergram

1. Calculation of the backprojection and integration over 180°

$$g(x, y) = \int\limits_0^{\pi} p_\gamma(\xi)\,d\gamma$$

2. Two-dimensional Fourier transform of the simple backprojection

$$g(x, y) \circ\!\!-\!\!-\!\!-\!\!-\!\!\bullet\, G(u, v)$$

3. Multiplication of the Fourier transform with the distance to the origin and two-dimensional inverse Fourier transform into the spatial domain

$$\sqrt{u^2 + v^2}G(u, v) \bullet\!\!-\!\!-\!\!-\!\!-\!\!\circ f(x, y) \tag{5.89}$$

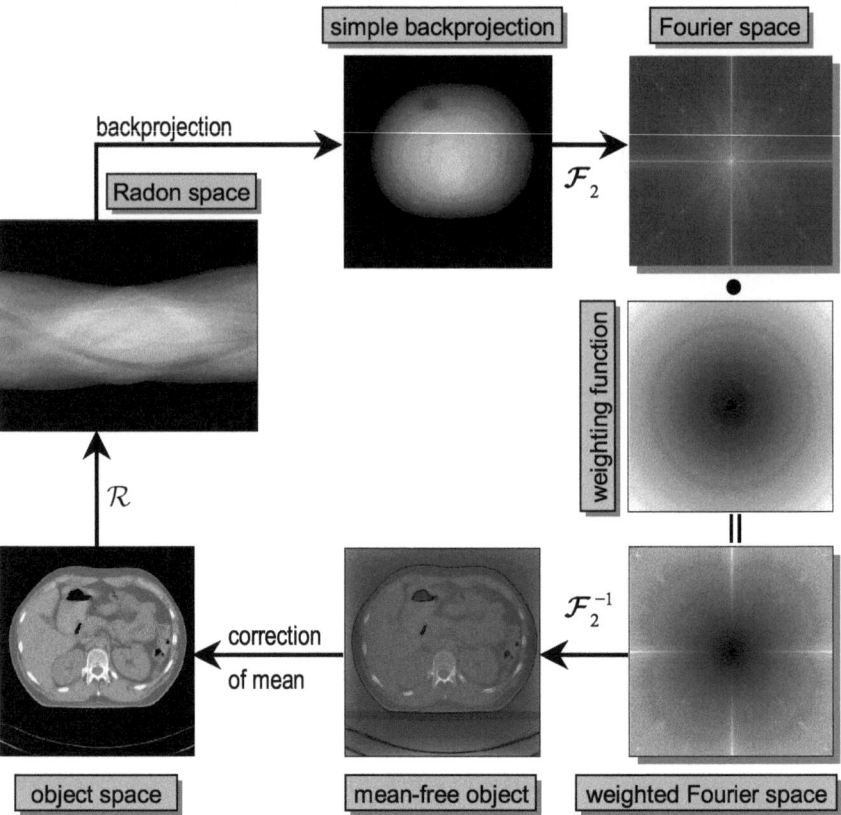

Fig. 5.30. Exemplary illustration of the filtered layergram process. When starting from the Radon space and creating a simple backprojection $g(x, y)$, the original image $f(x, y)$ is obtained by: Weighting the Fourier transform $G(u, v)$ of the simple backprojection with the distance function $|\mathbf{q}|$ in the frequency space; transforming it back to the spatial domain; and correcting the mean value of the image. The last step is necessary as the mean value of the image disappears by the distance weighting with $|\mathbf{q}|$ so that negative image values occur for the background

Obviously, the deconvolution filter $|\mathbf{q}|$ in (5.88) is a cone-shaped weighting function in the two-dimensional spectral domain. Since $g(x, y)$ has been obtained by "smearing" back the unfiltered projections – see (5.43) – that are accumulated layerwise over $180°$, the reconstruction follows Scheme 5.3, which is named a *filtered layergram*. Figure 5.30 illustrates the entire filtered layergram process exemplarily by means of an abdomen tomogram.

Please note that filtered backprojection and a filtered layergram are mathematically equivalent. However, in practice only filtered backprojection is implemented. This is due to the fact that the filtered backprojection reconstruction process can be started directly after measurement of the first projection profile. In this way, the reconstruction with filtered backprojection is faster, as data acquisition and image re-construction can be performed almost simultaneously. The filtered layergram process needs all projections before the Fourier space filtering can be started.

5.10
Filtered Backprojection and Radon's Solution

In this section, it is shown that the filtered backprojection using the high-pass filtered projection signals $h_\gamma(\xi)$ is equivalent to Radon's original solution. To do so one writes the filtering with $|q|$ as

$$|q| = q \operatorname{sign}(q) \tag{5.90}$$

where

$$\operatorname{sign}(q) = \begin{cases} 1 & \text{for } q > 0 \\ 0 & \text{for } q = 0 \\ -1 & \text{for } q < 0 . \end{cases} \tag{5.91}$$

Substituting (5.90) into (5.68) and multiplying the integrand of (5.68) by $(i2\pi^2/i2\pi^2) = 1$, one obtains

$$h_\gamma(\xi) = \frac{1}{2\pi^2} \int_{-\infty}^{+\infty} i2\pi q P_\gamma(q) \left(\frac{\pi}{i} \operatorname{sign}(q) \right) e^{2\pi i q \xi} \, dq . \tag{5.92}$$

Using the convolution theorem again, this can also be expressed by

$$h_\gamma(\xi) = \frac{1}{2\pi^2} \left\{ \mathcal{F}^{-1}\{i2\pi q P_\gamma(q)\} * \mathcal{F}^{-1}\left\{ \frac{\pi}{i} \operatorname{sign}(q) \right\} \right\} . \tag{5.93}$$

In detail, the inverse Fourier transform of the first term is given by

$$2\pi i q P_\gamma(q) \circ\!\!-\!\!\bullet \frac{dp_\gamma(\xi)}{d\xi} \tag{5.94}$$

and the second term can be derived as an analog to the Fourier transform of the $\operatorname{sign}(x)$ function given in ▶ Sect. 4.10, (4.111), i.e.,

$$\frac{\pi}{i} \operatorname{sign}(q) \bullet\!\!-\!\!\circ \frac{1}{\xi} . \tag{5.95}$$

Thus, one can write

$$h_\gamma(\xi) = \frac{1}{2\pi^2} \left\{ 1/\xi * \frac{dp_\gamma(\xi)}{d\xi} \right\} . \tag{5.96}$$

Formulating (5.96) explicitly as a convolution, one obtains

$$h_\gamma(\xi) = \frac{1}{2\pi^2} \int_{-\infty}^{+\infty} \frac{1}{\xi - \xi'} \frac{dp_\gamma(\xi')}{d\xi'} \, d\xi' . \tag{5.97}$$

As mentioned above one receives the image, i.e., the attenuation values, $f(x, y)$, by integrating[7] the filtered projections along all angles:

$$f(x, y) = \int_0^\pi h_\gamma(\xi) \, d\gamma. \tag{5.98}$$

Substituting (5.97) into (5.98) one obtains

$$f(x, y) = \frac{1}{2\pi^2} \int_0^\pi \int_{-\infty}^{+\infty} \frac{1}{\xi - \xi'} \frac{dp_\gamma(\xi')}{d\xi'} \, d\xi' \, d\gamma. \tag{5.99}$$

Here, the differential expressions $d\xi'$ are "cancelled" and one further substitutes

$$\xi' = \xi + R. \tag{5.100}$$

This leads to

$$f(x, y) = -\frac{1}{2\pi^2} \int_0^\pi \int_{-\infty}^{+\infty} \frac{1}{R} \, dp_\gamma(\xi + R) \, d\gamma. \tag{5.101}$$

Changing the order of integration then gives

$$f(x, y) = -\frac{1}{2\pi^2} \int_{-\infty}^{+\infty} \frac{1}{R} \, d\left(\int_0^\pi p_\gamma(\xi + R) \, d\gamma \right). \tag{5.102}$$

Since the variables define a polar coordinate system, on which the entire plane is integrated, the integration limits can be changed as follows

$$f(x, y) = -\frac{1}{2\pi^2} \int_0^{+\infty} \frac{1}{R} \, d\left(\int_0^{2\pi} p_\gamma(\xi + R) \, d\gamma \right). \tag{5.103}$$

Inserting the bracket term of (5.103) back into the Hessian normal form of a straight line

$$x \cos(\gamma) + y \sin(\gamma) = \xi, \tag{5.104}$$

one obtains

$$f(x, y) = -\frac{1}{2\pi^2} \int_0^{+\infty} \frac{1}{R} \, d\left(\int_0^{2\pi} p_\gamma(x \cos(\gamma) + y \sin(\gamma) + R) \, d\gamma \right). \tag{5.105}$$

The integration of the bracket term now results in the mean value defined by Radon

$$\bar{p}_r(R) = \frac{1}{2\pi} \int_0^{2\pi} p(x \cos(\gamma) + y \sin(\gamma) + R, \gamma) \, d\gamma, \tag{5.106}$$

---

[7] For every angle the filtered backprojection is "smeared" back.

defined as the mean over the projection values $p(\xi, \gamma)$, of the tangents of the circle with the center $\mathbf{r} = (x, y)^T$ and radius R, so that finally

$$f(x, y) = -\frac{1}{\pi} \int_0^{+\infty} \frac{1}{R} \, d\overline{p}_{\mathbf{r}}(R) \tag{5.107}$$

is obtained, which is exactly the expression for the inversion of the projection problem given by Johann Radon in 1917. Figure 5.31 schematically shows which values have been averaged in Radon's original publication.

To determine the projection, integrating along all points

$$(\xi \cos(\gamma) - \eta \sin(\gamma), \xi \sin(\gamma) + \eta \cos(\gamma)), \tag{5.108}$$

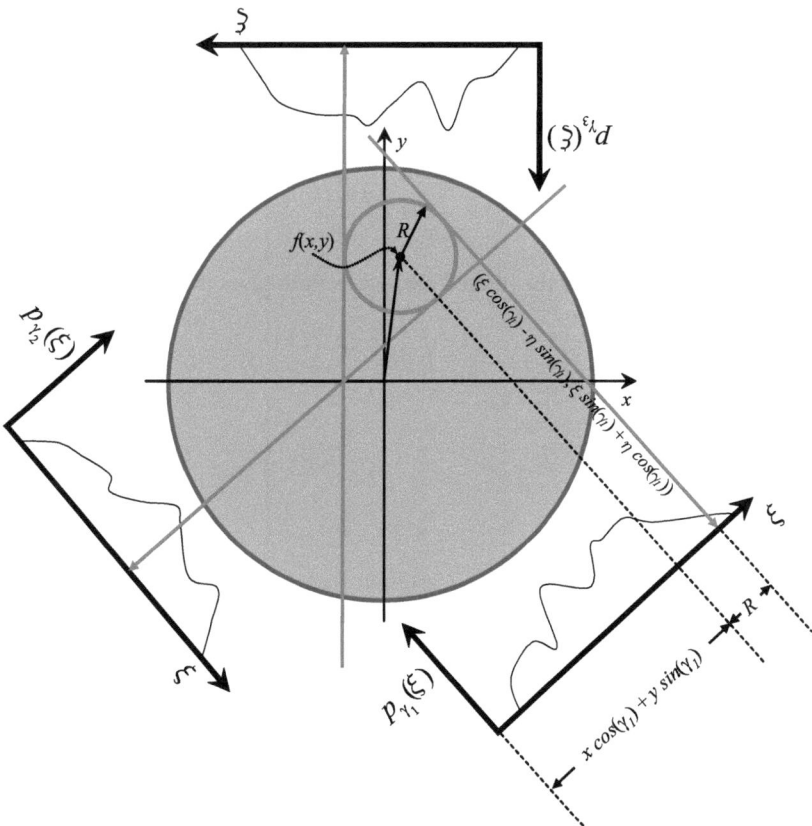

Fig. 5.31. In order to reconstruct the image value f at the point $(x, y)^T$, Johann Radon in 1917 first defined an average projection, which is obtained from the tangents of the circles with the radius R around $(x, y)^T$

which is along a straight line, can be written compactly as

$$f * \delta(\mathbf{L}) = \int f(\mathbf{r})\delta((\mathbf{r}^{\mathrm{T}} \cdot \mathbf{n}_\xi) - \xi)\, d\mathbf{r} \tag{5.109}$$

using the sift property of the δ-distribution.

In fact, the projection values at the points

$$(x \cos(\gamma) + y \sin(\gamma) + R, \gamma) \tag{5.110}$$

in the Radon space are averaged. In order to reconstruct the image value f at the point $(x, y)^{\mathrm{T}}$ the averaged values are weighted with the reciprocal value of the distance from that point and finally integrated. As this section started with (5.92), i.e., the filtered projection $h_y(\xi)$, this point of view is obviously equivalent to the filtered backprojection.

The filtered backprojection can be written very compactly when using the definition of the Hilbert transform introduced in ▸ Sect. 4.17.

With

$$\mathcal{H}\left\{\frac{dp_\gamma(\xi)}{d\xi}\right\} = \frac{1}{\pi} \int_{-\infty}^{+\infty} \frac{1}{\xi - \xi'} \frac{dp_\gamma(\xi')}{d\xi'}\, d\xi', \tag{5.111}$$

(5.97) can be written as

$$h_\gamma(\xi) = \frac{1}{2\pi}\mathcal{H}\left\{\frac{dp_\gamma(\xi)}{d\xi}\right\}. \tag{5.112}$$

This, along with (5.98) for the filtered backprojection, gives the expression[8]

$$f(x, y) = \frac{1}{2\pi} \int_0^\pi \mathcal{H}\left\{\frac{dp_\gamma(\xi)}{d\xi}\right\} d\gamma. \tag{5.113}$$

5.11
Cormack Transform

So far, only the vertical lines in the Cartesian Radon space have been considered, which, in terms of the measurement process, correspond to a complete projection signal acquired over an angle of $180°$. However, the previous section showing the relation between filtered backprojection and Radon's original work, excerpts of which were translated at the beginning of ▸ Chap. 5, also suggests a different view on data in the Radon space.

In this section, data arranged on horizontal lines in the Radon space are considered. This perspective leads to a decomposition into what is called the circular

---

[8] Strictly speaking, the derivation of p with respect to ξ is a partial derivative, so that (5.113) correctly reads $f(x, y) = \dfrac{1}{2\pi} \displaystyle\int_0^\pi \mathcal{H}\left\{\dfrac{\partial p_\gamma(\xi)}{\partial \xi}\right\}\Bigg|_{\xi = x\cos(\gamma) + y\sin(\gamma)} d\gamma.$

harmonics. As this method was proposed by Cormack, it is frequently referred to it as the Cormack transform.

In Fig. 5.32a once again an abdominal tomogram can be seen. The corresponding Cartesian Radon space reflecting the projection data from $\gamma = 0°$ to $\gamma = 180°$ is presented in Fig. 5.32b. The horizontal lines in this diagram belong to certain circles in the spatial domain whose tangents are the corresponding projection lines. This is schematically illustrated in Fig. 5.32c. In Fig. 5.32d, the Radon space data are given in polar coordinates. Obviously, the horizontal lines in the Cartesian Radon space belong to certain circles in the polar diagram. The upper half $(0 - \pi)$ of the circle drawn on Fig. 5.32a, c, and d corresponds to line 3 drawn on Fig. 5.32b, and the lower half of the circle $(\pi - 2\pi)$ corresponds to line 3'.

Due to the circular symmetry of the data acquisition system, every data ring $p_\gamma(\xi) = p(\xi, \gamma)$ is periodic in the projection angle γ, having period 2π. Thus, every ring can be written as a Fourier series

$$p(\xi, \gamma) = \sum_{n=-\infty}^{\infty} \mathfrak{p}_n(\xi) e^{in\gamma}, \qquad (5.114)$$

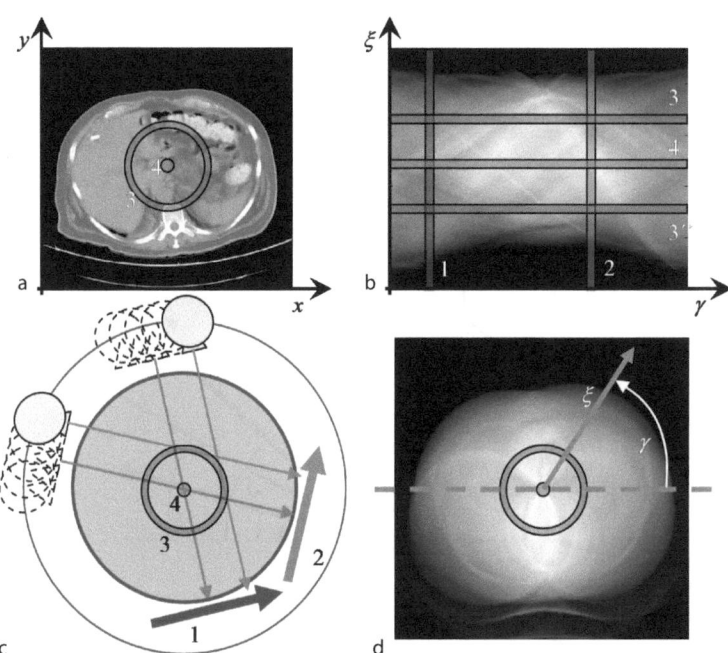

Fig. 5.32a–d. A decomposition into circular harmonics represents another view of Radon space data. Instead of the vertical lines 1 and 2, one changes the perspective to the horizontal lines 3 or 3' and 4 in the sinogram

where

$$p_n(\xi) = \frac{1}{2\pi} \int_0^{2\pi} p(\xi, \gamma) e^{-in\gamma} \, d\gamma \,. \tag{5.115}$$

The Fourier coefficients $p_n(\xi)$ of (5.115) are called circular harmonics.

The anatomical image in Fig. 5.32a can be viewed in this way, as every data ring $f(r, \delta)$ of the attenuation coefficients is 2π-periodic in the angle variable δ. To recall the relationship between (r, δ) and (ξ, γ), see Fig. 5.7. Hence, even in the tomogram, every ring can be represented as a Fourier series

$$f(r, \delta) = \sum_{n=-\infty}^{\infty} f_n(r) e^{in\delta}, \tag{5.116}$$

where

$$f_n(r) = \frac{1}{2\pi} \int_0^{2\pi} f(r, \delta) e^{-in\delta} \, d\delta \tag{5.117}$$

again represents the Fourier coefficients analogous to (5.115). In order to show the correspondence between the circular harmonics $p_n(\xi)$ and $f_n(r)$ in Fig. 5.33, the geometric situation is schematically drawn.

Once more the projection integral (5.10) is considered, whereas the integration path is extended to infinity and both paths – above and below the projection axis –

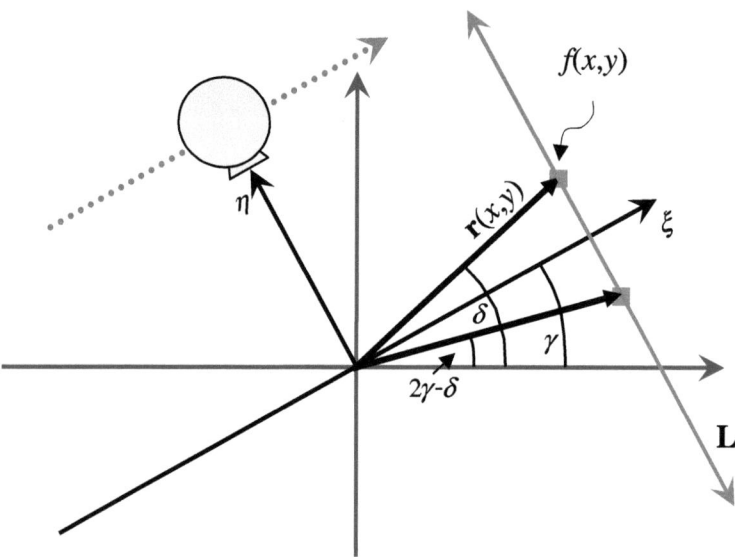

Fig. 5.33. Geometry of the integration path: For every point above the projection axis ξ, there is a mirror point with respect to the projection axis at angle $2\gamma - \delta$

are determined separately, i.e.,

$$p(\xi, \gamma) = \int_0^\infty f(r, \delta)\, d\eta + \int_{-\infty}^0 f(r, \delta)\, d\eta \,. \qquad (5.118)$$

Using the symmetry of Fig. 5.33, the two integrals in (5.118) can be combined to give

$$p(\xi, \gamma) = \int_0^\infty \{f(r, \delta) + f(r, 2\gamma - \delta)\}\, d\eta \,. \qquad (5.119)$$

Expressing the left- and right-hand sides of (5.119) by their Fourier expansions (5.114) and (5.116) respectively, one obtains

$$\sum_{n=-\infty}^\infty \mathfrak{p}_n(\xi)\, e^{in\gamma} = \int_0^\infty \left\{ \sum_{n=-\infty}^\infty \mathfrak{f}_n(r)\, e^{in\delta} + \sum_{n=-\infty}^\infty \mathfrak{f}_n(r)\, e^{in(2\gamma-\delta)} \right\} d\eta$$

$$= \int_0^\infty \left\{ \sum_{n=-\infty}^\infty \mathfrak{f}_n(r) \left(e^{in\delta} + e^{in(2\gamma-\delta)} \right) \right\} d\eta \qquad (5.120)$$

$$= \int_0^\infty \left\{ \sum_{n=-\infty}^\infty \mathfrak{f}_n(r)\, e^{in\gamma} \left(e^{in(\delta-\gamma)} + e^{-in(\delta-\gamma)} \right) \right\} d\eta \,.$$

Writing the cosine as its complex exponential functions

$$\cos(\alpha) = \frac{e^{i\alpha} + e^{-i\alpha}}{2} \,, \qquad (5.121)$$

(5.120) can be simplified to

$$\sum_{n=-\infty}^\infty \mathfrak{p}_n(\xi)\, e^{in\gamma} = 2 \int_0^\infty \left\{ \sum_{n=-\infty}^\infty \mathfrak{f}_n(r)\, e^{in\gamma} \cos(n(\delta - \gamma)) \right\} d\eta \,. \qquad (5.122)$$

Changing the coordinates (ξ, η) for the benefit of the polar presentation to $(r, \delta-\gamma)$, where

$$r = \sqrt{\xi^2 + \eta^2} \qquad (5.123)$$

and

$$\xi = r\cos(\delta - \gamma) \,, \qquad (5.124)$$

one can write the following after swapping the order of summation and integration

$$\sum_{n=-\infty}^\infty \mathfrak{p}_n(\xi)\, e^{in\gamma} = 2 \sum_{n=-\infty}^\infty e^{in\gamma} \int_{|\xi|}^\infty \frac{\mathfrak{f}_n(r)\cos\left(n\cos^{-1}(\xi/r)\right)}{\sqrt{r^2 - \xi^2}}\, r\, dr$$

$$= \sum_{n=-\infty}^\infty \left\{ 2 \int_{|\xi|}^\infty \frac{\mathfrak{f}_n(r)\cos\left(n\cos^{-1}(\xi/r)\right)}{\sqrt{r^2 - \xi^2}}\, r\, dr \right\} e^{in\gamma} \,. \qquad (5.125)$$

Note, as a consequence of (5.123), the integration along the variable r now has to start with $|\xi|$. Comparing the coefficients on both sides of (5.125), one can see that

$$p_n(\xi) = 2 \int_{|\xi|}^{\infty} \frac{f_n(r) \cos\left(n \cos^{-1}(\xi/r)\right)}{\sqrt{r^2 - \xi^2}} r \, dr . \tag{5.126}$$

Using the definition of the Chebyshev polynomials[9] of the first kind

$$T_n(x) = \cos\left(n \cos^{-1}(x)\right) , \tag{5.127}$$

(5.126) further simplifies to

$$p_n(\xi) = 2 \int_{|\xi|}^{\infty} \frac{f_n(r) T_n(\xi/r)}{\sqrt{r^2 - \xi^2}} r \, dr . \tag{5.128}$$

(5.128) is called the Cormack transform (following the notation given by Barrett and Swindell [1981]). Furthermore, with the help of the Abel transform introduced in ▸ Sect. 4.16.

$$p_n(\xi) = A\{f_n(r) T_n(\xi/r)\} \tag{5.129}$$

is obtained. However, for practical reasons, the inversion of (5.128) is important as $f_n(r)$ is of interest. From the Fourier expansion with the circular harmonics $f_n(r)$ of (5.117), the distribution of the attenuation coefficients $f(r, \delta)$ is then obtained. Cormack (1963) gave the following expression for this[10]

$$f_n(r) = -\frac{1}{\pi} \int_r^{\infty} \frac{p'_n(\xi) T_n(\xi/r)}{\sqrt{\xi^2 - r^2}} d\xi . \tag{5.130}$$

(5.130) is called the inverse Cormack transform (Barrett and Swindell 1981). Substituting (5.130) into (5.116), the image is finally obtained with

$$f(r, \delta) = -\frac{1}{\pi} \sum_{n=-\infty}^{\infty} \int_r^{\infty} \frac{p'_n(\xi) T_n(\xi/r)}{\sqrt{\xi^2 - r^2}} e^{in\delta} d\xi . \tag{5.131}$$

It is also possible to decompose the Fourier transform of the object $F(u, v)$ into its circular harmonics. Following the Fourier slice theorem, the Fourier transforms of the projections can be found in a polar representation under the projection angle γ.

---

[9] That these are actually polynomials in (ξ/r) results from the equation (Heuser 1992):
$\cos(n\varphi) = (\cos(\varphi))^n - \binom{n}{2}(\cos(\varphi))^{n-2}(\sin(\varphi))^2 + \binom{n}{4}(\cos(\varphi))^{n-4}(\sin(\varphi))^4 - + \dots$
[10] Formulating with Chebyshev polynomials of the first kind is problematical in practice because T_n increases exponentially for n outside $[-1, +1]$. However, a stable version of (5.130) can be found with Chebyshev polynomials of the second kind (Cormack 1964; Natterer 1999).

Hence, it is possible to write

$$P(q,\gamma) = \sum_{n=-\infty}^{\infty} \mathfrak{P}_n(q) e^{in\gamma} \tag{5.132}$$

where

$$\mathfrak{P}_n(q) = \frac{1}{2\pi} \int_0^{2\pi} P(q,\gamma) e^{-in\gamma} \, d\gamma . \tag{5.133}$$

(5.116) and (5.132) are radial expressions. Therefore, the Hankel transform introduced in ▶ Sect. 4.15 can be used to establish the connection of both expressions. It holds that

$$\mathfrak{P}_n(q) = i^n 2\pi \int_0^{\infty} \mathfrak{f}_n(r) J_n(2\pi qr) r \, dr \tag{5.134}$$

and

$$\mathfrak{f}_n(r) = i^n 2\pi \int_0^{\infty} \mathfrak{P}_n(q) J_n(2\pi qr) q \, dq . \tag{5.135}$$

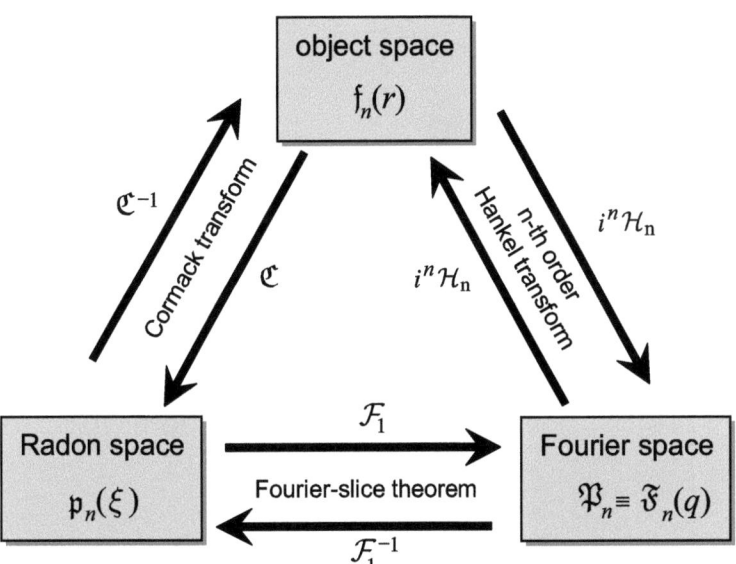

Fig. 5.34. Schematic overview of the relations among the spatial domain, the Radon space, and the Fourier space according to the Cormack transform (adapted from Barrett and Swindell [1981])

These are written in a more compact way as

$$\mathfrak{P}_n(q) = i^n \mathcal{H}_n\{\mathfrak{f}_n(r)\} \tag{5.136}$$

and

$$\mathfrak{f}_n(r) = i^n \mathcal{H}_n\{\mathfrak{P}_n(q)\} \tag{5.137}$$

respectively. These are – as a generalization of the definition given in ▶ Sect. 4.15 – the Hankel transforms of the nth order. Figure 5.34 schematically represents the quantities discussed in this section. For rotationally symmetric objects, the Cormack transform can be identified with the Abel transformation of ▶ Sect. 4.16.

6   Algebraic and Statistical Reconstruction Methods

Contents

6.1   Introduction 201
6.2   Solution with Singular Value Decomposition 207
6.3   Iterative Reconstruction with ART 211
6.4   Pixel Basis Functions and Calculation of the System Matrix 218
6.5   Maximum Likelihood Method 223

6.1
Introduction

Currently, the filtered backprojection (FBP) method, which has been discussed in ▸ Sect. 5.7, is the reconstruction algorithm of choice because it is very fast, especially on dedicated hardware. However, one disadvantage of FBP is that it essentially weights all X-rays equally. Since X-ray tubes produce a polychromatic spectrum, beam-hardening image artifacts arise in the reconstruction. Artifacts of this type are particularly dominant if metal objects are inside the patient, because FBP interprets the corresponding projection data as inconsistent. Here, algebraic and statistical reconstruction methods will serve as alternatives because artificial or inherent beam weighting reduces the influence of rays running through metal objects.

Algebraic and statistical methods for computed tomography (CT) image reconstruction are widely disregarded in clinical routine due to the substantial amount of inherent computational effort. On the other hand, the continuously growing computational power of today's standard computers has led to a rediscovery of these methods.

At the advent of CT, the first image reconstructions were carried out using algebraic reconstruction techniques (ART). As mentioned above, in today's clinical routine FBP is the working horse of CT image reconstruction due to the computational expense of ART[1]. However, ART is more instructive since it represents the reconstruction problem as a linear system of equations.

---

[1] Today, special iterative statistical techniques are widely used in nuclear diagnostic imaging in order to overcome the problems in the signal-to-noise ratio caused by poor photon statistics.

Generally, these methods are much easier to understand than the Fourier-based strategies discussed so far. Thereby, one takes the discrete nature of the practical realization of CT into account from the very beginning. Discretization of the projection, $p_\gamma(\xi)$, is technically dictated by the design of the detector array based on a set of discrete detector elements. The discretization of the tomographic image, i.e., the size and number, N, of pixels inside the field of view, has to be determined before image reconstruction can take place. Figure 6.1 shows the situation of spatially discrete attenuation values schematically. The tomographic image that has to be reconstructed, however, consists of a discrete array of unknown variables, f_j, with $j = \{1, \dots, N\}$, i.e., the unknown attenuation coefficients. The set of projections through the object can easily be modeled by a linear system of equations. Figure 6.1 motivates this set-up of a linear system of equations. Passing through tissue, the intensity of the X-ray beam is weakened according to the attenuation coefficients, $f_j = \mu_j$. If the image is small, the solution of the corresponding low dimensional linear system of equations can be obtained, for example, by Gaussian elimination (cf. Press et al. [1990]).

For each ray through the object, one obtains the already known projection sum defined in ▸ Sect. 5.2 and, furthermore, the following system of equations for the situation of the 2×2-pixel image in Fig. 6.1a.

$$f_1 + f_3 = p_1 , \tag{6.1}$$
$$f_2 + f_4 = p_2 , \tag{6.2}$$
$$f_1 + f_4 = p_3 , \tag{6.3}$$
$$f_1 + f_2 = p_4 , \tag{6.4}$$

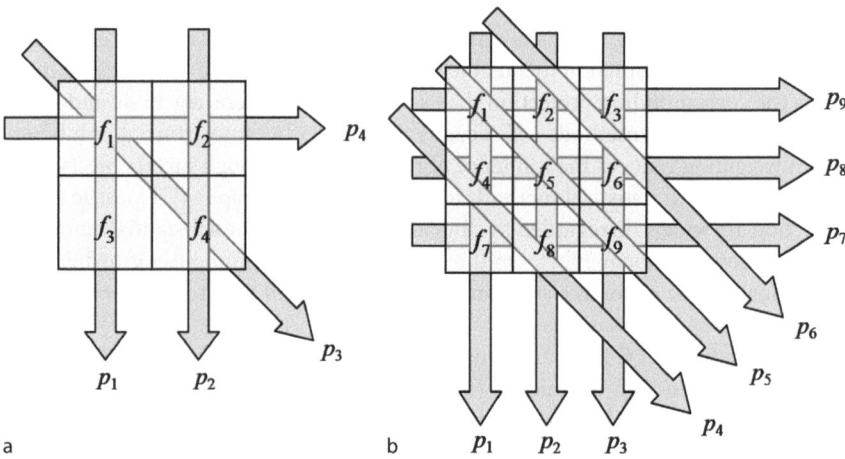

a   b p_1 p_2 p_3

Fig. 6.1. The principle of algebraic reconstruction is very simple. The set of projections results in a linear system of equations. In the left image all four unknown attenuation values can be determined exactly using four projections from three projection angles. If the grid that is to be reconstructed is finer, more projections have to be measured

Four equations are obtained with four unknown quantities that can be solved exactly, as long as the physical measuring procedure is not afflicted with noise and no linear dependencies occur – that is as long as the rank is 4 in the example above[2]. Now the image shall be refined spatially as demonstrated with the 3×3-pixel image in Fig. 6.1b. It is immediately clear that the number of required independent X-ray projections grows quadratically with the linear refinement of the image. In the given example, one obtains a solvable problem consisting of nine equations and nine unknown attenuation values. Looking at the diagonal projections, a difference is apparent compared with the horizontal or the vertical projection direction: The path length through each element of the object is obviously different. This circumstance must be taken into account in the set-up of the system of equations.

In contrast to the methods introduced in the previous chapter, in which the object is sampled with a δ-line, when using algebraic methods, one proceeds with the physically correct assumption that the X-ray beam has a certain width. When passing through tissue, one now has to take into account how much of the pixel that is to be reconstructed is passed through by the beam. For this purpose, one introduces weights that reflect the relation between the area that is illuminated by the beam and the entire area of the pixel. Figure 6.2 shows this ratio schematically. A beam of

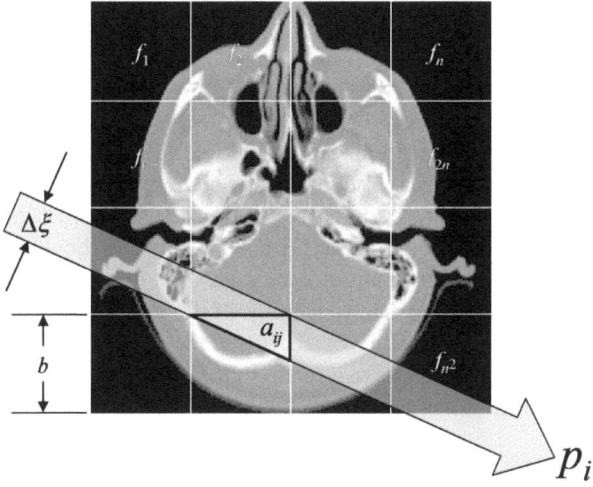

Fig. 6.2. The X-ray beam of width $\Delta\xi$ does not traverse all pixels of size b^2 equally when passing through the tissue. The area of the pixel section that has actually been passed through and that is to be reconstructed must be included in the system of equations as a weighting

---

[2] This is true if a diagonal projection through the object is involved. Two horizontal and two vertical projections alone would lead to an under-determined system of equations having a rank of 3.

width, $\Delta\xi$, passes through the tissue – again illustrated by cranial tomography. The pixel size is given by b^2. The weight a_{ij} is thus determined by the relation[3]

$$a_{ij} = \frac{\textit{illuminated area of pixel j by ray i}}{\textit{total area of pixel j}} \tag{6.5}$$

and lies in the interval $0 \le a_{ij} \le 1$.

From the generalizations of (6.2) to (6.4), one obtains the following system of equations:

$$\sum_{j=1}^{N} a_{ij}f_j = p_i , \tag{6.6}$$

in which the weights, a_{ij}, are taken into account. N is the number of the pixels, n^2, that are to be reconstructed, and $i = \{1,\ldots,M\}$ is the index of the projection, with $M = N_p D$ being the total number of projections of all detector elements, D, of the detector array in all projection directions, N_p. In an expanded form, one may write (6.6) as

$$\begin{aligned}
a_{11}f_1 + a_{12}f_2 + \cdots + a_{1N}f_N &= p_1 \\
a_{21}f_1 + a_{22}f_2 + \cdots + a_{2N}f_N &= p_2 \\
&\vdots \\
a_{M1}f_1 + a_{M2}f_2 + \cdots + a_{MN}f_N &= p_M .
\end{aligned} \tag{6.7}$$

Within nuclear diagnostic imaging, as in PET[4] and SPECT[5], in which the representation of the photon paths as a set of linear equations is used in a statistical approach, the weightings are to be interpreted as probabilities that gamma quanta from the area element[6] j are detected in projection i. This shows the strength of the algebraic and statistical techniques. Linear physical processes can be built into the imaging model via appropriate weightings of the projection equations. In this way, the imaging quality can be improved because the mathematical model that the reconstruction is based on can be tailored to match the real physical situation.

Writing all projections as column vector

$$\mathbf{p} = (p_1,\ldots,p_M)^T , \tag{6.8}$$

and writing the attenuation values that are to be reconstructed as a column vector as well (they were presented as an image matrix in the previous sections)

$$\mathbf{f} = (f_1,\ldots,f_N)^T , \tag{6.9}$$

---

[3] To be precise, this definition is only exact in the case of emission CT.
[4] PET = Positron Emission Tomography.
[5] SPECT = Single Photon Emission Computed Tomography.
[6] Physically, the gamma quanta emerge from a volume element.

the weightings are thus presented as an $M \times N$ matrix

$$\mathbf{A} = \begin{pmatrix} a_{11} & a_{12} & \cdots & a_{1N} \\ a_{21} & \cdot & & a_{2N} \\ \vdots & & \cdot & \vdots \\ a_{M1} & & & a_{MN} \end{pmatrix}, \tag{6.10}$$

such that the system of equations becomes

$$\mathbf{p} = \mathbf{A}\mathbf{f}, \tag{6.11}$$

where $\mathbf{A}$ can be understood as design matrix (Press et al. 1990). In CT, this matrix is also referred to as the system matrix (Toft 1996). Through a direct comparison with (5.21) the following duality between the presentation as matrix and the Radon transform can be identified:

$$\begin{matrix} \mathbf{p} & = \mathbf{A} & \mathbf{f} \\ \updownarrow & \updownarrow & \updownarrow \\ p_\gamma(\xi) & = & \mathcal{R}\{f(x,y)\} \end{matrix} \tag{6.12}$$

Therefore, vector $\mathbf{p}$ contains all values of the Radon space, which means it contains all values of the sinogram, and $\mathbf{f}$ is the vector that contains all gray values of the image grid, i.e., the attenuation coefficients. The mathematical difficulties in this view on the reconstruction problem can be summarized by the following points:

- The system of equations (6.11) can only be solved exactly under idealized physical conditions. In the present case, however, one has to deal with real data, i.e., data afflicted with noise. Therefore, even in the case $N = M$, only an approximate solution can be found for $\mathbf{f}$. Furthermore, for high-quality CT scanners it is true that $M > N$, which means that the number of projections is higher than the number of pixels that are to be reconstructed. Mathematically, this situation leads to an over-determined system of equations.
- Typically, the system matrix, $\mathbf{A}$, is almost singular, which means that it contains very small singular values such that the reconstruction problem is an ill-conditioned problem.
- $\mathbf{A}$ does not have a simple structure and so no fast inversion has been found so far. On the other hand, $\mathbf{A}$ is a sparse matrix as only $N^{1/2}$ pixels contribute to an entry in the Radon space (cf. Fig. 6.5).
- $\mathbf{A}$ is usually very large, so direct inversions are extremely time- and memory-intensive (cf. Fig. 6.5).

However, there are interesting advantages of the algebraic approach as well:

- Irregular geometries of scanners or missing data in the sinogram lead to severe difficulties in the direct reconstruction methods. In the matrix formalism, however, these geometric conditions can be considered and taken into account adequately.
- Finite detector widths and different detector sensitivities can be taken into account. Therewith, better modeling of the real physical measurement process can be obtained.
- Beams running through objects that potentially produce inconsistencies in the Radon space can be weighted appropriately.

The solution of (6.11) can generally be found by minimization of the following function

$$\chi^2 = |\mathbf{A}\mathbf{f} - \mathbf{p}|^2 . \tag{6.13}$$

There is always a solution for this optimization problem. The solution is called the *least squares minimum norm* or *pseudo solution*. One thereby searches for a matrix called the pseudo inverse $\mathbf{A}^+$ of $\mathbf{A}$, also called the *Moore–Penrose* inverse (Natterer 2001), with the following properties:

$$\begin{align} \mathbf{A}\mathbf{A}^+\mathbf{A} &= \mathbf{A} \\ \mathbf{A}^+\mathbf{A}\mathbf{A}^+ &= \mathbf{A}^+ \\ (\mathbf{A}\mathbf{A}^+)^T &= \mathbf{A}\mathbf{A}^+ \\ (\mathbf{A}^+\mathbf{A})^T &= \mathbf{A}^+\mathbf{A} \end{align} \tag{6.14}$$

In a certain sense, $\mathbf{A}^+$ represents the inverse matrix to the square matrix $\mathbf{A}$ such that

$$\widetilde{\mathbf{f}} = \mathbf{A}^+\mathbf{p} . \tag{6.15}$$

As the pseudo solution (6.15) is a compromise in the least squares sense, it is denoted by $\widetilde{\mathbf{f}}$.

With respect to the duality between the matrix presentation and the Radon transform equation,

$$\mathbf{g} = \mathbf{A}^T\mathbf{p} \tag{6.16}$$

presents the adjoint Radon transform of sinogram values to the image space. (6.16) represents the unfiltered backprojection that is analogous to (5.43).

(6.11) can be brought into standard form by multiplying with $\mathbf{A}^T$ from the left, thus

$$\mathbf{A}^T\mathbf{p} = \mathbf{A}^T\mathbf{A}\mathbf{f} . \tag{6.17}$$

This leads to the solution

$$\mathbf{f} = (\mathbf{A}^T\mathbf{A})^{-1}\mathbf{A}^T\mathbf{p} . \tag{6.18}$$

Interestingly, (6.18) can be discussed in the language of the Fourier-based reconstruction methods. In the sense of the adjoint Radon transform (6.16) mentioned above, here matrix $\mathbf{A}^{\mathrm{T}}$ represents the backprojection operator and, consequently, the operator $(\mathbf{A}^{\mathrm{T}}\mathbf{A})^{-1}$ represents the necessary filtering. As discussed in ▸ Sect. 5.9, it is indeed sensible to perform a simple backprojection prior to the filtering. Obviously, this leads to the layergram method so that (6.18) can be seen in duality to (5.88). However, it is also possible to represent the filtered backprojection discussed in ▸ Sect. 5.7 as a matrix equation.

Starting with (6.18), some rearrangements lead to the following result:

$$
\begin{aligned}
\mathbf{f} &= (\mathbf{A}^{\mathrm{T}}\mathbf{A})^{-1}\mathbf{A}^{\mathrm{T}}\mathbf{p} \\
&= (\mathbf{A}^{\mathrm{T}}\mathbf{A})^{-1}\mathbf{A}^{\mathrm{T}}(\mathbf{A}\mathbf{A}^{\mathrm{T}})(\mathbf{A}\mathbf{A}^{\mathrm{T}})^{-1}\mathbf{p} \\
&= (\mathbf{A}^{\mathrm{T}}\mathbf{A})^{-1}(\mathbf{A}^{\mathrm{T}}\mathbf{A})\mathbf{A}^{\mathrm{T}}(\mathbf{A}\mathbf{A}^{\mathrm{T}})^{-1}\mathbf{p} \\
&= \mathbf{A}^{\mathrm{T}}(\mathbf{A}\mathbf{A}^{\mathrm{T}})^{-1}\mathbf{p} \ .
\end{aligned}
\tag{6.19}
$$

Here, the term $(\mathbf{A}\mathbf{A}^{\mathrm{T}})^{-1}$ plays the role of the high-pass filter $|q|$ from (5.68). So the pseudo inverse $\mathbf{A}^{+}$ is given by

$$
\mathbf{A}^{+} = \mathbf{A}^{\mathrm{T}}(\mathbf{A}\mathbf{A}^{\mathrm{T}})^{-1} = (\mathbf{A}^{\mathrm{T}}\mathbf{A})^{-1}\mathbf{A}^{\mathrm{T}}
\tag{6.20}
$$

and is determined practically by singular value decomposition.

6.2
Solution with Singular Value Decomposition

Using singular value decomposition (SVD) to solve (6.11) needs the CT system matrix to represent the design matrix, $\mathbf{A}$, in the following system of equations:

$$
\left(\begin{array}{c} \mathbf{A} \end{array}\right)\left(\begin{array}{c} \mathbf{f} \end{array}\right) = \left(\begin{array}{c} \mathbf{p} \end{array}\right) .
\tag{6.21}
$$

Within SVD any $M \times N$ matrix $\mathbf{A}$ with $M \geq N$, can be decomposed into

$$
\mathbf{A} = \mathbf{U}\boldsymbol{\Sigma}\mathbf{V}^{\mathrm{T}} = \mathbf{U}\left(\mathbf{diag}\left(\sigma_j\right)\right)\mathbf{V}^{\mathrm{T}} ,
\tag{6.22}
$$

where $\mathbf{U}$ is an $M \times N$ orthogonal matrix and $\mathbf{V}$ is an $N \times N$ orthogonal matrix in the sense of their columns being orthonormal. $\boldsymbol{\Sigma}$ is a diagonal $N \times N$ matrix whose entries are the singular values σ_j. One obtains the pseudo inverse of $\mathbf{A}$ through

$$
\mathbf{A}^{+} = \mathbf{V}\left(\mathbf{diag}\left(\frac{1}{\sigma_j}\right)\right)\mathbf{U}^{\mathrm{T}} .
\tag{6.23}
$$

In that way, the solution of (6.21) can be found by

$$\begin{pmatrix} \tilde{\mathbf{f}} \end{pmatrix} = \begin{pmatrix} & \mathbf{V} & \end{pmatrix} \begin{pmatrix} 1/\sigma_1 & \cdot & 0 \\ & \cdot & \\ \cdot & \cdot & \cdot \\ 0 & \cdot & 1/\sigma_N \end{pmatrix} \begin{pmatrix} & \mathbf{U}^{\mathrm{T}} & \end{pmatrix} \begin{pmatrix} \mathbf{p} \end{pmatrix}. \tag{6.24}$$

To illustrate the principle, an X-ray illumination from four directions of an object is given schematically in Fig. 6.3. The image will be reconstructed as a 2×2 pixel image.

The system of equations belonging to Fig. 6.3 is the concretization of (6.21), thus

$$\begin{pmatrix} 0 & 1 & 1 & 0 \\ 1 & 0 & 1 & 0 \\ 0 & 1 & 0 & 1 \\ 1 & 0 & 0 & 1 \\ 0 & 0 & 1 & 1 \\ 1 & 1 & 0 & 0 \end{pmatrix} \begin{pmatrix} f_1 \\ f_2 \\ f_3 \\ f_4 \end{pmatrix} = \begin{pmatrix} 5 \\ 13 \\ 2 \\ 10 \\ 7 \\ 8 \end{pmatrix}. \tag{6.25}$$

Here, the *nearest neighbors method* was used to calculate the weights, a_{ij}, represented in the system matrix. Different methods for the determination of weights that are more appropriate to the physical situation will be explained in the next section. The system of equations (6.25) is over-determined, which means that there are more equations (projections, p_i) than unknown quantities (pixel values, f_j). Obviously, some of the rows in this system are linearly dependent. The number of linearly independent equations can be expressed by the number of non-zero singular values. This number gives the *rank* of the matrix, **A**.

With the singular value decomposition described above, one obtains the pseudo inverse that solves the system of equations. Matrix **A** hereafter contains four finite

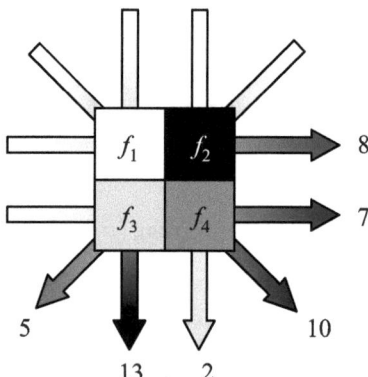

Fig. 6.3. Example of an object that is irradiated from four directions. Six projection values are acquired. The image will be reconstructed as a 2×2 pixel image. A linear black to white lookup table simulates weak to strong attenuation

singular values and is thus of the rank 4. The singular values are $\sigma_1 = 2.4495$ and $\sigma_2 = \sigma_3 = \sigma_4 = 1.4142$. With the decomposition presented in (6.23), one obtains the pseudo inverse

$$\mathbf{A}^+ = \begin{pmatrix} -0.1667 & 0.3333 & -0.1667 & 0.3333 & -0.1667 & 0.3333 \\ 0.3333 & -0.1667 & 0.3333 & -0.1667 & -0.1667 & 0.3333 \\ 0.3333 & 0.3333 & -0.1667 & -0.1667 & 0.3333 & -0.1667 \\ -0.1667 & -0.1667 & 0.3333 & 0.3333 & 0.3333 & -0.1667 \end{pmatrix} \tag{6.26}$$

so that (6.15)

$$\widetilde{\mathbf{f}} = \begin{pmatrix} 8 \\ 0 \\ 5 \\ 2 \end{pmatrix} = \mathbf{A}^+ \begin{pmatrix} 5 \\ 13 \\ 2 \\ 10 \\ 7 \\ 8 \end{pmatrix} \tag{6.27}$$

is the solution in the least squares sense. A comparison with Fig. 6.3 indicates that (6.27) does indeed provide the correct solution.

Unfortunately, in practice, the solution of the system of equations (6.21) cannot easily be obtained, because typically, the problem is ill-conditioned, i.e., very small singular values, σ_j, might occur. In consequence, even small measurement errors in the projections, $\mathbf{p}$, may induce large fluctuations in the reconstructed image, $\widetilde{\mathbf{f}}$. In these cases, the solution has to be stabilized with a technique called regularization. Press et al. (1990) suggested weighting the spectrum of the singular values appropriately. In the easiest case, $1/\sigma_j$ is replaced by a zero if σ_j exceeds a certain threshold.

Finally, the example of (6.25) will be considered again from the perspective of the adjoint reconstruction problem (6.16). To do so, one has to transpose the system matrix, $\mathbf{A}$, and write

$$\begin{pmatrix} g_1 \\ g_2 \\ g_3 \\ g_4 \end{pmatrix} = \begin{pmatrix} 0 & 1 & 0 & 1 & 0 & 1 \\ 1 & 0 & 1 & 0 & 0 & 1 \\ 1 & 1 & 0 & 0 & 1 & 0 \\ 0 & 0 & 1 & 1 & 1 & 0 \end{pmatrix} \begin{pmatrix} 5 \\ 13 \\ 2 \\ 10 \\ 7 \\ 8 \end{pmatrix} = \begin{pmatrix} 31 \\ 15 \\ 25 \\ 19 \end{pmatrix}. \tag{6.28}$$

As argued above, (6.28) presents the unfiltered backprojection. Figure 6.4 illustrates the method schematically, using the sample data from Fig. 6.3.

So far it has been shown that, in principle, the direct algebraic method leads to an exact image reconstruction using SVD. If the number of projections, M, and the number of pixels, N, to be reconstructed are small, the system of equations (6.11) can actually be solved directly. In the example (6.25) it has been chosen that $M = 6$ and $N = 4$. However, it must be clear that in modern CT scanners the real physical situation is given by a very large system of equations. Even computers of the latest generation are not prepared to solve the problem with such a method of brute force.

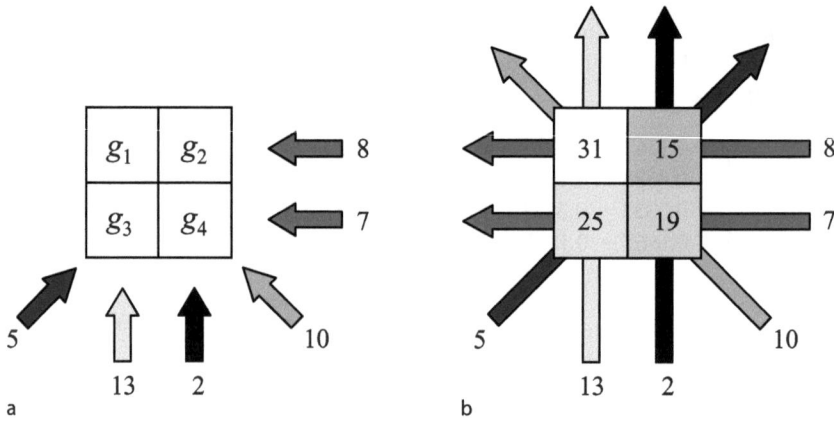

Fig. 6.4a,b. Schematic presentation of the adjoint problem $\mathbf{g} = \mathbf{A}^T\mathbf{p}$ to equation $\mathbf{p} = \mathbf{A}\mathbf{f}$. The adjoint reconstruction problem presents the unfiltered backprojection

A modern CT system typically has about 1.000 detectors in a single detector slice and acquires just as many projections in a 360°-rotation such that M is in the order of 10^6. For an image matrix of 512 × 512 pixels that is to be reconstructed, N is to the order of 10^5. In Fig. 6.5, it is illustrated that the system of equations (6.11) to be solved consists of 10^5 unknown quantities and 10^6 equations. That is why all practicable algorithms for the solution of the linear system of equations work with iterative concepts that do not need a direct inversion of matrix $\mathbf{A}$.

It should be noted that if the linear system of equations (6.11) could be represented in diagonal form, the solution could be directly determined. Indeed, the Fourier-based methods can be interpreted in such a way that the Fourier transform reduces the reconstruction problem to a diagonalized system of equations. That is why the Fourier methods are very efficient (Epstein 2003).

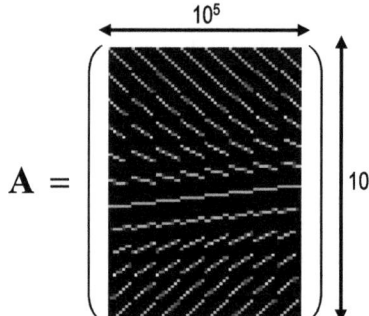

Fig. 6.5. The system matrix, $\mathbf{A}$, is a huge but sparse matrix. The structure of entries of $\mathbf{A}$ is given in *gray dots*. Since modern CT scanners produce an image matrix of at least 512 × 512 pixels, N is in the order of 10^5. Usually, a single-slice detector array consists of 10^3 detector elements and data are acquired from 10^3 projection angles. This leads to a matrix size of $10^5 \times 10^6$ elements

6.3
Iterative Reconstruction with ART

ART stands for *algebraic reconstruction technique*. In fact, Hounsfield used this method for the first CT image reconstructions. While doing so, as mentioned before, he was not acquainted with *Radon*'s publication in 1917. However, later it became clear that iterative ART is a reinvention of *Kaczmarz*'s method published in 1937 (Kaczmarz 1937).

The methods, which use iterative strategies to solve (6.21), usually start with the notion that the realization of an image, $\mathbf{f} = (f_1, \ldots, f_N)^{\mathrm{T}}$, presents a point in an N-dimensional solution space. Starting with an initial image, $\mathbf{f}^{(0)}$, a sequence of images, $\{\mathbf{f}^{(1)}, \mathbf{f}^{(2)}, \ldots\}$, is calculated iteratively that converges to the desired tomographic reconstruction.

In the first step, a forward projection,

$$\mathbf{p}^{(n)} = \mathbf{A}\mathbf{f}^{(n)}, \tag{6.29}$$

of the nth image approximation $\mathbf{f}^{(n)}$ is determined. The projection, $\mathbf{p}^{(n)}$, determined in the nth forward projection can then be compared with the actual measured projection, $\mathbf{p}$. The comparison between the determined and the measured projection yields correction terms that are applied to the nth image approximation, $\mathbf{f}^{(n)}$, resulting in the $(n+1)$th image approximation. This process is iteratively repeated such that with another forward projection, the projection $\mathbf{p}^{(n+1)}$ is determined.

Usually, iterative methods are structured into three categories (Schramm 2001):

- *Methods with correction of all object pixels at the same time:* In this case all of the corrections are determined in one step using all information contained in the projection data set. By doing so all object pixels are corrected at the same time. Iterative *Least Squares* techniques (ILST) belong to this category, as well as the *maximum likelihood* methods, which will be discussed below.
- *Methods with pixel-wise correction:* In this category all pixels of the nth iteration are corrected one after the other. For the determination of the correction term only the projection elements to which the relevant pixel made a contribution in the forward projection are consulted.
- *Methods with beam-wise correction:* Methods like ART operate by using the information of only one beam sum. Then the object pixels that made a contribution to the appropriate projection pixel are corrected. Afterward the determination of the next beam sum follows. ART is therefore sometimes called the ray-by-ray reconstruction method.

An example first published by Kak and Slaney (1988) and Rosenfeld and Kak (1982) is repeated here to illustrate the idea of the correction using the strategy of beam-wise methods. The N-dimensional solution space in which the realization of an image, $\mathbf{f} = (f_1, \ldots, f_N)^\mathrm{T}$, is optimized, is divided into M hyperplanes given by the system of equations (6.7). If there is a unique solution of (6.7) or (6.11), the intersection of all hyperplanes will result in a single point, namely $\mathbf{f} = (f_1, \ldots, f_N)^\mathrm{T}$, that represents the solution.

For illustration, the dimension of the problem is reduced to $N = M = 2$, which means the desired image consists of two pixels and two measured projections are available. The corresponding system of equations is

$$a_{11}f_1 + a_{12}f_2 = p_1$$
$$a_{21}f_1 + a_{22}f_2 = p_2 \tag{6.30}$$

In Fig. 6.6 the iterative solution scheme can be understood graphically.

As a starting point, one needs an initial image, $\mathbf{f}^{(0)}$, as a base for which the iteration shall be started. This image could result from a rudimentary backprojection, for example. However, an image that is equivalent to the zero vector, $\mathbf{f}^{(0)} = (0, 0, \ldots, 0)^\mathrm{T}$, serves the purpose equally. This vector is projected perpendicularly onto the first straight line that represents the first X-ray beam with the projection result, p_1, in order to obtain a new and improved image, $\mathbf{f}^{(1)}$. This image is then projected perpendicularly onto the second straight line. Therewith, one obtains an image that is improved with respect to $\mathbf{f}^{(1)}$ because $\mathbf{f}^{(2)}$ lies closer to the intersection point of the straight lines than its two predecessors. Note, if both straight lines (6.30)

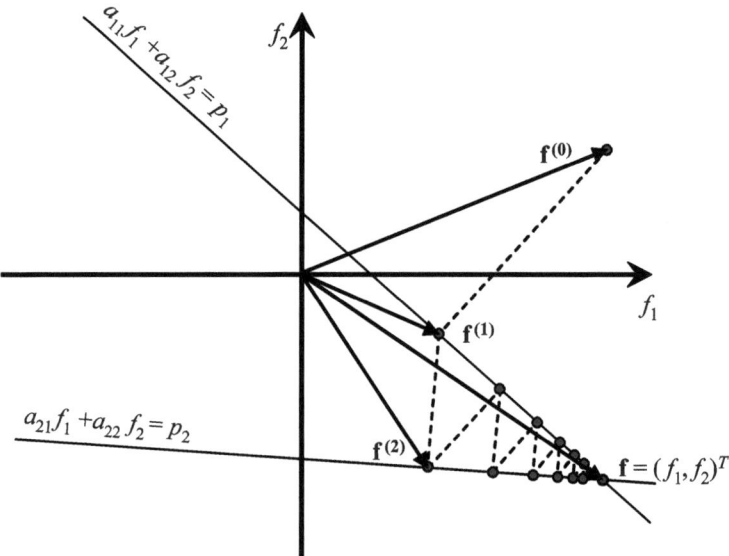

Fig. 6.6. Iterative solution of the system of equations (6.30) (adapted from Kak and Slaney [1988] and Rosenfeld and Kak [1982]). In the two-dimensional solution space each equation of the system (6.30) is represented by a *straight line*. The intersection point of the two lines gives the solution vector $\mathbf{f} = (f_1, f_2)^\mathrm{T}$, i.e., the desired pixels of the image

are perpendicular to each other, one will reach the intersection point within two iterations. In practice, this method almost always converges. The only exception is the case of parallel straight lines intersecting in infinity. However, physically this would mean that one has measured the same direction twice and therefore either no new spatial information can be gained (mathematically this means that the system of equations is singular because it is linearly dependent) or a different projection result is obtained. This result is caused by measurement noise, which is inconsistent.

The duality (6.12) between the matrix formalism and the backprojection can help to obtain an iteration equation. The basic operation, therefore, is the inner product between a certain row, i, of the system matrix $\mathbf{A}$, thus

$$\mathbf{a}_i = (a_{i1}, a_{i2}, \ldots, a_{iN}), \tag{6.31}$$

and the solution vector, thus the image

$$\mathbf{f} = (f_1, \ldots, f_N)^{\mathrm{T}}, \tag{6.32}$$

that is

$$p_i = \sum_{j=1}^{N} a_{ij} f_j, \tag{6.33}$$

as this is just the Radon transform, i.e., a projection value in the Radon space. The iteration equation then, is given by

$$\mathbf{f}^{(n)} = \mathbf{f}^{(n-1)} - \frac{\left(\mathbf{a}_i \mathbf{f}^{(n-1)} - p_i\right)}{\mathbf{a}_i (\mathbf{a}_i)^{\mathrm{T}}} (\mathbf{a}_i)^{\mathrm{T}} \tag{6.34}$$

(Kak and Slaney 1988).

This result can be obtained by simple linear algebra. Within the iteration step $\mathbf{f}^{(n-1)} \rightarrow \mathbf{f}^{(n)}$, one has to search the intersection point, $\mathbf{f}^{(n)}$, of the straight projection line

$$a_{i1} f_1 + a_{i2} f_2 = p_i \tag{6.35}$$

and the perpendicular straight dashed line drawn in Fig. 6.6. Since the projection line (6.35) is given in Hessian's normal form, one initially brings it to the slope-intercept form, i.e.,

$$f_2 = -\frac{a_{i1}}{a_{i2}} f_1 + \frac{p_i}{a_{i2}}. \tag{6.36}$$

(6.36) yields (a_{i2}/a_{i1}) as the slope of the perpendicular line. Together with the old image point, $\mathbf{f}^{(n-1)}$, one obtains

$$\frac{f_2 - f_2^{(n-1)}}{f_1 - f_1^{(n-1)}} = \frac{a_{i2}}{a_{i1}} \tag{6.37}$$

as the desired dashed perpendicular projection line in the point-slope form. Converting both equations to the same form leads to the following system of equations:

$$a_{i1}f_1 + a_{i2}f_2 = p_i$$

$$-a_{i2}f_1 + a_{i1}f_2 = -a_{i2}f_1^{(n-1)} + a_{i1}f_2^{(n-1)} \ . \tag{6.38}$$

For both new pixels $(f_1, f_2)^T$, which are the components of vector $\mathbf{f}^{(n)}$, it is consequently true that

$$f_1(a_{i1}^2 + a_{i2}^2) = a_{i1}p_i - a_{i1}a_{i2}f_2^{(n-1)} + a_{i2}^2 f_1^{(n-1)}$$

$$f_2(a_{i1}^2 + a_{i2}^2) = a_{i2}p_i - a_{i1}a_{i2}f_1^{(n-1)} + a_{i1}^2 f_2^{(n-1)} \tag{6.39}$$

or further in vector form

$$\binom{f_1}{f_2} = \frac{1}{a_{i1}^2 + a_{i2}^2} \binom{a_{i1}p_i - a_{i1}a_{i2}f_2^{(n-1)} + a_{i2}^2 f_1^{(n-1)}}{a_{i2}p_i - a_{i1}a_{i2}f_1^{(n-1)} + a_{i1}^2 f_2^{(n-1)}} \ . \tag{6.40}$$

Adding a zero to both components, i.e.,

$$\binom{f_1}{f_2} = \frac{1}{a_{i1}^2 + a_{i2}^2} \binom{a_{i1}p_i - a_{i1}a_{i2}f_2^{(n-1)} + a_{i2}^2 f_1^{(n-1)} + a_{i1}^2 f_1^{(n-1)} - a_{i1}^2 f_1^{(n-1)}}{a_{i2}p_i - a_{i1}a_{i2}f_1^{(n-1)} + a_{i1}^2 f_2^{(n-1)} + a_{i2}^2 f_2^{(n-1)} - a_{i2}^2 f_2^{(n-1)}} \ , \tag{6.41}$$

one can see that

$$\binom{f_1}{f_2} = \mathbf{f}^{(n-1)} + \frac{1}{a_{i1}^2 + a_{i2}^2} \binom{a_{i1}p_i - a_{i1}a_{i2}f_2^{(n-1)} - a_{i1}^2 f_1^{(n-1)}}{a_{i2}p_i - a_{i1}a_{i2}f_1^{(n-1)} - a_{i2}^2 f_2^{(n-1)}} \ . \tag{6.42}$$

Using the notation for inner products here, one obtains

$$\binom{f_1}{f_2} = \mathbf{f}^{(n-1)} + \frac{1}{\mathbf{a}_i(\mathbf{a}_i)^T} (p_i - \mathbf{a}_i \mathbf{f}^{(n-1)}) \binom{a_{i1}}{a_{i2}} \tag{6.43}$$

and finally

$$\binom{f_1}{f_2} = \mathbf{f}^{(n-1)} - \frac{\mathbf{a}_i \mathbf{f}^{(n-1)} - p_i}{\mathbf{a}_i(\mathbf{a}_i)^T} \binom{a_{i1}}{a_{i2}} \ . \tag{6.44}$$

This is the iteration given in (6.34) since $\mathbf{f}^{(n)} = (f_1, f_2)^T$. The key idea of ART can also be expressed in the following way: For projection i, the equation is fulfilled in the nth iteration (Toft 1996). That means

$$\mathbf{a}_i \mathbf{f}^{(n)} = \mathbf{a}_i \mathbf{f}^{(n-1)} - \frac{\mathbf{a}_i \mathbf{f}^{(n-1)} - p_i}{\mathbf{a}_i (\mathbf{a}_i)^T} \mathbf{a}_i (\mathbf{a}_i)^T = p_i . \tag{6.45}$$

Compared with filtered backprojection, the calculation expense of the iteration is a major disadvantage. Therefore, one's interest is to accelerate the convergence of the iteration equation. To do so, a heuristic relaxation parameter, λ, may be introduced into the iteration (6.34) to speed up the convergence so that the iteration equation reads

$$\mathbf{f}^{(n)} = \mathbf{f}^{(n-1)} - \lambda_n \frac{\left(\mathbf{f}^{(n-1)} \mathbf{a}_i - p_i \right)}{\mathbf{a}_i (\mathbf{a}_i)^T} (\mathbf{a}_i)^T . \tag{6.46}$$

The optimal value, λ, thereby depends on the iteration step n, the sinogram values, and the sampling parameters. However, it has been proven experimentally that a small shift away from the value $\lambda_n = 1$ can, indeed, increase the convergence speed (Herman 1980).

Looking at Fig. 6.6, the question arises as to why $\mathbf{f}^{(0)}$ is projected onto beam $i = 1$ rather than onto $i = 2$. The chosen projection beam is, in fact, selected randomly and it is a popular strategy to implement the index as an equally distributed random variable.

As in ▸ Chap. 5, the ART mode of operation shall be clarified by a specific example. Therefore, the example from Fig. 6.3 shall be consulted again. In summary, the following reconstruction procedure, which can be divided into four steps, is available as presented in Scheme 6.1.

Scheme 6.1 Algebraic reconstruction technique (ART)

1. Determination of an initial image:

$$\mathbf{f} = (0, 0, 0, 0) .$$

2. Calculation of forward projections based on the nth estimation

$$\mathbf{p}^{(n)} = \mathbf{A}\mathbf{f}^{(n)} .$$

3. Correction of the estimation (projection index i being distributed randomly)

$$\mathbf{f}^{(n)} = \mathbf{f}^{(n-1)} - \frac{\left(\mathbf{a}_i \mathbf{f}^{(n-1)} - p_i \right)}{\mathbf{a}_i (\mathbf{a}_i)^T} (\mathbf{a}_i)^T .$$

4. Iteration from step 2 when the method yields a change in consecutive image values larger than a fixed threshold. Otherwise: End of iteration.

For the first iteration, the result shall be written down explicitly here. In detail, and for the example from Fig. 6.3, this means that one first has to appoint the start image, $\mathbf{f} = (0,0,0,0)^T$, before calculating the forward projections (being $i = 2$

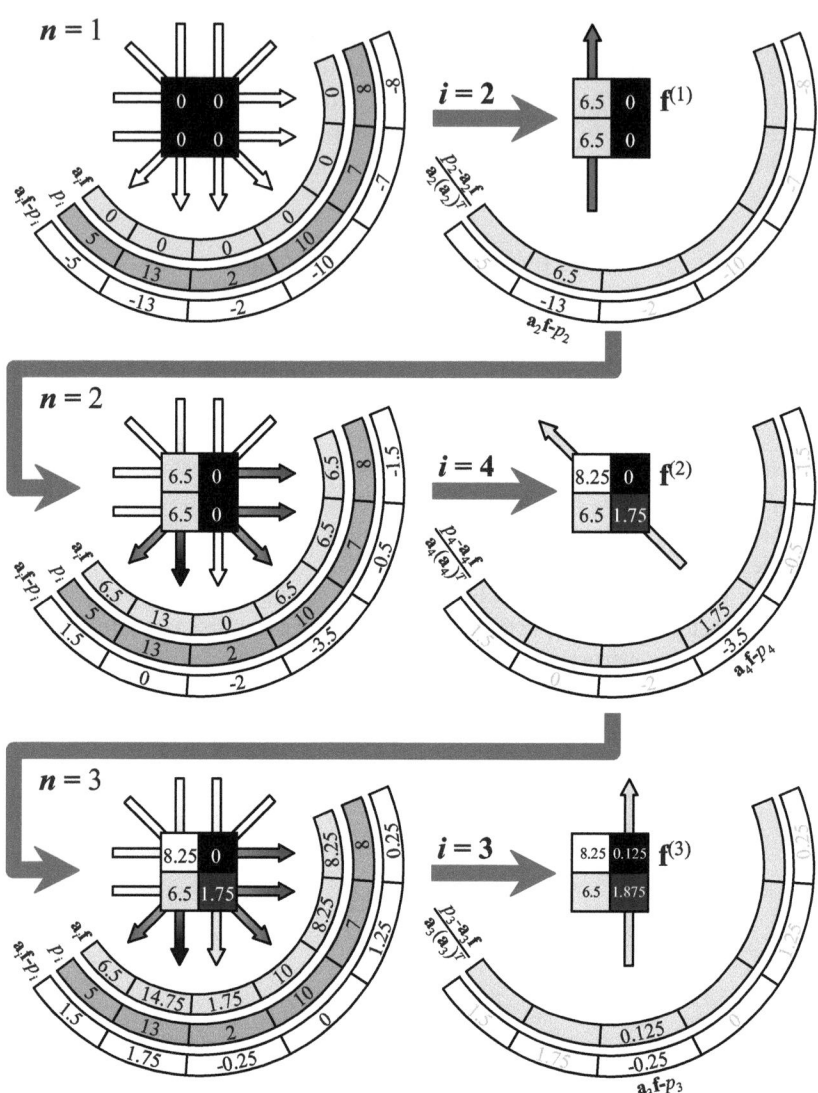

Fig. 6.7. The first three algebraic reconstruction technique (ART) iterations of the exemplary 2×2 pixel image of Fig. 6.3 using $\mathbf{f}^{(0)} = (0,0,0,0)^T$ as an initial image. The correct image values are $\mathbf{f} = (8,0,5,2)^T$

here). Comparison with the measured projection yields the correction term and the first approximation

$$\mathbf{f}^{(1)} = \begin{pmatrix} 6.5 \\ 0 \\ 6.5 \\ 0 \end{pmatrix} = \begin{pmatrix} 0 \\ 0 \\ 0 \\ 0 \end{pmatrix} + \frac{1}{(1\ 0\ 1\ 0)\begin{pmatrix} 1 \\ 0 \\ 1 \\ 0 \end{pmatrix}} \left(13 - (1\ 0\ 1\ 0)\begin{pmatrix} 0 \\ 0 \\ 0 \\ 0 \end{pmatrix} \right) \begin{pmatrix} 1 \\ 0 \\ 1 \\ 0 \end{pmatrix} . \tag{6.47}$$

For instance in Matlab[TM], the core content of the iteration can actually be programmed in only one line:

$$f = f - \left(\left(a(i,:) * f - p(i) \right) / \left(a(i,:) * a(i,:)' \right) \right) * a(i,:)' \right) . \tag{6.48}$$

Figure 6.7 illustrates the interaction scheme for the first three iterations using the exemplary 2×2 pixel image of Fig. 6.3. In the middle row, the second correction step $(n = 2)$, i.e., the result of (6.47), is shown. The second forward projection, $\mathbf{p}^{(1)} = (6.5, 13, 0, 6.5, 6.5, 6.5)$, has to be compared with the actual measured projections, $\mathbf{p} = (5, 13, 2, 10, 7, 8)$, in order to determine the next correction terms that will further refine the image. That way the differences between the forward projections and the measured projections decrease in each iteration. In the last row of Fig. 6.7, it can be seen that the correction terms have decreased significantly after only three iterations.

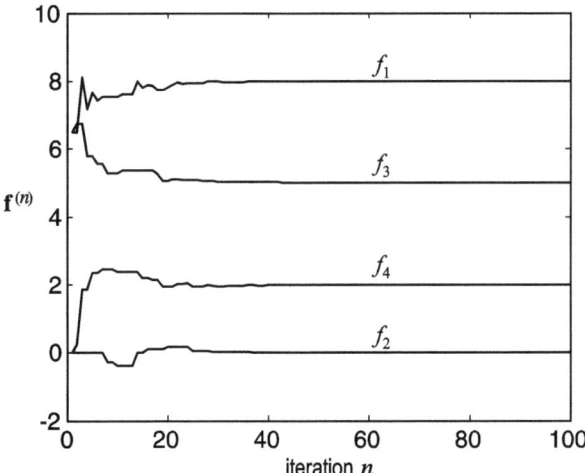

Fig. 6.8. Algebraic reconstruction technique convergence for the example from Fig. 6.7. The pixel values are plotted versus the iteration steps from $n = 1$ to $n = 100$

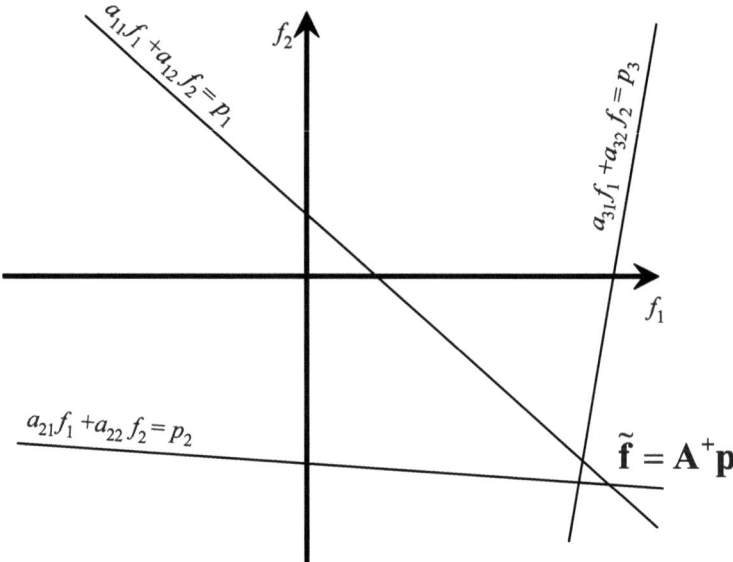

Fig. 6.9. Search for the pseudo solution, $\widetilde{\mathbf{f}} = \mathbf{A}^{+}\mathbf{p}$, which represents a solution of the over-determined system of equations in the least squares sense

Figure 6.8 shows the convergence behavior in the pixel values of the image. The image values, $\mathbf{f}^{(n)} = (f_1^{(n)}, f_2^{(n)}, f_3^{(n)}, f_4^{(n)})$, are plotted versus the iteration step, n. One can see that all image values converge very well. However, it should be mentioned that ART is noise-sensitive. The more the data are covered with noise, the worse the convergence behavior becomes (Dove 2001).

It is, incidentally, easy to understand that the system of equations (6.30) has exactly one solution: The intersection point of both straight projection lines. However, if only one more projection from another direction is added, usually it means that a clear intersection point cannot be found for real data as they are typically afflicted with noise and artifacts. Figure 6.9 illustrates such a situation. In this case, and according to (6.13) to (6.15), a pseudo solution, $\widetilde{\mathbf{f}}$, in this over-determined system of equations must be found.

6.4
Pixel Basis Functions and Calculation of the System Matrix

Now that the principle of the ART has become clear, the next question has to be answered: Where can the system matrix, $\mathbf{A}$, be obtained? In nuclear diagnostic imaging, the $\mathbf{A}$ entries are weights that are defined by (6.5) and that reflect the actual contribution of every pixel to the activity. However, before the contribution of a pixel can be determined, the definition of a pixel that has been given in the introduction of this chapter should be briefly discussed again here.

6.4.1
Discretization of the Image: Pixels and Blobs

The discretization of an image can be described by a (linear) series expansion approach in which a continuous image is approximated with a linear combination of a finite number of basis functions. If $f(x, y) = f(\mathbf{r})$ is the actual continuous spatial distribution of the attenuation coefficients, then the idea is to approximate the image $f(\mathbf{r})$ with coefficients

$$\mathbf{f} = (f_1, \ldots, f_N)^{\mathrm{T}} , \tag{6.49}$$

in the sense that

$$\mathbf{f}(\mathbf{r}) \approx \widehat{\mathbf{f}}(\mathbf{r}) = \sum_{j=1}^{N} f_j \phi_j (\mathbf{r} - \mathbf{r}_j) , \tag{6.50}$$

where N is the number of basis functions, $\phi_j(\mathbf{r})$, and attenuation coefficients, f_j, with $j = \{1, \ldots, N\}$ respectively. The vector $\mathbf{r} - \mathbf{r}_j$ is pointing from the center, $\mathbf{r}_j$, of basis function, j, to the current position, $\mathbf{r}$. If the same basis function is used for every coefficient, the index, j, can be omitted. In imaging systems, the basis functions are usually aligned on a rectangular grid; however, hexagonal or other patterns are possible as well.

There are at least two factors that affect the quality of the image approximation. The first one is the number of basis functions that are used to represent an image. A high number of coefficients results in a better approximation than a low number of basis functions. In practice, however, the number of coefficients is often limited and chosen according to the resolution of the data. Additionally, the number of image coefficients also depends on the type of the chosen basis function. For example, for SPECT and PET applications it has been shown in (Yendiki and Fessler 2004) that blob-based reconstructions need fewer image coefficients for equal image quality than voxel-based reconstructions.

The second aspect, which affects the image approximation, is the type of the basis function, $\phi_j(\mathbf{r})$. In general, basis functions can be categorized by the size of their spatial support, i.e., the size of the region of non-zero function values. While voxels, blobs, B-splines, and overlapping spheres have only small or local support, Fourier series are non-local or global basis functions.

In order to obtain reproducible results, it is necessary for the reconstructed image to be essentially independent of the orientation of the underlying grid. Therefore, basis functions that lead to shift and rotational invariance of the reconstruction should be preferred. Further, for X-ray CT, and some other imaging modalities, it is reasonable to constrain the image to non-negative values. Generally, it is sufficient that if all image coefficients are positive, then the resulting function will be positive as well.

The most common basis functions in digital imaging are pixels and voxels for two- and three-dimensional cases respectively. Both are based on rectangular functions. An exemplary alternative representation, called the blob, is given by a local basis function derived from a *Kaiser–Bessel* window introduced into the field of tomography by Lewitt (1990, 1992). A blob is defined as

$$\phi_{j_{m,a,\alpha}}\left(\|\mathbf{r}-\mathbf{r}_j\|\right)=\phi_{j_{m,a,\alpha}}(r)=\begin{cases}\dfrac{\sqrt{1-(r/a)^2}^{m}J_m\left(\sqrt{1-(r/a)^2}\right)}{J_m(\alpha)} & \text{for } 0\le r\le a \\ 0 & \text{otherwise}\end{cases}\quad,$$

$$(6.51)$$

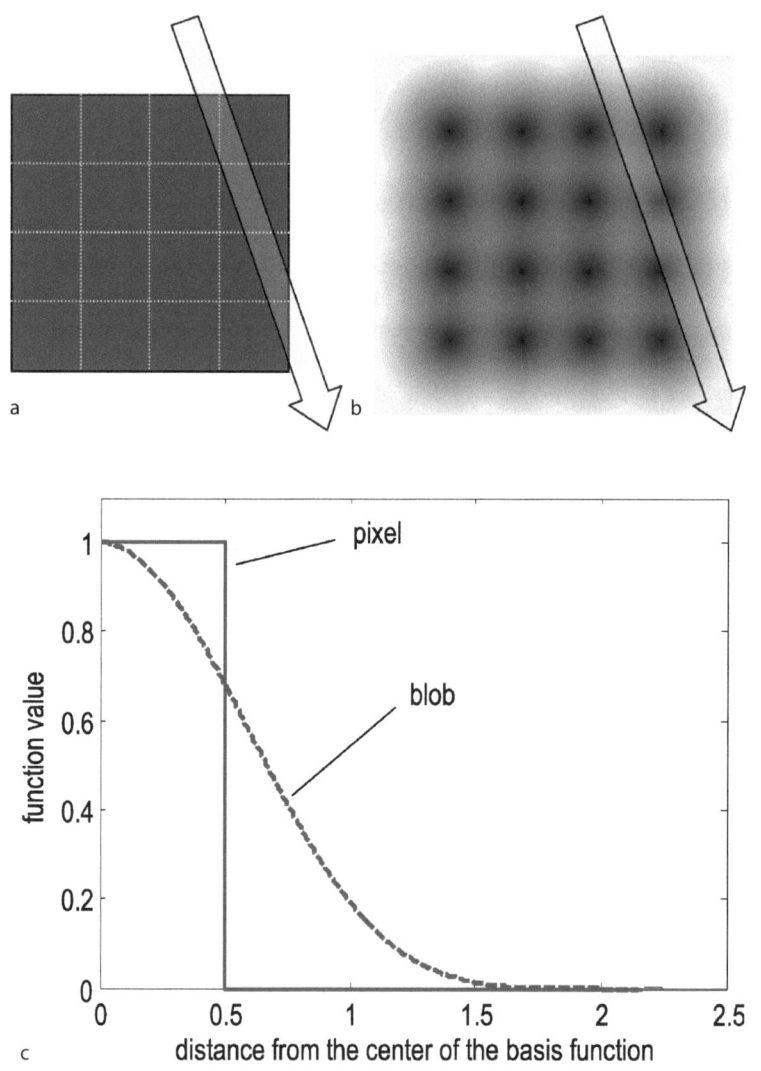

Fig. 6.10. Basis functions of the image: **a** pixels, **b** blobs, and **c** profile of the two different basis functions. The parameters for the blob shown here are $a=2$, $\alpha=10.4$, and $m=2$

where J_m is the modified Bessel function (Watson 1966) of order m; a is the radius of the basis function and α determines the shape of the blob. In contrast to pixels or voxels, blobs are rotationally (2D) or spherically (3D) symmetric basis functions. A two-dimensional comparison of pixels and blobs is shown in Fig. 6.10, together with the function profiles through their centers.

Aligned on a simple quadratic grid, adjacent blobs must overlap in order to cover the entire image. However, due to their overlap, a reconstruction based on blobs is inherently smoother than a reconstruction using pixels or voxels. Hence, the change over from pixels to blobs adds a global smoothness to the reconstruction, without changing the underlying reconstruction algorithm.

6.4.2
Approximation of the System Matrix in the Case of Pixels

In Fig. 6.11 the nearest neighbors method for the determination of the weights, a_{ij}, is presented for pixels schematically. The center of each pixel,

$$f_j = f_{kN+l} \,, \tag{6.52}$$

in the $(N = n^2)$ image shall have a quadratic neighborhood of the size b^2. The index of the single projections runs from $i = \{1, \ldots, M\}$, whereas $M = N_p D$ stands for the

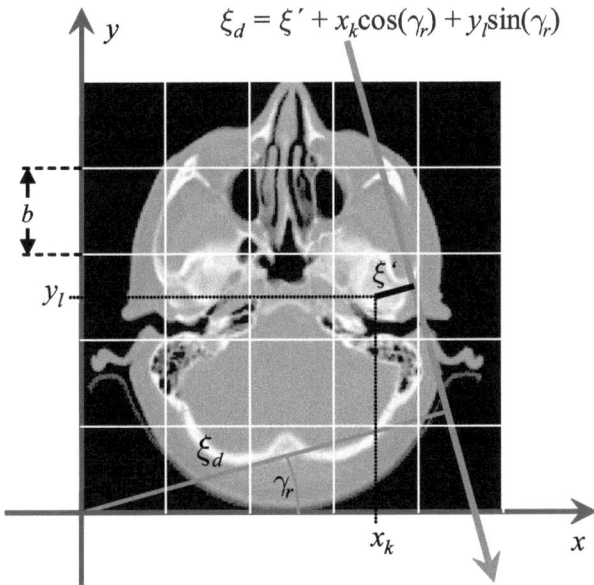

Fig. 6.11. Determination of the weights, $a_{i,j}$, with the nearest neighbors method. Here, the projection line is once again assumed to be a δ-line. However, this principle can be transferred easily to a line of the finite thickness, $\Delta\xi$. In contrast to Fig. 6.2 a finer grid is determined here

total number of projections of the array detectors, $d = \{1, \ldots, D\}$, and all projection directions $r = \{1, \ldots, N_p\}$, which means that

$$p_i = p_{rD+d} \tag{6.53}$$

holds true. The distance between the beam, ξ', and the pixel center can be determined by

$$\xi' = \xi_d - (x_k \cos(\gamma_r) + y_l \sin(\gamma_r)) . \tag{6.54}$$

Now, the following rule results for the element a_{ij}: If

$$|\xi' \cos(\gamma_r)| < \frac{b}{2} \text{ and } |\xi' \sin(\gamma_r)| < \frac{b}{2} \tag{6.55}$$

then

$$a_{ij} = a_{rD+d,kN+l} = b . \tag{6.56}$$

Of course, higher order strategies are possible too. For instance, one may take the beam length through pixel (x_k, y_l) (Toft 1996).

6.4.3
Approximation of the System Matrix in the Case of Blobs

The basis of tomographic imaging is the projection integral

$$p_i = \int_{L_i} f(\mathbf{r}) \, d\eta \tag{6.57}$$

where L_i is a line segment between the source and detector pixel, i. The integral represents the accumulated attenuation along a single beam if the width of the beam is infinitely small. Substituting the definition of $f(\mathbf{r})$ from (6.50) into (6.57) leads to

$$p_i = \int_{L_i} \sum_{j=1}^{N} f_j \phi_j(\mathbf{r} - \mathbf{r}_j) \, d\eta . \tag{6.58}$$

Reordering integration and summation leads to

$$p_i = \sum_{j=1}^{N} \left\{ \int_{L_i} \phi_j(\mathbf{r} - \mathbf{r}_j) \, d\eta \right\} f_j$$

$$= \sum_{j=1}^{N} a_{ij} f_j \tag{6.59}$$

where

$$a_{ij} = \int_{L_i} \phi_j(\mathbf{r} - \mathbf{r}_j) \, d\eta \tag{6.60}$$

describes the contribution of the jth basis function to the ith beam. If pixels are used, then a_{ij} is equal to the length of the intersection between beam i and pixel j as indicated in the subsection above. The coefficients a_{ij} are not necessarily restricted to simple line integrals. It is also possible to formulate more realistic system models, which include, for instance, scatter and complex sampling trajectories. Furthermore, it has been discussed in (Müller 2006) that a finite beam width of the X-ray can easily be incorporated into the model as well.

6.5
Maximum Likelihood Method

In the previous iterative reconstruction methods, one always started with projections that could be modeled as line integrals. The maximum likelihood method gives an alternative description. It is a statistical estimation method in which the image obtained is one that matches the measured projection values best, taking into account the measurement statistics of the real values. For a brief overview of the measurement statistics in X-ray systems, ▸ Sect. 2.6 should be consulted.

Formulated more precisely, the measuring procedure must be modeled as a stochastic process whose parameters, $\mathbf{f}^*$, have to be estimated through a given random sample, $\mathbf{p}$ (these are the projection values). In the following subsections it shall be distinguished whether the image reconstruction takes place for nuclear diagnostics (where $\mathbf{f}^*$ is the expectation value for the activity of the radioactive tracer) or for CT (where $\mathbf{f}^*$ is the expectation value for the attenuation of the X-ray quanta).

This is a completely different approach to image reconstruction than the direct methods described above. It is typically used in situations where the number of quanta on the detectors is quite small and the sinogram therefore contains a lot of noise. In these cases, noise can dominate the reconstructed image if the direct methods for image reconstruction are used. Furthermore, the statistical method may be used even in situations where projections are either missing or inconsistent. These are cases in which the filtered backprojection typically fails (Dempster 1977; Bouman and Sauer 1996).

As mentioned above, today, statistical methods are used in nuclear diagnostic imaging since PET and SPECT suffer from much worse statistics than does CT. As the filtered backprojection is faster than the iterative statistical methods and as the number of quanta is usually high in CT, there is no commercial push to transfer the iterative methods to the CT scanners of the current generation. However, this method must be presented here because, as with the continuously improving performance of computers, the iterative methods will, in the near future, also need a computing time that will be attractive for practical use. Additionally, statistical methods might possibly result in practical approaches that decrease the dose for CT. Low-dose imaging suffers from poor quantum statistics, such that statistical image reconstruction methods are appropriate.

6.5.1
Maximum Likelihood Method for Emission Tomography

The method presented by Shepp und Vardi (1982) is based on the assumption that the gamma quanta reaching the individual detector elements obey Poisson statistics. This is a direct consequence of the statistical properties of the radioactive decay and it can be motivated according to the processes explained in ▸ Sect. 2.6. The number of decays per time unit in the object pixel, f_j, represents a Poisson-distributed random variable with the expectation value f_j^*.

Thus, the probability of measuring a certain number, $\mathcal{N}_j = f_j$, of decays from a pixel with an expected activity of f_j^* can be modeled through

$$P(\mathcal{N}_j = f_j) = \frac{\left(f_j^*\right)^{f_j}}{f_j!} e^{-f_j^*} . \tag{6.61}$$

Every linear combination

$$\sum_{j=1}^{N} a_{ij} f_j = p_i \tag{6.62}$$

of the N pixels with the expectation value

$$\sum_{j=1}^{N} a_{ij} f_j^* = p_i^* , \tag{6.63}$$

is again Poisson-distributed and statistically independent, whereas the pixels are the statistically independent activities, f_j, so that for the M projection values $\mathcal{M}_i = p_i$ of (6.62), the Poisson distribution

$$P(\mathcal{M}_i = p_i) = \frac{\left(p_i^*\right)^{p_i}}{p_i!} e^{-p_i^*} \tag{6.64}$$

holds true. One obtains the joint probability of all statistically independent projection values, that is, the probability of observing projection $\mathbf{p}$ at the given expectation value $\mathbf{p}^*$, through multiplication of the single probabilities, so that

$$P(\mathbf{p}|\mathbf{p}^*) = \prod_{i=1}^{M} \frac{\left(p_i^*\right)^{p_i}}{p_i!} e^{-p_i^*} \tag{6.65}$$

holds true. By applying (6.63), one obtains

$$P(\mathbf{p}|\mathbf{p}^*) = \prod_{i=1}^{M} \frac{\left(\sum\limits_{j=1}^{N} a_{ij} f_j^*\right)^{p_i}}{p_i!} e^{-\sum\limits_{j=1}^{N} a_{ij} f_j^*} = P(\mathbf{p}|\mathbf{f}^*) . \tag{6.66}$$

It does not make sense to understand (6.66) as a function of the fixed projection measurement, $\mathbf{p}$. Rather, (6.66) is given as a function of the variable parameter $\mathbf{f}^*$, i.e., the expectation value of the spatial activity distribution

$$L(\mathbf{f}^*) = \prod_{i=1}^{M} \frac{\left(\sum_{j=1}^{N} a_{ij} f_j^* \right)^{p_i}}{p_i!} \, e^{-\sum_{j=1}^{N} a_{ij} f_j^*} . \tag{6.67}$$

(6.67) is the likelihood function. The key idea of the maximum likelihood method is to vary the expectation values of all activities, f_j^*, with $j = \{1, \dots, N\}$ in order to find the maximum of the likelihood functional, L. The distribution $\mathbf{f}^*$, for which L is maximal, is called the maximum likelihood solution of the statistical reconstruction problem and reflects the most probable solution.

Through the logarithm of L, (6.67) simplifies to

$$l(\mathbf{f}^*) = \ln\left(L(\mathbf{f}^*) \right) = \sum_{i=1}^{M} \left(p_i \ln\left(\sum_{j=1}^{N} a_{ij} f_j^* \right) - \ln(p_i!) - \sum_{j=1}^{N} a_{ij} f_j^* \right). \tag{6.68}$$

$l(\mathbf{f}^*)$ is called the *log likelihood function*. Taking the logarithm is possible since the logarithm is a strictly monotonous function and, therefore, the location of the maximum will not change. Thus, the optimization can formally be expressed by

$$\mathbf{f}^*_{\max} = \max_{\mathbf{f}^* \in \Omega} \{ l(\mathbf{f}^*) \} , \tag{6.69}$$

where Ω is set of possible solutions. A necessary condition for the maximum of $l(\mathbf{f}^*)$ is the vanishing of the first derivation

$$\frac{\partial l(\mathbf{f}^*)}{\partial f_r^*} = \sum_{i=1}^{M} \left(p_i \frac{\partial \left(\ln\left(\sum_{j=1}^{N} a_{ij} f_j^* \right) \right)}{\partial f_r^*} - a_{ir} \right)$$

$$= \sum_{i=1}^{M} \left(\frac{p_i a_{ir}}{\sum_{j=1}^{N} a_{ij} f_j^*} - a_{ir} \right) = \sum_{i=1}^{M} \frac{p_i a_{ir}}{\sum_{j=1}^{N} a_{ij} f_j^*} - \sum_{i=1}^{M} a_{ir} . \tag{6.70}$$

That means that, overall, the expression

$$\frac{\partial l(\mathbf{f}^*)}{\partial f_r^*} = \sum_{i=1}^{M} \frac{p_i a_{ir}}{\sum_{j=1}^{N} a_{ij} f_j^*} - \sum_{i=1}^{M} a_{ir} = 0 \tag{6.71}$$

has to result in zero.

However, a sufficient condition for a global maximum is the concavity of the function that is to be maximized. This condition is usually verified with the second derivative, which means application of the Hessian matrix with respect to the f_j^* with $j = \{1, \ldots, N\}$,

$$
\frac{\partial^2 l(\mathbf{f}^*)}{\partial f_r^* \partial f_s^*} = \sum_{i=1}^{M} \left(p_i a_{ir} \frac{\partial \left(\left(\sum_{j=1}^{N} a_{ij} f_j^* \right)^{-1} \right)}{\partial f_s^*} \right)
$$

(6.72)

$$
= \sum_{i=1}^{M} \left(\frac{-p_i a_{ir}}{\left(\sum_{j=1}^{N} a_{ij} f_j^* \right)^2} a_{is} \right) = - \sum_{i=1}^{M} \frac{p_i a_{ir} a_{is}}{\left(\sum_{j=1}^{N} a_{ij} f_j^* \right)^2} \; .
$$

The symmetric Hessian matrix is negative semi-definite, i.e., $l(\mathbf{f}^*)$ is concave and the optimal value obtained is a global maximum. As a consequence, the commonly referred to *Kuhn–Tucker* conditions for each j are fulfilled too (Shepp and Yardi 1982), i.e., it holds true that

$$
f_j^* \left. \frac{\partial l(\mathbf{f}^*)}{\partial f_j^*} \right|_{\mathbf{f}_{\max}^*} = 0 \text{ for all } j \text{ with } f_j^* > 0
$$

(6.73)

and

$$
\left. \frac{\partial l(\mathbf{f}^*)}{\partial f_j^*} \right|_{\mathbf{f}_{\max}^*} \leq 0 \text{ for all } j \text{ with } f_j^* = 0 \; .
$$

(6.74)

The first *Kuhn–Tucker* condition assures that the activity values f_j^* with $j = \{1, \ldots, N\}$ cannot become negative and, at the same time, it leads to an iteration scheme that will be derived in the following.

If (6.73) is applied to (6.71), one obtains

$$
f_r^* \frac{\partial l(\mathbf{f}^*)}{\partial f_r^*} = f_r^* \left(\sum_{i=1}^{M} \frac{p_i a_{ir}}{\sum_{j=1}^{N} a_{ij} f_j^*} - \sum_{i=1}^{M} a_{ir} \right) = 0 \; ,
$$

(6.75)

thus

$$
f_r^* \sum_{i=1}^{M} \frac{p_i a_{ir}}{\sum_{j=1}^{N} a_{ij} f_j^*} - f_r^* \sum_{i=1}^{M} a_{ir} = 0 \; ,
$$

(6.76)

and, therefore,

$$
f_r^* = \frac{f_r^*}{\sum_{i=1}^{M} a_{ir}} \sum_{i=1}^{M} \frac{p_i a_{ir}}{\sum_{j=1}^{N} a_{ij} f_j^*} \; .
$$

(6.77)

Obviously, the intersection point between a straight line of slope one and the functional on the right side of (6.77) must be found. The intersection point is obtained by the fixpoint iteration,

$$f_r^{*(n+1)} = \frac{f_r^{*(n)}}{\sum\limits_{i=1}^{M} a_{ir}} \sum\limits_{i=1}^{M} \frac{p_i a_{ir}}{\sum\limits_{j=1}^{N} a_{ij} f_j^{*(n)}} \tag{6.78}$$

which, at the same time, represents the iteration rule for the image reconstruction. (6.78) is called the *Expectation Maximation* (EM) algorithm. The idea is to assume $f_r^{*(n+1)}$ as the refined value for the activity at pixel r for every step $n \rightarrow n+1$. In a forward projection

$$p_i^{*(n)} = \sum\limits_{j=1}^{N} a_{ij} f_j^{*(n)} , \tag{6.79}$$

the expected projection values on the detector elements resulting from this are then calculated in the denominator of (6.78). These estimates are compared with the actual measured projection values, p_i. The resultant multiplicative correction factor

$$\frac{1}{\sum\limits_{i=1}^{M} a_{ir}} \sum\limits_{i=1}^{M} \frac{p_i a_{ir}}{\sum\limits_{j=1}^{N} a_{ij} f_j^{*(n)}} \tag{6.80}$$

for activity values $f_r^{*(n)}$ in all pixels, leads to a new and improved image, $f_r^{*(n+1)}$, with $r = \{1, \ldots, N\}$. The log likelihood function (6.68) increases with each step $n \rightarrow n+1$ and a maximum is found within a given uncertainty, if the iterative refinement step $|f_r^{*(n+1)} - f_r^{*(n)}|$ falls below a given threshold.

For initialization of the iteration, the average projection sum

$$f_j^{*(0)} = \frac{1}{M} \sum\limits_{i=1}^{M} p_i, \text{ for all } j = \{1, \ldots, N\} \tag{6.81}$$

turns out to be convenient. Figure 6.12 demonstrates the convergence for the example from Fig. 6.3. (6.78) represents a simple gradient method (Lange et al. 1987). This can be seen when adding a zero to the right side of (6.78), thus

$$f_r^{*(n+1)} = f_r^{*(n)} + \frac{f_r^{*(n)}}{\sum\limits_{i=1}^{M} a_{ir}} \left\{ \sum\limits_{i=1}^{M} \frac{p_i a_{ir}}{\sum\limits_{j=1}^{N} a_{ij} f_j^{*(n)}} - \sum\limits_{i=1}^{M} a_{ir} \right\} , \tag{6.82}$$

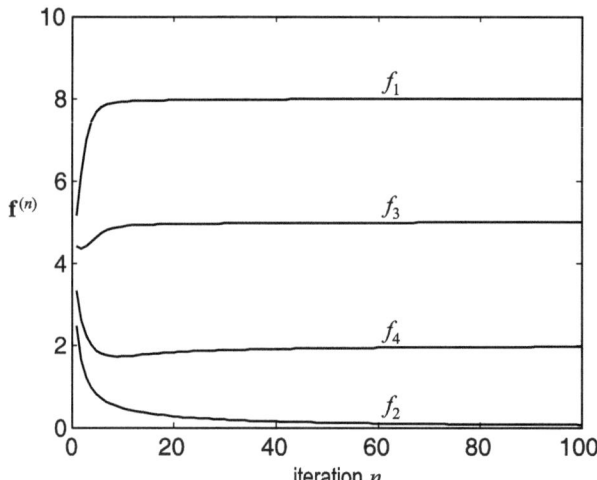

Fig. 6.12. Convergence of the maximum likelihood method for the example from Fig. 6.3. The single pixel values are shown in the iteration steps $n = 1$ to $n = 100$

in which the term in brackets is just equal to (6.71), meaning that

$$f_r^{*(n+1)} = f_r^{*(n)} + \frac{f_r^{*(n)}}{\sum\limits_{i=1}^{M} a_{ir}} \left\{ \frac{\partial l(\mathbf{f}^*)}{\partial f_r^*} \right\} , \qquad (6.83)$$

or in vector notation[7]

$$\mathbf{f}^{*(n+1)} = \mathbf{f}^{*(n)} + \mathbf{D}\left(\mathbf{f}^{*(n)}\right) grad\left(l(\mathbf{f}^*)\right) . \qquad (6.84)$$

Therefore,

$$\mathbf{D}(\mathbf{f}^{*(n)}) = \mathrm{diag} \left(\frac{f_r^{*(n)}}{\sum\limits_{i=1}^{M} a_{ir}} \right) \qquad (6.85)$$

is a diagonal matrix.

Contrary to iteration (6.34), no negative pixel values occur during iteration here. Scheme 6.2 once again summarizes the individual steps that are to be implemented for realization. This overview clarifies that the cost is definitely higher than that of filtered backprojection.

The most important properties of the maximum likelihood method resulting from the iteration formula (6.78) are:

---

[7] The *grad* operator is defined as a column vector here.

Scheme 6.2 Maximum likelihood method in nuclear diagnostics

1. Determination of a start image that is an initial estimation:

$$f_j^{*(0)} = \frac{1}{M} \sum_{i=1}^{M} p_i, \text{ for all } j = \{1, \ldots, N\} .$$

2. Determination of a normalization:

$$s_r = \sum_{i=1}^{M} a_{ir} .$$

3. Calculation of forward projections based on the $(n-1)$th estimation:

$$\mathbf{p}' = \mathbf{A}\mathbf{f}^{(n-1)} .$$

4. Calculation of the relative difference between the measured projection, p_i, and the forward projection based on the $(n-1)$th estimation:

$$p_i^{\text{rel}} = \frac{p_i}{p'_i} .$$

5. Backprojection of the relative difference of the projections into the image space:

$$\mathbf{f}^{\text{back}} = \mathbf{A}^{\text{T}} \mathbf{p}^{\text{rel}} .$$

6. For every pixel, the current estimation $(n-1)$ is multiplied with the backprojection of the relative difference and weighted with the normalization from step 2:

$$f_r^{(n)} = \frac{f_r^{(n-1)} f_r^{\text{back}}}{s_r} .$$

7. Go on with step 3 as long as the method comes up with differences larger than a chosen threshold in further iterations. Otherwise: End of iteration.

- Positivity, i.e., it holds true that

$$f_j^{*(n)} \geq 0 \tag{6.86}$$

for all pixels j and all iteration steps n, if $f_j^{*(0)} \geq 0$.

- The algorithm is scaling, i.e.,

$$\sum_{i=1}^{M} a_{ij} f_j^{*(n)} = \sum_{j=1}^{N} \left(f_j^{*(n)} \sum_{i=1}^{M} a_{ij} \right) = \sum_{i=1}^{M} p_i^{*(n)} \tag{6.87}$$

holds true for all iteration steps $n > 1$.

- Monotonous maximization of $L(\mathbf{f}^*)$, i.e.,

$$L(\mathbf{f}^{*(n+1)}) \geq L(\mathbf{f}^{*(n)}) \tag{6.88}$$

whereas the identity is only true for $L(\mathbf{f}^{*(n)}) = \max_{\mathbf{f}^* \in \Omega} \{L(\mathbf{f}^*)\}$.

6.5.2
Maximum Likelihood Method for Transmission CT

In transmission CT, the projection sum (6.62) cannot be measured directly because one does not deal with gamma quanta originating inside the body. The raw data, rather, are the X-ray quanta that are generated in the X-ray tube outside the body, attenuated exponentially when passing through the body. As shown in ▸ Sect. 2.6.1, the number of quanta generated in the X-ray tube is a Poisson-distributed random variable. Discussing the absorption or the scattering processes of X-ray quanta in a pixel, j, as random events, the number of quanta reaching the detector after passing through the pixel per time unit obeys Poisson statistics[8] as well.

The probability of an absorption and scattering process is proportional to the attenuation coefficient, f_j, which means that the attenuation of the intensity caused by pixel j – provided there is a constant X-ray path length – is given by

$$\frac{\Delta I}{I} \propto -f_j . \tag{6.89}$$

As in the previous section, it is again the absolute number of quanta measured by the detector that shows a Poisson distribution.

Based on the Beer–Lambert law of attenuation and the fact that the number of X-ray quanta is proportional to the radiation intensity, one can write

$$I_i \propto n_i^* = n_0\, e^{-\sum\limits_{j=1}^{N} a_{ij} f_j^*} . \tag{6.90}$$

The number of quanta, n_0, generated by the X-ray tube is constant and can be assumed to be known through calibration. f_j^* are the expectation values of the attenuation coefficients. In contrast to the section above, one has to formulate the maximum likelihood problem that is to be solved for transmission CT or, more specifically, for the number of X-ray quanta in detector i, i.e., n_i^*, and not directly for the image value, f_j^*. The probability of measuring a certain number, n_i, at an expected value for the quanta number, n_i^*, can be modeled as

$$P(n_i) = \frac{\left(n_i^*\right)^{n_i}}{n_i!} e^{-n_i^*} \tag{6.91}$$

where all n_i are again considered to be statistically independent.

One obtains the joint probability of all measured, statistically independent numbers for X-ray quanta, i.e., the probability of observing the set of numbers, **n**, at

---

[8] In ▸ Sect. 2.6.2 it has been shown that cascaded Poisson processes again obey a Poisson distribution.

the given expectation value, $\mathbf{n}^*$, through multiplication of the single probabilities, so that

$$P(\mathbf{n}|\mathbf{n}^*) = \prod_{i=1}^{M} \frac{(n_i^*)^{n_i}}{n_i!} e^{-n_i^*} \tag{6.92}$$

holds true. By applying (6.90), one obtains

$$P(\mathbf{n}|\mathbf{n}^*) = \prod_{i=1}^{M} \frac{\left(n_0 e^{-\sum\limits_{j=1}^{N} a_{ij}f_j^*} \right)^{n_i}}{n_i!} e^{-n_0 e^{-\sum\limits_{j=1}^{N} a_{ij}f_j^*}} = P(\mathbf{n}|\mathbf{f}^*) . \tag{6.93}$$

Since the number of measured X-ray quanta, $\mathbf{n}$, is not a variable, (6.93) should instead be understood as a function of the variable distribution of the expectation values of attenuation coefficients, $\mathbf{f}^*$, i.e.,

$$L(\mathbf{f}^*) = \prod_{i=1}^{M} \frac{\left(n_0 e^{-\sum\limits_{j=1}^{N} a_{ij}f_j^*} \right)^{n_i}}{n_i!} e^{-n_0 e^{-\sum\limits_{j=1}^{N} a_{ij}f_j^*}} . \tag{6.94}$$

(6.94) is the likelihood function for transmission CT (Andia 2000). Here, analogous to the previous section, the key idea of the maximum likelihood method is the variation of the expectation values of attenuation coefficients to find the optimal likelihood functional, L. Again, the distribution $\mathbf{f}^*$, for which L is a maximum, is called the maximum likelihood solution of the statistical reconstruction problem and reflects the most probable solution.

The logarithm of L simplifies (6.94) to

$$l(\mathbf{f}^*) = \ln\left(L(\mathbf{f}^*) \right)$$

$$= \sum_{i=1}^{M} \left(\ln\left(n_0 e^{-\sum\limits_{j=1}^{N} a_{ij}f_j^*} \right)^{n_i} - \ln(n_i!) - n_0 e^{-\sum\limits_{j=1}^{N} a_{ij}f_j^*} \right) \tag{6.95}$$

$$= \sum_{i=1}^{M} \left(n_i \ln(n_0) - n_i \sum_{j=1}^{N} a_{ij}f_j^* - \ln(n_i!) - n_0 e^{-\sum\limits_{j=1}^{N} a_{ij}f_j^*} \right) .$$

$l(\mathbf{f}^*)$ is the log likelihood function of transmission CT. Formally, one may write the optimization as

$$\mathbf{f}^*_{max} = \max_{\mathbf{f}^* \in \Omega} \{l(\mathbf{f}^*)\} , \tag{6.96}$$

where Ω denotes the set of possible solutions. A necessary condition for the maximum of $l(\mathbf{f}^*)$ is the vanishing of the first derivation

$$\frac{\partial l(\mathbf{f}^*)}{\partial f_r^*} = \sum_{i=1}^{M} \left(\frac{\partial \left(n_i \ln(n_0) - n_i \sum_{j=1}^{N} a_{ij} f_j^* - \ln(n_i!) - n_0 e^{-\sum_{j=1}^{N} a_{ij} f_j^*} \right)}{\partial f_r^*} \right)$$

$$= \sum_{i=1}^{M} \left(-\frac{\partial \left(n_i \sum_{j=1}^{N} a_{ij} f_j^* \right)}{\partial f_r^*} - \frac{\partial \left(n_0 e^{-\sum_{j=1}^{N} a_{ij} f_j^*} \right)}{\partial f_r^*} \right)$$

$$= \sum_{i=1}^{M} \left(-n_i a_{ir} + n_0 a_{ir} e^{-\sum_{j=1}^{N} a_{ij} f_j^*} \right) = n_0 \sum_{i=1}^{M} a_{ir} e^{-\sum_{j=1}^{N} a_{ij} f_j^*} - \sum_{i=1}^{M} n_i a_{ir} . \tag{6.97}$$

This means that the expression

$$\frac{\partial l(\mathbf{f}^*)}{\partial f_r^*} = n_0 \sum_{i=1}^{M} a_{ir} e^{-\sum_{j=1}^{N} a_{ij} f_j^*} - \sum_{i=1}^{M} n_i a_{ir} \tag{6.98}$$

has to become zero again.

However, the sufficient condition for a global maximum of $l(\mathbf{f}^*)$ is the concavity of the function that is to be maximized. This condition is usually verified with the second derivative. That means in the case of f_j^* with $j = \{1, \ldots, N\}$, the application of the Hessian matrix is

$$\frac{\partial^2 l(\mathbf{f}^*)}{\partial f_r^* \partial f_s^*} = \sum_{i=1}^{M} \left(\frac{\partial \left(n_0 a_{ir} e^{-\sum_{j=1}^{N} a_{ij} f_j^*} - n_i a_{ir} \right)}{\partial f_s^*} \right)$$

$$= \sum_{i=1}^{M} \left(n_0 a_{ir} e^{-\sum_{j=1}^{N} a_{ij} f_j^*} (-a_{is}) \right) = -n_0 \sum_{i=1}^{M} \left(a_{is} a_{ir} e^{-\sum_{j=1}^{N} a_{ij} f_j^*} \right) . \tag{6.99}$$

Here, too, the symmetric Hessian matrix is negative semi-definite, i.e., $l(\mathbf{f}^*)$ is concave and the optimal value obtained is a global maximum. As a consequence, the *Kuhn–Tucker* conditions for each j are fulfilled, too, which again leads to an iteration scheme. When applying (6.73) to (6.98), one obtains

$$f_r^* \frac{\partial l(\mathbf{f}^*)}{\partial f_r^*} = f_r^* \left(n_0 \sum_{i=1}^{M} a_{ir} e^{-\sum_{j=1}^{N} a_{ij} f_j^*} - \sum_{i=1}^{M} n_i a_{ir} \right) = 0 \qquad (6.100)$$

thus

$$f_r^* n_0 \sum_{i=1}^{M} a_{ir} e^{-\sum_{j=1}^{N} a_{ij} f_j^*} - f_r^* \sum_{i=1}^{M} n_i a_{ir} = 0 \qquad (6.101)$$

and finally

$$f_r^* = \frac{f_r^* n_0}{\sum_{i=1}^{M} n_i a_{ir}} \sum_{i=1}^{M} a_{ir} e^{-\sum_{j=1}^{N} a_{ij} f_j^*} . \qquad (6.102)$$

Obviously, the intersection point between a straight line of a slope of one and the functional on the right side of (6.102) must be found. The intersection point is obtained by the fixpoint iteration

$$f_r^{*(n+1)} = \frac{f_r^{*(n)} n_0}{\sum_{i=1}^{M} n_i a_{ir}} \sum_{i=1}^{M} a_{ir} e^{-\sum_{j=1}^{N} a_{ij} f_j^{*(n)}} , \qquad (6.103)$$

which also represents the iteration rule for image reconstruction. Recalling that the number, n_i, of X-ray quanta measured in the detector, i, is proportional to the intensity of the X-ray radiation,

$$n_i = n_0 e^{-\sum_{j=1}^{N} a_{ij} f_j} \qquad (6.104)$$

one may write (6.103) as

$$f_r^{*(n+1)} = f_r^{*(n)} \frac{\sum_{i=1}^{M} a_{ir} e^{-\sum_{j=1}^{N} a_{ij} f_j^{*(n)}}}{\sum_{i=1}^{M} a_{ir} e^{-p_i}} . \qquad (6.105)$$

However, according to Beer–Lambert's law, the measured projection values,

$$p_i = \sum_{j=1}^{N} a_{ij} f_j , \qquad (6.106)$$

appear as an argument of the exponential function in the denominator of (6.105). In (6.105), $f_r^{*(n+1)}$ is the refined value for the X-ray attenuation at pixel r for every step $n \rightarrow n+1$. From this, the number of expected X-ray quanta,

$$n_i^{(n)} \propto e^{-\sum\limits_{j=1}^{N} a_{ij} f_j^{*(n)}}, \tag{6.107}$$

measured by the detectors can be determined. These are compared with the actual measured X-ray quanta,

$$n_i \propto e^{-p_i} . \tag{6.108}$$

The ratio

$$\sum_{i=1}^{M} a_{ir} e^{-\sum\limits_{j=1}^{N} a_{ij} f_j^{*(n)}} \Bigg/ \sum_{i=1}^{M} a_{ir} e^{-p_i} \tag{6.109}$$

of (6.105) improves the image in every iteration. The log likelihood function (6.95) increases with each step $n \rightarrow n+1$ and a maximum is found within a given uncertainty if the iterative refinement step, $|f_r^{*(n+1)} - f_r^{*(n)}|$, falls below a given threshold.

In order to write (6.103) in the form[9]

$$\mathbf{f}^{*(n+1)} = \mathbf{f}^{*(n)} + \mathbf{D}(\mathbf{f}^{*(n)}) grad \left(l(\mathbf{f}^*) \right) \tag{6.84}$$

just as in the previous section, the diagonal matrix (6.85) of the emission tomography has to be replaced by the appropriate expression for transmission CT:

$$\mathbf{D}(\mathbf{f}^{*(n)}) = \mathrm{diag} \left(\frac{f_r^{*(n)}}{\sum\limits_{i=1}^{M} a_{ir} e^{-p_i}} \right) . \tag{6.110}$$

As shown by (Lange and Fessler 1995), (6.105) can be written as

$$
\begin{aligned}
f_r^{*(n+1)} &= f_r^{*(n)} + \frac{f_r^{*(n)}}{\sum\limits_{i=1}^{M} a_{ir} e^{-p_i}} \left\{ \frac{\partial l(\mathbf{f}^*)}{\partial f_r^*} \right\} \\
&= f_r^{*(n)} + \frac{f_r^{*(n)}}{\sum\limits_{i=1}^{M} a_{ir} e^{-p_i}} \left\{ \sum_{i=1}^{M} a_{ir} e^{-\sum\limits_{j=1}^{N} a_{ij} f_j^{*(n)}} - \sum_{i=1}^{M} a_{ir} e^{-p_i} \right\} .
\end{aligned}
\tag{6.111}
$$

Unfortunately, the simple fixpoint iteration (6.105) is often numerically unstable. Therefore, methods for regularization of the inverse problem have to be used. These methods shall be presented briefly in the next section.

---

[9] The *grad* operator is defined as a column vector here.

6.5.3
Regularization of the Inverse Problem

The maximum likelihood approach (6.96) in imaging is a very high dimensional inverse problem and is often revealed to be unstable, i.e., it may produce oscillations that complicate the search for optimal parameters, $\mathbf{f}^*$. The typical countermeasure to this problem is the introduction of a regularization term to the log likelihood function. This penalty term controls the compromise between spatial resolution and noise in the image. It should be noted here that this compromise has to be found for the method of the filtered backprojection as well: In filtered backprojection one has to find an appropriate deviation from the linear weighting of the spectrum of the projection integral in order to attenuate high frequencies. This regularization definitely reduces noise in the image; however, at the same time this is also true for the spatial resolution. This is a typical trade-off in ill-posed problems.

In terms of linear algebra, regularization improves the conditioning of the inverse problem, which potentially leads to faster convergence of the algorithm. Furthermore, it is possible to integrate a priori known, desired properties of the image into the regularization strategy.

The estimation of the image for the regularized problem is formally given as an additive extension of (6.96), i.e.,

$$\mathbf{f}^*_{max} = \max_{\mathbf{f}^* \in \Omega} \{\ln(L(\mathbf{f}^*)) + \ln(R(\mathbf{f}^*))\} \,, \tag{6.112}$$

where R represents the regularization functional and Ω denotes the set of possible solutions, which is, compared with (6.96), limited by the penalty term, R. Severe complications of this method are that one has to find an appropriate regularization functional, R, and a parameter that controls the strength of the penalty.

Bayes' view, which understands the regularization as an *a priori* model, shall briefly be described here. A typical Bayesian estimation method is the *maximum a posteriori* (MAP) method. For this method, two statistical models are required. One model describes the physical process of emission of gamma quanta (for emission CT), which means

$$p_i \text{ obeys Poisson} \left(\sum_{j=1}^{N} a_{ij} f_j^* \right), \tag{6.113}$$

or the transmission of X-ray quanta (in transmission CT), which means

$$n_i \text{ obeys Poisson} \left(n_0 \, e^{-\sum_{j=1}^{N} a_{ij} f_j^*} \right). \tag{6.114}$$

This first model is the original maximum likelihood term.

The other model, namely the *a priori* model, is the probability distribution of the original image. The reconstruction quality of the MAP method depends sensitively on the choice of an appropriate second model. The choice of the *a priori* model seems to be a big challenge, since detailed knowledge about the image that is to be reconstructed is required and has to be formulated mathematically as the functional $R(\mathbf{f}^*)$, called the *prior*, in (6.112). Fortunately, it can be shown (Green 1990) that knowledge is neither necessary on a large scale nor does one need complex knowledge about the image involved to obtain sufficient regularization. In order to suppress noise in the reconstructed image, it is reasonable to require that gray values of spatially neighboring pixels do not differ substantially in their mean. The a priori knowledge is hence limited to the direct neighborhood and can be formulated in a simple mathematical form.

In practice the image is very often modeled as a *Markoff random field* (MRF). For such stochastic processes, the conditional distribution, $P(f_n|f_1,\ldots,f_{n-1})$, only depends on the gray values of direct neighboring pixels[10]. Therefore, an important expression is the *Gibbs* distribution

$$R(\mathbf{f}^*) = \frac{1}{Z} e^{-\lambda^q \sum_{c \in C} V_c(\mathbf{f}^*)},\qquad(6.115)$$

because a random field is a Markoff random field if, and only if, the probability distribution obeys function (6.115). Z is a normalizing constant[11], $V_c(\mathbf{f}^*)$ is a potential function of a local group of pixels, c is commonly called a clique, and C denotes the set of all cliques (Lehmann et al. 1997). Examples of the possible cliques of a 4- and 8-neighborhood in a two-dimensional image are given in Fig. 6.13.

The parameter λ^q represents the regularization parameter[12] (the control factor for the influence of regularization) where $1 \le q \le 2$. In Andia (2000) several potential functions are given. A very simple function for $q = 2$ is the potential

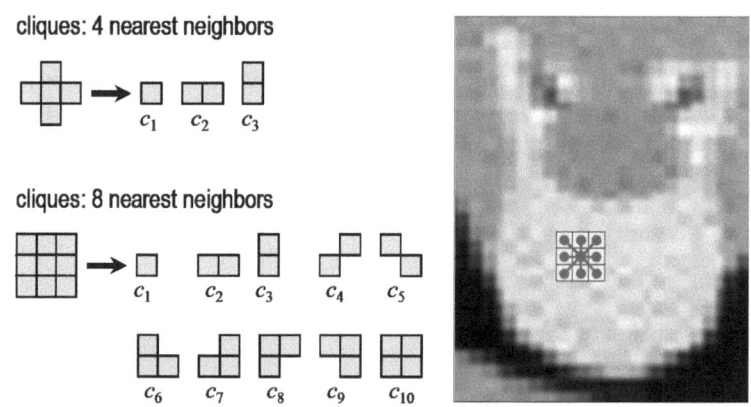

Fig. 6.13. The cliques of a 4- and 8-neighborhood for the definition of Gibbs potential

[10] In so far as the Markoff process is the stochastic equivalent to the differential equation (Lehmann et al. 1997).

[11] In physics Z is called the partition function.

[12] In physics λ^{-q} represents the temperature.

that penalizes the squares of the difference of neighboring pixels $d = (f_i - f_j)$, thus

$$V_{c_k}(d) = d^2 . \tag{6.116}$$

This potential, the commonly referred to Gaussian MRF, suppresses the noise, but also penalizes edges in the image. In order to be able to control the strength of the regularization, one defines

$$R(\mathbf{f}^*) = \frac{1}{Z} e^{-\lambda^q \sum\limits_{\{j,k\} \in C} w_{j,k}|f_j - f_k|^q} \tag{6.117}$$

as the Gibbs distribution of the generalized Gaussian MRF (GGMRF; Bouman and Sauer 1996). Here, $w_{j,k}$ is a weighting that assesses the respective neighborhood of the clique (e.g., $w_{j,k} = 1$ for orthogonal neighbors and $w_{j,k} = 2^{-1/2}$ for diagonal neighbors [Green 1990]). Additionally, smaller values of the exponent, q, of the potential function of (6.117) allow sharper edges within the image reconstruction.

Of course one can also find edge preserving regularizations by an appropriate choice of the potential function. The regularized log likelihood function for transmission CT that results from (6.117) reads

$$\tilde{l}(\mathbf{f}^*) = \sum_{i=1}^{M} \left(-n_i \sum_{j=1}^{N} a_{ij}f_j^* - n_0 e^{-\sum\limits_{j=1}^{N} a_{ij}f_j^*} \right) + \ln R(\mathbf{f}^*) , \tag{6.118}$$

whereas the partition function, Z, which depends on the regularization parameter, λ^q, does not have to be considered for optimization (Hebert and Leahy 1989). An iterative fixpoint solution can be found for (6.118) as well. Using the recipe of the section above, starting with (6.97), one obtains

$$\frac{\partial \tilde{l}(\mathbf{f}^*)}{\partial f_r^*} = \sum_{i=1}^{M} \left(-\frac{\partial \left(\sum\limits_{j=1}^{N} a_{ij}f_j^* \right)}{\partial f_r^*} - \frac{\partial \left(n_0 e^{-\sum\limits_{j=1}^{N} a_{ij}f_j^*} \right)}{\partial f_r^*} \right) + \frac{\partial \ln R(\mathbf{f}^*)}{\partial f_r^*} \tag{6.119}$$

$$= n_0 \sum_{i=1}^{M} a_{ir} e^{-\sum\limits_{j=1}^{N} a_{ij}f_j^*} - \sum_{i=1}^{M} a_{ir} - \frac{\partial \left(\lambda^q \sum\limits_{\{j,k\} \in C} w_{j,k} |f_j - f_k|^q \right)}{\partial f_r^*} ,$$

and finally

$$f_r^{*(n+1)} = f_r^{*(n)} \frac{\sum\limits_{i=1}^{M} a_{ir} e^{-\sum\limits_{j=1}^{N} a_{ij}f_j^{*(n)}}}{\sum\limits_{i=1}^{M} a_{ir} e^{-p_i} + \lambda^q q \sum\limits_{c_i \in C} w_{kj} \left| f_k - f_j^* \right|^{q-1} \text{sign}\left(f_k - f_j^* \right)} , \tag{6.120}$$

for the special GGMRF approach of (6.118).

6.5.4

Approximation Through Weighted Least Squares

Another method of stabilizing the optimization is to incorporate the reliability of the measured values. This is possible with the method of weighted least squares for which the functional

$$\mathbf{f}_{min}^* = \min_{\mathbf{f} \in \Omega} \left\{ \frac{1}{2} (\mathbf{Af}^* - \mathbf{p})^T \mathbf{C}^{-1} (\mathbf{Af}^* - \mathbf{p}) \right\} \tag{6.121}$$

has to be minimized (Andia 2000), and where Ω denotes the set of possible solutions. Contrary to (6.13), the inverse covariance matrix, $\mathbf{C}^{-1}$, which weights the measured values of low reliability adequately, is incorporated here.

The functional (6.121) that has to be optimized results from a Taylor expansion of the log maximum likelihood functional

$$l(\mathbf{f}^*) = \sum_{i=1}^{M} \left(n_i \ln(n_0) - n_i \sum_{j=1}^{N} a_{ij} f_j^* - \ln(n_i!) - n_0 \, e^{-\sum_{j=1}^{N} a_{ij} f_j^*} \right)$$

$$= \sum_{i=1}^{M} \left(n_i \ln(n_0) - n_i p_i^* - \ln(n_i!) - n_0 \, e^{-p_i^*} \right), \tag{6.95}$$

in which the expressions for the expected projection sums,

$$p_i^* = \sum_{j=1}^{N} a_{ij} f_j^* , \tag{6.122}$$

are used and the dependencies of p_i^* at

$$p_i = \ln\left(\frac{n_0}{n_i}\right), \tag{6.123}$$

are developed to the second order (Sauer and Bouman 1993), hence

$$l(\mathbf{f}^*) \approx \sum_{i=1}^{M} \left(n_i \ln(n_0) - n_i p_i - n_i(p_i^* - p_i) - \ln(n_i!) \right.$$

$$\left. - n_0 \left(e^{-p_i} - e^{-p_i}(p_i^* - p_i) + \frac{e^{-p_i}}{2}(p_i^* - p_i)^2 \right) \right)$$

$$= \sum_{i=1}^{M} \left(n_i \ln(n_0) - n_i p_i^* - \ln(n_i!) - n_i + n_i(p_i^* - p_i) - \frac{n_i}{2}(p_i^* - p_i)^2 \right)$$

$$= \sum_{i=1}^{M} \left(n_i \ln(n_0) - \ln(n_i!) - n_i - n_i p_i - \frac{n_i}{2}(p_i^* - p_i)^2 \right)$$

$$= \sum_{i=1}^{M} \left(n_i \ln(n_0) - \ln(n_i!) - n_i(1 + p_i) - \frac{n_i}{2}(p_i^* - p_i)^2 \right)$$

$$= -\frac{1}{2} \sum_{i=1}^{M} \left(n_i(p_i^* - p_i)^2 \right) + \sum_{i=1}^{M} \left(n_i \ln(n_0) - \ln(n_i!) - n_i(1 + p_i) \right). \tag{6.124}$$

Overall, this expression can be summarized as

$$l(\mathbf{f}^*) \approx -\frac{1}{2} \sum_{i=1}^{M} \left(n_i \left(\sum_{j=1}^{N} a_{ij} f_j^* - p_i \right)^2 \right) + \sum_{i=1}^{M} c(n_i) \qquad (6.125)$$

or, equivalently, in vector and matrix form as

$$l(\mathbf{f}^*) \approx -\frac{1}{2}(\mathbf{Af}^* - \mathbf{p})^T \mathbf{C}^{-1}(\mathbf{Af}^* - \mathbf{p}) + c(\mathbf{n}) \qquad (6.126)$$

so that the maximization of the log likelihood function (6.95) can be approximated by the minimization of the weighted least squares in (6.121). Therefor, $c(\mathbf{n})$ is independent of $\mathbf{f}^*$ and can be ignored in the optimization. This result holds analogously true for the maximum likelihood approach for emission CT.

Finally, the physical meaning of the covariance matrix should be explained. In the case of emission CT,

$$p_i = n_i \qquad (6.127)$$

is the number of quanta that is measured in the detector and

$$\mathbf{C} = \begin{pmatrix} n_1 & 0 & 0 & 0 & 0 \\ 0 & n_2 & 0 & 0 & 0 \\ 0 & 0 & . & . & . \\ 0 & 0 & . & . & . \\ 0 & 0 & . & . & n_M \end{pmatrix} \qquad (6.128)$$

is the covariance matrix. (6.128) is the central statement for the Poisson distribution of detected quanta, namely $\sigma_i^2 = \langle n_i \rangle$.

In the case of transmission CT, it holds true that

$$p_i = \ln\left(\frac{n_0}{n_i}\right) \qquad (6.129)$$

and

$$\mathbf{C} = \begin{pmatrix} \dfrac{1}{n_1} & 0 & 0 & 0 & 0 \\ 0 & \dfrac{1}{n_2} & 0 & 0 & 0 \\ 0 & 0 & . & . & . \\ 0 & 0 & . & . & . \\ 0 & 0 & . & . & \dfrac{1}{n_M} \end{pmatrix}. \qquad (6.130)$$

(6.130) implies that $\sigma_i^2 = 1/\langle n_i \rangle$. This means that the more quanta that are measured, the smaller the signal variance and, consequently, the stronger the data are weighted as reliable within the minimization (in ▸ Chap. 9 this circumstance will be discussed in detail). The reverse holds for emission CT. The approximated regularized maximum likelihood approach then reads

$$
\mathbf{f}^*_{\min} = \min_{\mathbf{f} \in \Omega} \left\{ \frac{1}{2} (\mathbf{Af}^* - \mathbf{p})^{\mathrm{T}} \mathbf{C}^{-1} (\mathbf{Af}^* - \mathbf{p}) + \lambda^q \sum_{\{j,k\} \in C} w_{j,k} \left| f_j - f_k \right|^q \right\} \tag{6.131}
$$

and can be optimized with, for example, the Newton–Raphson method (Fessler 1996; Bouman and Sauer 1996).

7 Technical Implementation

Contents

7.1   Introduction 241
7.2   Reconstruction with Real Signals 242
7.3   Practical Implementation of the Filtered Backprojection 255
7.4   Minimum Number of Detector Elements 258
7.5   Minimum Number of Projections 259
7.6   Geometry of the Fan-Beam System 261
7.7   Image Reconstruction for Fan-Beam Geometry............................. 262
7.8   Quarter-Detector Offset and Sampling Theorem 293

7.1
Introduction

The geometrical design of a CT scanner is relatively simple. As already described in the preliminary remarks on CT in ▸ Sects. 3.3 to 3.6, the development stages can be divided into four generations. Two of these generations – the first and the third one – are particularly interesting. The first generation – the pencil-beam concept – exactly reflects the Radon reconstruction process, since parallel beams define the slice plane (cf. Fig. 7.1, middle and right side). The second generation is just a temporary intermediate developmental step toward the third generation, the fan-beam concept. The third-generation scanners (Fig. 7.1, left side) are those most frequently implemented to date. For this reason, the focus of this chapter will be on the reconstruction mathematics based on fan-beam geometry.

The fourth generation is again a stage of evolutionary progress that has not yet been frequently implemented. From a mathematical point of view, this generation is identical to the third generation and will therefore not be considered here separately. The fast backprojection methods will not be considered here either. A good summary on modern reconstruction methods is given in Ingerhed (1999).

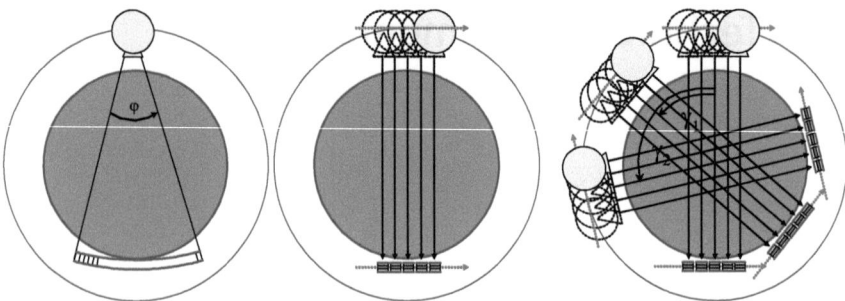

Fig. 7.1. The currently most frequently used third-generation scanner (*left*) compared with the first generation (*center and right*)

7.2
Reconstruction with Real Signals

The spectral weighting function, $|q|$, has the unit of a spatial frequency and is defined for all frequencies, so that one has to integrate

$$h_\gamma(\xi) = \int\limits_{-\infty}^{+\infty} P_\gamma(q)\,|q|\,e^{2\pi i q\xi}\,dq \tag{7.1}$$

along the whole frequency domain. As the real signal is spatially discrete and also spatially limited due to the limited number of detectors, the spectrum of the projection is repeated periodically as a result of the sampling process. In ▸ Sect. 4.18 it was reported that Shannon's sampling theorem ensures faultless sampling of a signal with a sampling frequency, $\Delta\xi = (2Q)^{-1}$, if the maximum available spatial frequency is less than Q. Due to this, weighting of the periodic spectrum with the ramp function, $|q|$, is only reasonable in the interval $[-Q, Q]$. Let the sampled projection signal be

$$p_\gamma(j\Delta\xi), \quad \text{with } j = 0, 1, \dots, D-1\,. \tag{7.2}$$

The sampling parameters $\Delta\xi$ (width of one single detector element) and D (number of the detector elements) can be easily changed in CT systems of the first generation by means of tube and detector feeds. In systems of the third generation, which have been most frequently used to date, $\Delta\xi$ is actually determined by the fixed detector width and D by the fixed number of detector elements. In this chapter it will be shown later on (cf. ▸ Sect. 7.8) that the sampling intervals can be synthetically halved by taking structural measures, such as a quarter detector shift or flying focus concept.

In principle, the Fourier transform of the projection

$$P_\gamma(q) = \int\limits_{0}^{D\Delta\xi} p_\gamma(\xi)e^{-2\pi i q\xi}d\xi \tag{7.3}$$

must now of course only be performed over the total detector length, i.e., over the interval $[0, D\Delta\xi]$. Outside the detector the projection is assumed to be zero. $P_\gamma(q)$ can now be discretized by

$$P_\gamma\left(k\frac{2Q}{D}\right) = \frac{1}{2Q}\sum_{j=0}^{D-1} p_\gamma\left(\frac{j}{2Q}\right)e^{-2\pi i(jk/D)}, \qquad (7.4)$$

where

$$P_\gamma(k\Delta q), \quad \text{with } k = 0, 1, \ldots, D-1 \qquad (7.5)$$

and

$$\Delta q = \frac{2Q}{D}. \qquad (7.6)$$

For the windowed high-pass filtered projection

$$h_\gamma(\xi) \approx \int_{-Q}^{+Q} P_\gamma(q)|q|e^{2\pi i q\xi}\,dq, \qquad (7.7)$$

the Fourier transform must be approximated. Obviously, only an approximation can be calculated because, from a mathematical point of view, the Fourier transform cannot be limited to $[-Q, Q]$ (cf. ▶ Chap. 4). This is due to the fact that the projection signal is in any case limited in the spatial domain by the finite length of the detector, as already described above.

However, the energy, which is physically contained in the high-frequency bands, can be neglected for a sufficiently high value of D, i.e., a sufficiently long detector array. Thus, a band limitation that is required to ensure a sufficient sampling process, only destroys a small amount of information such that discretization can be expressed as shown:

$$h_\gamma\left(j\frac{1}{2Q}\right) = h_\gamma(j\Delta\xi) \approx \frac{2Q}{D}\sum_{k=-D/2}^{D/2-1} P_\gamma\left(k\frac{2Q}{D}\right)\left|k\frac{2Q}{D}\right|e^{2\pi i(jk/D)}. \qquad (7.8)$$

This equation yields the filtered projection, $h_\gamma(j\Delta\xi)$, at the sampling points, $j\Delta\xi$, of the projection, $p_\gamma(\xi)$, by the inverse finite and discrete Fourier transform of the product of $|k\,2Q/D|$ and $P_\gamma(k\,2Q/D)$ in the frequency domain. The image to be reconstructed finally results from the discrete approximation of the integral

$$f(x, y) = \int_0^\pi h_\gamma(\xi)\,d\gamma$$

$$\approx \frac{\pi}{N_p}\sum_{n=1}^{N_p} h_{\gamma_n}(x\cos(\gamma_n) + y\sin(\gamma_n)), \qquad (7.9)$$

where γ_n for $n = 1, 2, \ldots, N_P$ are the angles of the measured projections since the projection angles are also just available as discrete values. The standardization in front of the sum in (7.9) is accomplished by the discretization of the angle element

$$d\gamma \rightarrow \Delta\gamma = \frac{\pi}{N_P} \, . \tag{7.10}$$

In practice, it does not seem to be useful to multiply the frequency domain for real data with a linearly increasing function, as shown in (7.8), since linear spectral weighting increases the noise in the high frequency band. Therefore, the reconstructed tomogram may finally be degraded by unacceptably noisy signals. This can be remedied by means of suitable data windowing.

7.2.1
Frequency Domain Windowing

The Fourier transform theory lists a multitude of frequency or spectral windows. These are superior to a sharp rectangular band limitation within the frequency interval $[-Q, Q]$ by means of the linearly increasing function given in (7.8).

Taking into account the window function in the frequency domain leads to the following approximation:

$$h_\gamma\left(j\frac{1}{2Q}\right) = h_\gamma(j\Delta\xi) \approx \frac{2Q}{D} \sum_{k=-D/2}^{D/2-1} P_\gamma\left(k\frac{2Q}{D}\right) \left|k\frac{2Q}{D}\right| W\left(k\frac{2Q}{D}\right) e^{2\pi i(jk/D)}, \tag{7.11}$$

where W is the corresponding window function. By the way, it makes no difference whether the sum in (7.11) runs from $-D/2$ to $D/2 - 1$ or from 0 to D, since the sampled signals are part of a periodic spectrum (cf. ▸ Sect. 4.18).

There are several proposals concerning the shape of the window function. From these, a selection of eight are briefly discussed, one by one, in the following paragraphs.

Taking into account the linear weighting of the whole spatial frequency domain,

$$G(q) = |q| \text{ for all } q \, , \tag{7.12}$$

this (theoretically) means no windowing; thus, $W(q) \equiv 1$ for all q. Ramachandran and Lakshminarayanan (1971) have proposed to cut the frequency band with a rectangle.

The rectangle

$$W(q) = \text{rect}(q) \, , \tag{7.13}$$

or

$$G(q) = |q| \, \text{rect}(q) \tag{7.14}$$

is in this case defined by a normalized interval $q = [-\varepsilon, \varepsilon]$ with $\varepsilon = 0.5$, just to give one example. This approach is above all used to prevent weighting with $|q|$ resulting in an excessively high noise increase in the high bands. In principle, the rectangular window must determine a unique interval for the periodic, discrete frequency band.

However, the sharp edges result in undesired sidelobes; thus, it seems to be useful to flatten the edges of the window intervals. This provides a certain degree of freedom, which is used in CT to implement a variety of windows ranging from kernels that emphasize edges to ones that yield rather smooth images.

In order to prevent an excessively sharp window edge, a cosine function (cosine I in Fig. 7.2) is applied at first to the rectangle, the argument of which ranges from $-\pi/4$ to $\pi/4$, i.e.,

$$W(q) = \text{rect}(q)\cos(\pi q/2) \tag{7.15}$$

or

$$G(q) = |q|\,\text{rect}(q)\cos(\pi q/2)\,. \tag{7.16}$$

Shepp and Logan (1974) have made a similar proposal, which is most commonly used today

$$W(q) = \text{rect}(q)\,\text{sinc}(q) \tag{7.17}$$

or

$$G(q) = |q|\,\text{rect}(q)\,\text{sinc}(q)\,. \tag{7.18}$$

The application of the *sinc* function to the rectangle proposed by *Shepp* and *Logan* yields a slightly smoother convolution kernel than that of *cosine I* in (7.16). However, as mentioned above, a multitude of windowing methods are described in the literature (Oppenheim and Schafer 1999) that are mainly aimed at making the window edge even smoother to prevent the formation of sidelobes with high energy in the convolution kernel. This may, for example, again be achieved by means of a cosine window; the argument of which ranges from $-\pi/2$ up to $\pi/2$

$$W(q) = \text{rect}(q)\cos(\pi q) \tag{7.19}$$

or

$$G(q) = |q|\,\text{rect}(q)\cos(\pi q)\,. \tag{7.20}$$

Here, attenuation in the frequency domain occurs without discontinuity such that the ripple content of the kernel can be better suppressed than with the above-mentioned functions. For many applications in which sharp edges in the spatial domain are not such a decisive factor, even smoother kernels are used. In this context the following window functions of *Hamming*, *Hanning*, and *Blackman*, respectively, should be mentioned. All of these are based on the rectangle function to which a properly selected term is added, consisting of a section of a cosine function with appropriate frequency and weighting factor. In fact, the spectrum of the cosine function partly compensates for the ripple content of the spectrum of the rectangle function.

The commonly referred to *Hamming* window is generated symmetrically around zero with the interval $q = [-\varepsilon, \varepsilon]$ where $\varepsilon = 0.5$

$$W(q) = \text{rect}(q)\,(0.54 - 0.46\cos(2\pi q)) \tag{7.21}$$

or

$$G(q) = |q|\,\text{rect}(q)\,(0.54 - 0.46\cos(2\pi q))\,. \tag{7.22}$$

Accordingly, what is called the *Hanning* window is generated by

$$W(q) = \text{rect}(q)\,(0.5 - 0.5\cos(2\pi q)) \tag{7.23}$$

or

$$G(q) = |q|\,\text{rect}(q)\,(0.5 - 0.5\cos(2\pi q))\,, \tag{7.24}$$

and what is known as the *Blackman* window by

$$W(q) = \text{rect}(q)\,(0.42 - 0.5\cos(2\pi q) + 0.08\cos(4\pi q)) \tag{7.25}$$

or

$$G(q) = |q|\,\text{rect}(q)\,(0.42 - 0.5\cos(2\pi q) + 0.08\cos(4\pi q))\,. \tag{7.26}$$

Huesman et al. (1977) describe further windows with considerable attenuation of the higher frequencies, such as the *Parzen* window, which is listed below as a final example.

$$W(q) = \begin{cases} \text{rect}(q)\left(1 - 6|q|^2(1 - |q|)\right) & \text{for } |q| \le \dfrac{\varepsilon}{2} \\ \text{rect}(q)\left(2(1 - |q|)^3\right) & \text{for } \dfrac{\varepsilon}{2} < |q| \le \varepsilon \end{cases} \tag{7.27}$$

or, respectively,

$$G(q) = \begin{cases} \text{rect}(q)\left(|q| - |q|6|q|^2(1 - |q|)\right) & \text{for } |q| \le \dfrac{\varepsilon}{2} \\ \text{rect}(q)\left(2|q|(1 - |q|)^3\right) & \text{for } \dfrac{\varepsilon}{2} < |q| \le \varepsilon\,. \end{cases} \tag{7.28}$$

Figure 7.2 illustrates the different window functions given in the frequency domain.

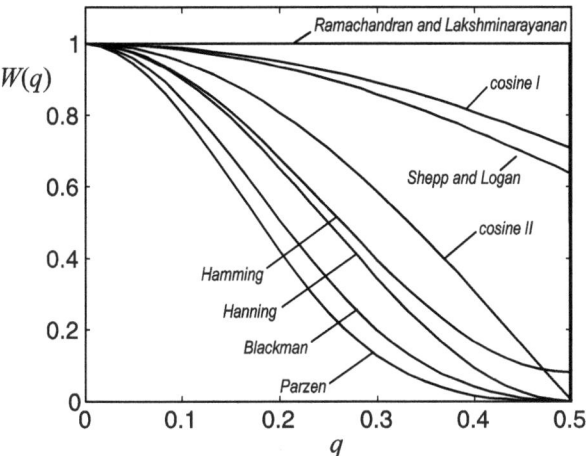

Fig. 7.2. Window functions used for band limitation by weighting the spectrum linearly

7.2.2
Convolution in the Spatial Domain

For many window functions, the kernels can be explicitly expressed in the spatial domain. As already described in ▶ Sect. 4.11, the filtering process in the frequency domain can equivalently be described in the spatial domain by a convolution. The convolution for continuous signals reads

$$h_\gamma(\xi) = \int\limits_{-\infty}^{+\infty} p_\gamma(z) g(\xi - z)\, dz \,. \tag{7.29}$$

In ▶ Sect. 5.7 the difficulties related to the integration of the Fourier integral with non-square integrable functions have already been pointed out. The function

$$g(\xi) = \lim_{\beta \to 0} g_\beta(\xi) = \lim_{\beta \to 0} \frac{\beta^2 - (2\pi\xi)^2}{(\beta^2 + (2\pi\xi)^2)^2} = -\frac{1}{(2\pi\xi)^2} \tag{7.30}$$

must be considered as a distribution with a positive δ-peak at $\xi = 0$. However, this limit consideration is not really useful in practice.

In this special case, the situation is different since in real systems band-limited signals have to always be considered such that the linear weighting function, which is of relevance in practice, reads

$$G(q) = |q| \operatorname{rect}(q) \,. \tag{7.31}$$

This weighting function has been defined above as an example on the normalized interval $q = [-\varepsilon, \varepsilon]$ with $\varepsilon = 0.5$. The band limitation describes the regularization of the reconstruction problem introduced in ▶ Sect. 6.5.3, by means of which the high frequencies are cut off. The regularization results in a sufficiently smooth filter in the spatial domain for which a closed formula is available that can be explicitly calculated.

The regularized, i.e., band-limited kernel in the spatial domain will only be calculated in detail as an example of the windowing process for the cut-off described by *Ramachandran* and *Lakshminarayanan*. For this purpose, the inverse Fourier transform,

$$g(\xi) = \int_{-\varepsilon}^{+\varepsilon} |q| e^{i2\pi\xi q} \, dq = -\int_{-\varepsilon}^{0} q e^{i2\pi\xi q} \, dq + \int_{0}^{+\varepsilon} q e^{i2\pi\xi q} \, dq \,, \tag{7.32}$$

of the ramp that is windowed by a symmetrical rectangle with a length of 2ε is calculated. The error, which arises from the fact that the high frequencies will be cut off, will then be negligible, if the frequency band of $P_y(q)$ is actually limited to the interval $q = [-\varepsilon, \varepsilon]$. This is often the case for real signals, since the projection represents an averaging process and thus acts as a low-pass filter. On simple radiographs, the poor image contrast is actually due to this averaging process.

Integral (7.32) is carried out via integration by parts, thus

$$\int q e^{aq} \, dq = q \frac{e^{aq}}{a} - \int \frac{e^{aq}}{a} \, dq$$
$$= q \frac{e^{aq}}{a} - \frac{e^{aq}}{a^2} + C = \left(\frac{aq-1}{a^2} \right) e^{aq} + C \tag{7.33}$$

so that the result for (7.32) is

$$g(\xi) = -\left[\left(\frac{2\pi i \xi q - 1}{(2\pi i \xi)^2} \right) e^{2\pi i \xi q} \right]_{-\varepsilon}^{0} + \left[\left(\frac{2\pi i \xi q - 1}{(2\pi i \xi)^2} \right) e^{2\pi i \xi q} \right]_{0}^{\varepsilon} . \tag{7.34}$$

If the limits are substituted, the above equation reads

$$g(\xi) = -\left(\frac{-1}{(2\pi i \xi)^2} \right) + \left(\frac{-2\pi i \xi \varepsilon - 1}{(2\pi i \xi)^2} \right) e^{-2\pi i \xi \varepsilon} + \left(\frac{2\pi i \xi \varepsilon - 1}{(2\pi i \xi)^2} \right) e^{2\pi i \xi \varepsilon} - \left(\frac{-1}{(2\pi i \xi)^2} \right) . \tag{7.35}$$

An evaluation of the bracketed expressions yields

$$g(\xi) = \frac{1}{(2\pi i \xi)^2} - \frac{\varepsilon}{2\pi i \xi} e^{-2\pi i \xi \varepsilon} - \frac{1}{(2\pi i \xi)^2} e^{-2\pi i \xi \varepsilon} + \frac{\varepsilon}{2\pi i \xi} e^{2\pi i \xi \varepsilon}$$
$$- \frac{1}{(2\pi i \xi)^2} e^{2\pi i \xi \varepsilon} + \frac{1}{(2\pi i \xi)^2} . \tag{7.36}$$

After sorting these expressions

$$g(\xi) = \frac{\varepsilon}{2\pi i \xi} e^{2\pi i \xi \varepsilon} - \frac{\varepsilon}{2\pi i \xi} e^{-2\pi i \xi \varepsilon} - \left(\frac{1}{(2\pi i \xi)^2} e^{2\pi i \xi \varepsilon} - \frac{2}{(2\pi i \xi)^2} + \frac{1}{(2\pi i \xi)^2} e^{-2\pi i \xi \varepsilon} \right), \tag{7.37}$$

the following *sine* or *sinc* expressions can be found in the above equation:

$$g(\xi) = \frac{\varepsilon}{\pi\xi} \left(\frac{e^{2\pi i \xi \varepsilon} - e^{-2\pi i \xi \varepsilon}}{2i} \right) - \left(\frac{1}{\pi\xi} \right)^2 \left(\frac{e^{\pi i \xi \varepsilon} - e^{-\pi i \xi \varepsilon}}{2i} \right)^2$$

$$= \frac{2\varepsilon^2 \sin(2\pi\xi\varepsilon)}{2\pi\xi\varepsilon} - \varepsilon^2 \left(\frac{\sin(\pi\xi\varepsilon)}{\pi\xi\varepsilon} \right)^2 \tag{7.38}$$

$$= 2\varepsilon^2 \, \text{sinc}(2\xi\varepsilon) - \varepsilon^2 \, \text{sinc}^2(\xi\varepsilon) \,.$$

With $\varepsilon = 1/(2\Delta\xi)$ this results in

$$g(\xi) = \frac{2\left(\frac{1}{2\Delta\xi}\right)^2 \sin(\pi\frac{\xi}{\Delta\xi})}{\pi\frac{\xi}{\Delta\xi}} - \left(\frac{1}{2\Delta\xi}\right)^2 \left(\frac{\sin(\pi\frac{\xi}{2\Delta\xi})}{\pi\frac{\xi}{2\Delta\xi}} \right)^2 \tag{7.39}$$

and with the double angle formula for the *cosine*

$$\cos(2x) = 1 - 2\sin^2(x) \,, \tag{7.40}$$

this can be changed to

$$\sin^2(x) = \frac{1}{2}\left(1 - \cos(2x)\right) \,, \tag{7.41}$$

which finally gives

$$g(\xi) = \frac{1}{2\left(\Delta\xi\right)^2} \left(\frac{\sin(\pi\frac{\xi}{\Delta\xi})}{\pi\frac{\xi}{\Delta\xi}} + \frac{\cos\left(\pi\frac{\xi}{\Delta\xi}\right) - 1}{\left(\pi\frac{\xi}{\Delta\xi}\right)^2} \right) \,. \tag{7.42}$$

In analogy to the previous calculation, the kernel of *Shepp and Logan* can be calculated by

$$g(\xi) = -\frac{2}{\pi^2\Delta\xi^2} \frac{1 - 2\frac{\xi}{\Delta\xi}\sin\left(\pi\frac{\xi}{\Delta\xi}\right)}{4\left(\frac{\xi}{\Delta\xi}\right)^2 - 1} \tag{7.43}$$

(this result was given by Morneburg [1995]). Expressions for the *Hanning* and *Hamming* windows are listed by Huesman et al. (1977). As a final example, the kernel of the *Parzen* window should be mentioned.

$$g(\xi) = \frac{\begin{array}{c} 24\pi\frac{\xi}{\Delta\xi}\cos\left(\pi\frac{\xi}{\Delta\xi}\right) - 96\sin\left(\pi\frac{\xi}{\Delta\xi}\right) - 48\pi\frac{\xi}{\Delta\xi}\cos\left(\pi\frac{\xi}{2\Delta\xi}\right) \\ + 384\sin\left(\pi\frac{\xi}{2\Delta\xi}\right) - 2\left(\pi\frac{\xi}{\Delta\xi}\right)^3 - 72\pi\frac{\xi}{\Delta\xi} \end{array}}{4\pi^5 \frac{\xi^5}{(\Delta\xi)^3}} \tag{7.44}$$

Figure 7.3 shows the corresponding band-limited filter in the frequency and spatial domains respectively for the prominent windowing functions discussed above.

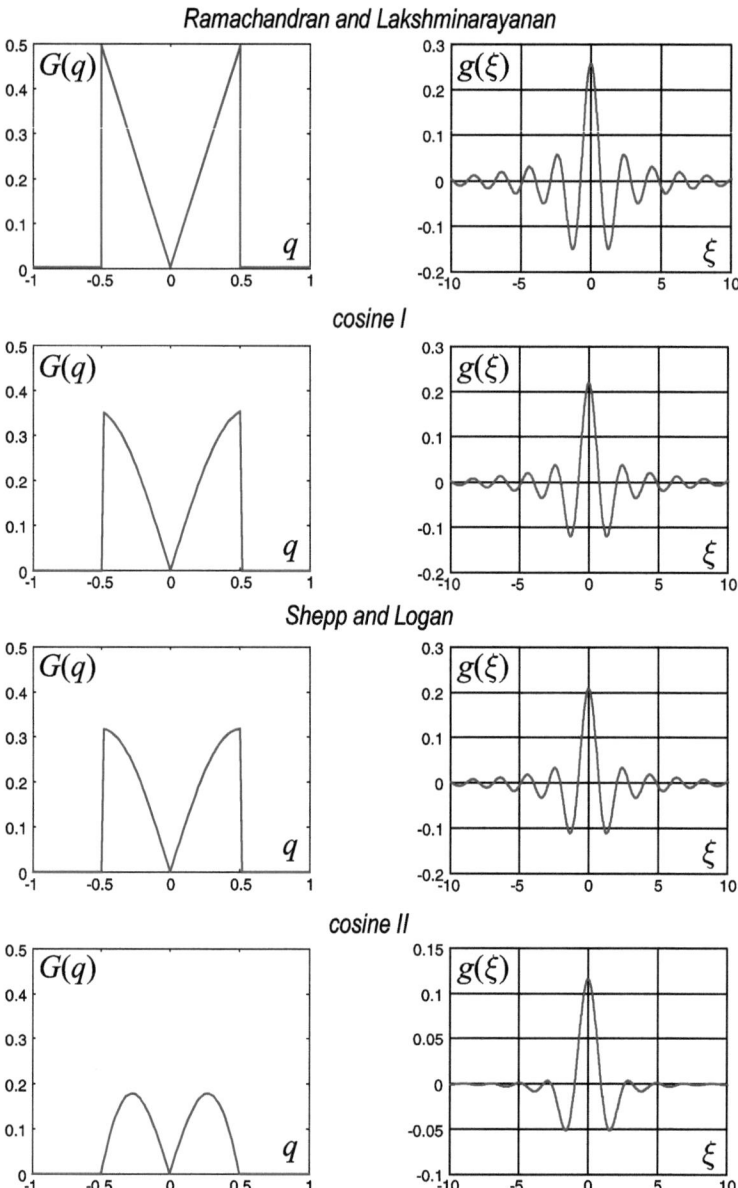

Fig. 7.3. Band-limited window functions of the filtered backprojection in the frequency domain (always *left-hand side*) with the corresponding Fourier transforms in the spatial domain, i.e., the corresponding convolution kernels (always *right-hand side*). The sampling interval is set to $\Delta\xi = 1$

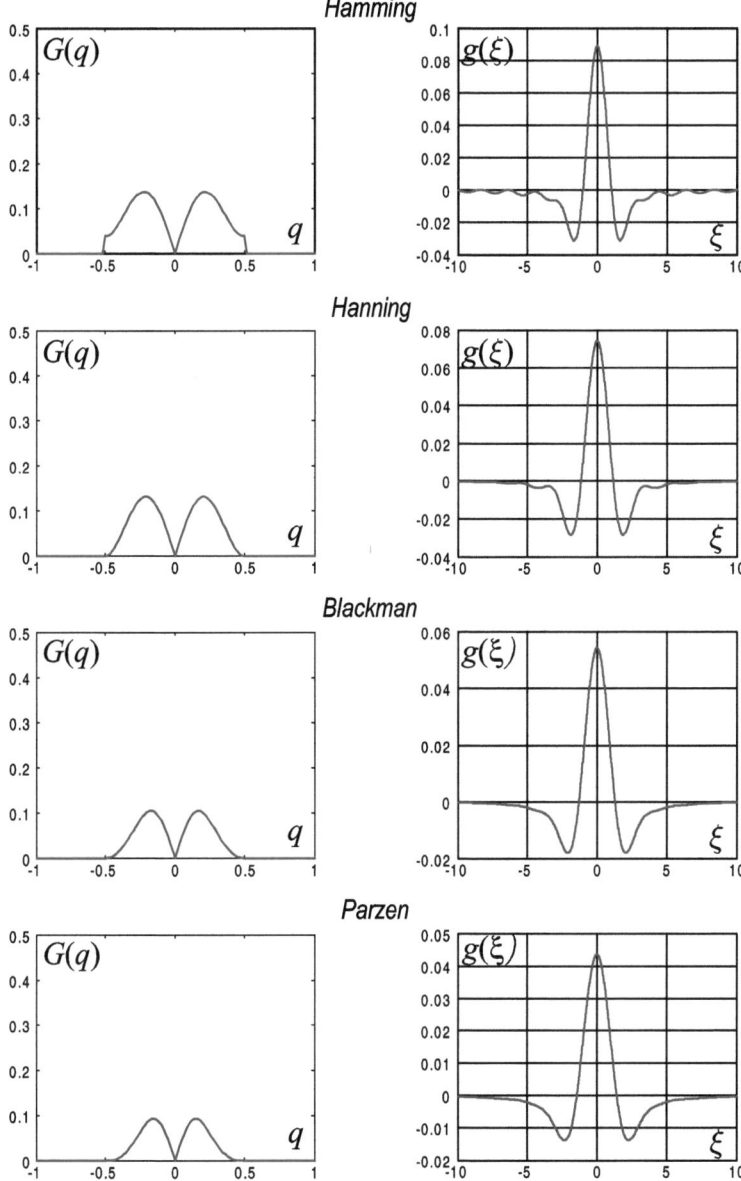

Fig. 7.3. Band-limited window functions of the filtered backprojection (continued)

The common aspect of the different windowing functions is that even spatial frequencies below the maximum interesting frequency are attenuated. For this reason, the spatial resolution of the reconstructed image decreases. These 'soft' convolution kernels are frequently used in nuclear medicine (Morneburg 1995).

The filter of *Ramachandran* and *Lakshminarayanan* provides the best spatial resolution in the reconstructed images. However, as already indicated above, noise in the reconstructed image is considerably increased with real data, i.e., with noisy projection signals. Furthermore, in regions with considerable contrast differences, the sharp limitation in the frequency domain results in oscillating intensity values in the spatial domain, which appear as artifacts in the image. With its overshoots, this phenomenon, commonly referred to as the *Gibbs* phenomenon, contaminates the point to be reconstructed with neighboring points of the reconstructed image.

If smoother windows are applied in the frequency domain, the number of oscillations in the spatial domain is reduced more and more. On the other hand, the central maximum of the convolution kernel becomes broader and smaller as the selected window becomes softer. A comparison between the filter proposed by *Ramachandran* and *Lakshminarayanan* with a maximum value

$$g(0) = \frac{1}{(2\Delta\xi)^2} \qquad (7.45)$$

and the filter proposed by *Parzen* with a maximal value of

$$g(0) = \frac{0.175}{(2\Delta\xi)^2} \qquad (7.46)$$

(cf. Huesman et al. [1977]) illustrates the conditions (cf. Fig. 7.3).

Artifacts in regions with a sharp contrast are prevented by using the kernels described and, furthermore, noise is considerably suppressed. However, the spatial resolution is impaired. Therefore, the selection of the filter is always a trade-off between resolution and noise in the reconstructed image. The selection of the filter to be used will always depend on the clinical application. A multitude of filters, for convenience subdivided into anatomical groups, are implemented on modern CT scanners. These filters have to be selected by the operator.

The kernel proposed by *Shepp* and *Logan* is nowadays accepted as a standard kernel. Regarding the main maximum, it is almost identical to the ideal filter proposed by *Ramachandran* and *Lakshminarayanan*. The *sinc* function, which is additionally applied to the rectangle, counteracts the sidelobes such that the artifacts are significantly reduced with just a slightly reduced spatial resolution.

7.2.3
Discretization of the Kernels

The continuously defined convolution

$$h_\gamma(\xi) = \int\limits_{-\infty}^{+\infty} p_\gamma(z) g(\xi - z) \, dz \,, \qquad (7.47)$$

can be expressed in its discrete form as follows:

$$h_\gamma\left(j\frac{1}{2\varepsilon}\right) = h_\gamma(j\Delta\xi) = \Delta\xi \sum_{n=-D/2}^{D/2-1} p_\gamma(n\Delta\xi)g((j-n)\Delta\xi) . \tag{7.48}$$

Evaluating the inverse transform of the windowed ramp at the sampling points (which are determined by the window width) results in the discretized version of the kernels $g(n)$.

If the definition of the *sinc* function,

$$\text{sinc}(x) = \frac{\sin(\pi x)}{\pi x} , \tag{7.49}$$

is applied to the kernel of *Ramachandran* and *Lakshminarayanan* (7.38), this becomes

$$g(\xi) = 2\varepsilon^2 \text{sinc}(2\varepsilon\xi) - \varepsilon^2 \text{sinc}^2(\varepsilon\xi) . \tag{7.50}$$

Sampling at the points $\xi = n\Delta\xi$, combined with the replacement of the sampling rate by $\varepsilon = 1/(2\Delta\xi)$, yields the expression

$$g(n\Delta\xi) = \frac{2}{(2\Delta\xi)^2} \text{sinc}(n) - \frac{1}{(2\Delta\xi)^2} \text{sinc}^2\left(\frac{n}{2}\right)$$

$$= \frac{1}{4\Delta\xi^2}\left(2\text{sinc}(n) - \text{sinc}^2\left(\frac{n}{2}\right)\right) . \tag{7.51}$$

From this, the discrete form of the convolution kernel can be found,

$$g(n\Delta\xi) = \begin{cases} \dfrac{1}{(2\Delta\xi)^2} & \text{for } n = 0 \\ 0 & \text{for } n \text{ even}, (n \neq 0), \\ -\dfrac{1}{(n\pi\Delta\xi)^2} & \text{for } n \text{ odd}, \end{cases} \tag{7.52}$$

since both *sinc* terms disappear with even n. For odd n, only the first *sinc* term is always zero.

The discretized kernel of *Shepp* and *Logan* can be calculated analogously by

$$g(n\Delta\xi) = -\frac{2}{\pi^2\Delta\xi^2}\frac{1-2n\sin(\pi n)}{4(n)^2-1} . \tag{7.53}$$

For integers, n, this yields

$$g(n\Delta\xi) = -\frac{2}{(\pi\Delta\xi)^2}\frac{1}{4n^2-1} . \tag{7.54}$$

Figure 7.4 shows the discretized kernels for *Ramachandran* and *Lakshminarayanan*, *Shepp* and *Logan*, *Hamming*, as well as the *Blackman* filters. For all the filters, only

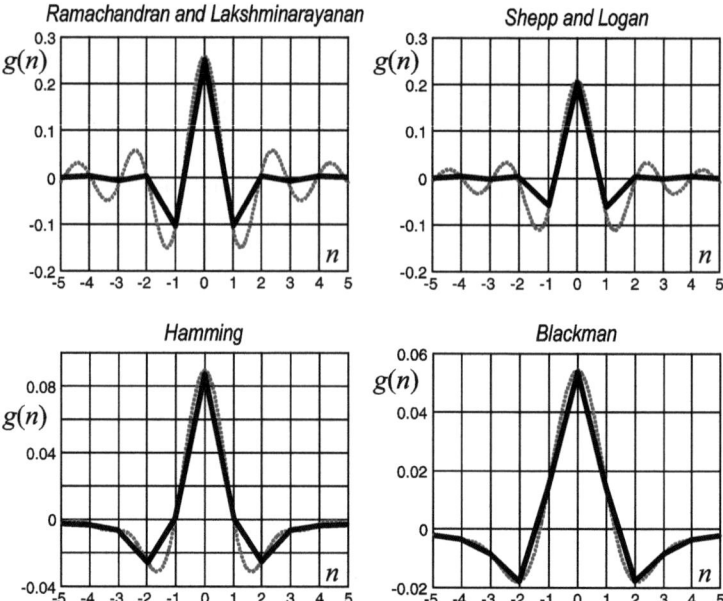

Fig. 7.4. Discretized band-limited kernels. The corresponding interpolated values of the kernels are shown as *fine dotted lines* in the background

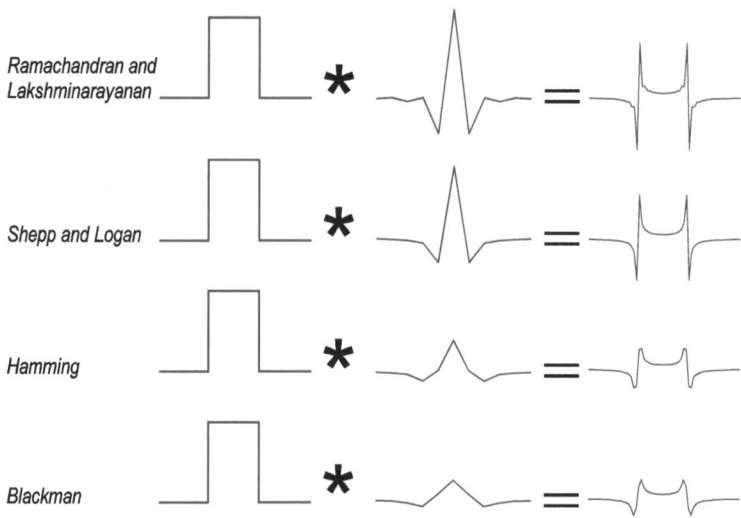

Fig. 7.5. Impact of the different convolution kernels upon a rectangular signal. The kernels become smoother from top to bottom

the value for $n = 0$ is positive. Further kernels are explicitly described in Huesman et al. (1977).

For the other integer values of n, the kernel data are negative or zero. Due to the negative kernel values, blurring of the gray values of an object into neighboring regions is compensated for. Figure 7.5 shows the result of the discretized kernels applied to a rectangular profile. In this figure, the kernels and the convolution results are uniformly normalized and thus directly comparable. Obviously, the filters become smoother from top to bottom.

7.3
Practical Implementation of the Filtered Backprojection

The concept of the discrete filtered backprojection will once again be formulated in this section. For this purpose, let the projection signal sampled at a projection angle γ_n be

$$p_{\gamma_n}(j\Delta\xi) \text{ with } j = -D/2, \ldots, 0, \ldots, D/2 - 1 \text{ and } n = 1, \ldots, N_p , \qquad (7.55)$$

where D is the number of detector elements, which are densely packed with a distance of $\Delta\xi$ from each other. N_p is the number of projections or views.

7.3.1
Filtering of the Projection Signal

One considers the projections, $p_{\gamma_n}(j\Delta\xi)$, measured under the projection angles, γ_n, and applies a high-pass filter at each γ_n. This corresponds to a multiplication of $P_{\gamma_n}(k\Delta q)$ with $|k\Delta q|$ in the frequency domain. The periodicity of the signal $P_{\gamma_n}(k\Delta q)$ must be taken into account here (cf. Fig. 7.6). In Matlab^{TM} the D frequency values are, for example, distributed such that the first value corresponds to

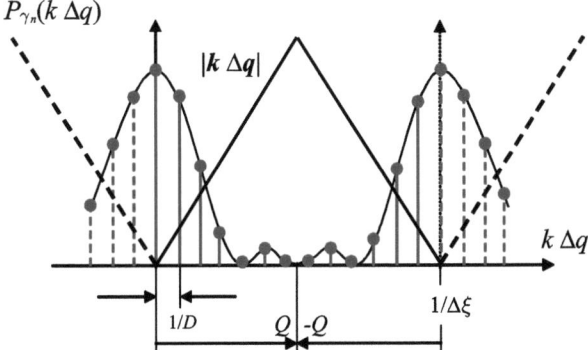

Fig. 7.6. Due to the sampling process, the spectrum of the projection signal is a periodic one. This must be taken into account when the signal is actually filtered in the frequency domain

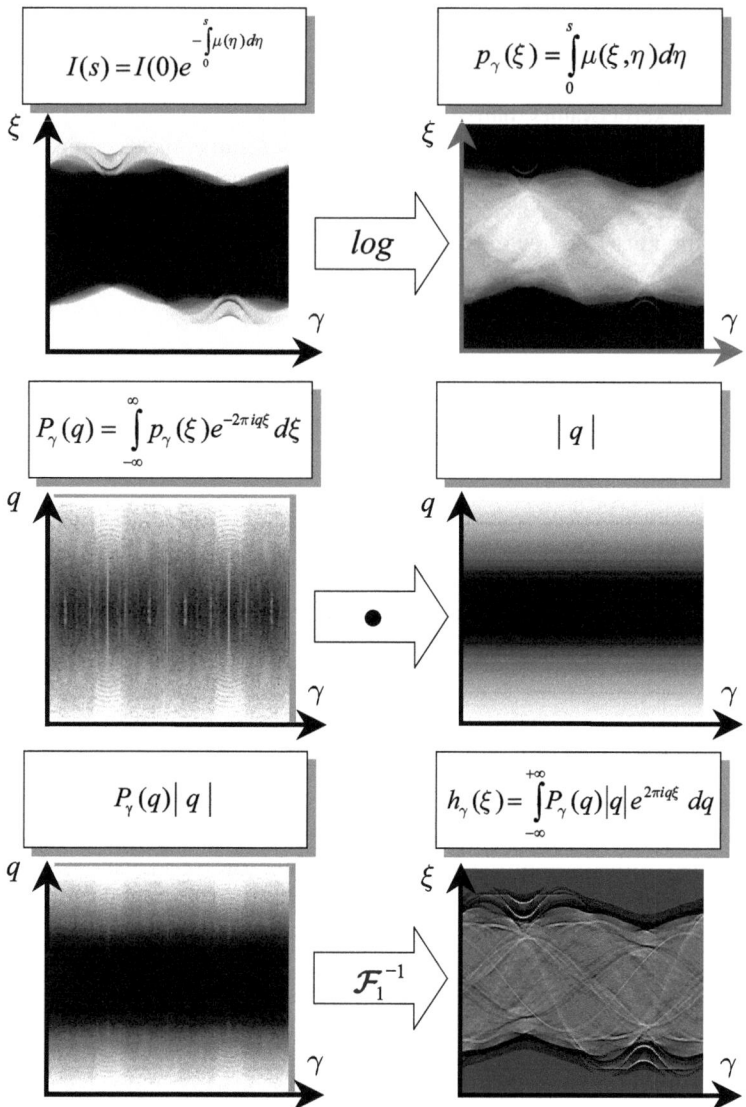

Fig. 7.7. Processing chain of the raw data. As a first step, the measured intensity is converted by logarithmation into the Radon space, the sinogram. Along the detector variable, ξ, one performs the one-dimensional Fourier transform for each projection angle, γ, and then weights each projection by multiplying it with the linear ramp $|q|$ in the frequency domain. The one-dimensional inverse Fourier transform along the q coordinate then yields the required filtered projection

the zero frequency quantity (i.e., the DC part at frequency $q = 0$). Then, the positive values range up to $k = D/2 - 1$. One finds the negative share of the spectral values seamless, i.e., the value $k = D/2$ corresponds to the frequency $q = -Q$. If one wants

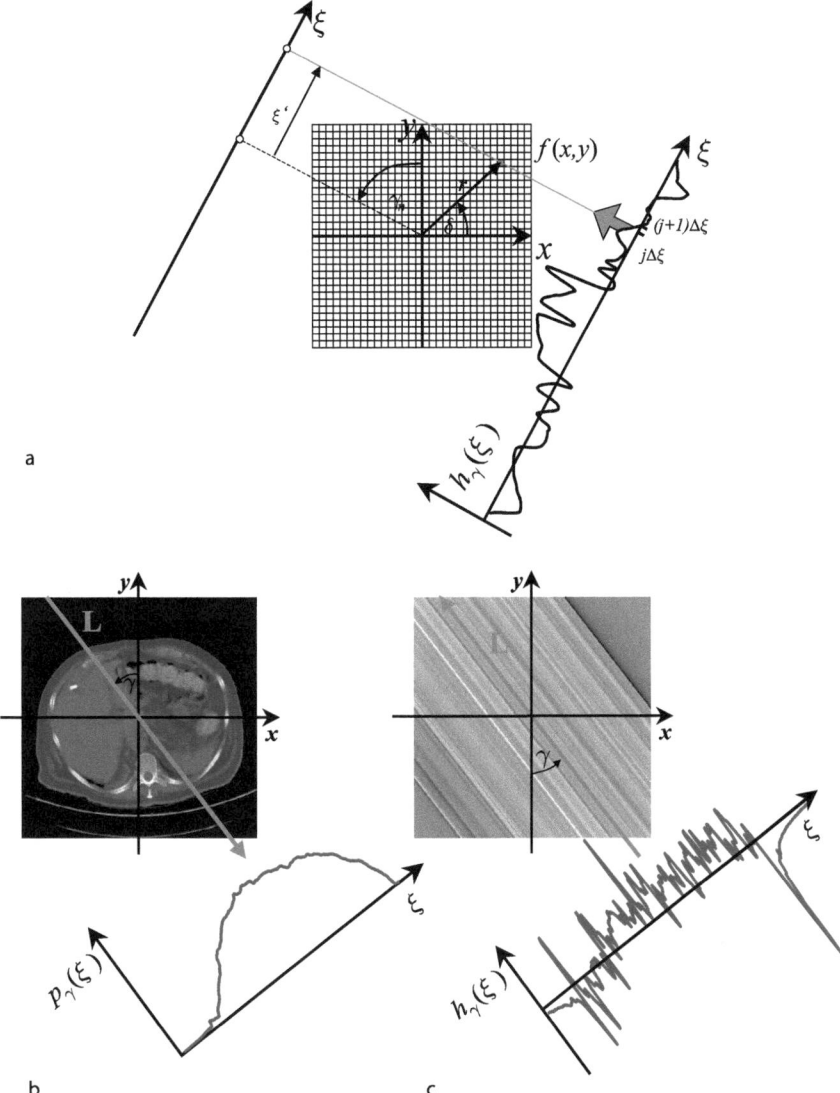

Fig. 7.8. a In general, the value of the filtered projection signal to be backprojected must be calculated by interpolation since the vector **r** points toward the grid position $(x, y)^{\mathrm{T}}$, which yields a value ξ' in the projection on the ξ'-axis ranging between the values $j\Delta\xi$ and $(j+1)\Delta\xi$. **b, c** show the projection and the related filtered backprojection under a single projection angle $\gamma = 40°$. The complete Radon space data are illustrated in the previous Fig. 7.7

to implement the multiplication of $P_{\gamma_n}(k\Delta q)$ with $|k\Delta q|$ in the frequency domain, the different intervals must be properly distinguished using Matlab^{TM}. Figure 7.7 illustrates, once again, the whole signal processing chain from the measured intensity to the filtered projection, based on real CT data of an abdominal section (cf. Fig. 7.8).

7.3.2
Implementation of the Backprojection

For an angle γ_n, every pixel in the (x, y) plane corresponds to the projection onto the detector axis ξ via the relation

$$\xi' = \left(\mathbf{r}^T \cdot \mathbf{n}_\xi\right) = x\cos(\gamma_n) + y\sin(\gamma_n) \ . \tag{7.56}$$

The high-pass filtered projection signal, $h_{\gamma_n}(\xi)$, has to be plotted versus the projection axis, ξ, for every angle, γ_n. Along line L, which is again given in the Hessian normal form (7.56) by all data points $\mathbf{r} = (x, y)^T$ satisfying $(\mathbf{r}^T\mathbf{n}_\xi) = \xi'$, the corresponding values $h_{\gamma_n}(\xi)$ are now to be plotted into the spatial domain of the data points $(x, y)^T$ for all ξ. It can easily be seen that the expression $\xi' = x\cos(\gamma_n) + y\sin(\gamma_n)$ does not necessarily result in a value ξ, where $h_{\gamma_n}(j\Delta\xi)$ is directly available. Figure 7.8a illustrates this situation. The corresponding $h_{\gamma_n}(\xi')$ can then be derived, for instance, by means of interpolation (e.g., nearest neighbor, linear, cubic, spline, etc.).

Thus, the filtered projection is "smeared" in a reverse direction over the spatial domain under the projection angle γ_n, i.e., along the original X-ray beams. Figure 7.8b shows the original projection for a tomogram of the abdomen with the projection angle $\gamma = 40°$. Figure 7.8c provides a single corresponding "smearing" in reverse direction of the filtered projection signal. This has to be done for all angles, γ_n, and these backprojections must consecutively be added.

7.4
Minimum Number of Detector Elements

The geometrical situation, which is found during the sampling process, is illustrated in Fig. 7.9. Without loss of generality, here the pencil-beam geometry of the first-generation CT scanners is used, because this geometry is easier to survey. The object space is illuminated along parallel X-ray lines with a constant angular spacing, $\Delta\gamma$. The spacing of the D sampling elements of the detector array is denoted by $\Delta\xi$.

In ▸ Sect. 4.18 it has already been explained that the maximum available frequency q_{max} in the data spectrum must be less than half the sampling rate $(\Delta\xi)^{-1}$, i.e.,

$$q_{max} < \frac{1}{2\Delta\xi} \ . \tag{7.57}$$

This is the *Nyquist* criterion. Thus, the spatial sampling interval obeys

$$\Delta\xi = \frac{1}{D\Delta q} < \frac{1}{2q_{max}} \ . \tag{7.58}$$

The corresponding polar Radon space represents the line integrals of the scan with the geometrical location of the sampling points in the spatial domain. The diameter of the Radon space therefore directly corresponds to the measurement field diameter (MFD) of the spatial domain. If $\Delta\xi$ is determined by q_{max}, then a desired MFD yields a minimum number of detectors

$$q_{max} < \frac{D\Delta q}{2} = \frac{D}{2 MFD} = \frac{1}{2\Delta\xi} \Rightarrow D_{min} > 2\, MFD\, q_{max}\; . \tag{7.59}$$

If it is assumed that the object to be reconstructed is a body with a rectangular projection profile, then the spectrum is again determined by a *sinc* function (the first root of which is to represent the maximum available relevant frequency).

If the object has the minimum diameter d_{min}, then the estimation of the maximum available frequency yields $q_{max} = (d_{min})^{-1}$. This rule of thumb will be discussed again in ▸ Sect. 7.8. Consequently, the sampling distance is $\Delta\xi = (D\Delta q)^{-1} < 0.5 d_{min}$. This results in the estimation

$$D_{min} > \frac{2\, MFD}{d_{min}} \tag{7.60}$$

for the minimum number of detectors in one projection.

7.5
Minimum Number of Projections

The Fourier slice theorem ensures that the Fourier transforms of the projections are able to fill up the entire frequency domain of the object. However, Fig. 7.9 shows that the distance of neighboring radial data lines in the Radon domain, (ξ, γ), just amounts to $\xi\Delta\gamma$. Therefore, the maximum distance of band-limited projections is $0.5\, MFD\, \Delta\gamma$. For an object with a maximum diameter d_{max}, the maximum distance in the frequency domain is of course limited.

With (7.60) it has been argued that

$$\frac{D_{min}}{2\, MFD} > q_{max}\; . \tag{7.61}$$

The number of projections is generally determined by

$$N_p = \frac{\pi}{\Delta\gamma}\; . \tag{7.62}$$

If, in addition, one takes into account the distance between the data points in the frequency domain of the individual projections, which is determined by the total length of the detector array

$$\Delta q = \frac{1}{D_{min}\Delta\xi} = \frac{1}{MFD}\; , \tag{7.63}$$

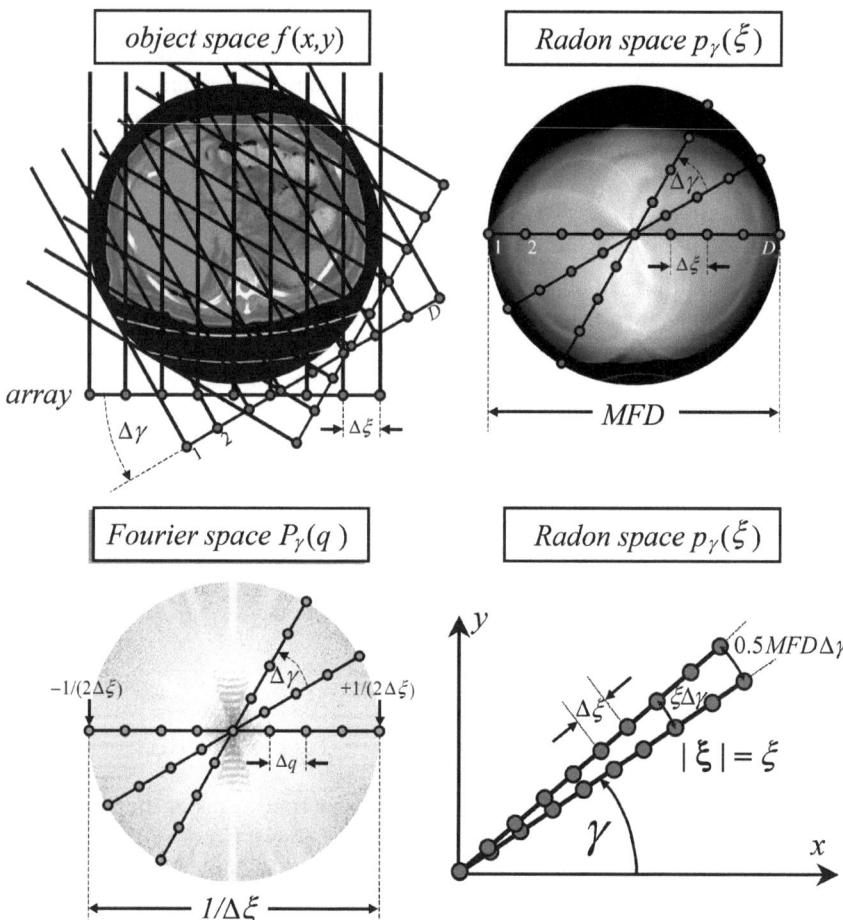

Fig. 7.9. Geometrical location of the sampling points in the object space, Radon space, and frequency domain. The number of required sampling points (and thus the number of detector elements D and the number of projection directions N_p) is estimated by Shannon's sampling theorem

and if it is further assumed that the maximum distance,

$$0.5\,MFD\,\Delta\gamma \approx \Delta\xi\,, \tag{7.64}$$

perpendicular to the ξ data line in the (ξ, γ) space should approximately correspond to the distance between the data points in the Radon domain of a single projection (cf. Fig. 7.9, lower right side), then the estimation

$$N_p = \frac{\pi}{\Delta\gamma} \approx \pi \frac{MFD}{2\Delta\xi} \tag{7.65}$$

holds for the minimum number of projections. According to (7.63) the right-hand side of the estimation (7.65) thus reads

$$\pi \frac{MFD}{2\Delta\xi} = \pi \frac{\Delta\xi D_{min}}{2\Delta\xi} \, , \qquad (7.66)$$

so that finally one obtains

$$N_p = \frac{\pi}{\Delta\gamma} \approx \pi \frac{D_{min}}{2} \, . \qquad (7.67)$$

As a practicable rule of thumb one finds that the requirements stipulated in the sampling theorem are met if

$$N_p \approx D_{min} \, . \qquad (7.68)$$

Due to the typically rectangular sensitivity profile of detector elements, a detector quarter shift resulting in a duplication of the spatial resolution in a 360° rotation of the sampling unit – which will be discussed in detail in ▸ Sect. 7.8 – must still be taken into account.

7.6
Geometry of the Fan-Beam System

This chapter, first of all, describes the geometry of a CT scanner of the third generation. Due to the latest developments in the field of the two-dimensional flat panel detectors, it must be assumed that the geometry of the third generation – that means a sampling unit consisting of an X-ray source and a detector array mounted on the same rotation disk – will make its way. For single-slice detector arrays located on a circular arc with the center in the X-ray focus inside the X-ray tube, the geometrical variables outlined in Fig. 7.10 play an important role.

Strictly speaking, the CT scanners of the third generation must still be subdivided into those with curved detector arrays (with equidistant angles between the detector elements) and plane detector arrays (with equidistant detector spacing). The corresponding derivations for image reconstruction with both detector types are described in the following sections.

In these systems the X-ray source moves along a circle trajectory determined by the coordinates $(- FCD \sin(\theta), FCD \cos(\theta))$ – as long as the CT sampling unit rotates counterclockwise, i.e., in a mathematically positive sense. Here FCD denotes the focus center distance. The angle θ is the projection angle defined by the central beam. In the curved detector arrays all detector elements have the same focus center distance (FDD) from the source.

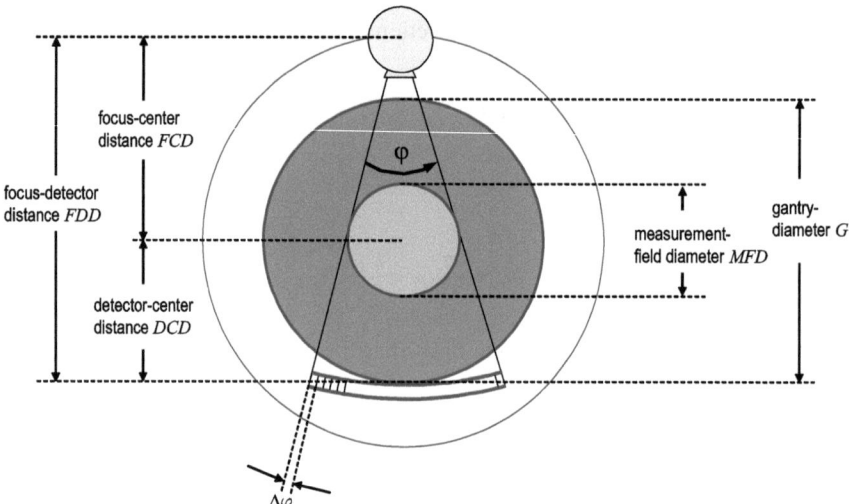

Fig. 7.10. Fan-beam geometry in CT scanners of the third generation. Curved detector arrays with constant $\Delta\varphi$ are very important in clinical practice

7.7
Image Reconstruction for Fan-Beam Geometry

The conceptional philosophies of the different CT generations have already been pointed out in ▸ Chap. 3. The differences in the reconstruction methods between the first and the third generation will be discussed in detail in this section. ▸ Chap. 5 described the reconstruction methods for the geometry of a pencil-beam system, i.e., for parallel X-ray beams within one projection angle. Within this geometry, the reconstruction methods are easier to understand than those for fan-beam geometry.

A serious practical disadvantage of pencil-beam geometry is, in fact, the rather difficult mechanical data acquisition process with separate rotary and shift steps, which would result in an unacceptably high data acquisition time in clinical practice. In CT scanners of the third generation, there is just one single synchronous rotation of the X-ray tube and the detector array, both of which are combined on a sampling unit, i.e., a rotating inner disk. With this construction the acquisition time is considerably reduced since the modern slip-ring technology enables a continuous rotation of the inner CT sampling unit without the previously required pauses for linear offset.

This chapter will now discuss the mathematical differences in the reconstruction methods with respect to the transition from pencil-beam to fan-beam geometry. Figures 7.11 and 7.12 provide the difference between the Radon spaces of a parallel-beam and a fan-beam system.

In order to ensure easily traceable projection results, the following approach is applied to a phantom consisting of two squares of different sizes and one circle. All objects have the same attenuation properties, thus simulating identical material.

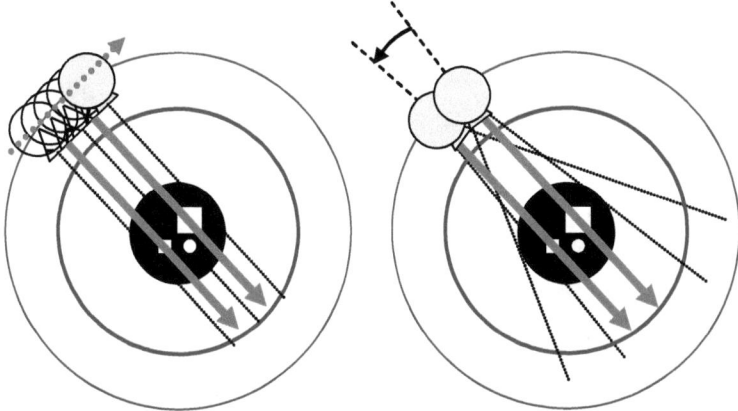

Fig. 7.11. Comparison of parallel-beam and fan-beam geometry for the simple phantom consisting of three objects. The corresponding longest X-ray beam paths through the diagonal of the square objects are symbolized by thicker *red lines*. While these paths are found under a single projection angle with parallel-beam geometry, in fan-beam geometry these rays are obviously located in fans of different projection angles

On the upper right-hand side of Fig. 7.12, a polar representation of the Radon space with parallel-beam geometry is shown. A direct comparison of the two projection geometries will be based on the Cartesian Radon spaces, which have already been introduced. The Cartesian representation of the projection values that is equivalent to the polar representation is shown on the lower left side of Fig. 7.12. Here, parallel-beam geometry is given in a range from $0°$ up to $360°$.

The data in the Cartesian Radon space are rather simple to interpret. The projections start at $0°$ (12 o'clock position of the tube) and, subsequently, the X-ray tube is moved in the counterclockwise sense, i.e., in a mathematically positive direction, around the phantom. First of all, at $0°$ the projection of the circle is added to the projection of the large square. The projection values of the small square are separated in the projection at $0°$ from the values of the other objects. Because the rotation is positive in a mathematical sense, the Radon value with the highest score is reached at $135°$ (and then later in the opposite direction, which means at $315°$).

The detector array central value, $p_{135°}(0) = p_{315°}(0)$, is the peak value, because the longest path through the diagonals of the two square objects, arranged one behind another, has been traveled on the corresponding line of the X-ray beam. At first glance, the Radon representations of the data are the same for both parallel-beam and fan-beam geometry. In particular, it becomes evident that for both representations the value for $p_{135°}(0) = p_{315°}(0)$ is a peak value. The central projection is actually identical in both Radon spaces. However, differences arise with increasing distance from the central projection.

The dashed lines in Fig. 7.12 connecting prominent points are intended to direct the eyes. These points can be easily interpreted with parallel-beam geometry at an angle of $45°$. In ▶ Sect. 5.8 the projection data profiles for a square object have al-

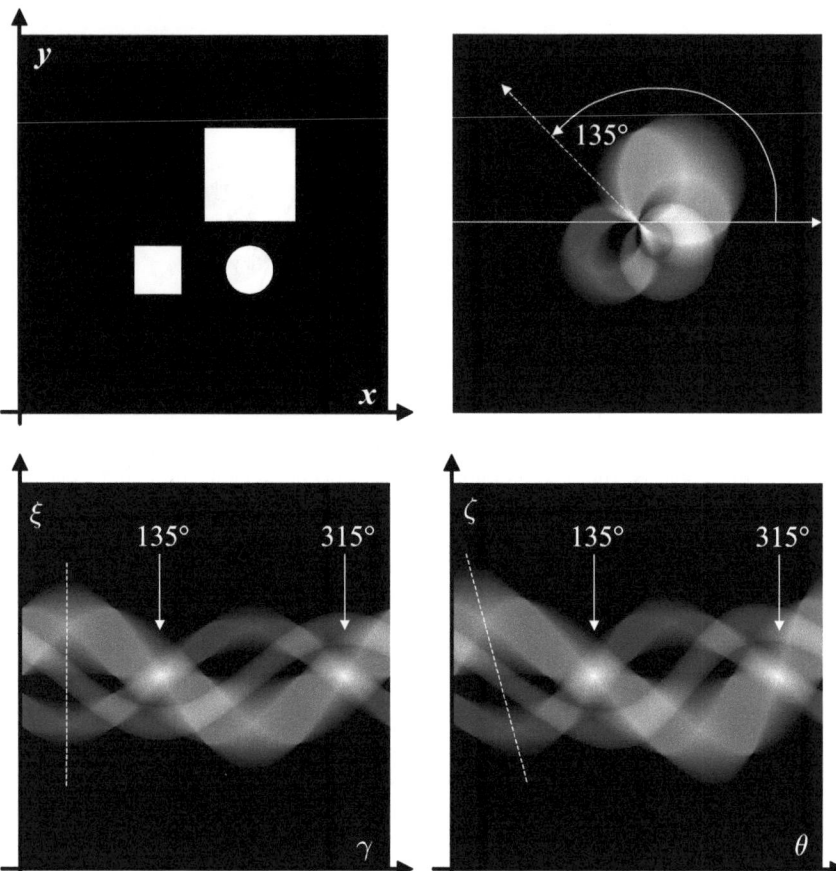

Fig. 7.12. Comparison between the Radon spaces of parallel-beam and fan-beam geometry. The *upper left-hand side* shows a simple phantom. It consists of two *squares* of different sizes and a *circle*. All objects have identical attenuation coefficients, which are homogeneously distributed. On the *upper right* and *lower left* the Radon space of parallel-beam geometry with its polar or Cartesian representation can be seen respectively. On the *lower right* the projection values in the Radon space of the phantom with fan-beam geometry are shown. The two *lower* pictures are identical only at first glance, particularly because the highest attenuation values at 135° and 315° are located at the same point. With an increasing distance from the central projection, however, there are indeed differences. A *dashed line* connecting prominent positions of the diagonal lines passing through the objects in the two Radon space representations is intended to direct the eyes

ready been discussed. With an angle of 0° one finds a uniform intensity throughout the rectangle profile and at 45° a corresponding triangle profile.

In the Radon space of parallel-beam geometry, one finds the triangle profile of both square objects at an angle of 45°. With fan-beam geometry, this point appears earlier for the larger square and later for the smaller square. Figure 7.11 illustrates this phenomenon. The beams with the longest paths through the diagonal lines of the square objects are located inside one projection angle with parallel-beam geometry. However, within fan-beam geometry one never finds the beams with the longest paths inside one fan, but in different fans with slightly shifted projection angles.

7.7.1
Rebinning of the Fan Beams

Figure 7.13 schematically outlines the transition from fan-beam to pencil-beam geometry. The rotation of the central beam of the fan is considered about the angle $\theta = \theta_2 - \theta_1$. During the rotation of the sampling unit the X-ray focus and the detector form concentric circles, which generally have different radii. It is obviously possible to find an X-ray beam in the projection, $\phi_{\theta_2}(\zeta)$, that is parallel to one within the fan of beams, $\phi_{\theta_1}(\zeta)$. Synthetic pencil-beam geometry may thus be reconstructed by rebinning the corresponding beams of the fans with different projection angles and by recombining them.

When rebinning the fan beams, the geometrical boundary conditions must of course be carefully considered. Parameters are the fan, the aperture angle, or the length of the detector array. Figure 7.14 shows that the boundary beams of pencil-beam geometry may not be generated by rebinning with a limited aperture angle of the fan because they simply do not exist with other projection angles.

Figure 7.15 illustrates that the maximum achievable distance between the boundary beams of synthesized pencil-beam geometry, this means the distance between the first beam $\phi_{\theta_1}(\zeta_1)$ of the projection with the angle θ_1 and the last beam $\phi_{\theta_2}(\zeta_D)$ of the projection with the angle θ_2, is directly determined by the aperture angle of the fan. This is obviously due to the fact that the related maximum change of the projection angle, $\theta = \psi_{max} + \psi_{min}$, exactly corresponds to the aperture angle, φ. In

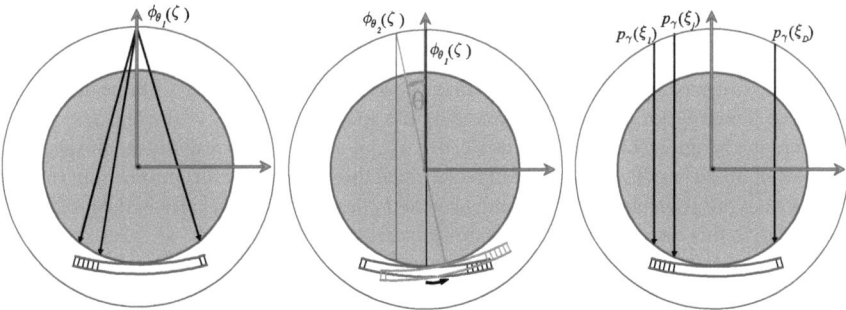

Fig. 7.13. *Left*: CT scanner with fan-beam geometry. X-ray focus and detector array rotate synchronously around a common center – the iso-center in the measurement field. *Center*: After one rotation of the sampling unit about the angle θ the beams in the new fan, which are parallel to the beams of the old fan, are to be found. *Right*: Pencil-beam geometry may be synthesized by rebinning the projection values from different projection angles

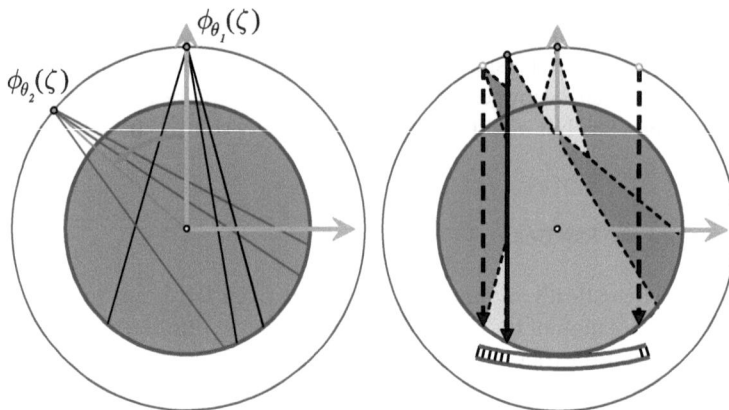

Fig. 7.14. On the *left* of the picture it is shown that the maximum change of the projection angle, θ, is not independent of the aperture angle of the fan for the rebinning process. On the *right* the two outer parallel beams (*dashed lines*) cannot be generated by rebinning with a limited aperture angle

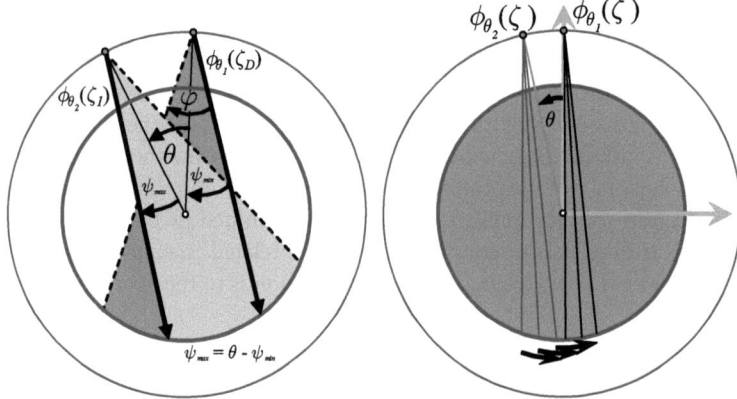

Fig. 7.15. The maximum achievable distance between the outer boundary beams of the synthesized pencil-beam geometry is limited by the aperture angle, φ, of the fan. The related maximum change of the projection angle, $\theta = \psi_{\max} + \psi_{\min}$, just corresponds to the aperture angle. On the *right* of the picture it can be seen that the pairwise parallel beams from two fan projections are available with the existing geometry only in the central area of the projection. However, in this case they always have the same angular displacement

▸ Sect. 7.8 it will be demonstrated that the angle $\psi_{\min}$ is not necessarily equal to the angle $\psi_{\max}$ because asymmetrical detector arrangements are preferred in practice.

The pairwise parallel beams from two different projection angles of the fans are only found in the central area of the projection with the same angular displacement. It should be mentioned that of course two beams of a single fan projection may never be related to the same projection angle of the synthesized pencil-beam

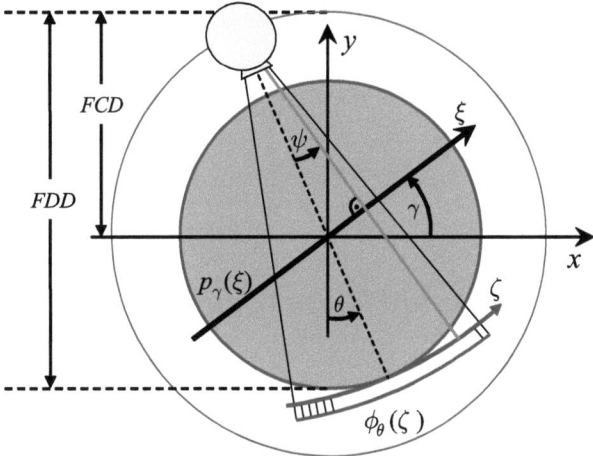

Fig. 7.16. The geometrical conditions at the transition from pencil-beam to fan-beam geometry and vice versa. Any value of the projection, $\phi_\theta(\zeta)$, on the curved detector, which is characterized by the arc angle, ψ, and thus finally by the arc length, $\zeta = FDD\,\psi$, must be related to a virtual linear detector. This is how a parallel projection, $p_\gamma(\xi)$, can be synthesized

projection. In order to be able to calculate which beam ζ in the fan projection $\phi_\theta(\zeta)$ corresponds to which beam ξ in the parallel projection $p_\gamma(\xi)$, both systems have to be compared with the same figure (cf. Fig. 7.16). In order to be able to measure both projections as a function of a line section (detector arc length, ζ, or linear detector length, ξ) one will have to differentiate here between the angle, ψ, and the corresponding position on the detector array, $\zeta = FDD\,\psi$.

In practice only the fan-beam projection $\phi_\theta(\zeta)$ is measured and thus known. The projection angle for the parallel projection is – as already mentioned in ▸ Sect. 5.2 – the angle, γ, between the virtual linear detector axis, ξ, and the x-axis. Due to the curvature of the detector array, the projection angle in fan-beam geometry is the angle, θ, between the central beam of the fan and the y-axis.

Considering Fig. 7.16, one will now have to find the relationship between the piercing point, ξ, of the virtual linear detector array under the projection angle, γ, with the projection point, ψ (ψ is an angle coordinate), of the real curved detector array under the projection angle, θ. This obviously obeys

$$\xi = FCD \sin(\psi) \tag{7.69}$$

and, in analogy,

$$\zeta = FDD\,\psi. \tag{7.70}$$

Taking into account the curvature of the detector, one finds the angular relationship

$$\gamma = \theta + \psi \tag{7.71}$$

or, equivalently,

$$\psi = \arcsin\left(\frac{\xi}{FCD}\right) \tag{7.72}$$

and

$$\theta = \gamma - \psi. \tag{7.73}$$

This directly yields

$$\phi_\theta(\zeta) = \phi_\theta(FDD\ \psi) = p_{\theta+\psi}(FCD\sin(\psi)) \tag{7.74}$$

or, inversely,

$$p_\gamma(\xi) = \phi_{\gamma-\psi}(\zeta)$$
$$= \phi_{\gamma-\arcsin\left(\frac{\xi}{FCD}\right)}\left(FDD\arcsin\left(\frac{\xi}{FCD}\right)\right). \tag{7.75}$$

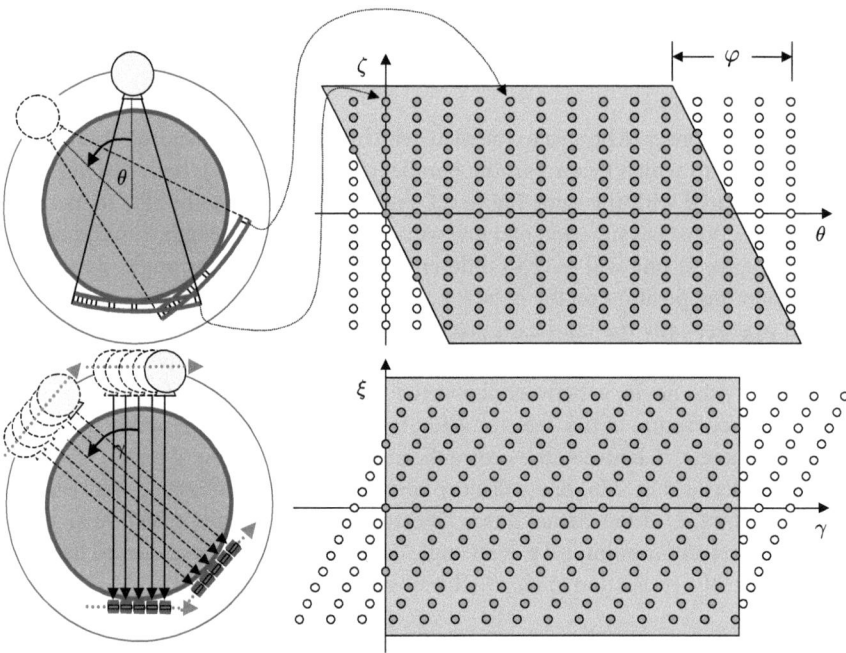

Fig. 7.17. Location of the data in the Radon spaces of the corresponding geometries for small aperture angles of the fan such that the *sine* or the *arcsine* are not yet perceptible. The *circles* reflect the sampling points of the projection values on the curved detector array, $\phi_\theta(\zeta)$, (*upper graphic*) or an imaginary parallel projection, $p_\gamma(\xi)$, (*lower graphic*) respectively. According to (7.69) through (7.75), it is evident that the angle interval in which projections are acquired in fan-beam geometry must be larger than π by the aperture angle, φ, of the fan. This ensures that a complete set of values can be interpolated in the Radon space of the parallel projection

Figures 7.17 and 7.18 show the location of the data in the Radon space schematically. It is thus possible to calculate a parallel-beam projection based on the fan-beam projection by rebinning. However, it immediately becomes obvious that the measured fan projection must be interpolated. In particular, two facts have to be taken into account in this context. First of all, the calculated data points of the parallel projection are not equidistant and, unfortunately, the measuring interval required for a real parallel projection

$$\gamma \in [0, \pi] \tag{7.76}$$

does not suffice to interpolate a complete Radon space derived from the fan-beam projection. In fact, data must be measured within the angle interval

$$\theta \in [\psi_{min}, \pi + \psi_{max}] \; . \tag{7.77}$$

Compared with parallel-beam geometry, the required measuring interval of fan-beam geometry is increased by the aperture angle of the fan, i.e., by $\varphi = \psi_{max} + \psi_{min}$. In Fig. 7.17 the data required for the reconstruction are highlighted in blue.

Certain boundary conditions must be taken into account to ensure fast rebinning of the sample values. Let $\Delta\theta$ and $\Delta\gamma$ be the angular intervals in which a fan-beam or a parallel-beam profile is measured respectively. Let these intervals obey

$$\Delta\theta = \Delta\gamma = \Delta\varphi \; . \tag{7.78}$$

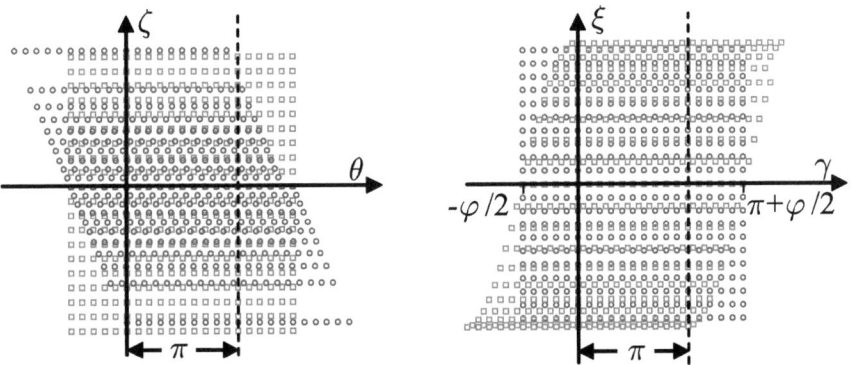

Fig. 7.18. Locations of the data points in the Radon spaces with respect to the different geometries even for a very large (in practice not feasible) aperture angle of the fan. On the *left*, the regular grid of the (ζ, θ) Radon space is indicated by *squares*. Diagonal to it, the data points necessary for the calculation of the Radon space of the synthetic parallel-beam system overlaid by *circles* can be seen. On the *right* the regular grid of the (ξ, γ) Radon space to be calculated (indicated by *circles*) can be seen. Diagonal to it, the actual data measured by the fan-beam system, which has been overlaid with *squares*, are shown. The characteristic *sine* function stands out

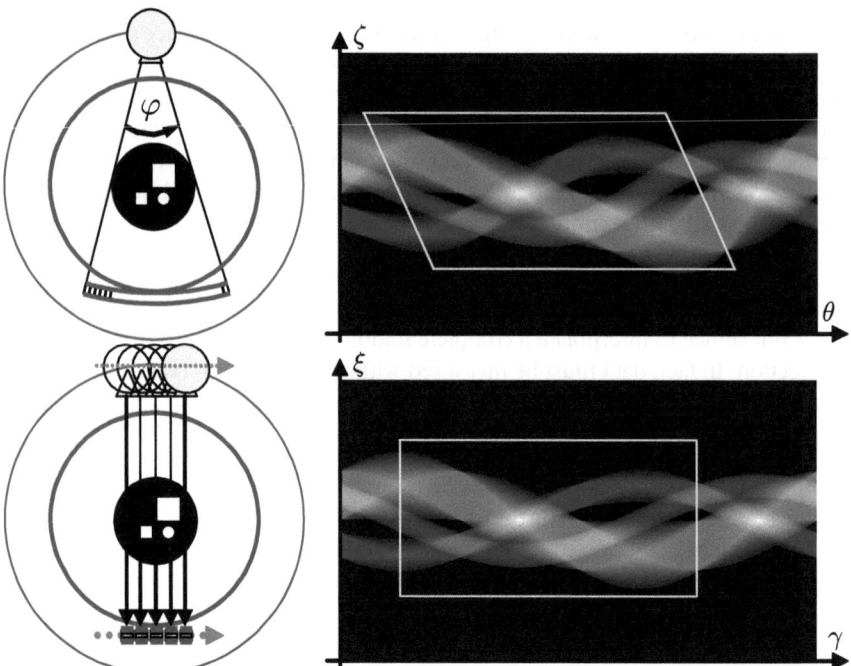

Fig. 7.19. The Radon spaces for fan-beam geometry (*top*) and parallel-beam geometry (*bottom*) are illustrated by a simple phantom over 360°

Then, one may write for integer numbers n, m

$$\phi_{m\Delta\varphi}(FDD\ n\Delta\varphi) = p_{(m+n)\Delta\varphi}(FCD\sin(n\Delta\varphi)) \ . \tag{7.79}$$

According to (7.79) the nth beam in the mth fan projection is equal to the nth beam in the $(m+n)$th parallel projection. The resultant, generated parallel beams are of course not equidistant and, therefore, an interpolation on the ξ-axis is necessary. Figure 7.19 once again shows the relationship between parallel-beam and fan-beam geometry of the data in the Radon space of the software phantom introduced in Fig. 7.11.

7.7.2
Complementary Rebinning

From a physical point of view, there is no difference in the overall attenuation of the X-ray beam caused by the tissue on its way forth and back, so the following equality applies:

$$p_\gamma(\xi) = p_{\gamma\pm\pi}(-\xi) \ . \tag{7.80}$$

Within parallel-beam geometry the object is therefore entirely sampled after half a rotation. For fan-beam geometry, however, an analog relationship can be established, since two beams are identical if their angles obey

$$\theta_2 = \theta_1 + 2\psi_1 \pm \pi \,. \tag{7.81}$$

In this case one finds

$$\psi_2 = -\psi_1 \,. \tag{7.82}$$

Figure 7.20 illustrates these angular conditions.

If the above fact is taken into account in (7.69) through (7.75), then Fig. 7.20 can be used to find

$$\begin{aligned} \phi_\theta(\zeta) &= p_{\theta+\psi}\left(FCD\sin(\psi)\right) \\ &= p_{\theta+\psi\pm\pi}\left(-FCD\sin(\psi)\right) \\ &= \phi_{\theta+2\psi\pm\pi}(-\zeta) \,. \end{aligned} \tag{7.83}$$

This is how a complementary X-ray source can be constructed. Figure 7.21 schematically shows the formation of the complementary X-ray source.

One may now use this knowledge concerning sampling geometry to further fill up the Radon space. Based on the interval (7.77), the complementary projection

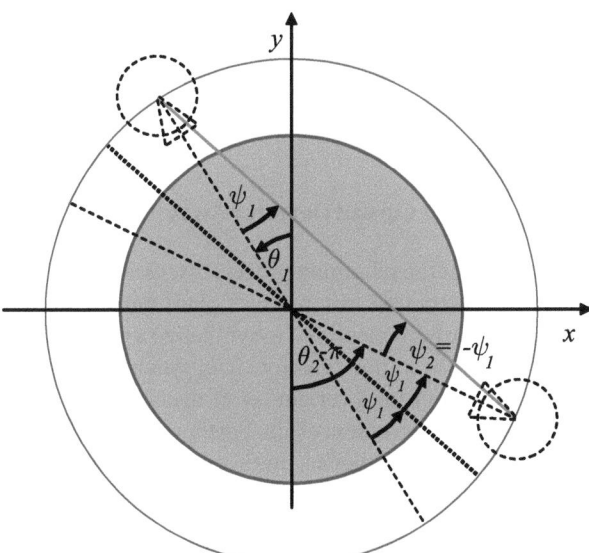

Fig. 7.20. In fan-beam geometry there are beams within the fans of the second half of the full rotation where the paths back and forth are identical with respect to their overall attenuation

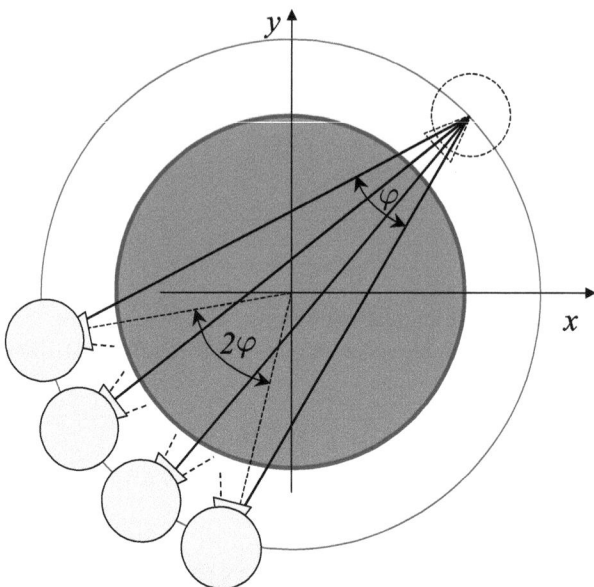

Fig. 7.21. Construction of a complementary X-ray source resulting from the physical symmetry of the attenuation path of the radiation

angle interval is obviously

$$2\pi - \psi_{\max} \leq \theta \leq 2\pi - \psi_{\min} . \tag{7.84}$$

7.7.3
Filtered Backprojection for Curved Detector Arrays

A practical disadvantage of the rebinning process discussed in the previous section is the fact that the reconstruction may not take place until a sufficient number of fan-beam projections have been measured to synthesize a parallel-beam projection. The reconstruction process can not, therefore, run synchronously to the sampling process, but rather must wait for a certain set of measured data. Hence, it represents a bottleneck in the reconstruction chain. Furthermore, rebinning of the Radon space generally requires a non-negligible, time-consuming interpolation; therefore, one might alternatively also consider how a direct filtered backprojection for fan-beam geometry might look. For this purpose one usually starts from the filtered backprojection of pencil-beam geometry (Horn 1979; Kak and Slaney 1988; Schlegel and Bille 2002), i.e.,

$$f(x, y) = \int\limits_{0}^{\pi} \int\limits_{-\infty}^{+\infty} P_{\gamma}(q)\, e^{2\pi i q \xi} |q|\, dq\, d\gamma$$

$$= \int\limits_{0}^{\pi} \int\limits_{-\infty}^{+\infty} P_{\gamma}(q)\, e^{2\pi i q (x\cos(\gamma) + y\sin(\gamma))} |q|\, dq\, d\gamma \qquad (7.85)$$

$$= \int\limits_{0}^{\pi} \int\limits_{-\infty}^{+\infty} P_{\gamma}(q)\, e^{2\pi i q (\mathbf{r}^{\mathrm{T}} \cdot \mathbf{n}_{\xi})} |q|\, dq\, d\gamma \; .$$

Here, one replaces the frequency representation of the projection $P_{\gamma}(q)$ by the explicit Fourier transform of the spatial projection signal $p_{\gamma}(\xi)$, i.e.,

$$f(x, y) = \int\limits_{0}^{\pi} \int\limits_{-\infty}^{+\infty} \left(\int\limits_{-\infty}^{+\infty} p_{\gamma}(\xi')\, e^{-2\pi i q \xi'}\, d\xi' \right) e^{2\pi i q \xi} |q|\, dq\, d\gamma \; . \qquad (7.86)$$

Changing the integration order one finds

$$f(x, y) = \int\limits_{0}^{\pi} \int\limits_{-\infty}^{+\infty} \int\limits_{-\infty}^{+\infty} p_{\gamma}(\xi')\, e^{-2\pi i q \xi'}\, e^{2\pi i q (\mathbf{r}^{\mathrm{T}} \cdot \mathbf{n}_{\xi})} |q|\, d\xi'\, dq\, d\gamma$$

$$\qquad (7.87)$$

$$= \int\limits_{0}^{\pi} \int\limits_{-\infty}^{+\infty} \int\limits_{-\infty}^{+\infty} p_{\gamma}(\xi')\, e^{2\pi i q (\mathbf{r}^{\mathrm{T}} \cdot \mathbf{n}_{\xi} - \xi')} |q|\, d\xi'\, dq\, d\gamma \; .$$

At this point, one concentrates on the integration over the frequency variable q, i.e.,

$$f(x, y) = \int\limits_{0}^{\pi} \int\limits_{-\infty}^{+\infty} p_{\gamma}(\xi') \left(\int\limits_{-\infty}^{+\infty} e^{2\pi i q (\mathbf{r}^{\mathrm{T}} \cdot \mathbf{n}_{\xi} - \xi')} |q|\, dq \right) d\xi'\, d\gamma \; . \qquad (7.88)$$

The bracket term in integral (7.88),

$$g\left(\mathbf{r}^{\mathrm{T}} \cdot \mathbf{n}_{\xi} - \xi' \right) = \int\limits_{-\infty}^{+\infty} e^{2\pi i q (\mathbf{r}^{\mathrm{T}} \cdot \mathbf{n}_{\xi} - \xi')} |q|\, dq \; , \qquad (7.89)$$

is the filter, or convolution kernel in the spatial domain, which must be considered as a distribution (Schlegel and Bille 2002). This is based upon the fact that the integral (7.89) cannot be directly evaluated due to difficulties related to the convergence that have been already mentioned in ▸ Sect. 5.7. However, as a distribution within another integral, (7.89) may lead to meaningful results. Substituting the left side of (7.89) into (7.88) leads to the following convolution:

$$f(x, y) = \int\limits_{0}^{\pi} \int\limits_{-\infty}^{+\infty} p_{\gamma}(\xi') g\left(\mathbf{r}^{\mathrm{T}} \cdot \mathbf{n}_{\xi} - \xi'\right) d\xi' d\gamma$$

$$= \frac{1}{2} \int\limits_{0}^{2\pi} \left(\int\limits_{-\infty}^{+\infty} p_{\gamma}(\xi') g\left(\mathbf{r}^{\mathrm{T}} \cdot \mathbf{n}_{\xi} - \xi'\right) d\xi' \right) d\gamma ,$$

(7.90)

which is defined in the second line of (7.90) for a backprojection over 360°.

At this point, the transformation of the coordinates to fan-beam geometry is to be carried out, which means that one has to change the coordinates:

$$(\xi', \gamma) \rightarrow (\zeta, \theta) .$$

(7.91)

When the integration variables are substituted, one is first and above all interested in the change in the infinitesimal area element, $d\xi \, d\gamma$, during the coordinate transformation. As already discussed in ▶ Sect. 5.7 and Fig. 5.24, the new area element must be multiplied by the Jacobian. This means that the area element, $d\xi \, d\gamma$, is determined in the new coordinates by $J \, d\zeta \, d\theta$ where J is given by

$$J \equiv \det\left(\frac{\partial(\xi', \gamma)}{\partial(\zeta, \theta)}\right) = \begin{vmatrix} \dfrac{\partial \xi'}{\partial \zeta} & \dfrac{\partial \gamma}{\partial \zeta} \\[2mm] \dfrac{\partial \xi'}{\partial \theta} & \dfrac{\partial \gamma}{\partial \theta} \end{vmatrix} .$$

(7.92)

Direct substitution of the transformation equations (7.69) through (7.71) yields

$$J = \begin{vmatrix} \dfrac{\partial\left(FCD \sin\left(\frac{\zeta}{FDD}\right)\right)}{\partial \zeta} & \dfrac{\partial\left(\theta + \frac{\zeta}{FDD}\right)}{\partial \zeta} \\[4mm] \dfrac{\partial\left(FCD \sin\left(\frac{\zeta}{FDD}\right)\right)}{\partial \theta} & \dfrac{\partial\left(\theta + \psi\right)}{\partial \theta} \end{vmatrix}$$

$$= \begin{vmatrix} \dfrac{FCD}{FDD} \cos\left(\dfrac{\zeta}{FDD}\right) & \dfrac{1}{FDD} \\[4mm] 0 & 1 \end{vmatrix}$$

(7.93)

$$= \dfrac{FCD}{FDD} \cos\left(\dfrac{\zeta}{FDD}\right) = \dfrac{FCD}{FDD} \cos\left(\psi\right) .$$

Thus, the infinitesimal area element is replaced during the coordinate transformation (7.91) with

$$d\xi' d\gamma \rightarrow \dfrac{FCD}{FDD} \cos\left(\dfrac{\zeta}{FDD}\right) d\zeta \, d\theta .$$

(7.94)

If the arc angle, ψ, is used instead of the arc length, ζ, one analogously finds

$$d\xi' \, dy \rightarrow FCD \cos(\psi) \, d\psi \, d\theta \,. \tag{7.95}$$

To understand how the inner product $(\mathbf{r}^{\mathrm{T}} \mathbf{n}_\xi)$ appears in the new coordinates, one may define the points on the backprojection lines in polar coordinates. Figure 7.22 illustrates the geometrical situation. The point $\mathbf{r} = (x, y)^{\mathrm{T}}$ is given by the equations

$$\begin{aligned} x &= r \cos(\delta) \\ y &= r \sin(\delta) \end{aligned} \,. \tag{7.96}$$

With

$$\mathbf{n}_\xi = \begin{pmatrix} \cos(\gamma) \\ \sin(\gamma) \end{pmatrix}, \tag{7.97}$$

the inner product determined by the addition law for the *cosine*,

$$\cos(\alpha \pm \beta) = \cos(\alpha) \cos(\beta) \mp \sin(\alpha) \sin(\beta)\,, \tag{7.98}$$

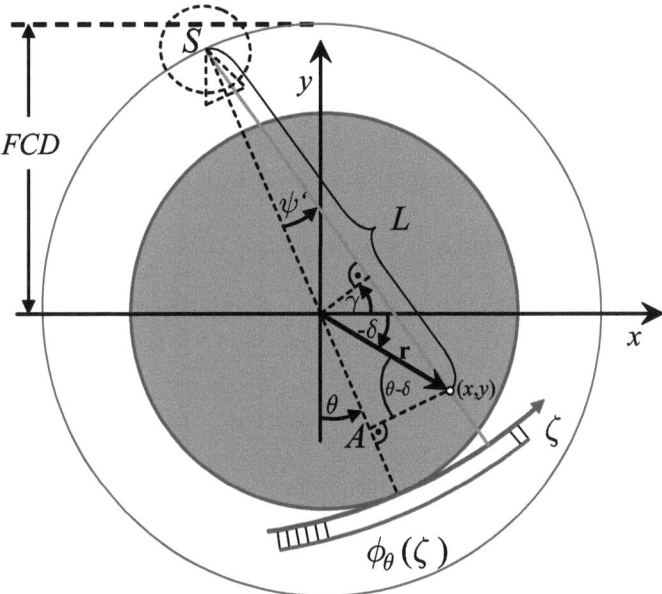

Fig. 7.22. New variables are introduced for the filtering process into the coordinate system of the fan beam. L is the distance between the point $\mathbf{r} = (x, y)^{\mathrm{T}}$ and the X-ray source. The point $\mathbf{r} = (x, y)^{\mathrm{T}}$ is characterized by its distance to the center of rotation, r, and the angle δ

reads

$$
\begin{aligned}
\mathbf{r}^{\mathrm{T}} \cdot \mathbf{n}_\xi &= (r\cos(\delta), r\sin(\delta)) \begin{pmatrix} \cos(\gamma) \\ \sin(\gamma) \end{pmatrix} \\
&= r\cos(\delta)\cos(\gamma) + r\sin(\delta)\sin(\gamma) \\
&= r\cos(\gamma - \delta) \,.
\end{aligned}
\tag{7.99}
$$

With the new coordinates (7.90) thus reads

$$
\begin{aligned}
&f(r, \delta) \\
&= \frac{1}{2} \int_0^{2\pi} \left(\int_{\psi_{\min}}^{\psi_{\max}} \phi_\theta(FDD\,\psi)g\left(r\cos(\gamma - \delta) - FCD\sin(\psi)\right)FCD\cos(\psi)\,\mathrm{d}\psi \right) \mathrm{d}\theta
\end{aligned}
\tag{7.100}
$$

and, taking into account the angle relationship (7.73), this yields

$$
\begin{aligned}
&f(r, \delta) \\
&= \frac{1}{2} \int_0^{2\pi} \left(\int_{\psi_{\min}}^{\psi_{\max}} \phi_\theta(FDD\,\psi)g\left(r\cos(\theta + \psi - \delta) - FCD\sin(\psi)\right)FCD\cos(\psi)\,\mathrm{d}\psi \right) \mathrm{d}\theta \,.
\end{aligned}
\tag{7.101}
$$

This, the argument of the kernel,

$$
g = g\left(r\cos(\theta + \psi - \delta) - FCD\sin(\psi)\right),
\tag{7.102}
$$

shall be examined more closely. Using the addition law (7.98), one may write

$$
\begin{aligned}
r\cos(\theta + \psi - \delta) - FCD\sin(\psi) &= r\cos(\theta - \delta)\cos(\psi) - r\sin(\theta - \delta)\sin(\psi) \\
&\quad - FCD\sin(\psi) \\
&= r\cos(\theta - \delta)\cos(\psi) \\
&\quad - (r\sin(\theta - \delta) + FCD)\sin(\psi) \,.
\end{aligned}
\tag{7.103}
$$

As indicated in Fig. 7.22, L denotes the distance between the X-ray source and the point $\mathbf{r} = (x, y)^{\mathrm{T}}$. If one considers the rectangular triangle formed by the points $\mathbf{S}$, $\mathbf{A}$, and $\mathbf{r}$, then simple trigonometry yields

$$
L\cos(\psi') = FCD + r\sin(\theta - \delta)
\tag{7.104}
$$

and

$$
L\sin(\psi') = r\cos(\theta - \delta) \,.
\tag{7.105}
$$

In Fig. 7.22, it can be seen that

$$L = \sqrt{(FCD + r\sin(\theta - \delta))^2 + (r\cos(\theta - \delta))^2} \qquad (7.106)$$

or

$$L = \sqrt{(FCD - x\sin(\theta) + y\cos(\theta))^2 + (x\cos(\theta) + y\sin(\theta))^2} \,. \qquad (7.107)$$

Here, ψ' is a certain angle of the fan arc angle coordinate, ψ, determined by the point (r, δ) and the projection angle, θ. This special angle is determined by

$$\psi' = \arctan\left(\frac{r\cos(\theta - \delta)}{FCD + r\sin(\theta - \delta)}\right) \qquad (7.108)$$

or

$$\psi' = \arctan\left(\frac{x\cos(\theta) + y\sin(\theta)}{FCD - x\sin(\theta) + y\cos(\theta)}\right). \qquad (7.109)$$

Thus, substituting (7.104) and (7.105) into (7.103) or (7.102) respectively, one finds

$$g = g\left(L\sin(\psi')\cos(\psi) - L\cos(\psi')\sin(\psi)\right). \qquad (7.110)$$

With the addition law,

$$\sin(a - b) = \sin(a)\cos(b) - \cos(a)\sin(b), \qquad (7.111)$$

the argument of the kernel, g, reduces to

$$g = g\left(L\sin(\psi' - \psi)\right). \qquad (7.112)$$

If one substitutes this expression into (7.101), one finds

$$f(r, \delta) = \frac{1}{2}\int_0^{2\pi}\left(\int_{\psi_{\min}}^{\psi_{\max}} p_{\theta + \psi}\left(FCD\sin(\psi)\right)g\left(L\sin(\psi' - \psi)\right)FCD\cos(\psi)\,d\psi\right)d\theta$$

$$(7.113)$$

or

$$f(r, \delta) = \frac{1}{2}\int_0^{2\pi}\left(\int_{\psi_{\min}}^{\psi_{\max}} \phi_\theta\left(FDD\,\psi\right)g\left(L\sin(\psi' - \psi)\right)FCD\cos(\psi)\,d\psi\right)d\theta. \qquad (7.114)$$

The inner integral again represents a convolution. Since the function g in (7.112) is the spatial convolution kernel of the linear frequency ramp, thus

$$g(\xi) = \int_{-\infty}^{+\infty} |q| e^{2\pi i q \xi} \, dq \, , \tag{7.115}$$

one may substitute here the arguments found for fan-beam geometry, so that

$$g(L\sin(\psi)) = \int_{-\infty}^{+\infty} |q| e^{2\pi i q (L \sin(\psi))} \, dq \, . \tag{7.116}$$

The substitution

$$q' = \frac{qL\sin(\psi)}{\psi} \tag{7.117}$$

then yields

$$g(L\sin(\psi)) = \left(\frac{\psi}{L\sin(\psi)} \right)^2 \int_{-\infty}^{+\infty} |q'| e^{2\pi i \psi q'} \, dq' \, . \tag{7.118}$$

The integral term in (7.118) is just the desired Fourier transform of the linear frequency ramp. However, instead of the linear spatial detector variable, ξ, the angle coordinate, ψ, plays the main spatial part here. The following relationship thus applies:

$$g(L\sin(\psi)) = \left(\frac{\psi}{L\sin(\psi)} \right)^2 g(\psi) \, , \tag{7.119}$$

so that (7.114) finally reads

$$f(r, \delta) = \int_0^{2\pi} \frac{1}{L^2} \left(\int_{\psi_{\min}}^{\psi_{\max}} \phi_\theta(FDD\,\psi) \tilde{g}(\psi' - \psi) \, FCD\cos(\psi) \, d\psi \right) d\theta \, , \tag{7.120}$$

where

$$\tilde{g}(\psi) = \frac{1}{2} \left(\frac{\psi}{\sin(\psi)} \right)^2 g(\psi) \, . \tag{7.121}$$

In analogy to the filtered backprojection in ▸ Sect. 5.7, (5.67), one may now write the short expression

$$f(r,\delta) = \int_0^{2\pi} \frac{1}{L^2} h_\theta(\psi)\, d\theta .$$ (7.122)

(7.122) can be interpreted in such a way that the reconstruction with fan-beam geometry is a weighted filtered backprojection, where

$$h_\theta(\psi) = \tilde{\phi}_\theta(\psi) * \tilde{g}(\psi)$$ (7.123)

with

$$\tilde{\phi}_\theta(\psi) = \phi_\theta(FDD\,\psi)\,FCD\cos(\psi) .$$ (7.124)

In analogy to ▸ Sect. 5.7, Scheme 5.2, again the reconstruction instructions should be given in an overview scheme. The direct fan-beam reconstruction may be subdivided into three steps as shown in Scheme 7.1.

Scheme 7.1 Filtered backprojection in fan-beam geometry for the curved detector array

1. Pre-weighting of the fan-beam projection signal

$$\tilde{\phi}_\theta(\psi) = \phi_\theta(FDD\,\psi)\,FCD\cos(\psi) .$$

2. Filtering of the projection signal. The ramp filter

$$\tilde{g}(\psi) = \frac{1}{2}\left(\frac{\psi}{\sin(\psi)}\right)^2 g(\psi)$$

is a modified version of the filter for the parallel projection $g(\psi)$. Combined with the pre-weighting process described in the first step, the signal for backprojection results here from the convolution in the angular domain,

$$h_\theta(\psi) = \tilde{\phi}_\theta(\psi) * \tilde{g}(\psi) .$$

3. Finally, the backprojection is defined over the full angle of 2π:

$$f(r,\delta) = \int_0^{2\pi} \frac{1}{L^2} h_\theta(\psi)\, d\theta .$$

It can be seen as a filtered backprojection that is converging onto the source point in the X-ray tube. Therefore, the signal is weighted by the reciprocal squared distance between the source and the actual pixel.

7.7.4
Filtered Backprojection for Linear Detector Arrays

The reconstruction process based on the filtered backprojection with data of a linear instead of a curved detector array can be derived in analogy to the steps given in the previous chapter. Presently, in clinical practice linear arrays are of secondary importance. In addition, the geometrical conditions regarding the circular curved arrays are simpler and better adapted to reconstruction geometry because all X-ray paths within the fan are of the same length. However, it will be worthwhile understanding the geometrical situation of the backprojection based on linear detector arrays, because this is similar geometry to the case of cone-beam CT (described in ▸ Sect. 8.5), which is employed with a flat-panel detector matrix.

In a curved detector array the arc angle corresponds to a circular arc length, with the center located in the X-ray source, and the individual detector elements arranged in an equidistant angular pattern. Contrary to the above, one can just assume equidistant detectors in a linear detector array, since the corresponding angular change is non-linear. To illustrate this, Fig. 7.23 shows the geometrical situation.

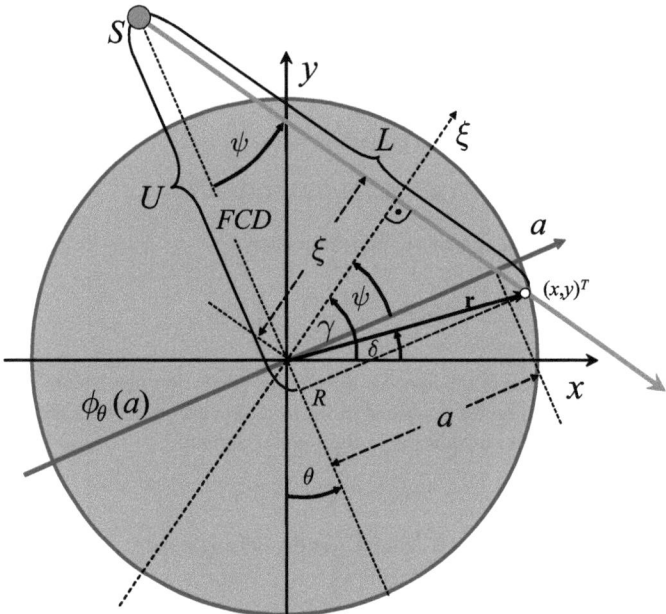

Fig. 7.23. For filtering in the coordinate system of a fan beam with linear detector array, it will be necessary to introduce new variables. L is the distance between a point, $\mathbf{r} = (x, y)^{\mathrm{T}}$, and the X-ray focus. The point $\mathbf{r} = (x, y)^{\mathrm{T}}$ is characterized by its distance, r, to the center of rotation and by the angle δ. The variable a marks the location of a virtual linear detector in the center of the measuring field. $\phi_\theta(a)$ is the measured projection value on the virtual detector and θ denotes the projection angle of the center beam of the fan-beam projection

$\phi_\theta(a)$ is the measured projection value, where the variable a denotes the position on the linear detector and θ the projection angle of the center beam. Many authors do not refer to the real detector, but rather to a virtual detector placed in the origin. This concept will also be used here. Similar to the geometry of the curved array, L is the distance between a point, $\mathbf{r} = (x, y)^T$, and the focus inside the X-ray source. The point $\mathbf{r} = (x, y)^T$ is characterized by its distance, r, from the center of rotation and by the angle δ.

If one assumes parallel-beam geometry for the point $\mathbf{r}$, one would find the parallel projection $p_\gamma(\xi)$ under the projection angle γ at detector position ξ. The ξ coordinate is shown as a dashed line in Fig. 7.23 and, by definition, it runs perpendicular to the related X-ray beam. In analogy to (7.69) and (7.71), the relationship with the fan-beam coordinates obeys

$$\xi = a \cos(\psi) = a \frac{FCD}{\sqrt{a^2 + FCD^2}} \tag{7.125}$$

and

$$\gamma = \theta + \psi = \theta + \arctan\left(\frac{a}{FCD}\right) . \tag{7.126}$$

At this point, the transform of the equation of the filtered backprojection,

$$f(x, y) = \frac{1}{2} \int_0^{2\pi} \left(\int_{-\infty}^{+\infty} p_\gamma(\xi) g(\mathbf{r}^T \cdot \mathbf{n}_\xi - \xi) \, d\xi \right) d\gamma , \tag{7.90}$$

into fan-beam geometry has to be carried out again, i.e., one has to substitute the coordinates of parallel-beam geometry with those of fan-beam geometry:

$$(\xi, \gamma) \rightarrow (a, \theta) . \tag{7.127}$$

Again, the new infinitesimal area element must be multiplied with the Jacobian for integration. This means that the old area element, $d\xi \, d\gamma$, is determined in the new coordinates by $J \, da \, d\theta$ where J is given by

$$J \equiv \det\left(\frac{\partial(\xi, \gamma)}{\partial(a, \theta)} \right) = \begin{vmatrix} \dfrac{\partial \xi}{\partial a} & \dfrac{\partial \gamma}{\partial a} \\ \dfrac{\partial \xi}{\partial \theta} & \dfrac{\partial \gamma}{\partial \theta} \end{vmatrix} . \tag{7.128}$$

If one directly substitutes the transformation equations (7.125) and (7.126) one finds

$$
J = \begin{vmatrix} \dfrac{\partial\left(a\frac{FCD}{\sqrt{a^2+FCD^2}}\right)}{\partial a} & \dfrac{\partial\left(\theta + \arctan\left(\frac{a}{FCD}\right)\right)}{\partial a} \\[3mm] \dfrac{\partial\left(a\frac{FCD}{\sqrt{a^2+FCD^2}}\right)}{\partial \theta} & \dfrac{\partial\left(\theta + \psi\right)}{\partial \theta} \end{vmatrix} \tag{7.129}
$$

$$
= \begin{vmatrix} \left(\dfrac{FCD}{\sqrt{a^2 + FCD^2}}\right)^3 & \dfrac{FCD}{a^2 + FCD^2} \\[3mm] 0 & 1 \end{vmatrix} = \left(\dfrac{FCD}{\sqrt{a^2 + FCD^2}}\right)^3 = \cos(\psi)^3 \, .
$$

Thus, the infinitesimal surface element is changed by the coordinate transformation (7.127) by

$$
d\xi \, d\gamma \rightarrow \left(\frac{FCD}{\sqrt{a^2 + FCD^2}}\right)^3 da \, d\theta \, . \tag{7.130}
$$

If, in addition, one again takes advantage of the relationship

$$
\mathbf{r}^{\mathrm{T}} \cdot \mathbf{n}_\xi = r \cos(\gamma - \delta) \, , \tag{7.99}
$$

one finds the expression

$$
f(r, \delta) = \frac{1}{2} \int\limits_{-\arctan\left(\frac{a}{FCD}\right)}^{2\pi - \arctan\left(\frac{a}{FCD}\right)} \left\{ \int\limits_{-a_{\min}}^{+a_{\max}} p_{\theta + \psi}\left(a\frac{FCD}{\sqrt{a^2 + FCD^2}}\right) \cdot \ldots \right.
$$

$$
\ldots \cdot g\left[r \cos\left(\theta + \arctan\left(\frac{a}{FCD}\right) - \delta\right) - \frac{a \, FCD}{\sqrt{a^2 + FCD^2}} \right] \tag{7.131}
$$

$$
\left. \times \left(\frac{FCD}{\sqrt{a^2 + FCD^2}}\right)^3 da \right\} d\theta
$$

for the filtered backprojection in the new coordinates. Within this expression, the following simplifications can be implemented immediately. The limits of integration of the new angle coordinate θ had to be changed as a direct consequence of the substitution. However, since the integration over the full angle 2π is invariant to a constant phase shift, it is also possible to integrate in the case of this new variable from 0 to 2π. Furthermore, the projection value, $p_\gamma(\xi)$, can be much easier expressed by means of the new detector variables since the coordinate transform yields

$$
p_\gamma(\xi) \rightarrow p_{\theta + \psi}\left(a\frac{FCD}{\sqrt{a^2 + FCD^2}}\right) = \phi_\theta(a) \, . \tag{7.132}
$$

(7.131) is thus simplified to read

$$f(r,\delta) = \frac{1}{2} \int_0^{2\pi} \left\{ \int_{-a_{min}}^{+a_{max}} \phi_\theta(a) g \left[r \cos \left(\theta + \arctan \left(\frac{a}{FCD} \right) - \delta \right) - \frac{a\, FCD}{\sqrt{a^2 + FCD^2}} \right] \cdots \right.$$

$$\left. \cdots \left(\frac{FCD}{\sqrt{a^2 + FCD^2}} \right)^3 da \right\} d\theta \,.$$

(7.133)

For convenience, one substitutes the argument of the convolution kernel g with

$$\psi = \arctan \left(\frac{a}{FCD} \right) , \tag{7.134}$$

which can easily be understood from Fig. 7.23. Then, addition law (7.98) can be used to calculate the argument of g with the following expression:

$$r \cos(\theta + \psi - \delta) - \frac{a\, FCD}{\sqrt{a^2 + FCD^2}}$$

$$= r \cos(\theta - \delta) \cos(\psi) - r \sin(\theta - \delta) \sin(\psi) - \frac{a\, FCD}{\sqrt{a^2 + FCD^2}} \,. \tag{7.135}$$

With the expressions – which also may be read from Fig. 7.23 –

$$\cos(\psi) = \frac{FCD}{\sqrt{a^2 + FCD^2}} \tag{7.136}$$

and

$$\sin(\psi) = \frac{a}{\sqrt{a^2 + FCD^2}} , \tag{7.137}$$

(7.135) is further modified to read

$$r \cos(\theta + \psi - \delta) - \frac{a\, FCD}{\sqrt{a^2 + FCD^2}}$$

$$= r \cos(\theta - \delta) \frac{FCD}{\sqrt{a^2 + FCD^2}} - r \sin(\theta - \delta) \frac{a}{\sqrt{a^2 + FCD^2}} - \frac{a\, FCD}{\sqrt{a^2 + FCD^2}} \tag{7.138}$$

$$= r \cos(\theta - \delta) \frac{FCD}{\sqrt{a^2 + FCD^2}} - (r \sin(\theta - \delta) + FCD) \frac{a}{\sqrt{a^2 + FCD^2}} \,.$$

In analogy to (7.106) some authors (e.g., Turbell [2001]) introduce the distance

$$U = FCD + r\sin(\theta - \delta)$$
(7.139)

or in analogy to (7.107)

$$U = FCD - x\sin(\theta) + y\cos(\theta)$$
(7.140)

(cf. Fig. 7.23)[1]. Similar to (7.108), one considers, with the coordinate a, the special position on the virtual detector determined by the point (r, δ) and the projection angle, θ. This very detector position is denoted a', so that

$$\tan(\psi') = \frac{a'}{FCD} = \frac{r\cos(\theta - \delta)}{FCD + r\sin(\theta - \delta)}$$
(7.141)

and thus

$$a' = FCD\frac{r\cos(\theta - \delta)}{FCD + r\sin(\theta - \delta)} = FCD\frac{r\cos(\theta - \delta)}{U}$$
(7.142)

or

$$a' = FCD\frac{x\cos(\theta) + y\sin(\theta)}{FCD - x\sin(\theta) + y\cos(\theta)} = FCD\frac{x\cos(\theta) + y\sin(\theta)}{U}.$$
(7.143)

Hence, the argument of g in (7.133) can be simplified to read

$$r\cos(\theta + \psi - \delta) - \frac{a\,FCD}{\sqrt{a^2 + FCD^2}} = U\frac{a'}{\sqrt{a^2 + FCD^2}} - U\frac{a}{\sqrt{a^2 + FCD^2}}$$
(7.144)

so that (7.133) is given by

$$f(r, \delta) = \frac{1}{2}\int_0^{2\pi}\left\{\int_{-a_{\min}}^{+a_{\max}}\phi_\theta(a)g\left[(a' - a)\frac{U}{\sqrt{a^2 + FCD^2}}\right]\left(\frac{FCD}{\sqrt{a^2 + FCD^2}}\right)^3 da\right\}d\theta .$$
(7.145)

The inner integral again represents a convolution. One should remember that the function

$$g(\xi) = \int_{-\infty}^{+\infty}|q|\,e^{2\pi i q\xi}\,dq$$
(7.146)

---

[1] Occasionally, the distance is also defined as the ratio $U_{\mathrm{KS}} = \dfrac{FCD + r\sin(\theta - \delta)}{FCD}$ (Kak and Slaney 1988).

is the spatial convolution kernel of the linear frequency ramp. Here, one substitutes the new variables found for fan-beam geometry with the linear detector, thus

$$g\left((a'-a)\frac{U}{\sqrt{a^2+FCD^2}}\right) = \int_{-\infty}^{+\infty} |q| e^{2\pi i q\left((a'-a)\frac{U}{\sqrt{a^2+FCD^2}}\right)} dq .$$ (7.147)

The substitution

$$q' = q\frac{U}{\sqrt{a^2+FCD^2}}$$ (7.148)

then yields

$$g\left((a'-a)\frac{U}{\sqrt{a^2+FCD^2}}\right) = \frac{a^2+FCD^2}{U^2} \int_{-\infty}^{+\infty} |q'| e^{2\pi i q'(a'-a)} dq'$$
$$= \frac{a^2+FCD^2}{U^2} g(a'-a)$$ (7.149)

and (7.145) thus finally reads

$$f(r,\delta) = \frac{1}{2} \int_0^{2\pi} \left\{ \int_{-a_{min}}^{+a_{max}} \phi_\theta(a) \frac{a^2+FCD^2}{U^2} g(a'-a) \left(\frac{FCD}{\sqrt{a^2+FCD^2}}\right)^3 da \right\} d\theta$$

$$= \frac{1}{2} \int_0^{2\pi} \frac{1}{U^2} \left\{ \int_{-a_{min}}^{+a_{max}} \phi_\theta(a) g(a'-a) \left(\frac{FCD^3}{\sqrt{a^2+FCD^2}}\right) da \right\} d\theta$$

(7.150)

$$= \frac{1}{2} \int_0^{2\pi} \frac{FCD^2}{U^2} \left\{ \int_{-a_{min}}^{+a_{max}} \phi_\theta(a) \left(\frac{FCD}{\sqrt{a^2+FCD^2}}\right) g(a'-a) da \right\} d\theta$$

$$= \frac{1}{2} \int_0^{2\pi} \frac{FCD^2}{U^2} \left\{ \left(\phi_\theta(a)\frac{FCD}{\sqrt{a^2+FCD^2}}\right) * g(a) \right\} d\theta .$$

In analogy to the filtered backprojection in ▸ Sect. 5.7, one may now write the short expression

$$f(r,\delta) = \int_0^{2\pi} \frac{FCD^2}{U^2} h_\theta(a) d\theta ,$$ (7.151)

where

$$h_\theta(a) = \frac{1}{2}\left(\phi_\theta(a)\frac{FCD}{\sqrt{a^2 + FCD^2}}\right) * g(a) \,.$$ (7.152)

The reconstruction instructions may again be divided into three steps listed in Scheme 7.2.

Scheme 7.2 Filtered backprojection with fan-beam geometry for the linear detector array

1. Coordinate transformation for the ramp filter: The required ramp filter

$$g\left((a' - a)\frac{U}{\sqrt{a^2 + FCD^2}}\right) = \frac{a^2 + FCD^2}{U^2}g(a' - a)$$

    must be exactly adapted to the form of the filter for the parallel projection by coordinate transform.
2. Filtering of the projection signal: The signal for backprojection results from the convolution

$$h_\theta(a) = \frac{1}{2}\left(\phi_\theta(a)\frac{FCD}{\sqrt{a^2 + FCD^2}}\right) * g(a)$$

    in the spatial domain of the linear detector.
3. The backprojection over the angle 2π results from

$$f(r, \delta) = \int_0^{2\pi} \frac{FCD^2}{U^2}h_\theta(a)\,d\theta \,,$$

    which means that the projection of the high-pass filtered detector signal in the reverse direction converges toward the focus of the X-ray source. While backprojecting, the signal must be weighted with the squared reciprocal variable U. In this case, U is the projection of the distance between the focus of the X-ray source and the current point onto the center beam of the fan.

7.7.5
Discretization of Backprojection for Fan-Beam Geometry

When dealing with the technical implementation of image reconstruction with fan-beam geometry, one has to answer the same question as described in ▸ Sect. 7.2.3

regarding the discretization of the kernels. This is due to the fact that, here also, one has a sampling process with a finite number of sampling points, either in equiangular steps, $\Delta\varphi$, for the curved detector array (cf. Fig. 7.10) or in equidistant steps, Δa, for the linear detector array.

As the single linear detector arrangements are of secondary importance, discretization will be discussed here in detail for the curved detector array only. Let the projection signal be sampled under a projection angle θ_n with a curved detector array having the form

$$\phi_{\theta_n}(j\Delta\varphi) \text{ with } j = -D/2, \ldots, 0, \ldots, D/2 - 1 \text{ and } n = 1, \ldots, N_p , \qquad (7.153)$$

where D is again the (even) number of detector elements. These detector elements are densely packed and located within an equiangular distance of $\Delta\varphi$ on a circle segment where the circle center is the focus of the X-ray source.

As in the previous section, the center beam is chosen as a reference beam and the projection angles in the fan therefore range from $\psi_{\min}$ to $\psi_{\max}$; the index in (7.153) has been chosen such that $\phi_{\theta_n}(0)$ with $j = 0$ is the center beam passing through the iso-center of rotation. Here, N_p is the number of projections or views.

While the backprojection in parallel-beam geometry runs along parallel beams, the projection in the reverse direction has to take place along the fan beams, which converge at the focus of the X-ray source. If Figs. 7.8a and 7.24 are compared, it can be seen that both concepts are so far analogous. Also, exactly as pointed out for parallel-beam geometry before, one cannot expect that in the calculation of the

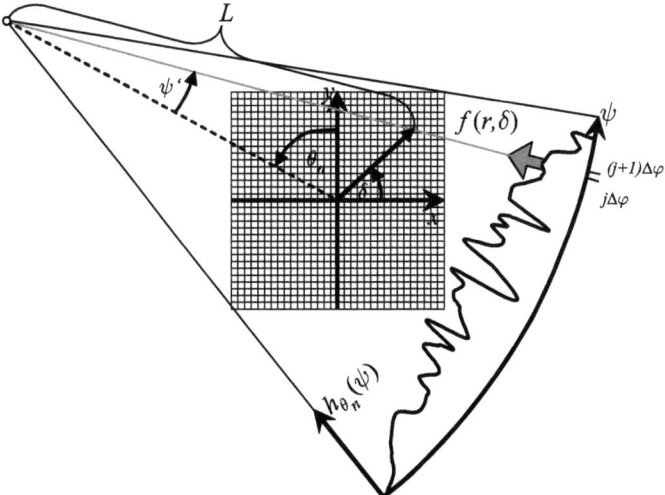

Fig. 7.24. Similar to parallel-beam geometry (cf. Fig. 7.8) the value of the filtered projection signal to be backprojected must be obtained by interpolation since the vector **r** points toward the integer grid position $(x, y)^T$, the projection of which on the ψ'-axis (or the ζ-axis) generally yields a value in the interval $[j\Delta\varphi, (j + 1)\Delta\varphi]$

contribution of $h_{\theta_n}(\psi)$ to a certain image pixel, $f(r,\delta)$, the required value for the angle ψ corresponds directly to an actual, measured angle, $j\Delta\varphi$. Instead, one has to interpolate between the sampled values.

As described in ▸ Sect. 7.2.3 for parallel-beam geometry, the kernels must also be discretized here. (7.121) shows that – compared with parallel-beam geometry – one has to use a slightly modified kernel for the filtering process. For convenience, one may therefore use the discretization results found in ▸ Sect. 7.2.3. The discrete sequence of the kernel (7.121) is given by

$$\tilde{g}(j\Delta\varphi) = \frac{1}{2}\left(\frac{j\Delta\varphi}{\sin(j\Delta\varphi)}\right)^2 g(j\Delta\varphi)\,. \tag{7.154}$$

As $g(j\Delta\varphi)$ is the discretized form of the kernel for parallel-beam geometry, a comparison with (7.52) yields the following sequence

$$\tilde{g}(j\Delta\varphi) = \begin{cases} \dfrac{1}{8\,(\Delta\varphi)^2} & \text{for } j = 0 \\[2ex] 0 & \text{for } j \text{ even, } (j \neq 0) \\[2ex] -\dfrac{1}{2\pi^2 \sin^2(j\Delta\varphi)} & \text{for } j \text{ odd}. \end{cases} \tag{7.155}$$

Similar to the concept for parallel-beam geometry, one may use the windowing technique so that the final result must read

$$h_{\theta_n}(j\Delta\varphi) = \tilde{\phi}_{\theta_n}(j\Delta\varphi) * \tilde{g}(j\Delta\varphi) * w(j\Delta\varphi)\,, \tag{7.156}$$

where w is the window function for which examples were given in ▸ Sect. 7.2.2. For the *Shepp–Logan* window this results in

$$\tilde{g}_{\mathrm{SL}}(j\Delta\varphi) = \frac{1}{2\pi^2\Delta\varphi}\left(\mathrm{ctg}\left(\left(j+\frac{1}{2}\right)\Delta\varphi\right) - \mathrm{ctg}\left(\left(j-\frac{1}{2}\right)\Delta\varphi\right)\right)\,. \tag{7.157}$$

Due to the bracketed terms, (7.157) may also be called the cotangent kernel (Kak and Slaney 1988).

Similar to the previous section, one may divide the reconstruction instructions into three steps as listed in Scheme 7.3. For the discretized version the arguments of the expressions must be replaced accordingly by the discretization (7.153) (Kak and Slaney 1988).

Scheme 7.3 Discretized filtered backprojection in fan-beam geometry

1. Pre-weighting of each individual fan projection acquired under angle θ_n

$$\tilde{\phi}_{\theta_n}(j\Delta\varphi) = \phi_{\theta_n}(FDD\,j\Delta\varphi)\,FCD\cos(j\Delta\varphi) \tag{7.158}$$

2. Filtering of the projection signal: The modified ramp filter

$$\tilde{g}(j\Delta\varphi) = \frac{1}{2}\left(\frac{j\Delta\varphi}{\sin(j\Delta\varphi)}\right)^2 g(j\Delta\varphi) \tag{7.159}$$

will also be sampled in equiangular steps. The discretized convolution thus reads

$$h_{\theta_n}(j\Delta\varphi) = \tilde{\phi}_{\theta_n}(j\Delta\varphi) * \tilde{g}(j\Delta\varphi) \tag{7.160}$$

3. In analogy to (7.9), the backprojection over the full angle of 2π with angular distances of $2\pi/N_p$ is given by

$$
\begin{aligned}
f(r,\delta) &= \int_0^{2\pi} \frac{1}{L^2} h_\theta(\psi)\,d\theta \\
&\approx \frac{2\pi}{N_p} \sum_{n=1}^{N_p} \frac{1}{L^2} h_{\theta_n}(\psi)\,.
\end{aligned}
\tag{7.161}
$$

This can also be seen as a filtered backprojection that is converging onto the focus point in the X-ray tube. Therefore, the signal is weighted by the reciprocal squared distance between the source and the actual pixel.

An example, which can be easily understood, will illustrate the intermediate results of filtered backprojection using fan-beam geometry and compare them with those of parallel-beam geometry. This example consists of a phantom with a central square of homogeneous attenuation. Figure 7.25 first shows the Radon spaces of the object. In addition to the known Cartesian representation of the projection values, $p_\gamma(\xi)$, this figure also shows the Radon space with its polar coordinates on the upper right-hand side. The Cartesian Radon space of the fan-beam geometry of the square is shown on the lower right-hand side. The intensity values are the projections $\phi_\theta(\zeta)$.

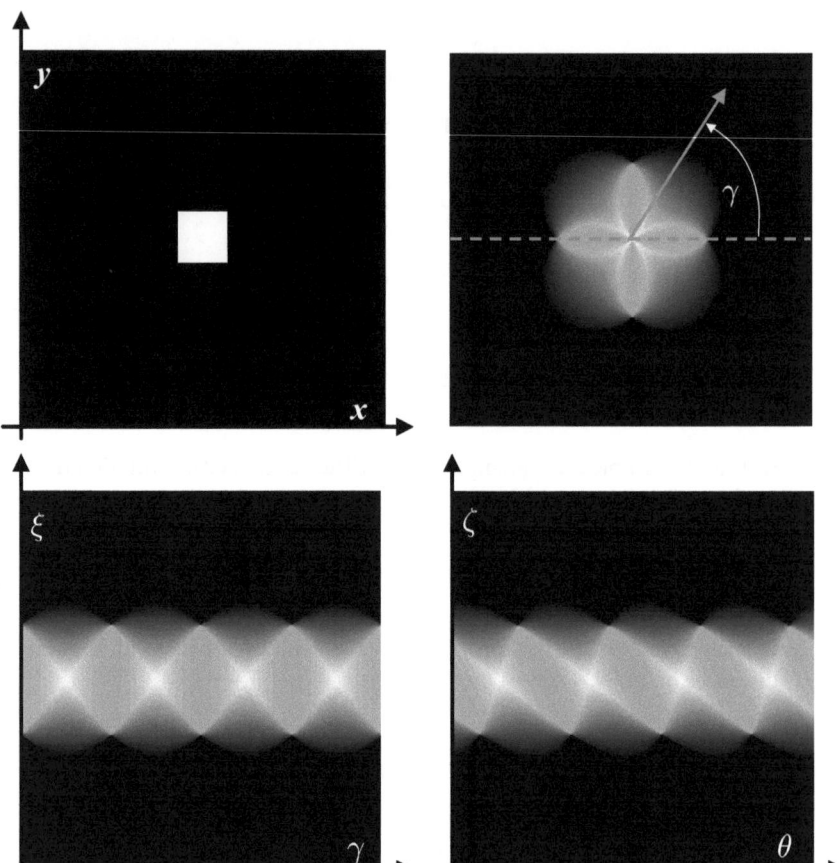

Fig. 7.25. To illustrate the differences in the Radon spaces between parallel-beam geometry and fan-beam geometry for the same object, once again a very simple phantom presented on the *upper left* is used here. The square object has homogeneously distributed attenuation coefficients. On the *upper right* and *lower left*, the Radon spaces of parallel-beam geometry with both their polar and Cartesian representations are depicted. Both representations are equivalent. On the *lower right*, the projection values in the Radon space of the phantom with fan-beam geometry are illustrated. The difference between the two pictures on the lower side is that the data in the Radon space of fan-beam geometry are tilted

As expected, a direct comparison of the Cartesian Radon space representations results in a tilting of the data, $\phi_\theta(\zeta)$, with fan-beam geometry compared with the data, $p_y(\xi)$, of parallel-beam geometry. The data rebinning concept explained in ▸ Sect. 7.7.1 can also be readily understood here. However, the projection data, $\phi_\theta(\zeta)$, will now be used to calculate a direct reconstruction according to (7.158) through (7.161).

The pictures displayed in Fig. 7.26 show the direct reconstructions with a successively increased number, N_p – cf. (7.161) second line – of the projections, $\phi_\theta(\zeta)$. In this case, N_p runs through the values $\{1, 3, 10, 25, 100, 180\}$. Similar to parallel-beam geometry, the true structure of the object is not displayed with a small number of views, N_p. Particularly striking is the fact that fan-beam geometry is directly reflected in the backprojections. This is, thus, also a very intuitive method. The individual projections look like the light cones of a pocket lamp.

Figure 7.27 once again directly compares the filtered backprojection with the two geometries – fan-beam and parallel-beam – taking $N_p = 3$ as an example. In the three-dimensional representation the specific filter response on the rectangle is readily visible for both geometries. This is described by (7.115) through (7.118). In parallel one can also recognize the quadratic decay of the single projection in a reverse direction with increasing distance to the source. This behavior is due to the $1/L^2$ normalization given in (7.161).

The linear detector array is discretized in the same way as the curved detector array. As $g(j\Delta a)$ in (7.150) is exactly the discretized form of the kernel for parallel-

Fig. 7.26. Backprojection with fan-beam geometry. Successive reconstruction of the tomographic phantom image based on the projection data. The number of projections is exemplarily increased image by image: $N_p = \{1, 3, 10, 25, 100, 180\}$. The reconstruction principle is again easy to understand intuitively. The backprojected data look like the light cones of a pocket lamp

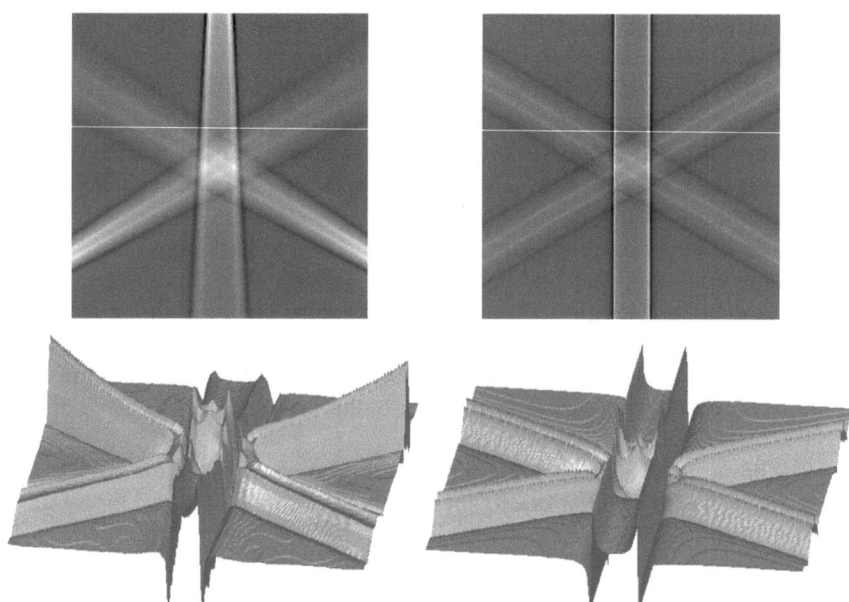

Fig. 7.27. Comparison between the filtered backprojection with fan-beam geometry (*left-hand column*) and the filtered backprojection of parallel-beam geometry (*right-hand column*). The length-dependent scaling according to (7.161) is readily visible with fan-beam geometry. However, the characteristic filter response to the rectangle is visible in both representations

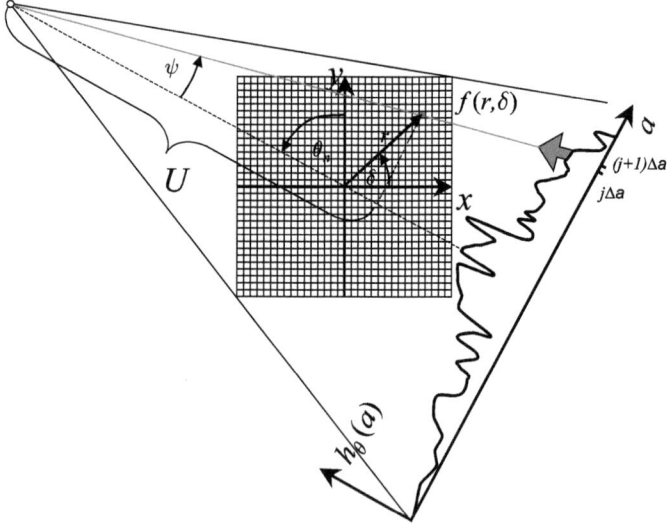

Fig. 7.28. Similar to parallel-beam geometry (cf. Fig. 7.8), the value of the filtered projection signal to be backprojected must be calculated by interpolation. This is necessary since the vector **r** points toward the integer grid position $(x, y)^{\mathrm{T}}$, the projection of which on the a-axis generally yields a value in the interval $[j\Delta a, (j + 1)\Delta a]$

beam geometry, one may directly apply (7.52) to find the sequence

$$g(j\Delta a) = \begin{cases} \dfrac{1}{4(\Delta a)^2} & \text{for } j = 0 \\[2mm] 0 & \text{for } j \text{ even, } (j \neq 0) \\[2mm] -\dfrac{1}{\pi^2 j^2 (\Delta a)^2} & \text{for } j \text{ odd} \end{cases} \qquad (7.162)$$

for practical implementation. Figure 7.28 shows the principle of backprojection with the required interpolation between individual detector positions.

7.8
Quarter-Detector Offset and Sampling Theorem

As described in the previous chapters, a reliable reconstruction of an object is ensured if any point of the object has been radiographed from all sides. For parallel-beam geometry, the object will of course just have to be illuminated over an angular range of 180°. This is because – due to the physical properties of the X-ray beam – the attenuation is independent of whether the beam passes through the object in a forward (L_1) or in a reverse direction (L_2).

Therefore,

$$\begin{aligned} f * \delta(L_1) &= \int_{\mathbb{R}} f(\mathbf{r}) \delta \left((\mathbf{r}^T \cdot \mathbf{n}_\xi) - \xi_1 \right) d\mathbf{r} \\ &= p_{\gamma_1}(\xi_1) = p_{\gamma_2}(\xi_2) \\ &= \int_{\mathbb{R}} f(\mathbf{r}) \delta \left((\mathbf{r}^T \cdot (-\mathbf{n}_\xi)) - \xi_2 \right) d\mathbf{r} \\ &= f * \delta(L_2) \end{aligned} \qquad (7.163)$$

actually just yields a mirror-inverted projection, which means

$$p_\gamma(\xi) = p_{\gamma + \pi}(-\xi) . \qquad (7.164)$$

Figure 7.29 illustrates this issue once again. The information obtained by acquiring data over angles exceeding 180° is thus redundant.

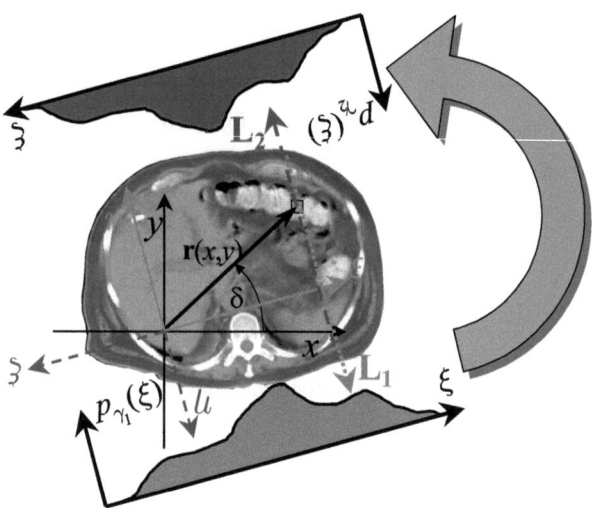

Fig. 7.29. If one illuminates an object under the angles γ and $\gamma + \pi$, then one finds a mirror-inverted projection with a symmetrical detector array, i.e., $p_\gamma(\xi) = p_{\gamma+\pi}(-\xi)$ and thus redundant information. But if the detector array is adjusted slightly asymmetrically, it is possible that the opposite projection samples the space precisely in such a way that the total sampling rate is doubled. One may easily understand that this occurs when the detector array is set off by one-quarter of the detector width

Nevertheless, all modern CT scanners create projections covering 360°. This initially seems to be absurd because – from a mathematical point of view – this redundant information is not necessary, i.e., the radiation dose may be doubled unnecessarily. Therefore, the radiation dose must be lowered if the 360° scan is carried out by choosing a reduced X-ray intensity. Noise must then be reduced in a further processing step where statistically independent noise contributions of centrally opposed projections are averaged.

Practical applications, however, use a different approach since the sampling theorem has consequences on the sampling process in CT. As mentioned earlier, the measurement of the attenuation values with a detector array of a third-generation scanner is, of course, a sampling process. Figure 7.30 shows the sampling situation with an array consisting of D detectors (enumerated from 1 to D here), each having a constant detector width, $\Delta\xi$. The detectors are arranged in an array as densely as possible. The spatial separation of the individual elements by the anti-scatter grid will not be examined here.

Even though the integration along the X-rays represents an averaging process filtering the projection signal like a low-pass, it must nevertheless be ensured that projection aliasing due to undersampling cannot occur. From a physical point of view the object is spatially continuously illuminated in the X-ray fan beam. Therefore, high-contrast edges such as transition zones at a bone/soft-tissue interface generate a large bandwidth. Hence, this yields a spatially, rapidly varying, continuous projection sampled with the detector array.

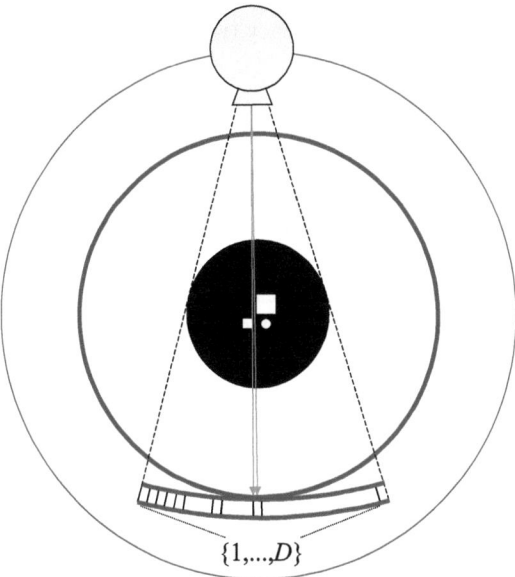

Fig. 7.30. A detector array with D detectors acquires the integral attenuation values. The phantom in the middle of the measuring field has sharp-edged structures and thus produces higher spatial frequencies than the spatial frequency of the detector array. The requirements stipulated in Shannon's sampling theorem are thus not met

As every detector element has a finite width, $\Delta\xi$, with constant sensitivity the measured values are averaged over this width, i.e., the measuring process already includes a natural low pass. Overall, however, the projection signal is a step function[2]. The profile of a single detector element is thus a rectangle function with a width, $\Delta\xi$. As described in ▶ Sect. 4.10, the frequency representation of a rectangle in the spatial domain is the *sinc* function, i.e.,

$$R(u) \propto \frac{\sin(\pi\Delta\xi u)}{\pi\Delta\xi u} . \qquad (7.165)$$

As a consequence, the measured signal is thus not band-limited so band overlap problems may arise. The essential energy contributions are included in the *sinc* function up to the first root. One therefore assumes that the highest available frequency (which is relevant in practice) is determined by the first root, given by

$$\pi\Delta\xi u = \pi . \qquad (7.166)$$

---

[2] A similar problem will also have to be discussed in the next chapter dealing with the secondary reconstruction of a three-dimensional image from a stack.

This means that the maximum frequency (which is again relevant in practice) is given by

$$u = \frac{1}{\Delta \xi} \, . \tag{7.167}$$

However, Shannon's sampling theorem states that the distance between the sampling points must be $\Delta \xi / 2$, i.e., that they must be twice as densely arranged than is actually possible due to the structural design.

According to Shannon's theorem, it would also be possible to filter with a low pass over two detector widths to prevent band overlap problems. However, the related deterioration of the modulation transfer function (MTF) of the system (cf. ▸ Sect. 9.2) would not be acceptable in clinical practice.

In fact, actual implementations do not filter over two detector elements, but increase the spatial sampling frequency with an elegant technical modification of the detector.

The spatial sampling rate is doubled by setting the detector array off by one quarter of the detector width. This is readily illustrated in Fig. 7.31: It can be seen that the central beams are located side-by-side at a distance of half the detector width after one rotation by 180°.

For the off-center beams, it is assumed that the space is sampled in the measuring field on concentric circles around the iso-center of rotation. With a symmetrical detector arrangement, one just needs a 180° rotation around the object to image

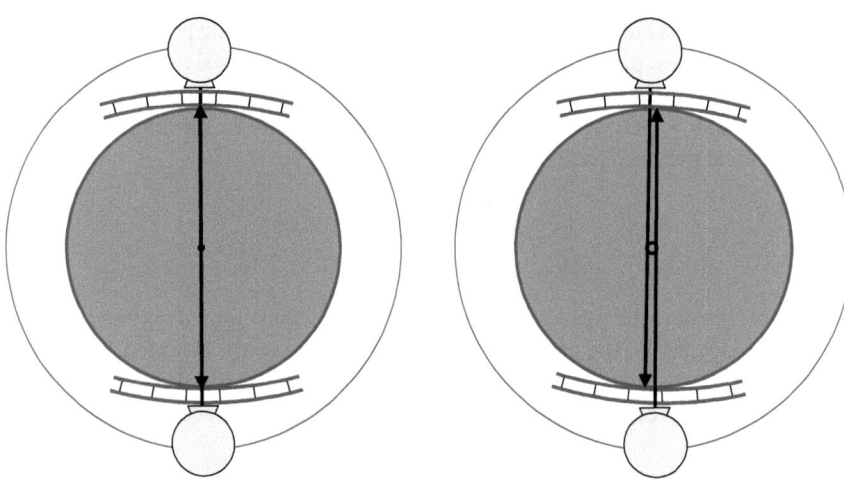

Fig. 7.31. Quarter detector shift in a CT scanner of the third generation. *Left*: Without an offset, the way forth and back yields the same projection value due to the symmetrical attenuation and thus no new information is obtained. *Right*: Taking advantage of an asymmetrical detector arrangement on the way forth and back of the X-ray beam, a sampling pattern can be achieved such that the desired double sampling rate is generated

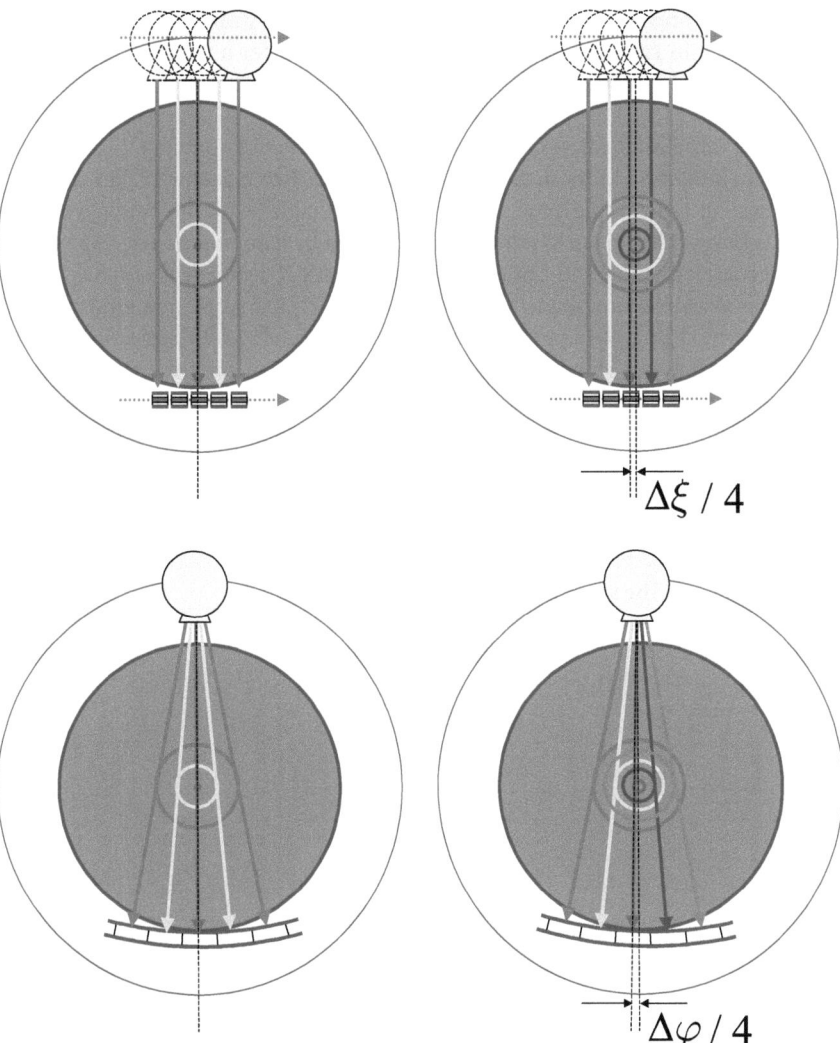

Fig. 7.32. Duplication of the sampling rate by a quarter detector shift to meet the requirements stipulated in Shannon's sampling theorem. The differences in radii of the concentric sampling circles are half as much with the quarter detector shift (*right-hand images*) than with the symmetrical detector arrangement (*left-hand images*). The principle is the same for parallel-beam geometry (*the two upper pictures*) as for fan-beam geometry (*two lower images*). Contrary to parallel-beam geometry, the distances between the sampling circles become smaller and smaller with an increasing angular distance from the central beam with fan-beam geometry

any point of the object from all sides. By setting off the detector array by one quarter of the detector width, the X-ray tube has to be rotated by 360° around the object to illuminate any point of the object under any angle. However, this additional technical effort results in the availability of supplementary sampling circles. The local difference in radii of those is just half as much as for a symmetrical detector arrangement. The spatial sampling rate has thus been doubled – as requested by Shannon.

Figure 7.32 also illustrates that the described concept of doubling the resolution can be put into effect for both fan-beam and parallel-beam geometry. However, the differences in radii are become smaller for larger angles with fan-beam geometry, whereas the parallel-beam system is characterized by a homogeneous resolution.

In general, this method can be used successfully if a sufficient number of projections is available for a single 360° rotation (with typical resolution limits ranging from 6 up to 13 line pairs per centimeter) approximately 1,000 projections are required per rotation (Morneburg 1995), and if the patient movement is negligibly small. This is due to the fact that it takes a certain time to carry out a 360° rotation and the complementary beams from the corresponding opposite side double the spatial sampling process.

Figure 7.33 illustrates the rebinning of the sample points in the Radon space. With the quarter detector shift, the sample points of the angle intervals on the backside $[\pi, 2\pi]$ are just located between the sample points of the angle intervals on the front side $[0, \pi]$. The spatial sampling rate is thus doubled.

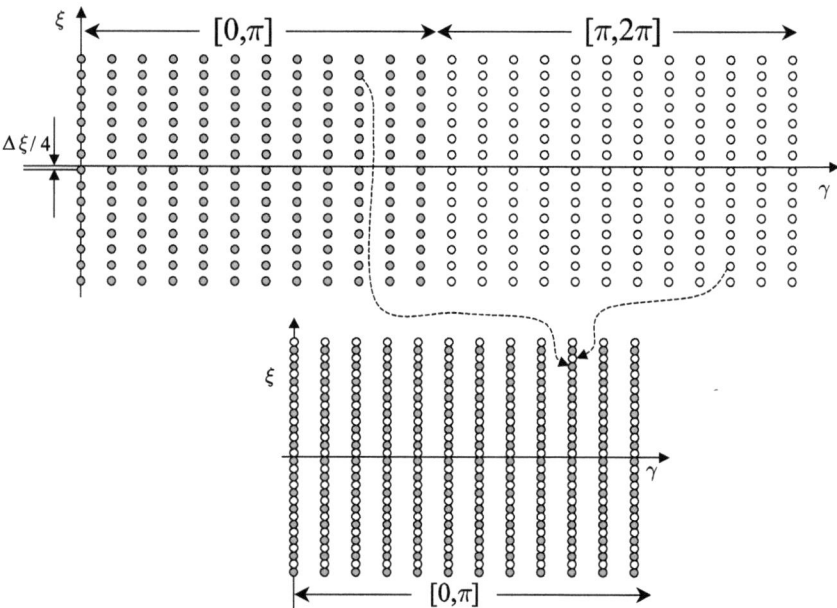

Fig. 7.33. Sampling points in the Radon space (of pencil-beam geometry) with a quarter detector shift. If one correctly rebins the sample points of the angle interval $[\pi, 2\pi]$ into the interval on the front side $[0, \pi]$, the spatial sampling rate is doubled

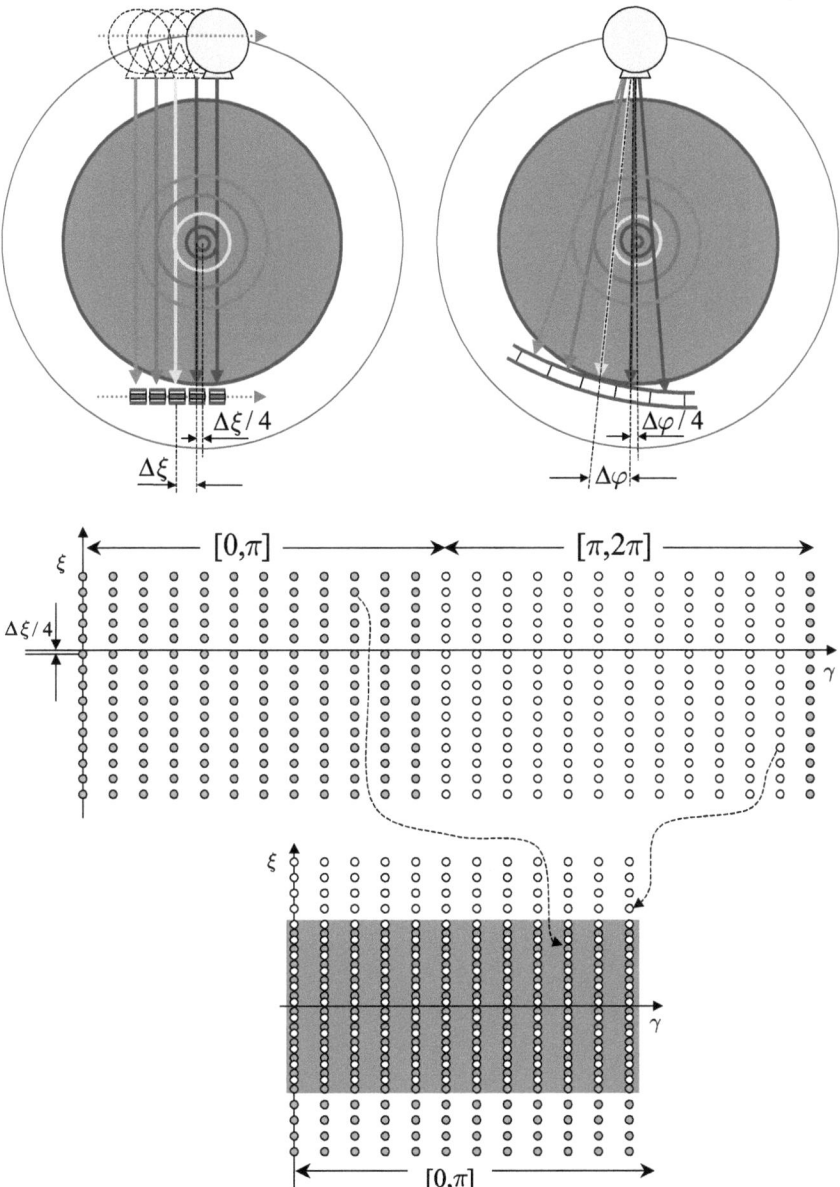

Fig. 7.34. Sampling points in the Radon space (of pencil-beam geometry) with a quarter detector shift and with a supplementary integer detector element offset. If the sample points are properly sorted from the angle interval $[\pi, 2\pi]$ into the interval on the front side $[0, \pi]$, one can see that the spatial sampling rate is only doubled in the central area. This area is highlighted in *gray*. At the same time, one observes a synthetic extension of the detector arrays whereby aliasing problems may occur due to the single spatial resolution

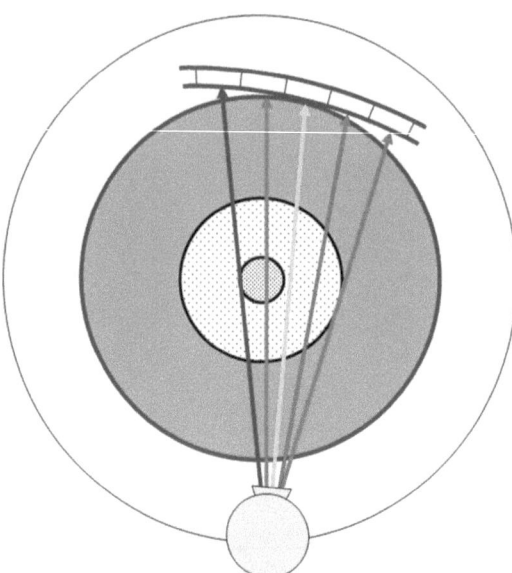

Fig. 7.35. As the detectors are positioned asymmetrically, measurement fields with different spatial resolutions are obtained. Due to the quarter detector shift, the resolution is twice as good in the inner measurement field than in the outer measurement field

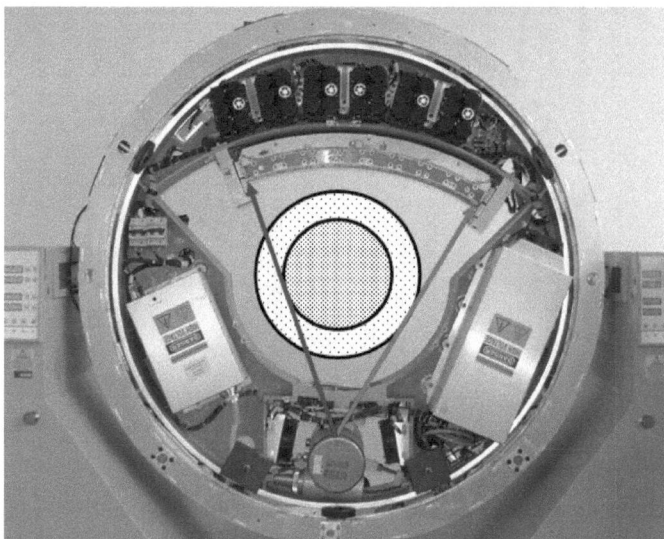

Fig. 7.36. Compact tomograph Philips Tomoscan M. For the outer measurement field it immediately becomes obvious why projections must be measured over 360°. As a result, this measuring method does indeed yield twice the resolution in the inner area

In compact low-cost CT systems, in which the engineers try to reduce the number of detector elements, the shift of the detector array is increased even more so that finally two different reconstruction areas are achieved: one inner area that has twice the spatial resolution of the outer area, due to the quarter detector shift. As the outer area of the measuring field usually covers the extremities, the corresponding loss in resolution does not play a major role in diagnostics.

Figure 7.34 illustrates the location of the sample points in the Radon space. The asymmetrical arrangement of the detector array results in a parallel offset of all points along the detector array axis. The points in the angle interval $[\pi, 2\pi]$ are sorted into the angle interval $[0, \pi]$ such that a central area with twice the sampling rate is generated (in Fig. 7.34, bottom, highlighted in gray). The detector array has thus been artificially extended. However, one only obtains a low spatial resolution in the extended area so that aliasing problems may occur.

However, given that – as already mentioned above – the area that is of diagnostic interest is in general positioned in the center of the measurement field, this effect plays a minor role. Figures 7.35 and 7.36 show the two resolution areas schematically on a compact Philips Tomoscan M^{TM} CT scanner. Another model of a compact scanner with asymmetrical detector array is the Siemens Somatom AR.T/C^{TM} scanner equipped with what is called a 4/5 detector.

8 Three-Dimensional Fourier-Based Reconstruction Methods

Contents

8.1	Introduction	303
8.2	Secondary Reconstruction Based on 2D Stacks of Tomographic Slices	304
8.3	Spiral CT	309
8.4	Exact 3D Reconstruction in Parallel-Beam Geometry	321
8.5	Exact 3D Reconstruction in Cone-Beam Geometry	336
8.6	Approximate 3D Reconstructions in Cone-Beam Geometry	366
8.7	Helical Cone-Beam Reconstruction Methods	394

8.1
Introduction

In this chapter, three-dimensional reconstruction methods will be discussed. The chapter starts in ▸ Sects. 8.2 and 8.3 with a description of pseudo three-dimensional methods that are based on stacks of two-dimensional images. The spiral CT method described in ▸ Sect. 8.3 also follows this pseudo three-dimensional approach, although in this case data acquisition is actually closer to a true three-dimensional reconstruction method. In this context, the term *true* means that the third dimension of the images is not the result of a secondary reconstruction in which the actual, i.e., primary, reconstruction only refers to two-dimensional Radon transforms.

Methods based on three-dimensional Radon transforms will be discussed in ▸ Sects. 8.4 and 8.5. In fact, the exact reconstruction methods based on cone-beam geometry are mathematically more demanding than the two-dimensional methods discussed in ▸ Chaps. 5 and 7. Although there have been studies investigating the use of algebraic reconstruction techniques[1] (cf. ▸ Chap. 6) related to cone-beam geometry, scientific interest is nowadays focused on Fourier methods, which will be exclusively addressed in this chapter.

Figure 8.1 demonstrates the tremendous development in the field of X-ray-based imaging to date. Figure 8.1a shows an excellent projection image in which a multitude of details of the anatomy are already visible. However, in this image

---

[1] ART or the statistical reconstruction methods discussed in ▸ Chap. 6 can be implemented in 3D in a straightforward manner.

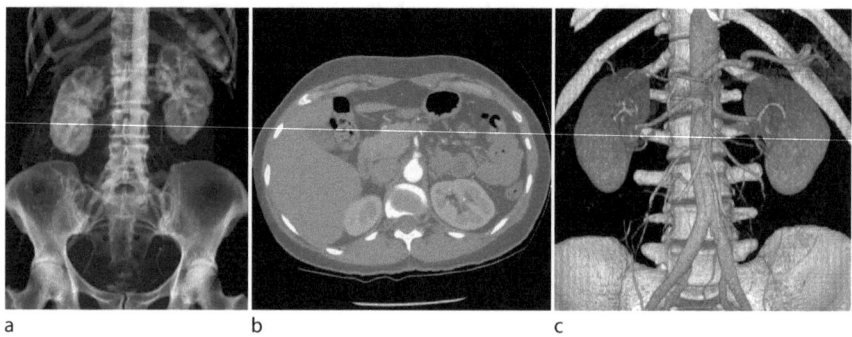

a   b   c

Fig. 8.1. Example of an abdominal situation illustrating the progress in imaging – starting with **a** a simple X-ray picture, then **b** the tomographic reconstruction, and in **c** the volume- or cone-beam CT reconstruction (courtesy of General Electric Medical Systems: Pfoh [2002])

the diagnosis is impaired due to the typical projective superimpositions. Figure 8.1b shows a tomogram slice, which brought along a revolutionary improvement of contrast and thus of diagnostic insight. The steps to volume- or cone-beam CT are no less revolutionary. With CT scanner prototypes equipped with special flat-panel detectors it is possible to show that a resolution of approximately 100 μm can be achieved. Figure 8.1c shows the result of such a volume scan. This high-resolution, isotropic volume representation is realized with a detector element density of approximately 5.000 elements per cm^2 (Pfoh 2002).

8.2
Secondary Reconstruction Based on 2D Stacks of Tomographic Slices

In the previous chapters, only two-dimensional images have been considered. However, today, medical applications of computed tomography (CT) are mainly related to three-dimensional imaging. In a first step, a stack of two-dimensional slices must be acquired. A conventional technique is given by a sequential procedure where the patient, laying on the patient table, is moved slightly in the axial direction of the scanner, i.e., the scanner's z-axis. The table then stops and a complete raw data set of a single slice is measured. Figure 8.2 shows a representative result of this process. The stack of tomogram slices is subsequently used to compute the three-dimensional representation of the depicted anatomy. This procedure is called *secondary reconstruction*.

As for the surface visualization method, it is necessary to select a gray value threshold representing the surface. In this context one has of course a certain degree of freedom, as different objects may be visualized as long as their Hounsfield values are clearly different from one another. The gray-value iso-surface will then be illuminated with a virtual light source so that the corresponding light reflections can be computed to display the result.

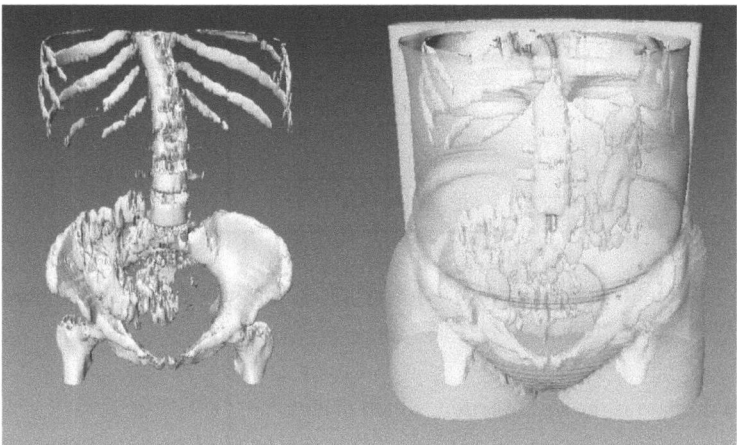

Fig. 8.2. Three-dimensional visualization of the slice stack. For visualization purposes a so-called surface rendering method is used based on a gray value iso-surface representation. If one selects a sufficiently high gray value threshold for the iso-surface, so that virtually only the bone structures are segmented, then the soft tissue parts obviously vanish (*left*). If one selects another threshold representing the soft tissue and the skin, then these objects can be visualized as well (*right*)

This rather simple surface segmentation method fails in all those cases in which distinct objects have very similar gray values. Therefore, it is hardly possible to identify an appropriate threshold that allows these objects to be clearly distinguished. Moreover, if the gray values vary in different areas of certain organs so that they cannot all be represented by a single threshold value, this method fails as well. In these unfortunately prevalent cases, one has to apply more intelligent and potentially also interactive segmentation methods. This wide range of different segmentation methods cannot be discussed here. A corresponding introduction is available, for example, in Lehmann et al. (1997). However, this problem will once again be addressed in ▸ Sect. 9.7.9. There it will be explained how and under which basic constraints the stack of slices is measured. Figure 8.3 shows the patient table feed as a function of the measuring time. This yields the above-mentioned stacks of slices, which, arranged side by side, result in the desired volume image.

From a signal theoretical point of view this sequential slice acquisition must again be interpreted as a sampling process. In this case, the samples are taken along the z-axis. The Nyquist criterion, which was introduced in ▸ Sect. 4.18, must of course also be taken into account in this context.

The intensity profile of the radiation in a single slice to be reconstructed, which is rectangular under ideal conditions, is restricted by a slot-like collimator in the third-generation scanners. Now let the axial width of this rectangle be Δz. As already discussed in ▸ Sect. 4.10, the frequency representation of a rectangle in the

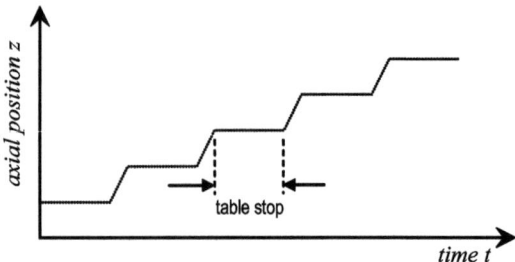

Fig. 8.3. The measurement position as a function of time. Raw data set measurements are taken each time the table stops

spatial domain is the *sinc* function, which means

$$Z(w) \propto \frac{\sin(\pi \Delta z w)}{\pi \Delta z w} . \tag{8.1}$$

Taking into account that the essential energy contributions of the *sinc* function (8.1) are included in the frequency interval up to the first root at

$$\pi \Delta z w = \pi , \tag{8.2}$$

then the maximum frequency of the signal relevant in practice amounts to

$$w = \frac{1}{\Delta z} . \tag{8.3}$$

In order to meet the requirements stipulated in Shannon's sampling theorem, the distance between the sampling points must therefore be $\Delta z/2$. This means that with a slice thickness of Δz the table feed may be as much as $\Delta z/2$. At least two slices must therefore be measured per slice thickness. Figure 8.4 shows the thickness of the slice

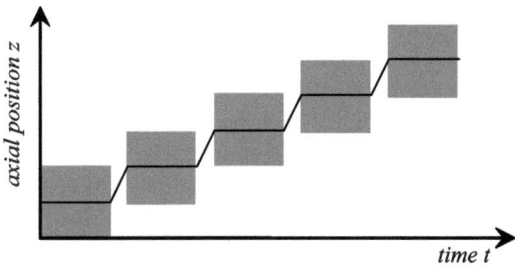

Fig. 8.4. The slice thicknesses in each measuring position as a function of measuring time. The values are always measured during the stop period of the table feed

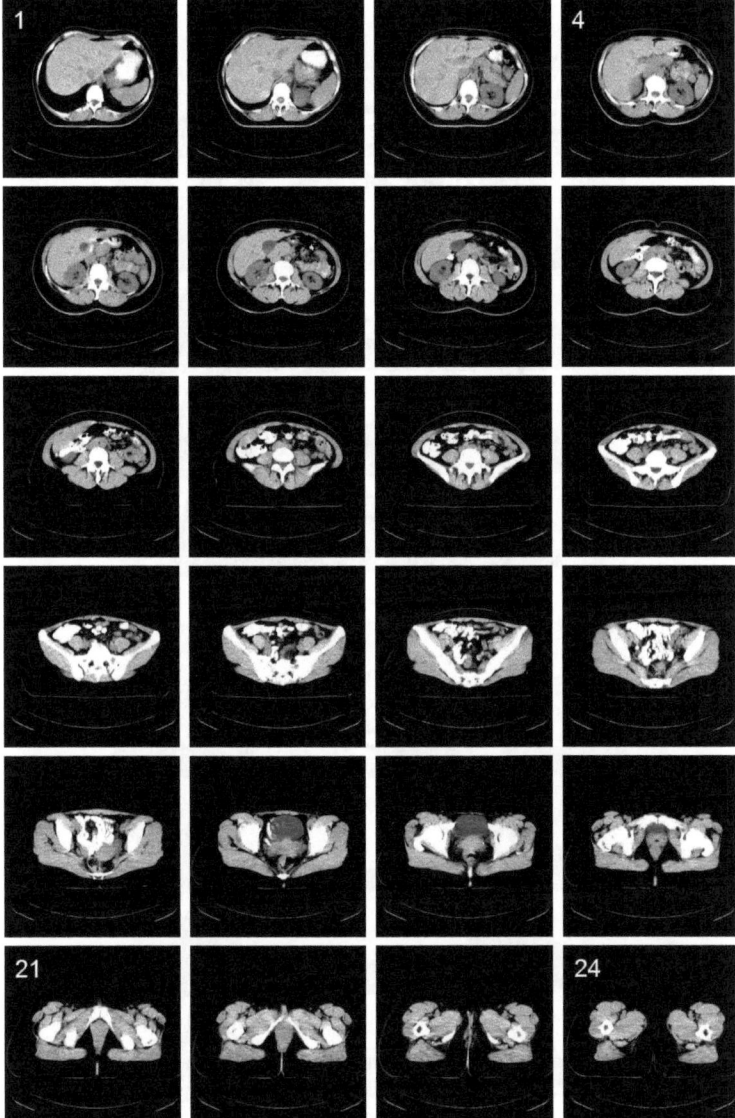

Fig. 8.5. Slice stack from abdomen tomography. A sequence of 24 successive slices is shown line by line

at any measuring position as a function of the time. Typical slice thicknesses range from 1 mm up to 10 mm.

This condition from signal theory is hardly taken into account in practice. In fact, many radiologists just place the slice images of a sequence side by side on their viewing light box to evaluate the radiographs quickly; thus, aliasing plays a sec-

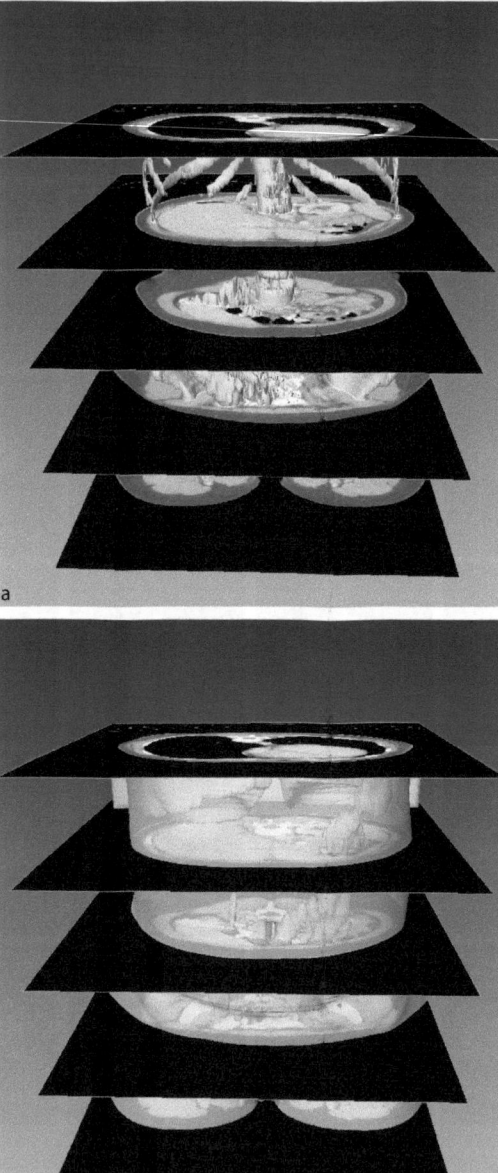

Fig. 8.6a,b. The individual images of a slice stack are arranged one after another. This arrangement results in a volumetric representation and therefore represents the basis for CT to be considered as a three-dimensional modality. A surface rendering technique has been used here

ondary part. Figure 8.5 shows such a sequence of 24 axial slices of the abdominal region.

As already discussed above, the slices of the stack shown in Fig. 8.5 must be placed on top of each other to provide a three-dimensional representation. Figure 8.6 illustrates this procedure schematically with five single slices being arranged using the same three-dimensional data set. The tomograms are stacked one by one with exactly matching pixels[2]. In general, the slice thickness considerably exceeds the pixel size within one slice. This means that the voxels[3] are anisotropic, a fact that must be taken into account when algorithms of three-dimensional image processing are used, for example, by isotropizing the data set by interpolation. Nevertheless, errors might occur during visualization and data evaluation if the above-mentioned sampling theorem is violated during data acquisition. A corresponding example is illustrated in ▸ Sect. 9.7.5.

Figure 8.6a shows five single slices together with a surface representation of the skeleton. For this purpose, the selected gray-value threshold is high enough to suppress any kind of soft tissue in the representation. In Fig. 8.6b the skin and some organs have been segmented by applying another, smaller gray-value threshold. Correspondingly, they are visualized as transparent material.

8.3
Spiral CT

A first step toward a true volume image is the so-called spiral CT method, which was proposed by *Willi Kalender* in 1989 on the annual RSNA[4] conference (Kalender et al. 1989). The inadequacies of the simple slice stack produced by conventional CT are easy to understand. Due to the preset collimation each slice has a certain width, which is also referred to as slice thickness. Within this slice thickness the intensity is weighted with its sensitivity profile – given by the source intensity distribution inside the collimation and the detector sensitivity profile – and then averaged.

This averaging process is a problem in all those cases in which the object is characterized by boundaries, which are angulated with respect to the axial slice, i.e., where structures to be displayed quickly change in the direction of the table feed. In these cases, the averaging process results in a step-like slice stack so that the structure to be displayed has a staircase-like appearance. This artifact formation will

---

[2] This only is true, if the patient holds his breath during data acquisition. Otherwise, motion artifacts appear (cf. ▸ Sect. 9.6.3). Sophisticated image processing methods can be employed to correct for these artifacts in a post-processing step.

[3] Voxel is an artificial word composed of *volume* (*x*) *element*.

[4] RSNA: Radiological Society of North America.

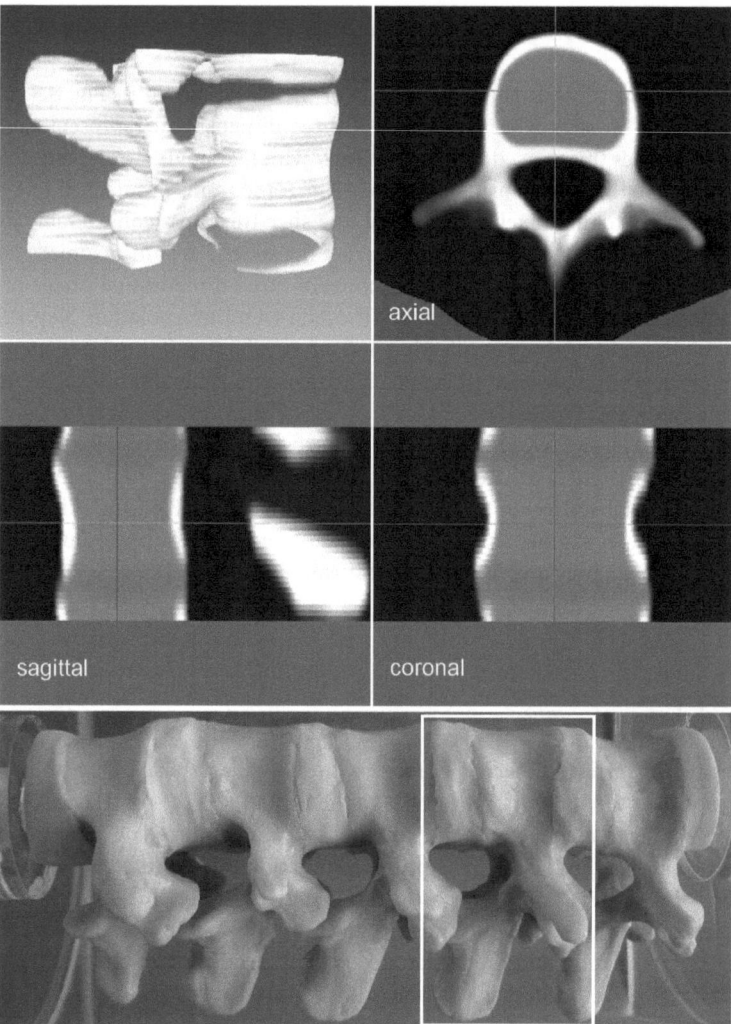

Fig. 8.7. Three-dimensional representation of a part of the vertebral phantom. Orthogonal reformatting, as well as volume rendering, is displayed. The slice thickness selected for the measurement is 1.5 mm. The step-like pattern of the three-dimensional rendering and of the calculated coronal and sagittal sections can be clearly distinguished

once again be discussed in ▸ Sect. 9.7.2. Figure 8.7 shows an example of this effect by means of a volume rendering[5] and MPR model[6] of a vertebral phantom.

---

[5] *Volume rendering*: Each voxel is assigned to a physical light reflectance and scattering. In the computer, this "*data fog*" is illuminated with a virtual light source and the optical events at the voxels are simulated.

[6] MPR: Multi-planar reformatting.

The development of slip-ring technology, which has already been briefly discussed in ▸ Sect. 3.8, made it possible to rotate the sampling unit, i.e., the tube and
detector array system, continuously. If the table feed is kept constant during the rotation, then the X-ray source rotates around the patient on a spiral path[7]; strictly
speaking, this movement describes a helical trajectory. This orbit interpretation is
based on a coordinate system attached to the patient table, since the X-ray source
does of course still run along a circular path. The spiral path only arises from the
patient's view. With this concept, the complete projection data acquisition of an object is possible, and the scan time for a volume could be considerably reduced in
comparison with conventional tomography.

Indeed, it is remarkable that spiral CT technology works at all. An essential requirement for the reconstruction methods described in ▸ Chap. 5 is the completeness of the raw data. This means that an object in the measurement field can only be
reconstructed if all points of the object are illuminated from all sides – i.e., over 180°
(cf. Fig. 8.8 left side). This condition is the reason why artifact-free conventional CT
scans of the heart are practically impossible because the heart motion shifts parts
out of the slice to be reconstructed while the sampling unit rotates around the heart.
Thus, the projection data to be used for the reconstruction process do not fulfill the
consistency condition. Rather, the differences between the projection data of a complete cycle should only be caused by the change of perspective. The reconstructed
tomogram is thus impaired by motion artifacts, which will be described in detail in
▸ Sect. 9.6.3.

In spiral CT scanners, the motion of the objects to be reconstructed is in fact
the decisive innovation compared with conventional CT scanners. Figure 8.8, right-
hand side, illustrates that the object to be examined is no longer scanned in a single plane. The reason for this is that, due to the continuous patient table feed, the
source trajectory is not a closed circular orbit. Therefore, a complete set of raw data
is not available for the reconstruction process – the data are inconsistent in terms of
▸ Chap. 5. In Fig. 8.8, bottom, the raw data space of spiral tomography of the head
is shown. During the patient table feed of 12 cm, projection data of six full cycles of
the sampling unit have been acquired. It can easily be seen how the information in
the sinogram changes while the patient is moved along the z-axis.

Figure 8.9 shows the conceptual difference between the source trajectories for
the conventional and the spiral CT method for a larger volume. The conventional
scanning scheme is shown on the left-hand picture. The tube moves along circular
paths, which are traversed completely and pushed over the patient successively with
discrete stops. Due to the finite collimation, the thickness of the slices is determined
by the width of the circular path. On the right-hand side it can be seen how the
circular path of the tube is combined with the continuous linear table feed in the
spiral CT scanner and thus arranges the data on a helical trajectory.

A decisive difference between the two methods is readily visible in Fig. 8.9. In
conventional tomography there are areas that are not illuminated at all with the

---

[7] One should actually name the orbit a *screw path* because a spiral is a purely 2D structure
in which the radius of the cycling trajectory point is linearly changed.

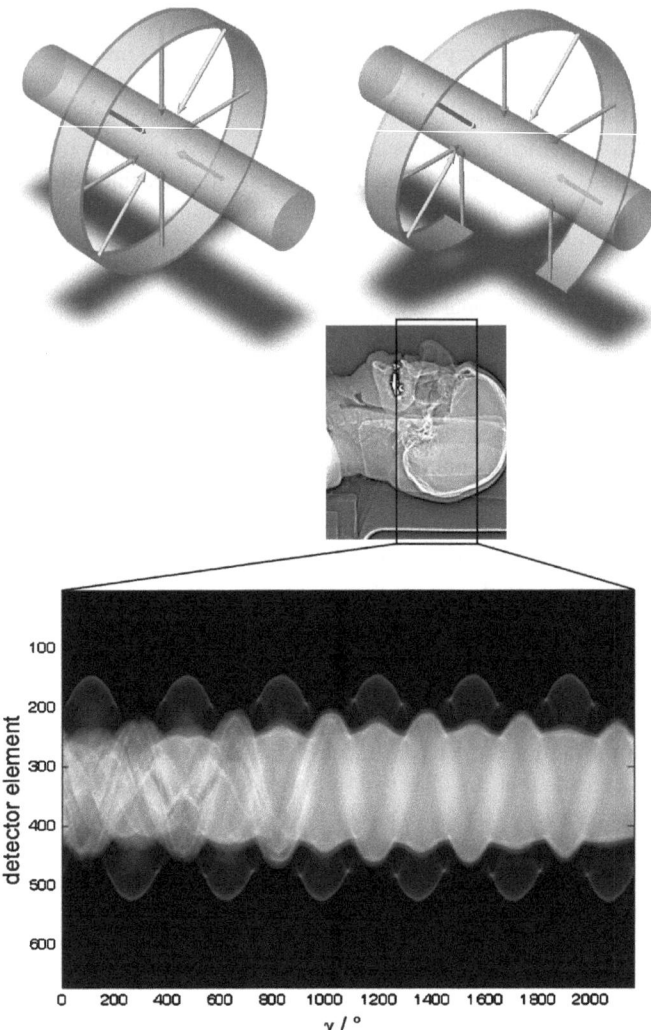

Fig. 8.8. Completeness of the data in conventional CT (*left*). Due to the continuous table feed, the raw data acquired with spiral CT are not complete. The path of the X-ray source is not a circular orbit. *Below*: Sinogram raw data of the head acquired with a spiral acquisition protocol

ratio of table feed and slice width illustrated above. Information required for image reconstruction is thus not available from these areas. The errors related to three-dimensional visualization have already been discussed above.

This is different in spiral CT scanners because – as already described above – these scanners are characterized by continuous data acquisition, i.e., all points along the z-axis are illuminated at least once. Therefore, at least partial information from

Fig. 8.9. Comparison of data acquisition between conventional CT and spiral CT. *Left*: Due to stop points of the table feed, the data are measured on separate circles in conventional CT. *Right*: The combination of the continuous rotation of the sampling unit and the continuous table feed results in a helical arrangement of the raw projection data in spiral CT scanners

each slice, which can be subsequently and arbitrarily selected, will be available for the reconstruction process. And, in fact, information is even available for smaller objects from different projection directions over the entire width of the slice due to the specific slice sensitivity profile determined by the chosen collimation.

The key idea governing the reconstruction process of the spiral CT method is based on the assumption that the missing data of one slice can be completed by interpolation. If this has been done, then the two-dimensional reconstruction procedures described in ▸ Chap. 5 are again available without any restriction. Figure 8.10 shows the simplest principle of a slice interpolation. The helical rise, i.e., the path along which the table is moving during one 360° rotation of the sampling unit, will be denoted with s here.

One may now select an arbitrary slice position, z_r, because no preferred axial position regarding the data basis exists due to the constant table feed. For the selected slice there initially is only a single projection angle, γ_r, for which the pro-

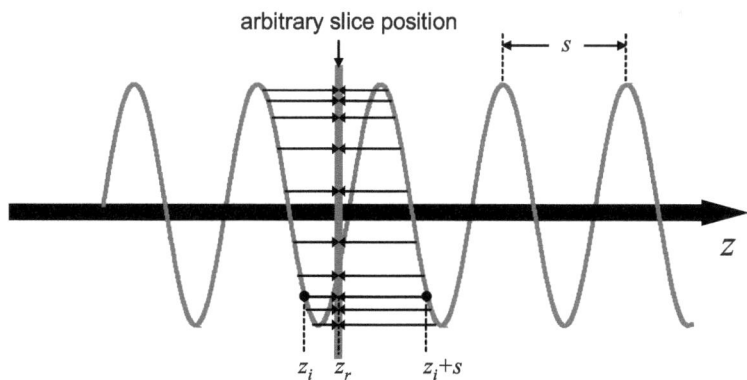

Fig. 8.10. Principle of slice interpolation in spiral CT. The corresponding neighboring projection data, which have actually been measured, contribute to the projection data of the selected slice by interpolation

jection data set $p_{\gamma_r}(\xi)$ is available. The projection data $p_\gamma(\xi)$ of all other projection angles must be provided accordingly by means of interpolation. For this purpose, the data that have not been measured under the other required projection angles in the selected slice position z_r, must be interpolated on the basis of the closest neighboring angles of the helical trajectory that have actually been measured.

The simplest case of such an interpolation scheme is a so-called linear interpolation. For this purpose, one first identifies each position z with a projection angle, so that

$$\gamma_i = \frac{z_i}{s} 360° \tag{8.4}$$

and thus

$$\gamma_i + 360° = \frac{z_i + s}{s} 360° . \tag{8.5}$$

In fact, the actual interpolation is a sum of the left-hand $p_{\gamma_i}(\xi)$ and right-hand $p_{\gamma_i+360°}(\xi)$ neighboring projection, each weighted with the corresponding distance to the selected slice. That is

$$p_{\gamma_r}(\xi) = (1 - \alpha)p_{\gamma_i}(\xi) + \alpha p_{\gamma_i+360°}(\xi) , \tag{8.6}$$

where the weight α is determined by

$$\alpha = \frac{z_r - z_i}{s} . \tag{8.7}$$

The linear interpolation over an angle of 360° already yielded remarkable success when it was introduced by *Willi Kalender*. This interpolation is abbreviated by 360°LI, where LI denotes linear interpolation. Nevertheless, it should be discussed whether the interpolation might yield even better reconstruction results if the acquired data basis could be measured more densely. Figure 8.11 illustrates that the raw spiral projection data of an acquisition interval over 720° are required for an interpolation of a single slice with the 360°LI method.

On the other hand, it is desired to achieve a higher density of the underlying projection data without having to increase the radiation dose. To realize this, one may take advantage of the fact that the attenuation of X-ray beams along the beam path is invariant to direction reversal. This is the reason why projection data must generally only be acquired over an angle interval of 180° in parallel-beam geometry. This means that with each measured projection profile $p_\gamma(\xi)$, one also knows $p_{\gamma+180°}(\xi)$. With this idea, a second helix is obtained, which, compared with the basic helix, has been shifted by 180°. This supplementary helix is shown in Fig. 8.11.

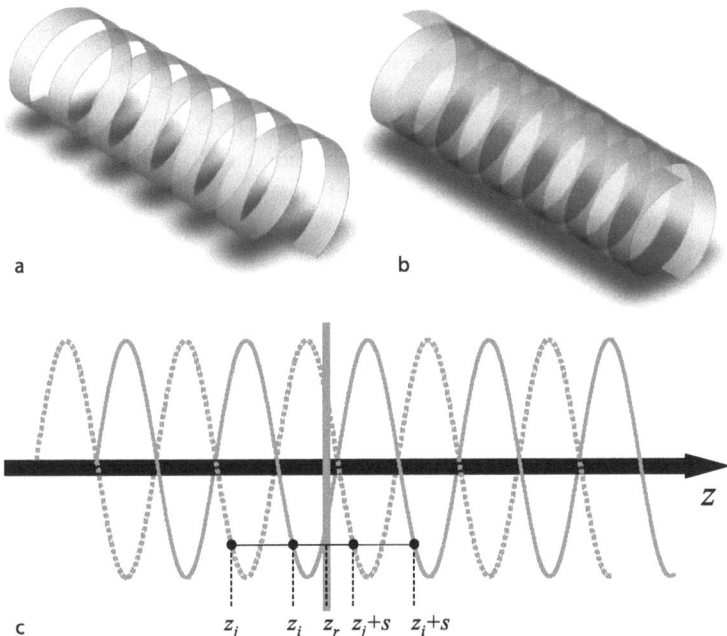

Fig. 8.11. Principle of slice interpolation in spiral CT. Due to the invariance of the attenuation to the reversal of the beam path, one obtains a supplementary helix, which is shifted by $180°$ compared with the basic helix. **a** Simple spiral with pitch $p = 2$, **b** supplementary helix filling the gap, and **c** interpolation with data from the supplementary helix. The complementary trajectory is indicated with a *dashed line*

The supplementary helix is generated regardless of whether parallel-beam geometry is considered or the fan-beam geometry of the third-generation tomographs, which are used mostly nowadays. In ▸ Sect. 7.7.2, it has already been mentioned that an inverse fan can be synthesized by complementary rebinning, which also describes a supplementary helix.

The benefit of the complementary helix is easy to understand. The data points for the interpolation are much closer to each other without increasing the radiation dose. Figure 8.11c illustrates that now the points z_i and $z_j + s$ are the nearest points for the approximation of the projection data point z_r. The frequently applied $180°\,\mathrm{LI}$ method uses the supplementary helix in a linear interpolation scheme. Overall, raw projection data from an acquisition interval of $2(180° + \varphi)$ are required to carry out the interpolation in a fan-beam system with the $180°$ LI method. In ▸ Sect. 7.7, it has already been discussed that one has to measure by the aperture angle φ of the fan beam beyond $180°$ if fan-beam geometry is used instead of parallel-beam geometry to acquire a complete Radon space.

In addition, one may carry out a higher order interpolation if the points z_j and $z_i + s$ indicated in Fig. 8.11c are considered as well. However, this issue will not be discussed in detail here. A survey of the interpolation algorithms used in CT today is available in Kalender (2000). There, the advantages and disadvantages of the corresponding interpolations are discussed in detail. This particularly also applies to devices with multi-line detectors and to methods with ECG triggering.

The specific scan protocol parameters of these methods will be briefly discussed in the following paragraphs. Contrary to conventional CT, one has to consider the table speed v_t as a function of the rotation frequency $1/T_{rot}$ of the sampling unit. This yields the table feed per rotation or the helical rise:

$$s = v_t\,T_{rot}\ .\tag{8.8}$$

If one includes the width or thickness of the X-ray fan d defined by the collimator, then these parameters are usually combined to define the new scan protocol parameter *pitch*

$$p = \frac{s}{d}\ .\tag{8.9}$$

Here, this parameter is defined for a system with a single detector array. Figure 8.12 illustrates the importance of the dimensionless pitch factor, which describes the table feed per rotation as a function of the slice collimation. Spiral acquisition trajectories with the pitch factors $p = 1$, $p = 1.5$, and $p = 2$ (from left to right) are shown. The table feed per rotation can obviously exceed the value of the slice thickness. Correspondingly, it has to be discussed whether or not the sampling theorem is violated in this situation.

In order to ensure that Shannon's sampling theorem is not violated, overlapping slices have to be acquired for the three-dimensional reconstruction in conventional CT as already described in the previous section. This requirement does not necessarily have to be met in spiral CT scanners because – in principle – a sequence of arbitrarily fine slices can be reconstructed from the spiral. Overlapping slices can therefore also be reconstructed retrospectively with this method. Figure 8.13 shows

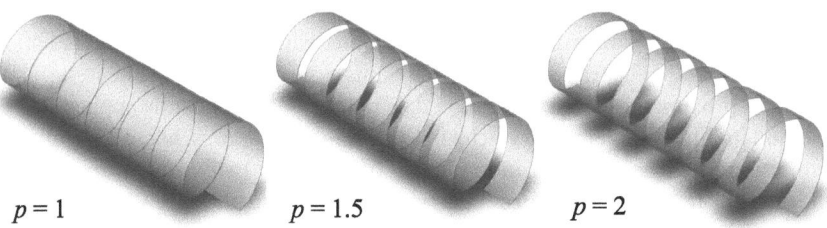

$p = 1$ $p = 1.5$ $p = 2$

Fig. 8.12. Impact of the dimensionless pitch factor (table feed per rotation in units of the slice thickness) in spiral CT. Different pitch factors p ($p = 1$, $p = 1.5$, and $p = 2$) are illustrated from left to right. A pitch factor $p > 1$ reduces the acquisition time since a larger volume in units of the slice thickness can be measured. The dose is obviously smaller for $p = 2$ than for $p = 1$

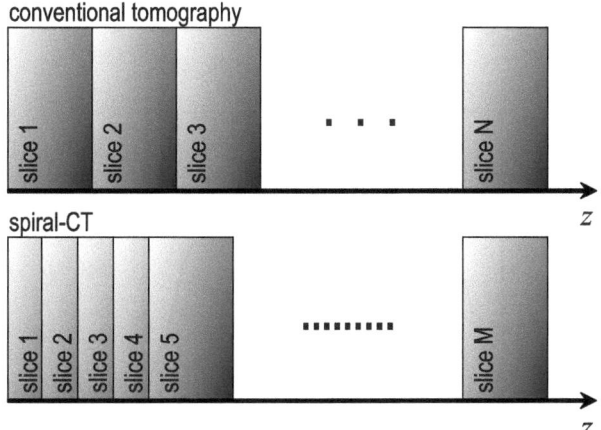

Fig. 8.13. A closer sequence of the images in the slice stack can be achieved by retrospective computation of the single slices with spiral CT ($M \gg N$). Therefore, staircase artifacts (cf. ▸ Sect. 9.7.2) can be reduced within the three-dimensional representation (adapted from Kalender [2000])

schematically the number of slices in spiral CT compared with conventional CT. In this context, one should take into account the fact that the slices are getting closer to each other, but that, on the other hand, the thickness measured for each slice is still determined by the slice thickness selected with the collimator.

Nevertheless, there are still some restrictions concerning the selection of the pitch factor. The pitch factor, p, typically ranges from 1 to 2. First of all, it should be clarified why the method actually works for $p = 1$. As discussed above, the projection data are, according to (8.6), interpolated from neighboring real projections on the helix. If one assumes that a very small, δ-shaped object has to be measured, and if one further assumes that a theoretical slice thickness can be adjusted with the sampling unit having a δ-shaped sensitivity profile, then the spiral CT method would not be successful at all. This is because the object would actually only be visible as a projection under a single angle. As a consequence, the Radon space for this object, approximated with the spiral CT interpolation scheme, would include inconsistent projection data, since no projection outside of the source orbit point at z_r contains information of the δ-object.

However, in real systems one neither measures such small objects nor is the sampling unit provided with a δ-shaped sensitivity profile. As long as a real object is located inside the profile of one slice thickness while the patient table moves forward, the corresponding neighboring projections also provide information on this object. Therefore, an interpolation finally makes sense. But why does a pitch factor $p > 1$ actually work? In order to answer this question one may once again have a look at Fig. 8.11. Due to the virtual supplementary helix, which has been shifted by 180°, it is theoretically possible to track all object points for a sufficiently long sampling period in the projections even with a pitch factor of up to $p = 2$. Data loss

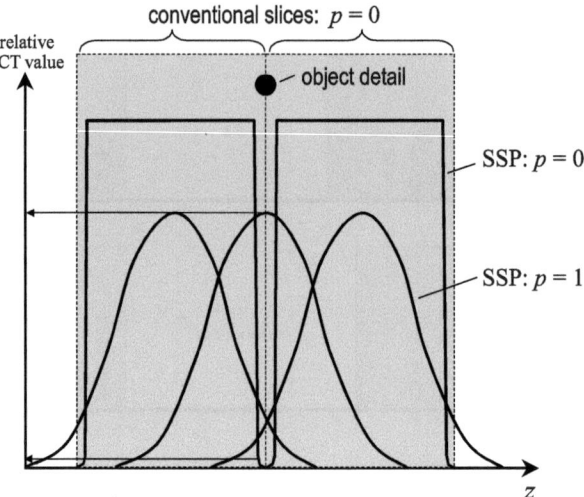

Fig. 8.14. Advantage of spiral CT regarding imaging of small structures. The slice sensitivity profiles (SSP) for conventional and spiral CT are represented versus the axial z-axis. If very small details are located exactly on the border of the conventional slices, then the image intensity may be very small due to the steeply descending sensitivity profile. In spiral CT, the maximum value of the slice sensitivity profile is generally smaller, but, on the other hand, also softer and wider. As arbitrarily selected slice positions can be reconstructed retrospectively, the structure detail can subsequently be positioned exactly in the area of the maximum value of the spiral sensitivity profile (adapted from Bushberg et al. [2002])

resulting in artifacts may arise for a pitch factor of $p > 2$, especially if rather small structures are scanned. The pitch factors selected in practice range from $1 < p < 1.8$.

Figure 8.14 schematically illustrates why conventional CT may under certain circumstances result in poorer image quality for small details than spiral CT. If a small structure is located exactly on the edge of two neighboring slices in a conventional tomogram with non-overlapping, densely arranged slices, then the intensity is equally distributed between the two slices even with a perfectly rectangular slice sensitivity profile (SSP). Unfortunately, in practice, the SSP is only approximately rectangular so that the measured response of the system is further weakened as a function of the steepness of the profile at the slice border.

On the other hand, a single-slice sensitivity profile of each reconstructed slice is broader and flatter in spiral CT than in conventional CT. This is due to the convolution of the conventional profile with the table feed function. The higher the pitch factor the broader and flatter the slice sensitivity profile. Contrary to conventional tomography, in which the spatial location of the structure details in relation to the location of the slices is random and can only be corrected by a supplementary measurement with an additional radiation dose, it is possible to position the structure detail exactly in the center of the SSP in spiral CT. This can be achieved by computing the slices at any slice position retrospectively.

It should be emphasized that the potential of the spiral CT method to compute arbitrarily dense overlapping slices does not mean that the spatial resolution in the axial direction can be increased arbitrarily. The axial resolution is always determined by the slice thickness given by the collimator (cf. Fig. 8.13). However, it is of course possible to reconstruct images every 1 mm for a helix scanned with a fan-beam thickness of 10 mm. But this does not finally result in a resolution of 1 mm. In analogy to Bushberg et al. (2002), one should therefore distinguish between a 1-mm *sampling pitch* and 10-mm *sampling aperture*.

The demand for ever faster CT scanners has accelerated the development of multi-line detector arrays, which also run along a helical path. Figure 8.15 schemat-

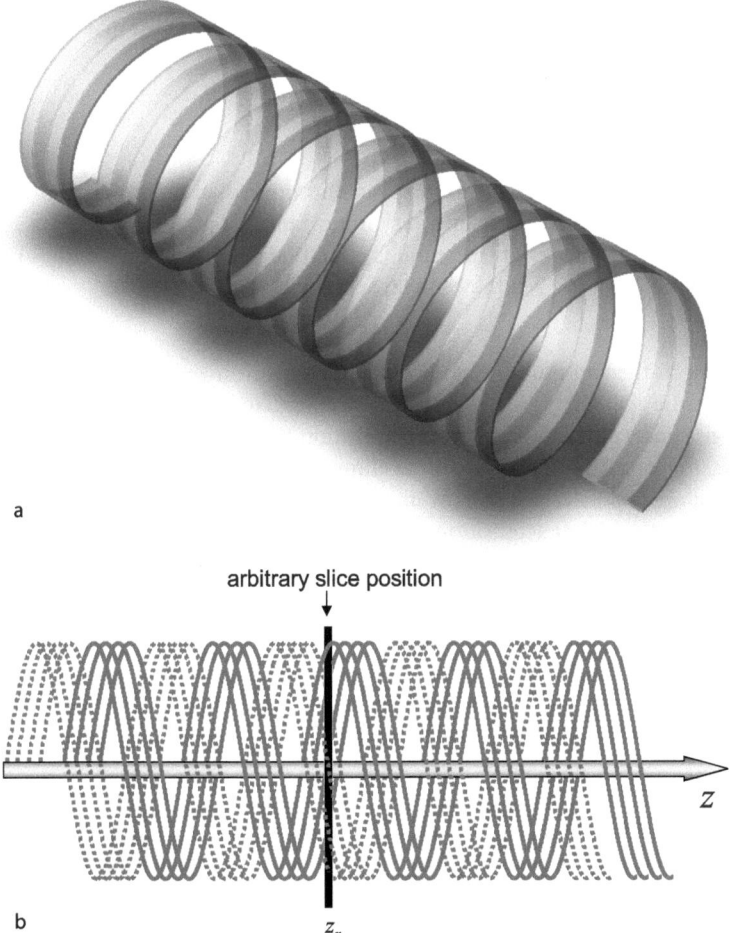

Fig. 8.15. a Helical orbits of a 4-line X-ray detector system. **b** Due to the concept of the inverse fan corresponding supplementary helices are shifted by 180° for each real orbit. These are shown with *dashed lines*

ically shows the corresponding paths of the actual detector lines as well as the corresponding supplementary helical paths for a 4-line detector array shifted by 180°. In this context, it is remarkable that only a single X-ray focus is used resulting in an X-ray cone where mathematical analysis is computationally expensive, especially in the case of a large cone aperture angle. However, good results can be achieved with an approximate reconstruction method for small apertures in the axial direction ($< 2°$). A discussion of the cone-beam reconstruction method will be the subject of the following sections of this chapter.

Figure 8.15 exemplarily illustrates that the neighboring points used to interpolate the projection data for the slice at the axial position, z_r, are even closer to each other than in the geometrical situation shown in Fig. 8.11. In order to evaluate this fact with respect to image quality and radiation dose, one has to consider the number of lines M of the detector array related to the pitch factor so that

$$p = \frac{s}{Md} \, . \tag{8.10}$$

Both the increased data acquisition speed and the reduction of radiation dose for pitch factors larger than one play a very important role in this context.

The main advantages of the spiral CT method can be summarized as listed in the following (Gay and Matthews 1998; Seeram 2001):

- Due to the continuous data acquisition synchronized with the table feed into the *gantry* it is possible to measure complete organs and larger volumes on a short time scale. Shorter acquisition times result from both, a pitch factor, $p > 1$, and, of course, the elimination of the table feed stop points that are characteristic of conventional CT.
- Due to the higher acquisition speed the number of artifacts associated with patient motion is reduced.
- Contrary to conventional CT, a complete data set can be acquired. Staircasing artifacts (cf. ▶ Sect. 9.7.2) in the three-dimensional reconstruction can thus be eliminated.
- Slices can be reconstructed retrospectively at any axial position. Very small structures can also be visualized in this way.

However, it must be taken into account that the X-ray tube is subjected to an extreme load due to this continuous acquisition process. This particularly applies to the thermal stress. The reader of this book can finally evaluate the imaging quality of the spiral method by comparing the reconstructions shown in Figs. 8.7 and 8.16. Figure 8.16 shows isotropic voxels with an edge length of 0.5 mm.

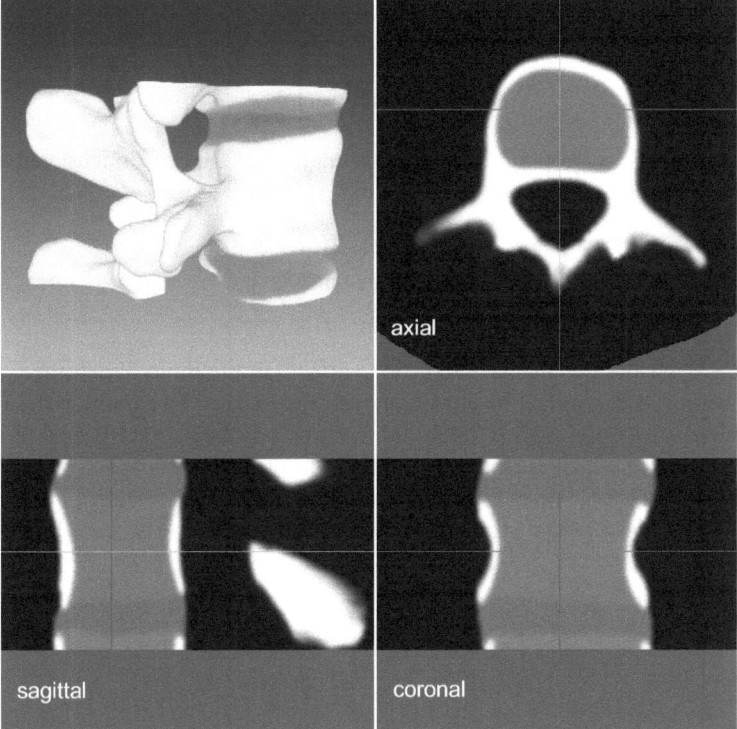

Fig. 8.16. The success of the spiral CT method can be readily understood if again the vertebral phantom shown in Fig. 8.7 is used for evaluation. A uniform, i.e., the isotropic resolution of 0.5 mm has been computed retrospectively. Staircasing artifacts, as occurring in Fig. 8.7, are considerably reduced here

8.4
Exact 3D Reconstruction in Parallel-Beam Geometry

The following sections will no longer deal with secondary reconstructions based on slice stacks, but rather with "true" three-dimensional reconstruction methods.

8.4.1
3D Radon Transform and the Fourier Slice Theorem

To define the Radon transform in the three-dimensional space corresponding to (5.18), one first has to study how the 1D integration line defined along the X-ray beam is now converted into a 2D surface. The projection integrals are thus converted into surface integrals. This means that a three-dimensional Radon value is a surface integral in the three-dimensional object space.

In analogy to the definition of the unit vector

$$\mathbf{n}_\xi = \begin{pmatrix} \cos(\gamma) \\ \sin(\gamma) \end{pmatrix} , \tag{5.11}$$

determining the corresponding projection angle γ for the two-dimensional reconstruction method (cf. Fig. 8.17a), the vector

$$\mathbf{n}_\xi = \begin{pmatrix} \cos(\gamma) \sin(\vartheta) \\ \sin(\gamma) \sin(\vartheta) \\ \cos(\vartheta) \end{pmatrix} \tag{8.11}$$

suitably describes the analog "projection surface" in the three-dimensional space. Each surface **A** can in fact be unambiguously determined by a point in the Radon space if one interprets this point as the root point of the surface normal of **A**, the prolongation of which runs through the origin of the Radon space. Figure 8.17b shows the surface **A**, which is the key element of the Radon transform in the three-dimensional description. Analogous to the description in the two-dimensional space, the vector

$$\boldsymbol{\xi} = \begin{pmatrix} \xi_x \\ \xi_y \\ \xi_z \end{pmatrix} = \xi \cdot \mathbf{n}_\xi = \begin{pmatrix} \xi \cos(\gamma) \sin(\vartheta) \\ \xi \sin(\gamma) \sin(\vartheta) \\ \xi \cos(\vartheta) \end{pmatrix} \tag{8.12}$$

can be defined, which unambiguously determines the Radon value.

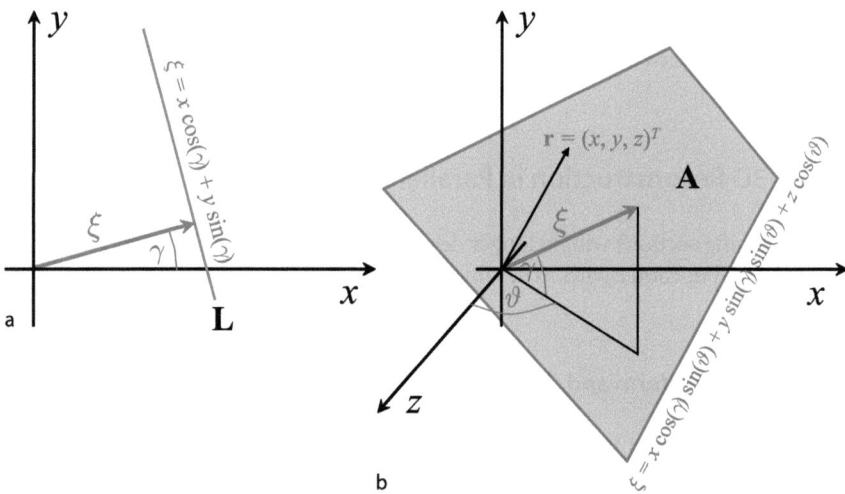

Fig. 8.17. a Geometrical description of the path of integration along a line **L** for the two-dimensional reconstruction problem. **b** Description of an analog integration surface **A** for the three-dimensional reconstruction problem

A point $\mathbf{r} = (x, y, z)^T$ is located in the plane $\mathbf{A}$, if the inner product

$$\xi = (\mathbf{r}^T \cdot \mathbf{n}_\xi) = x\cos(\gamma)\sin(\vartheta) + y\sin(\gamma)\sin(\vartheta) + z\cos(\vartheta) \qquad (8.13)$$

is constant. If $\mathbf{A}$ is a surface in the object space, then the sampling process can be described by

$$f(x, y, z) * \delta(\mathbf{A}) = \int_{\mathbb{R}^3} f(\mathbf{r})\delta(\mathbf{r} - \mathbf{A})\,d\mathbf{r} \qquad (8.14)$$

i.e., in full analogy to (5.16) or

$$f * \delta(\mathbf{A}) = \int_{\mathbf{r} \in \mathbf{A}} f(\mathbf{r})\,d\mathbf{r} . \qquad (8.15)$$

Due to the sift property of the δ-distribution, (8.14) yields all points $\mathbf{r}$ of the object space located on the surface $\mathbf{A}$. As the surface $\mathbf{A}$ is determined by $(\mathbf{r}^T\mathbf{n}_\xi) = \xi$, the projection integral may again be expressed by

$$
\begin{aligned}
f * \delta(\mathbf{A}) &= \int f(\mathbf{r})\delta((\mathbf{r}^T \cdot \mathbf{n}_\xi) - \xi)\,d\mathbf{r} \\
&= \int_{-\infty}^{\infty}\int_{-\infty}^{\infty}\int_{-\infty}^{\infty} f(x, y, z)\delta(x\cos(\gamma)\sin(\vartheta) \\
&\quad + y\sin(\gamma)\sin(\vartheta) + z\cos(\vartheta) - \xi)\,dx\,dy\,dz \\
&= p(\xi)
\end{aligned}
\qquad (8.16)
$$

i.e., in full analogy to (5.18). If the two-dimensional notation is consequently extended, the projection now changes with the two projection angles, γ and ϑ, so that

$$p = p(\xi, \gamma, \vartheta) = p_{\gamma,\vartheta}(\xi) . \qquad (8.17)$$

$p_{\gamma,\vartheta}(\xi)$ is called the Radon transform of the three-dimensional object. One may write

$$f(x, y, z) \; \circ\!\!\xrightarrow{\mathcal{R}_3}\!\!\bullet \; f * \delta(\mathbf{A}) = p_{\gamma,\vartheta}(\xi) \qquad (8.18)$$

or

$$p_{\gamma,\vartheta}(\xi) = \mathcal{R}_3\{f(x, y, z)\} . \qquad (8.19)$$

The three-dimensional Radon space consists of the surface integral values at the points (ξ, γ, ϑ) indicated by their spherical coordinates. The Fourier-slice theorem, which has already played a key role when we discussed the inversion of the Radon transform in ▸ Sect. 5.3 may also be formulated in three dimensions. For this purpose, one initially computes the Fourier transform of the Radon space again for all ξ of the corresponding projection angles γ and ϑ

$$P_{\gamma,\vartheta}(q) = \int_{-\infty}^{\infty} p_{\gamma,\vartheta}(\xi) e^{-2\pi i q \xi} d\xi, \tag{8.20}$$

where the projection values $p_{\gamma,\vartheta}(\xi)$ in (8.20) result from the projection integral extended into the two-dimensional space (cf. (5.28)), i.e.,

$$
\begin{aligned}
P_{\gamma,\vartheta}(q) &= \int_{-\infty}^{\infty} \left\{ \int_{-\infty}^{\infty} \int_{-\infty}^{\infty} \mu(\eta,\sigma,\xi) \, d\sigma \, d\eta \right\} e^{-2\pi i q \xi} d\xi \\
&= \int_{-\infty}^{\infty} \int_{-\infty}^{\infty} \int_{-\infty}^{\infty} \mu(\eta,\sigma,\xi) \, e^{-2\pi i q \xi} d\xi \, d\sigma \, d\eta .
\end{aligned}
\tag{8.21}
$$

Figure 8.18 shows the relationship between the fixed patient coordinate system and the rotating sampling system – consisting of the integration surface **A** – used for the description of the attenuation coefficients in the three-dimensional object space.

In the (x, y, z)-patient coordinate system, which still is a Cartesian but nevertheless a fixed system that has been rotated by γ and ϑ and shifted by the distance ξ

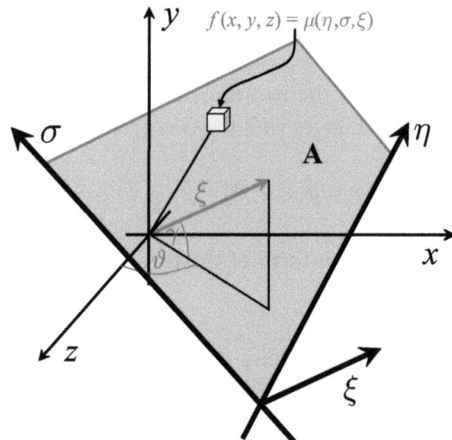

Fig. 8.18. Relation between the fixed (x, y, z) patient coordinate system and the coordinate system (ξ, η, σ) of the rotating surface **A**

one may write

$$P_{\gamma,\vartheta}(q) = \int\limits_{-\infty}^{\infty}\int\limits_{-\infty}^{\infty}\int\limits_{-\infty}^{\infty} \mu(\eta(x,y,z),\sigma(x,y,z),\xi(x,y,z))\,e^{-2\pi i q(\mathbf{r}^T\cdot\mathbf{n}_\xi)}\,dx\,dy\,dz$$

$$= \int\limits_{-\infty}^{\infty}\int\limits_{-\infty}^{\infty}\int\limits_{-\infty}^{\infty} f(x,y,z)\,e^{-2\pi i q(\mathbf{r}^T\cdot\mathbf{n}_\xi)}\,dx\,dy\,dz\,. \tag{8.22}$$

Since both systems are of Cartesian nature, the integration element $d\xi\,d\sigma\,d\eta$ can be directly replaced by $dx\,dy\,dz$, because the corresponding *Jacobian* is equal to one.

On the other hand, the Cartesian formulation of the Fourier transform of the three-dimensional object in the fixed (x,y,z)-patient coordinate system is determined by

$$F(u,v,w) = \int\limits_{-\infty}^{\infty}\int\limits_{-\infty}^{\infty}\int\limits_{-\infty}^{\infty} f(x,y,z)\,e^{-2\pi i(xu+yv+zw)}\,dx\,dy\,dz\,. \tag{8.23}$$

Regarding the changes in the coordinate systems it should be recalled here that a point in the frequency domain with the spherical coordinates (q,γ,ϑ) is characterized by the Cartesian frequency coordinates

$$u = q\cos(\gamma)\sin(\vartheta)$$
$$v = q\sin(\gamma)\sin(\vartheta) \tag{8.24}$$
$$w = q\cos(\vartheta)\,.$$

Therefore, the exponent in (8.23) may read

$$F(u,v,w) =$$

$$= \int\limits_{-\infty}^{\infty}\int\limits_{-\infty}^{\infty}\int\limits_{-\infty}^{\infty} f(x,y,z)\,e^{-2\pi i(xq\cos(\gamma)\sin(\vartheta)+yq\sin(\gamma)\sin(\vartheta)+zq\cos(\vartheta))}\,dx\,dy\,dz$$

$$= \int\limits_{-\infty}^{\infty}\int\limits_{-\infty}^{\infty}\int\limits_{-\infty}^{\infty} f(x,y,z)\,e^{-2\pi i q(x\cos(\gamma)\sin(\vartheta)+y\sin(\gamma)\sin(\vartheta)+z\cos(\vartheta))}\,dx\,dy\,dz$$

$$= \int\limits_{-\infty}^{\infty}\int\limits_{-\infty}^{\infty}\int\limits_{-\infty}^{\infty} f(x,y,z)\,e^{-2\pi i q(\mathbf{r}^T\cdot\mathbf{n}_\xi)}\,dx\,dy\,dz\,. \tag{8.25}$$

The integrand in the last line of (8.25) corresponds to the integrand of (8.22) so that the three-dimensional version of the Fourier-slice theorem can be summarized with

$$
\begin{aligned}
F(u(q, \gamma, \vartheta), v(q, \gamma, \vartheta), w(q, \gamma, \vartheta)) \\
= F(q \cos(\gamma) \sin(\vartheta), q \sin(\gamma) \sin(\vartheta), q \cos(\vartheta)) \\
= F_{\text{spherical}}(q, \gamma, \vartheta) \\
= P_{\gamma, \vartheta}(q) \, .
\end{aligned}
\tag{8.26}
$$

8.4.2
Three-Dimensional Filtered Backprojection

Starting from the inverse three-dimensional Fourier transform of the object spectrum

$$
f(x, y, z) = \int\limits_{-\infty}^{\infty} \int\limits_{-\infty}^{\infty} \int\limits_{-\infty}^{\infty} F(u, v, w) \, e^{2\pi i (xu + yv + zw)} \, du \, dv \, dw \,,
\tag{8.27}
$$

spherical coordinates given by the system (8.24) are introduced here. As already indicated in the previous chapters, the infinitesimal volume element $du \, dv \, dw$ must be replaced by $J \, dq \, dy \, d\vartheta$. The *Jacobian J* is here determined by

$$
J \equiv \det \left(\frac{\partial(u, v, w)}{\partial(q, \gamma, \vartheta)} \right) =
\begin{vmatrix}
\dfrac{\partial u}{\partial q} & \dfrac{\partial v}{\partial q} & \dfrac{\partial w}{\partial q} \\[2mm]
\dfrac{\partial u}{\partial y} & \dfrac{\partial v}{\partial y} & \dfrac{\partial w}{\partial y} \\[2mm]
\dfrac{\partial u}{\partial \vartheta} & \dfrac{\partial v}{\partial \vartheta} & \dfrac{\partial w}{\partial \vartheta}
\end{vmatrix}
\tag{8.28}
$$

$$
=
\begin{vmatrix}
\cos(\gamma) \sin(\vartheta) & \sin(\gamma) \sin(\vartheta) & \cos(\vartheta) \\
-q \sin(\gamma) \sin(\vartheta) & q \cos(\gamma) \sin(\vartheta) & 0 \\
q \cos(\gamma) \cos(\vartheta) & q \sin(\gamma) \cos(\vartheta) & -q \sin(\vartheta)
\end{vmatrix}
= q^2 \sin(\vartheta) \, .
$$

If one substitutes the new infinitesimal volume element in spherical coordinates one finds the inverse Fourier transform

$$
f(x, y, z) = \int\limits_{\vartheta=0}^{\pi} \int\limits_{\gamma=0}^{2\pi} \int\limits_{q=0}^{\infty} F(q \cos(\gamma) \sin(\vartheta), q \sin(\gamma) \sin(\vartheta), q \cos(\vartheta)) \cdot \ldots
$$

$$
\ldots \cdot e^{2\pi i q (x \cos(\gamma) \sin(\vartheta) + y \sin(\gamma) \sin(\vartheta) + z \cos(\vartheta))} q^2 \sin(\vartheta) \, dq \, dy \, d\vartheta \,.
\tag{8.29}
$$

According to the Fourier slice theorem (8.26) it holds that

$$F(q\cos(\gamma)\sin(\vartheta), q\sin(\gamma)\sin(\vartheta), q\cos(\vartheta)) = P_{\gamma,\vartheta}(q),\qquad(8.30)$$

so that

$$f(x, y, z) = \int\limits_{\vartheta=0}^{\pi}\int\limits_{\gamma=0}^{2\pi}\int\limits_{q=0}^{\infty} q^2 \sin(\vartheta) P_{\gamma,\vartheta}(q)$$
$$\cdot\, e^{2\pi i q(x\cos(\gamma)\sin(\vartheta)+y\sin(\gamma)\sin(\vartheta)+z\cos(\vartheta))}\, dq\, dy\, d\vartheta.\qquad(8.31)$$

For the inner integral, i.e., integration with respect to the frequency variable q, one may find the abbreviated expression

$$f(x, y, z) = \int\limits_{\gamma=0}^{2\pi}\int\limits_{\vartheta=0}^{\pi}\left\{\int\limits_{q=0}^{\infty} q^2 P_{\gamma,\vartheta}(q) e^{2\pi i q\xi}\, dq\right\}\sin(\vartheta)\, d\vartheta\, d\gamma$$

$$= \frac{1}{2}\int\limits_{\gamma=0}^{2\pi}\int\limits_{\vartheta=0}^{\pi}\left\{\int\limits_{q=-\infty}^{\infty} q^2 P_{\gamma,\vartheta}(q) e^{2\pi i q\xi}\, dq\right\}\sin(\vartheta)\, d\vartheta\, d\gamma\qquad(8.32)$$

$$= \frac{1}{2}\int\limits_{\gamma=0}^{2\pi}\int\limits_{\vartheta=0}^{\pi} h_{\gamma,\vartheta}(\xi)\sin(\vartheta)\, d\vartheta\, d\gamma.$$

As mentioned above, for a fixed point $\mathbf{r} = (x, y, z)^{\mathrm{T}}$ and fixed projection angles γ and ϑ the projection surface $\mathbf{A}$ is determined by ξ. Obviously, in three dimensions,

$$h_{\gamma,\vartheta}(\xi) = \int\limits_{-\infty}^{\infty} q^2 P_{\gamma,\vartheta}(q) e^{2\pi i q\xi}\, dq\qquad(8.33)$$

is the filtered projection signal obtained by multiplying the Fourier transform of $p_{\gamma,\vartheta}(\xi)$ with q^2 in the frequency domain. The backprojection, i.e., smearing back the filtered projection in a reverse direction, is given in the last line of (8.32).

8.4.3
Filtered Backprojection and Radon's Solution

Similar to ▶ Sect. 5.10 it can also be shown here that the filtered backprojection using $h_{\gamma,\vartheta}(\xi)$ of (8.33) can be expressed in terms of the original solution by Johann Radon in 1917 (cf. introduction of ▶ Chap. 5). Radon's solution can be obtained even more easily than demonstrated in ▶ Sect. 5.10 because, thanks to the square of q in (8.33), the complicated Fourier transform of the absolute value function is not involved here. If one starts from (8.33) and takes the generalization of rule (4.84) for the inverse Fourier transform of the nth derivative of the spatial function $f(x)$

$$\int_{-\infty}^{+\infty} (\mathrm{i}2\pi u)^n F(u)\, \mathrm{e}^{\mathrm{i}2\pi ux}\, du = \frac{d^n f(x)}{dx^n}\,, \tag{8.34}$$

into account, then (8.33) reads

$$
\begin{aligned}
h_{y,\vartheta}(\xi) &= -\frac{1}{4\pi^2} \int_{-\infty}^{\infty} (\mathrm{i}2\pi q)^2 P_{y,\vartheta}(q)\, \mathrm{e}^{2\pi \mathrm{i} q\xi}\, dq \\
&= -\frac{1}{4\pi^2} \frac{\partial^2 P_{y,\vartheta}(\xi)}{\partial \xi^2}\,,
\end{aligned}
\tag{8.35}
$$

whereby the partial derivative indicates that the projection integral is a variable depending on ξ and the projection angles y and ϑ. The reconstruction formula can then be derived from (8.32) into which (8.35) is substituted, so that

$$
\begin{aligned}
f(x,y,z) &= \frac{1}{2} \int_{y=0}^{2\pi} \int_{\vartheta=0}^{\pi} \left\{ \int_{q=-\infty}^{\infty} q^2 P_{y,\vartheta}(q)\, \mathrm{e}^{2\pi \mathrm{i} q\xi}\, dq \right\} \sin(\vartheta)\, d\vartheta\, dy \\
&= \frac{1}{2} \int_{y=0}^{2\pi} \int_{\vartheta=0}^{\pi} \left\{ \frac{1}{-4\pi^2} \frac{\partial^2 P_{y,\vartheta}(\xi)}{\partial \xi^2} \right\} \sin(\vartheta)\, d\vartheta\, dy \\
&= -\frac{1}{8\pi^2} \int_{y=0}^{2\pi} \int_{\vartheta=0}^{\pi} \frac{\partial^2 P_{y,\vartheta}(\xi)}{\partial \xi^2} \sin(\vartheta)\, d\vartheta\, dy\,.
\end{aligned}
\tag{8.36}
$$

Here, it has to be integrated over the entire unit sphere S, since the infinitesimal surface element of the unit sphere is, as outlined in Fig. 8.19,

$$dS = \sin(\vartheta)\, d\vartheta\, dy\,; \tag{8.37}$$

thus, the Radon inversion equation can be given in the compact form

$$f(x,y,z) = -\frac{1}{8\pi^2} \iint_S \frac{\partial^2 P_{y,\vartheta}(\xi)}{\partial \xi^2}\, dS. \tag{8.38}$$

Because one integrates over all directions n_ξ a more formal expression, is given by

$$f(x,y,z) = -\frac{1}{8\pi^2} \iint_S \frac{\partial^2 P_{y,\vartheta}(\xi)}{\partial \xi^2}\, dn_\xi \tag{8.39}$$

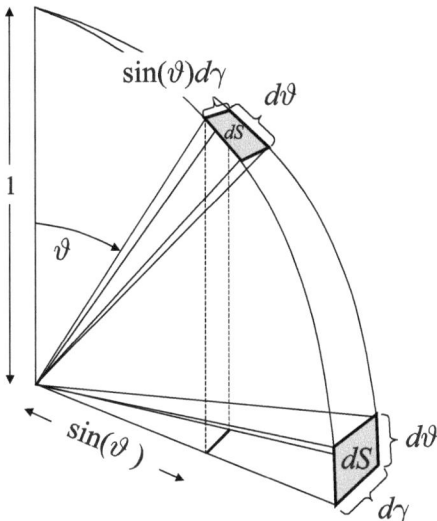

Fig. 8.19. The infinitesimal surface element dS of the unit sphere S depends on the polar angle ϑ with d$S = \sin(\vartheta)\,\mathrm{d}\vartheta\,\mathrm{d}\gamma$. Formally, the surface can be described by the infinitesimal surface normal in the direction of $\mathbf{n}_\xi$ so that one has to integrate over all infinitesimal directions $\mathrm{d}\mathbf{n}_\xi$

(Deans 1983; Natterer and Wübbeling 2001). Natterer and Wübbeling (2001) pointed out that the use of the three-dimensional inversion formula (8.38) must be seen as local tomography, because the calculation of f at the point $(x, y, z)^{\mathrm{T}}$ only requires the surface integration in a rather close neighborhood of the point $(x, y, z)^{\mathrm{T}}$. More detailed considerations concerning the mathematics of local tomography are available in Ramm and Katsevich (1996).

8.4.4
Central Section Theorem

A variant of the Fourier slice theorem is called the central section theorem. This modification assumes that two-dimensional parallel projections $p_{\alpha,\theta}(a, b)$ of the three-dimensional object $f(x, y, z)$ are available. Figure 8.20 outlines the underlying projection geometry.

In order to be able to describe the projection of the point $\mathbf{r}$, assigned with the attenuation value $f(\mathbf{r})$, along the straight lines defined by the actual paths of the X-ray beams in the three-dimensional space, it will be referred to here as the vectorial point slope form

$$\mathbf{r} = \boldsymbol{\xi} + \eta\mathbf{n}_\eta \tag{8.40}$$

of a straight line. The vector $\boldsymbol{\xi}$ is a distance vector, perpendicular to the detector normal $\mathbf{n}_\eta$, pointing to the projection line. This means that $\boldsymbol{\xi}$ lies within the detector

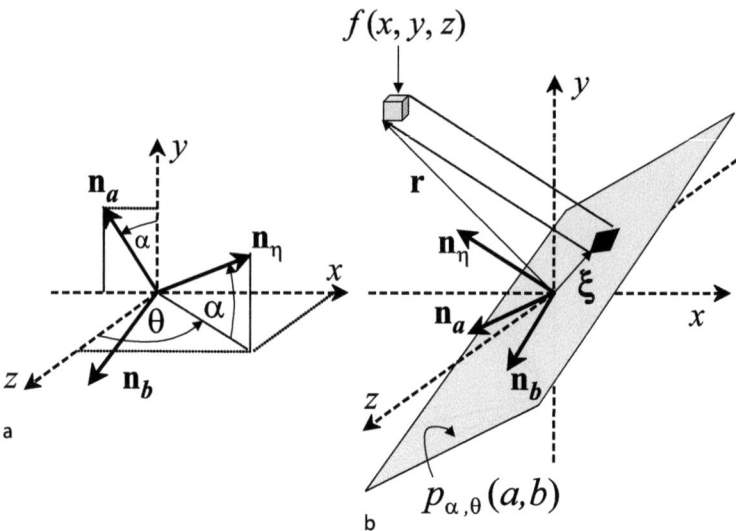

Fig. 8.20. a Orientation of the detector surface in the three-dimensional space using the detector surface normal $\mathbf{n}_\eta$, running parallel to the direction of the X-ray beams. The rotational degree of freedom around the detector normal is restricted by the unit vectors $\mathbf{n}_a$ and $\mathbf{n}_b$ along the orthogonal principal axes of the detector. It is important that $\mathbf{n}_a$ is always positioned such that $z = 0$. **b** Projection of a point $\mathbf{r} = (x, y, z)^T$ in the three-dimensional space along a line parallel to the detector surface normal $\mathbf{n}_\eta$ (i.e., along the actual direction of the X-ray beams) onto the detector surface. Here, the orientation in space is determined by the unit vectors $\mathbf{n}_a$ and $\mathbf{n}_b$. $p_{a,\theta}(a, b)$ denotes the projection values, i.e., the line integrals along the X-ray projection paths, which are measured at the point of intersection (a, b) on the detector plane

plane, i.e.,

$$\boldsymbol{\xi} = a\mathbf{n}_a + b\mathbf{n}_b \, , \tag{8.41}$$

such that the vector

$$\mathbf{r} = a\mathbf{n}_a + b\mathbf{n}_b + \eta\mathbf{n}_\eta \tag{8.42}$$

describes in its parameterized form the points on the projection line, which is perpendicular to the detector surface. The detector surface normal

$$\mathbf{n}_\eta = \begin{pmatrix} \cos(\alpha)\sin(\theta) \\ \sin(\alpha)\sin(\theta) \\ \cos(\theta) \end{pmatrix} \tag{8.43}$$

is unambiguously determined by the spherical angles α and θ. Figure 8.20a shows the position of the surface normal $\mathbf{n}_\eta$ in the fixed (x, y, z) coordinate system.

With the definition introduced above, the detector plane still has a rotational degree of freedom around the detector normal. Uniqueness is achieved by positioning the detector vector $\mathbf{n}_a$ into the plane spanned by x and y, i.e., so that its z component is zero; thus,

$$\mathbf{n}_a = \begin{pmatrix} -\sin(\alpha) \\ \cos(\alpha) \\ 0 \end{pmatrix} . \qquad (8.44)$$

This means, that

$$\mathbf{n}_b = \begin{pmatrix} -\cos(\alpha)\cos(\theta) \\ -\sin(\alpha)\cos(\theta) \\ \sin(\theta) \end{pmatrix} \qquad (8.45)$$

is also uniquely determined.

Using the parametric representation (8.42) the X-ray projection can now be formulated as the line integral

$$p_{\alpha,\theta}(a,b) = \int_{-\infty}^{\infty} f(a\mathbf{n}_a + b\mathbf{n}_b + \eta\mathbf{n}_\eta)\,d\eta . \qquad (8.46)$$

As illustrated in Fig. 8.20b, $p_{\alpha,\theta}(a,b)$ is the measured X-ray projection value on the detector plane. Toft (1996) designates (8.46) as a hybrid or generalized Radon transform for lines in the three-dimensional space, although – strictly speaking – $p_{\alpha,\theta}(a,b)$ does not represent a Radon value (cf. ▸ Sect. 8.4.1). Similar to the two-dimensional situation, the hybrid Radon transform can be represented as a convolution with δ-distributions. This means

$$p_{\alpha,\theta}(a,b) = \int_{\eta'=-\infty}^{\infty} \int_{b'=-\infty}^{\infty} \int_{a'=-\infty}^{\infty} f(a'\mathbf{n}_a + b'\mathbf{n}_b + \eta'\mathbf{n}_\eta)$$

$$\cdot \, \delta(a-a')\delta(b-b')\,da'\,db'\,d\eta' \qquad (8.47)$$

$$= \int_{\eta'=-\infty}^{\infty} f(a\mathbf{n}_a + b\mathbf{n}_b + \eta'\mathbf{n}_\eta)\,d\eta' .$$

The central section theorem – just like the Fourier slice theorem – actually relates the Fourier transform of the projection to the Fourier transform of subspaces of the three-dimensional space. The Fourier transform of $p_{\alpha,\theta}(a,b)$ can be expressed as

$$P_{\alpha,\theta}(q,p) = \int_{-\infty}^{\infty}\int_{-\infty}^{\infty} p_{\alpha,\theta}(a,b)\,e^{-2\pi i(aq+bp)}\,da\,db , \qquad (8.48)$$

where q and p are the frequency variables belonging to the spatial detector coordinates a and b respectively. Taking into account the definition of the hybrid Radon transform from (8.46), one may write

$$P_{\alpha,\theta}(q,p) = \int_{-\infty}^{\infty}\int_{-\infty}^{\infty}\int_{-\infty}^{\infty} f(a\mathbf{n}_a + b\mathbf{n}_b + \eta\mathbf{n}_\eta)\,e^{-2\pi i(aq+bp)}\,da\,db\,d\eta . \qquad (8.49)$$

As in the two-dimensional case, the expression has to be transformed to the fixed patient coordinate system, i.e.,

$$(a, b, \eta)^{\mathrm{T}} \rightarrow (x, y, z)^{\mathrm{T}} . \tag{8.50}$$

This is feasible with a single rotation, since

$$\mathbf{r} = \begin{pmatrix} x \\ y \\ z \end{pmatrix} = (\mathbf{n}_a \ \mathbf{n}_b \ \mathbf{n}_\eta) \begin{pmatrix} a \\ b \\ \eta \end{pmatrix} \tag{8.51}$$

(cf. (8.42)) can be expressed by substituting (8.44), (8.45), and (8.43) into (8.51). This yields

$$\mathbf{r} = \begin{pmatrix} -\sin(\alpha) & -\cos(\alpha)\cos(\theta) & \cos(\alpha)\sin(\theta) \\ \cos(\alpha) & -\sin(\alpha)\cos(\theta) & \sin(\alpha)\sin(\theta) \\ 0 & \sin(\theta) & \cos(\theta) \end{pmatrix} \begin{pmatrix} a \\ b \\ \eta \end{pmatrix} = \mathbf{Q}\boldsymbol{\eta} , \tag{8.52}$$

where $\mathbf{Q}$ is an orthogonal matrix, so that

$$\begin{pmatrix} a \\ b \\ \eta \end{pmatrix} = \mathbf{Q}^{-1}\mathbf{r} = \mathbf{Q}^{\mathrm{T}}\mathbf{r}$$

$$= \begin{pmatrix} -\sin(\alpha) & \cos(\alpha) & 0 \\ -\cos(\alpha)\cos(\theta) & -\sin(\alpha)\cos(\theta) & \sin(\theta) \\ \cos(\alpha)\sin(\theta) & \sin(\alpha)\sin(\theta) & \cos(\alpha) \end{pmatrix} \begin{pmatrix} x \\ y \\ z \end{pmatrix} . \tag{8.53}$$

The *Jacobian* is thus $J = 1$ and the infinitesimal volume element is to be transformed into

$$da\,db\,d\eta \rightarrow dx\,dy\,dz . \tag{8.54}$$

According to (8.53) the three components of the vector $(a, b, \eta)^{\mathrm{T}}$ obey

$$a = -x\sin(\alpha) + y\cos(\alpha) , \tag{8.55}$$
$$b = -x\cos(\alpha)\cos(\theta) - y\sin(\alpha)\cos(\theta) + z\sin(\theta) \tag{8.56}$$

and

$$\eta = x\cos(\alpha)\sin(\theta) + y\sin(\alpha)\sin(\theta) + z\cos(\theta) . \tag{8.57}$$

Using the unit vectors (8.43) through (8.45), one may write in short

$$a = \mathbf{r}^{\mathrm{T}} \cdot \mathbf{n}_a, \ b = \mathbf{r}^{\mathrm{T}} \cdot \mathbf{n}_b \text{ and } \eta = \mathbf{r}^{\mathrm{T}} \cdot \mathbf{n}_\eta . \tag{8.58}$$

(8.49) thus reads in the fixed patient coordinate system

$$P_{\alpha,\theta}(q,p) = \int\limits_{-\infty}^{\infty} \int\limits_{-\infty}^{\infty} \int\limits_{-\infty}^{\infty} f(x,y,z)\, e^{-2\pi i r^T \cdot (n_a q + n_b p)}\, dx\, dy\, dz\ . \qquad (8.59)$$

Obviously, the frequency vector

$$\mathbf{q} = q\mathbf{n}_a + p\mathbf{n}_b \qquad (8.60)$$

in the three-dimensional frequency domain with the Cartesian representation $\mathbf{v} = (u,v,w)^{\mathrm{T}}$ has the same orientation as

$$\boldsymbol{\xi} = a\mathbf{n}_a + b\mathbf{n}_b\ , \qquad (8.41)$$

in the three-dimensional spatial domain with the Cartesian representation $\mathbf{r} = (x,y,z)^{\mathrm{T}}$.

Figure 8.21a shows how the distance vector $\boldsymbol{\xi}$ defined in (8.41) can be identified with the Radon variable. The vector $\boldsymbol{\xi}$ is expressed with its spherical coordinates by the distance ξ and the angles γ and ϑ with respect to the fixed (x,y,z) patient coordinate system, exactly as described above in ▸ Sect. 8.4.1. Figure 8.21b illustrates the corresponding integration surface $\mathbf{A}$ of the Fourier slice theorem. This surface is perpendicular to the detector plane because the distance vector $\boldsymbol{\xi}$ lies in this very plane.

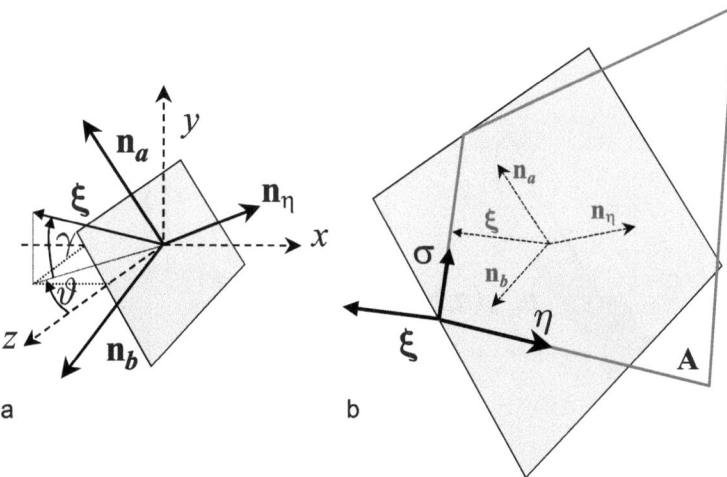

a   b

Fig. 8.21. a The distance vector in (8.41) can be identified with the Radon coordinates $\boldsymbol{\xi}$, which are determined by the distance ξ and the angles γ and ϑ in their spherical coordinates. The corresponding integration surface $\mathbf{A}$ is shown in **b**. Obviously, the surface $\mathbf{A}$ is perpendicular to the detector

(8.59) means that a (q, p) plane in the three-dimensional frequency domain (with the Cartesian coordinates u, v, and w) of the object $f(x, y, z)$ to be reconstructed, is identical with the two-dimensional Fourier transform $P_{\alpha, \theta}(q, p)$ of the corresponding projection value or hybrid Radon transform $p_{\alpha, \theta}(a, b)$ on the (a, b) detector plane. The orientation of the detector plane corresponds to the orientation of the plane in the frequency domain, since the Fourier transform is rotationally variant. In compact form, one may write

$$\mathcal{F}_2\{p_{\alpha, \theta}(a, b)\} = P_{\alpha, \theta}(q, p) = F(q\mathbf{n}_a + p\mathbf{n}_b) = \mathcal{F}_3\{f(\mathbf{r})\}|_{\mathbf{r} = \boldsymbol{\xi} = a\mathbf{n}_a + b\mathbf{n}_b} \cdot \quad (8.61)$$

Figure 8.22 shows the central section theorem schematically. On the lower left side, the vertical projection of a single point $f(x, y, z)$ onto an arbitrarily angulated detector plane can be seen in relation to a fixed (x, y, z) patient coordinate system. As mentioned above, the plane of all line integrals perpendicular to the detector plane is the hybrid Radon transform $p_{\alpha, \theta}(a, b)$. A two-dimensional Fourier transform then yields $P_{\alpha, \theta}(q, p)$. Due to the rotational variance of the Fourier transform, this plane is to be sorted into a fixed (u, v, w) frequency space under the angulation of

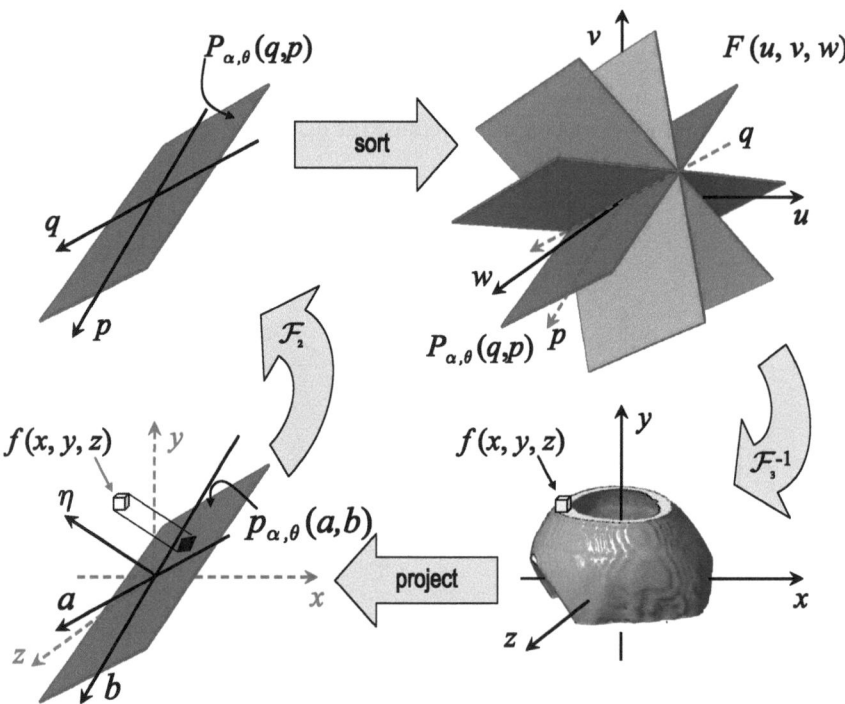

Fig. 8.22. Central section theorem. If all two-dimensional Fourier transforms (*upper left*) of three-dimensional parallel projections (*lower left*) are correctly sorted into the three-dimensional Fourier space (*upper right*), then a three-dimensional inverse Fourier transform yields the original object (*lower right*)

the measurement. If the projections have been measured for all angulations, then the object can finally be reconstructed by means of the inverse three-dimensional Fourier transform

$$f(x,y,z) = \int\limits_{-\infty}^{\infty} \int\limits_{-\infty}^{\infty} \int\limits_{-\infty}^{\infty} F(u,v,w)\, e^{2\pi i(xu+yv+zw)}\, du\, dv\, dw\,. \tag{8.62}$$

8.4.5
Orlov's Sufficiency Condition

In the mid-1970s, Orlov (1975) had already answered the question as to how many parallel projections must be measured in order to reliably reconstruct the object – it must be noted that this question initially was answered in the field of crystallography (cf. Orlov [1975]). According to (8.46) a set of parallel projections must be described by the hybrid Radon transform

$$p_{\alpha,\theta}(a,b) = \int\limits_{-\infty}^{\infty} f(\mathbf{r})\, d\eta = \int\limits_{-\infty}^{\infty} f(\boldsymbol{\xi} + \eta\mathbf{n}_\eta)\, d\eta\,, \tag{8.63}$$

where $\mathbf{n}_\eta \in \Omega \subset S$ and Ω is the set of directions for which the parallel projections are actually measured. S again is the unit sphere of all directions and $\boldsymbol{\xi}$ is located in the projection plane described by the surface normal vector $\mathbf{n}_\eta$; therefore, obviously, $(\boldsymbol{\xi}^T\mathbf{n}_\eta) = 0$. According to *Orlov's* sufficiency condition, $F(\mathbf{v})$, with $\mathbf{v} = (u,v,w)^T \in \mathbb{R}^3$, can be computed, if there is at least one projection direction $\mathbf{n}_\eta \in \Omega$ according to (8.63) for which $(\mathbf{v}^T\mathbf{n}_\eta) = 0$. Figure 8.23 shows a data acquisition situation in which the vector $\mathbf{v}$ cannot be reconstructed.

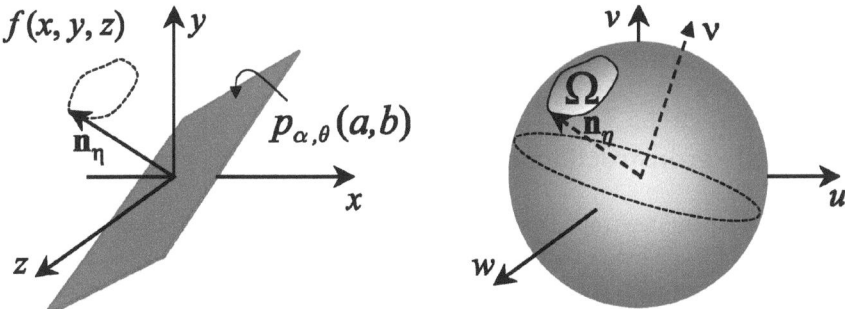

Fig. 8.23. *Orlov's* sufficiency condition. A requirement to be fulfilled by the measurement of parallel projections of all directions Ω (*left*) is that at least one direction must be perpendicular to the frequency vector $\mathbf{v}$. The whole frequency domain can actually be reconstructed if it is not possible to find a great circle on the directional unit sphere S that does not intersect the area of the measured projections, Ω. The picture on the right side shows a great circle, which does not touch Ω. The values $F(\mathbf{v})$ at the single frequency coordinate $\mathbf{v}$ cannot be reconstructed with this measurement

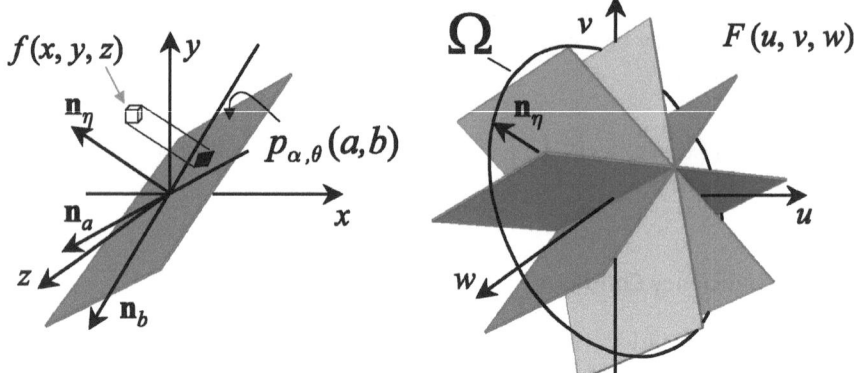

Fig. 8.24. The requirements stipulated in *Orlov*'s sufficiency condition are met in the case of the continuous 2π rotation of the detector surface. This is due to the fact that the set of all directions, Ω, in which the parallel projections were measured, forms a great circle of the directional unit sphere S. Therefore, any other great circle on S intersects Ω

This condition finally implies that the object $f(\mathbf{r})$ can only be reconstructed if the set Ω of the measured directions of the parallel projections is intersected by any great circle on the unit sphere, S. For Fig. 8.24, illustrating the central section theorem, now the question can be answered as to whether the continuous rotation of the projection direction $\mathbf{n}_\eta$ around the vector $\mathbf{n}_a$ is a complete measurement in terms of Orlov's sufficiency condition.

This is obviously the case, since the vector of the projection direction $\mathbf{n}_\eta$ describes a great circle on the directional unit sphere S with this rotation. It thus is impossible to find another great circle on S that does not intersect Ω. Due to Orlov's sufficiency condition, it is in fact obvious why the whole object cannot be reliably reconstructed with the tomosynthesis method described in ▸ Sect. 3.2.

8.5
Exact 3D Reconstruction in Cone-Beam Geometry

The development step from parallel-beam to fan-beam tomography, which was discussed at the beginning of ▸ Chap. 7, resulted in two major advantages. First, the increase in acquisition speed that resulted from the fact that it was no longer necessary to stop for the parallel shift of the detector system while the sampling unit was rotating around the patient. The second advantage was better utilization of the tube heat capacity (cf. ▸ Sect. 2.2.3). Using only a fine pencil X-ray beam by pinhole collimation represents an enormous waste of the generated radiation. As the limited thermal capacity of the X-ray tubes to date still is a major technical problem, the measurements could be carried out much more effectively by using fan-beam technology.

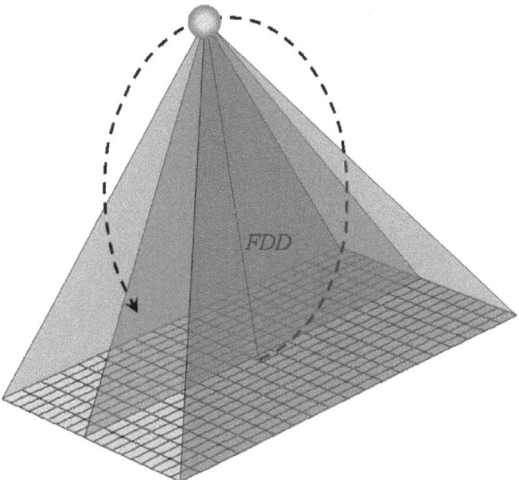

Fig. 8.25. Cone-beam illumination of a flat-panel detector. Such arrangements are frequently used in technical applications such as material testing in micro-CT. Medical applications more frequently employ cylindrical detector arrays. However, there are already a few proto-types working with flat-panel detectors in clinical applications

If one consequently follows the reasoning for fan-beam geometry, it seems to be useful not to employ only a collimated fan, but the entire cone beam generated by the X-ray source. This could technically not be realized for a long time, because a fast, high-quality, flat-panel detector system was not available[8]. However, two-dimensional detector systems for CT scanners have recently been developed so that nowadays all manufacturers of CT scanners offer multi-slice CT (MSCT) systems.

Figure 8.25 schematically shows the cone-beam geometry of an imaging system with a planar detector (flat-panel detector, cf. ▸ Sect. 2.5.3). The planar detector geometry was first of all applied for technical applications in micro-CT, since the required CCD chips were only available for this geometry. Anyway, Fig. 8.26 shows a prototype of a so-called volume CT (VCT) developed by General Electric Medical Systems.

This type of CT scanner provides spatial resolutions, which were never attained before and were rather associated with micro-CTs. However, this improvement in image quality is achieved at the expense of demanding reconstruction mathematics, which will be described in the following sections. Figure 8.27 shows cone-beam projections over an angular interval of 210°. The image actually shows the three-dimensional projection data of a wasp, which was scanned using micro-CT. Con-

---

[8] A major technical problem arises from the fact that the immense quantity of data must quickly be transferred from the rotating sampling unit to the processing system on the gantry or the scanner backend. These data must be transmitted via an interface with a large bandwidth. For this purpose optical slip-rings are used in modern scanners.

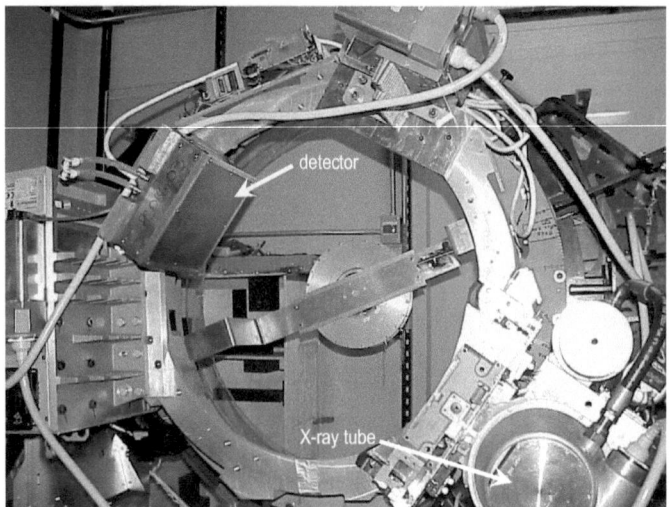

Fig. 8.26. Prototype of so-called volume CT in the GE research laboratories. A flat-panel detector measuring approximately 41 cm × 41 cm with a total number of 2,048 × 2,048 pixels is arranged opposite to the X-ray tube on the sampling unit. This detector technology is described in ▸ Sect. 2.5.3

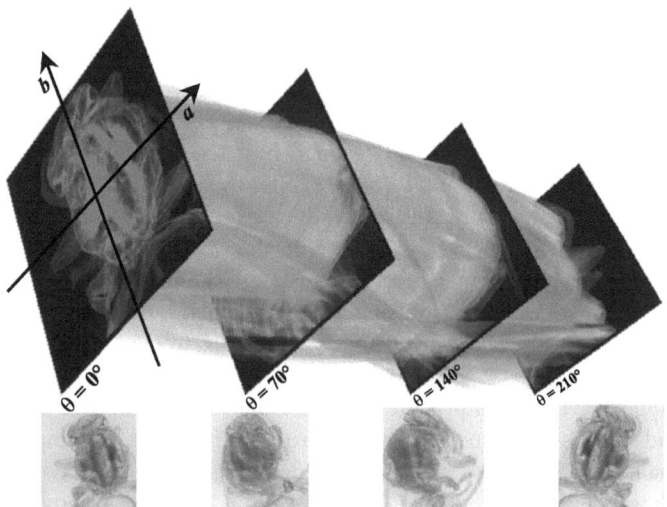

Fig. 8.27. Three-dimensional sinogram of cone-beam micro-CT. The raw projection data of a wasp are shown over an angle interval of $\theta = [0°, 210°]$. The detector coordinates are denoted (a, b). Single projections are displayed in steps of $70°$

trary to two-dimensional CT, the three-dimensional sinogram does not represent a Radon space. This problem will be discussed in detail in the following section.

8.5.1
Key Problem of Cone-Beam Geometry

As already shown in Fig. 8.18, three-dimensional reconstruction methods require integration over two-dimensional surfaces to obtain the required data $p_{\gamma,\vartheta}(\xi)$ in the Radon space (ξ, γ, ϑ).

If one has to deal with parallel-beam geometry, the surface integration can be split into two line integrals, the first of which is along the X-ray beams

$$\mathbf{X}^{\mathrm{P}} \left\{ \mu_{\gamma,\vartheta}(\sigma, \xi) \right\} = \int\limits_{-\infty}^{\infty} \mu(\eta, \sigma, \xi)\, \mathrm{d}\eta \;. \tag{8.64}$$

This equation is called the X-ray transform[9]. Subsequently, the second integration is along the direction of the detector, σ:

$$\begin{aligned} p_{\gamma,\vartheta}(\xi) &= \int\limits_{-\infty}^{\infty} \mathbf{X}^{\mathrm{P}} \left\{ \mu_{\gamma,\vartheta}(\sigma, \xi) \right\} \mathrm{d}\sigma \\ &= \int\limits_{-\infty}^{\infty} \int\limits_{-\infty}^{\infty} \mu(\eta, \sigma, \xi)\, \mathrm{d}\eta\, \mathrm{d}\sigma \;. \end{aligned} \tag{8.65}$$

(8.65) can readily be understood with Fig. 8.28a.

Unfortunately, the simple separation of the surface integral into two line integrals makes it impossible to find the required Radon values for the diverging X-ray beams, which almost always occur in practice.

If one first integrates along the direction of the X-ray beams for the case illustrated in Fig. 8.28b, i.e., if the X-ray transform, $\mathbf{X}\{\cdot\}$, is carried out in a first step, then the adaptation to the geometrical situation requires the introduction of polar coordinates (r, φ). On the detector this yields the projection values

$$\mathbf{X}^{\mathrm{c}} \left\{ \mu_{\gamma,\vartheta}(\varphi, \xi) \right\} = \int\limits_{-\infty}^{\infty} \mu(r, \varphi, \xi)\, \mathrm{d}r \;. \tag{8.66}$$

---

[9] Indexing of $\mathbf{X}\{\cdot\}$ with $\mathbf{p}$ or $\mathbf{c}$ denotes the X-ray parallel- or cone-beam transform respectively.

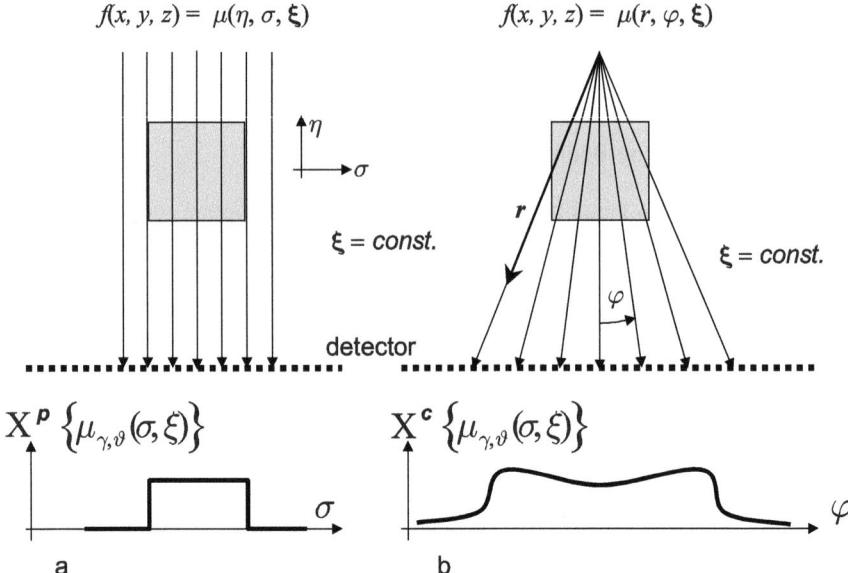

Fig. 8.28. Difference between parallel-beam and cone-beam geometry. **a** The parallel line integrals can easily be converted into a surface integral. **b** This cannot so easily be done with cone-beam geometry

If these projection values are integrated along the detector, this results in the second integration

$$
\tilde{p}_{\gamma,\vartheta}(\xi) = \int\limits_{-\pi/2}^{\pi/2} \mathbf{X}^c \left\{ \mu_{\gamma,\vartheta}(\varphi, \xi) \right\} d\varphi
$$

$$
= \int\limits_{-\pi/2}^{\pi/2} \int\limits_{-\infty}^{\infty} \mu(r, \varphi, \xi) \, dr \, d\varphi \ .
$$

(8.67)

To compare the result of parallel-beam geometry (8.65) with the result of cone-beam geometry (8.67), (8.65) must also be transformed into the polar coordinates, i.e.,

$$
p_{\gamma,\vartheta}(\xi) = \int\limits_{-\infty}^{\infty} \int\limits_{-\infty}^{\infty} \mu(\eta, \sigma, \xi) \, d\eta \, d\sigma
$$

$$
= \int\limits_{-\pi/2}^{\pi/2} \int\limits_{-\infty}^{\infty} \mu(r, \varphi, \xi) r \, dr \, d\varphi \ ,
$$

(8.68)

where the *Jacobian* yields the variable r in the new infinitesimal area element $r \, dr \, d\varphi$.

This means that the X-ray transforms

$$\mathbf{X}^{\mathrm{c}}\left\{\mu_{\gamma,\vartheta}(\varphi,\xi)\right\} \neq \mathbf{X}^{\mathbf{P}}\left\{\mu_{\gamma,\vartheta}(\sigma,\xi)\right\} \tag{8.69}$$

of the two geometries are not in line with each other; therefore, the data $\tilde{p}_{\gamma,\vartheta}(\xi)$ do not represent the Radon values required for the reconstruction process in cone-beam geometry. This means

$$\tilde{p}_{\gamma,\vartheta}(\xi) \neq p_{\gamma,\vartheta}(\xi) . \tag{8.70}$$

8.5.2
Method of Grangeat

As shown in the previous section, there is no direct way to obtain the Radon transform in cone-beam systems. However, Grangeat (1990) has found a technique to compute a derivative of the Radon transform from the X-ray transform[10]. One variation of the results will be discussed here. The main issues discussed in this section are adapted from a presentation by Jacobson (1996).

In a first step, the Radon transform is differentiated with respect to the direction of ξ, so that the expression

$$\frac{\partial}{\partial\xi}p_{\gamma,\vartheta}(\xi) = \frac{\partial}{\partial\xi}\int\limits_{-\pi/2}^{\pi/2}\int\limits_{-\infty}^{\infty}\mu(r,\varphi,\xi)r\,dr\,d\varphi \tag{8.71}$$

can be obtained. Here, the order of the differentiation and the integration may be changed, since one does not integrate over the radial Radon component ξ. This results in

$$\frac{\partial}{\partial\xi}p_{\gamma,\vartheta}(\xi) = \int\limits_{-\pi/2}^{\pi/2}\int\limits_{-\infty}^{\infty}\frac{\partial}{\partial\xi}\mu(r,\varphi,\xi)r\,dr\,d\varphi . \tag{8.72}$$

To understand the meaning of the derivative $\partial/\partial\xi$, one has to study the geometrical conditions in the cone-beam projection. Figure 8.29 shows the corresponding three-dimensional geometry. In fact, one does not refer to a special source trajectory. Initially, the Radon data should be computed for an arbitrarily selected source detector situation from the projection data on the detector, which means from the cone-beam X-ray transform (8.66).

O denotes the detector origin and **S** denotes the X-ray source position. One now considers the use of a virtual detector situated in the iso-center of the measurement field in order to simplify the geometrical approach. The results for other detector positions, especially those for situations found in practice, can then easily be derived from the intercept theorem.

The Radon sphere of the point **S** has the diameter $FCD = \|\mathbf{S} - \mathbf{O}\|$ and the detector surface is the tangential surface of the Radon sphere at the point **O**. That means that the vector **S–O** points in the direction of the detector normal.

---

[10] A literature survey dealing with the three-dimensional cone-beam reconstruction is available in (Grangeat 1997).

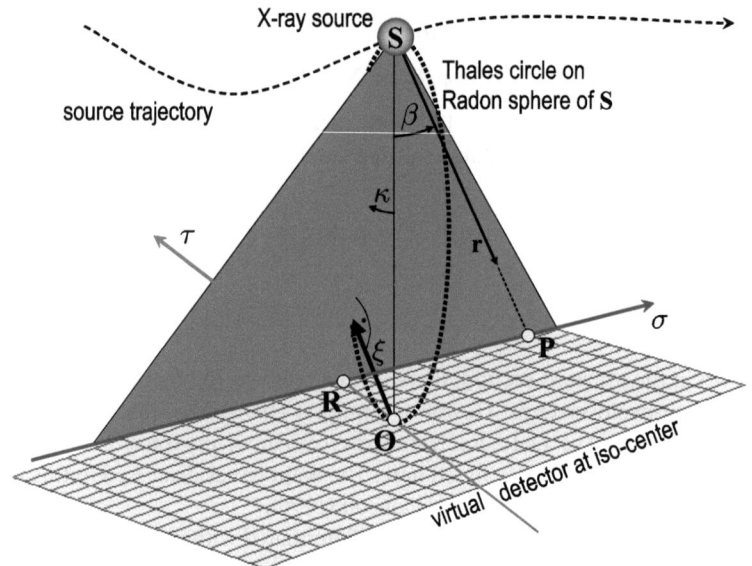

Fig. 8.29. For an arbitrary trajectory of the X-ray source **S**, the derivative of the X-ray transform must be calculated from the detector data. For this purpose, a virtual detector is placed in the iso-center **O** of the measurement field. An arbitrary fan of the X-ray cone-beam is shown shaded in *red*

As described in ▸ Sect. 8.4.1,

$$\boldsymbol{\xi} = \begin{pmatrix} \xi_x \\ \xi_y \\ \xi_z \end{pmatrix} = \xi \cdot \mathbf{n}_\xi = \begin{pmatrix} \xi \cos(\gamma) \sin(\vartheta) \\ \xi \sin(\gamma) \sin(\vartheta) \\ \xi \cos(\vartheta) \end{pmatrix} \qquad (8.12)$$

unequivocally determines the Radon value $p_{\gamma,\vartheta}(\xi) = \mathcal{R}_3\{f(x,y,z)\}$, which is required for the reconstruction process. This is due to the fact that the X-ray fan – i.e., the integration surface through the point[11] **r** to be reconstructed – as described in ▸ Sect. 8.4.1, is unambiguously determined by the vector **ξ**. **ξ** is perpendicular to this X-ray fan, as illustrated with red color in Fig. 8.29 and it has its origin in **O**.

The foot-point of **ξ** on the X-ray fan and the points **O** and **S** form a rectangular triangle on a great circle of the Radon sphere of **S** according to the theorem of Thales. The central beam of the fan is tilted off the detector normal direction **S–O** by the angle κ. The vector **r** and the detector normal enclose the angle β.

The central beam of the fan, which is perpendicular to **ξ**, hits the detector at point **R**. The whole fan beam intersects the detector surface along the σ-axis. The intersection point **P** of the beam through the point **r** to be reconstructed can therefore be found on the σ-axis. The line connecting the detector origin **O** and point

---

[11] In this consideration the vector **r** has its foot-point in the X-ray source.

R defines a radial detector τ-axis through **O**, which is always perpendicular to the X-ray fan-detector cutting line, σ. Figure 8.30 once again illustrates the described geometrical situation with a different perspective from that shown in Fig. 8.29.

The derivative $\partial/\partial\xi$ of the Radon transform proposed in (8.71) appears in the projection on the virtual flat-panel detector as a radial derivative. As the detector data comprise the only physically measured reconstruction data basis, one thus first has to study the consequences related to the variation of ξ with regard to the radial component ξ in (8.71) on the detector. The infinitesimal change of the vector, ξ, by $d\xi$ results in a parallel shift of the σ-axis by $d\tau$ on the detector. However, the exact size of the infinitesimal shift must be computed here[12]. Figures 8.31a and b illustrate the geometrical conditions in the X-ray fan plane (a) or perpendicular to this plane, respectively, i.e., in the plane (b) defined by τ and **S–O**.

The point **r** to be reconstructed has the polar coordinates (r, φ) for a pre-defined fixed Radon vector ξ in the corresponding X-ray fan. As illustrated in Fig. 8.31a the projection of the radius **r** onto the central beam of the X-ray fan is thus $r\cos(\varphi)$. This projection appears again in Fig. 8.31b, which is a perpendicular view to the above figure.

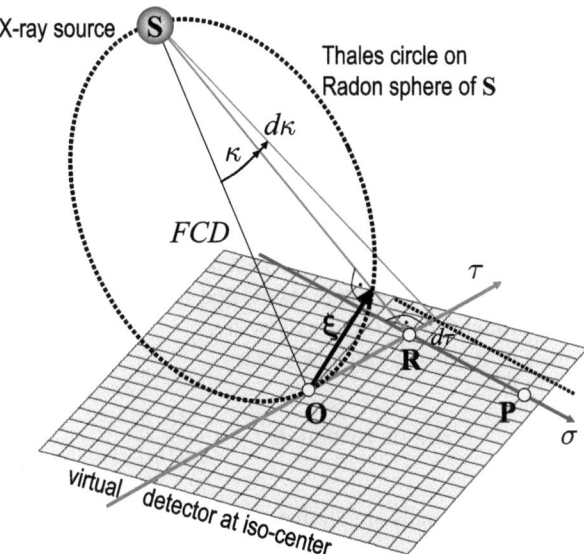

Fig. 8.30. A planar X-ray fan from the cone-beam is represented as a linear line of intersection along the σ-axis on the virtual detector in the iso-center **O** of the measuring field. The projection of the Radon vector ξ on the detector defines a radial τ-axis through **O**, which is perpendicular to the σ-axis

---

[12] As – by definition – the vector of ξ always ends on the Radon sphere, a change in the length with constant angles γ and ϑ results in the fact that the new vector points to the Radon data of other detector positions in space. This has to be accounted for in the case of the outline for an algorithmic implementation.

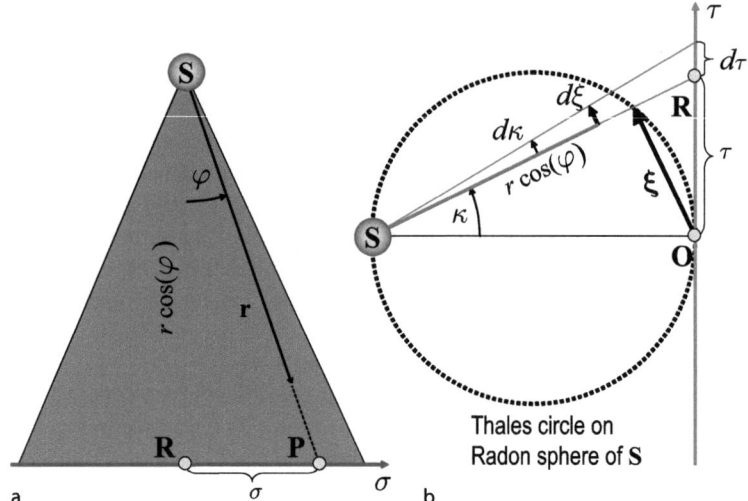

Fig. 8.31. To compute the derivative of the Radon transform from the detector data, i.e., from the X-ray transform, the geometric situation inside **a** an arbitrary X-ray fan of the X-ray cone-beam and **b** the geometry within the surface that is perpendicular to this X-ray fan are illustrated

The variation $d\xi$ of $\boldsymbol{\xi}$ in the direction of the radial Radon coordinate ξ, which was proposed in the first equation of this section, yields an infinitesimally small vector, which can be shifted in parallel along a line $\mathbf{R} - \mathbf{S}$. This actually defines the infinitesimal change of the fan tilt angle κ by $d\kappa$ in the rectangular triangle defined by the catheti $d\xi$ and $r\cos(\varphi)$ respectively. That means, the shorter the vector $\mathbf{r}$, the larger the infinitesimal angular change $d\kappa$ related to $d\xi$. This, in fact, determines the infinitesimal parallel shift of the σ-axis by $d\tau$.

Furthermore, Fig. 8.31b also illustrates that

$$\sin(d\kappa) = \tan(d\kappa) = \frac{d\xi}{r\cos(\varphi)} \tag{8.73}$$

or, because $d\kappa$ is infinitesimally small, that

$$d\kappa = \frac{d\xi}{r\cos(\varphi)} . \tag{8.74}$$

Additionally, Fig. 8.31b shows that κ depends on ξ. Therefore, if a function $v(\kappa(\xi))$ is differentiated with respect to ξ, the chain rule

$$\frac{\partial v}{\partial \xi} = \frac{\partial v}{\partial \kappa}\frac{\partial \kappa}{\partial \xi} \tag{8.75}$$

must be applied. The derivation of κ with respect to ξ can be replaced by an adaptation of (8.74)

$$\frac{\partial \kappa}{\partial \xi} = \frac{1}{r \cos(\varphi)} \, , \tag{8.76}$$

so that

$$\frac{\partial v}{\partial \xi} = \frac{\partial v}{\partial \kappa} \frac{1}{r \cos(\varphi)} \, . \tag{8.77}$$

For convenience, the function v is defined as the integrand in (8.71), i.e., the attenuation values weighted by the distance to the X-ray source

$$v = \mu(r, \varphi, \boldsymbol{\xi}) \cdot r \, . \tag{8.78}$$

Then, (8.77) and (8.78) can be substituted into (8.72) to find the expression

$$\frac{\partial}{\partial \xi} p_{\gamma, \vartheta}(\xi) = \int\limits_{-\pi/2}^{\pi/2} \int\limits_{-\infty}^{\infty} \frac{\partial}{\partial \kappa} \frac{1}{\cos(\varphi)} \mu(r, \varphi, \boldsymbol{\xi}) \, dr \, d\varphi \, , \tag{8.79}$$

in which the order of integration and differentiation can again be changed so that one finds

$$\frac{\partial}{\partial \xi} p_{\gamma, \vartheta}(\xi) = \frac{\partial}{\partial \kappa} \int\limits_{-\pi/2}^{\pi/2} \frac{1}{\cos(\varphi)} \left[\int\limits_{-\infty}^{\infty} \mu(r, \varphi, \boldsymbol{\xi}) \, dr \right] d\varphi \, . \tag{8.80}$$

The integration along all X-ray beams yields the X-ray transform from (8.80) in the brackets of (8.66), i.e., we can finally find the expression

$$\frac{\partial}{\partial \xi} p_{\gamma, \vartheta}(\xi) = \frac{\partial}{\partial \kappa} \int\limits_{-\pi/2}^{\pi/2} \frac{1}{\cos(\varphi)} \mathbf{X}^{c} \left\{ \mu_{y, \vartheta}(\varphi, \xi) \right\} d\varphi \, , \tag{8.81}$$

which is the main result of Grangeat (1990).

In fact, (8.81) represents a fundamental relationship between the first derivative of the Radon transform and the X-ray transform within cone-beam geometry. The key step is the substitution (8.75) because the right-hand side of (8.79) is relieved from the explicit r dependence. This is the only way to represent the integration over r – the bracketed term in (8.80) – as an X-ray transform, which has actually been measured.

However, the differentiation with respect to the tilt angle κ is rather inconvenient, because in practice one actually has to deal with the principal axes a and b of the flat-panel detector, which is shown in Fig. 8.32. From Fig. 8.31b, it can be read that

$$\tau = \|\mathbf{S} - \mathbf{O}\| \tan(\kappa) = FCD \tan(\kappa) \,. \tag{8.82}$$

The differentiation with respect to the tilt angle κ yields

$$\frac{d\tau}{d\kappa} = \frac{FCD}{\cos^2(\kappa)} \,. \tag{8.83}$$

If one substitutes relationship (8.83) into (8.81), one obtains

$$\frac{\partial}{\partial \xi} p_{\gamma,\vartheta}(\xi) = \frac{FCD}{\cos^2(\kappa)} \frac{\partial}{\partial \tau} \int\limits_{-\pi/2}^{\pi/2} \frac{1}{\cos(\varphi)} \mathbf{X}^c \left\{ \mu_{\gamma,\vartheta}(\varphi, \tau(\xi)) \right\} d\varphi \,. \tag{8.84}$$

For the second detector coordinate, it can be seen from Fig. 8.31a that

$$\sigma = \|\mathbf{S} - \mathbf{R}\| \tan(\varphi) \,. \tag{8.85}$$

The differentiation with respect to the fan angle φ yields

$$\frac{d\sigma}{d\varphi} = \frac{\|\mathbf{S} - \mathbf{R}\|}{\cos^2(\varphi)} \,. \tag{8.86}$$

Furthermore, Fig. 8.31a shows that

$$\cos(\varphi) = \frac{\|\mathbf{S} - \mathbf{R}\|}{\|\mathbf{S} - \mathbf{P}\|} \tag{8.87}$$

and thus

$$d\varphi = \frac{\cos(\varphi)}{\|\mathbf{S} - \mathbf{P}\|} d\sigma \tag{8.88}$$

can be substituted into (8.81) so that, finally,

$$\frac{\partial}{\partial \xi} p_{\gamma,\vartheta}(\xi) = \frac{1}{\cos^2(\kappa)} \frac{\partial}{\partial \tau} \int\limits_{-\infty}^{\infty} \frac{FCD}{\|\mathbf{S} - \mathbf{P}\|} \mathbf{X}^c \left\{ \mu_{\gamma,\vartheta}(\sigma, \tau(\xi)) \right\} d\sigma \tag{8.89}$$

can be found.

Figure 8.29 illustrates the meaning of the weighting introduced in (8.89). While integrating along the σ-axis of the detector, each detector point must be pre-weighted with its inverse distance to the X-ray source. This pre-weighting step can be expressed by β, i.e., the angle with respect to the detector normal direction $\mathbf{S}$–$\mathbf{O}$, so that

$$\frac{\partial}{\partial \xi} p_{\gamma,\vartheta}(\xi) = \frac{1}{\cos^2(\kappa)} \frac{\partial}{\partial \tau} \int\limits_{-\infty}^{\infty} \cos(\beta) \mathbf{X}^c \left\{ \mu_{\gamma,\vartheta}(\sigma, \tau(\xi)) \right\} d\sigma \,. \tag{8.90}$$

In fact, the detector data can be pre-weighted for all detector positions before the integration is carried out. Of course, (8.90) only makes sense if the surfaces, defined by the Radon coordinates ξ, actually contain the X-ray source. Otherwise, the distances $\|\mathbf{S} - \mathbf{P}\|$ are not defined. This is the so-called *Tuy–Smith* condition, which will be discussed in detail in ▸ Sect. 8.6.1.

8.5.3
Computation of the First Derivative on the Detector

(8.90) means that it has to be integrated along the straight lines in the direction of the detector σ coordinate, namely, the detector values differentiated in the direction of the detector τ coordinate. Here, the approach proposed by *Grangeat* will be used, which starts from the fact that the order of integration and differentiation may be changed in (8.90). This leads to

$$\frac{\partial}{\partial \xi} p_{\gamma,\vartheta}(\xi) = \frac{1}{\cos^2(\kappa)} \int_{-\infty}^{\infty} \frac{\partial}{\partial \tau} X_w^c \left\{ \mu_{\gamma,\vartheta}(\sigma, \tau(\xi)) \right\} d\sigma, \tag{8.91}$$

where

$$X_w^c \left\{ \mu_{\gamma,\vartheta}(\sigma, \tau(\xi)) \right\} = \cos(\beta) X^c \left\{ \mu_{\gamma,\vartheta}(\sigma, \tau(\xi)) \right\} \tag{8.92}$$

denotes the weighted X-ray projection.

Figure 8.32 illustrates that the (σ, τ) coordinate system is rotated by the angle δ with respect to the (a, b) principal detector axis system.

The radial differentiation, which means the partial derivation of the weighted detector values in the direction of the τ coordinate, can be split into two terms. This can be done when using a description in components of the principal coordinate system, i.e.,

$$\frac{\partial}{\partial \tau} X_w^c \left\{ \mu_{\gamma,\vartheta}(\sigma, \tau(\xi)) \right\} = G_a(\sigma, \tau) \sin(\delta) + G_b(\sigma, \tau) \cos(\delta), \tag{8.93}$$

where the partial derivation in the direction of the principal detector axes reads

$$G_a(\sigma, \tau) = \frac{\partial}{\partial a} X_w^c \left\{ \mu_{\gamma,\vartheta}(\sigma, \tau(\xi)) \right\} \text{ and } G_b(\sigma, \tau) = \frac{\partial}{\partial b} X_w^c \left\{ \mu_{\gamma,\vartheta}(\sigma, \tau(\xi)) \right\}. \tag{8.94}$$

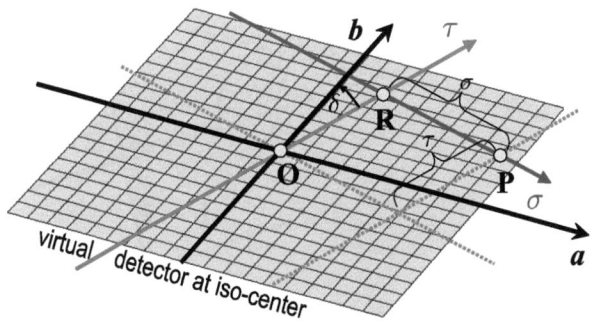

Fig. 8.32. The (σ, τ) system is rotated by the angle δ with respect to the coordinate system of the detector principal axes a and b

8.5.4
Reconstruction with the Derivative of the Radon Transform

(8.90) provides an interesting relationship between the integration along radial derivatives on the flat-panel detector and the first derivative of the Radon transform with respect to the radial Radon coordinate. However, which advantage one may derive from this relationship with regard to object reconstruction remains to be clarified. For this purpose, the geometrical situation will once again be studied in Fig. 8.33.

The key idea of *Grangeat* is that (8.91) yields a value on the Radon sphere of the parallel projection, which is actually located at the point $\boldsymbol{\xi} = (\xi, \gamma, \vartheta)$. However, this approach does not enable direct access to the Radon values, only to their first radial derivative. These values are assembled on a meridian surface $\mathbf{V}_y$. The polar orientation of $\mathbf{V}_y$ just corresponds to the two Radon coordinates (ξ, ϑ).

All values $\partial p_{y,\vartheta}(\xi)/\partial\xi$ are thus collected on different meridian surfaces, $\mathbf{V}_y$, which are inclined by the corresponding angle y. For the complete object reconstruction

$$f(x, y, z) = -\frac{1}{8\pi^2} \iint_S \frac{\partial^2 p_{y,\vartheta}(\xi)}{\partial\xi^2} \, dS , \qquad (8.38)$$

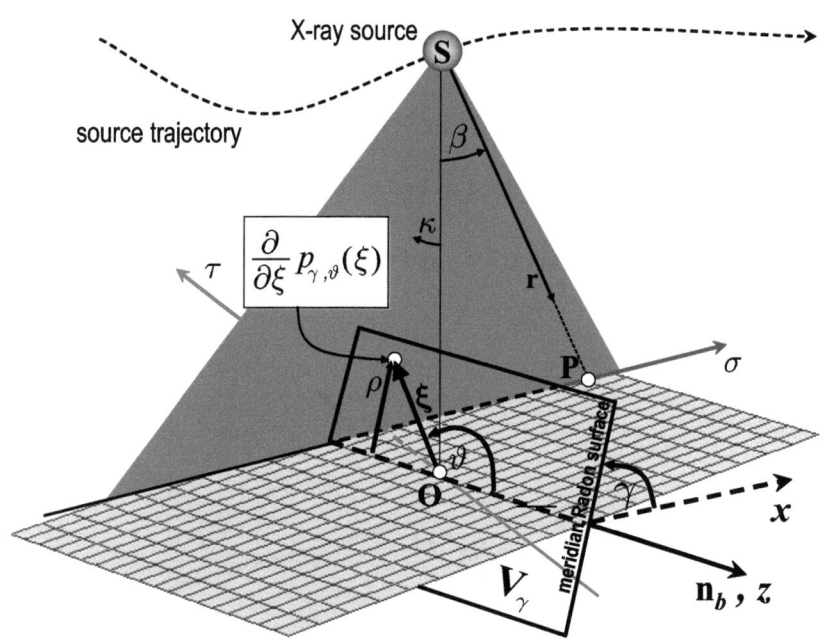

Fig. 8.33. The cone-beam projection does not provide a direct access to the Radon transform required for the reconstruction process. However, (8.91) yields a radially differentiated value of the Radon transform on the corresponding point situated on the Radon sphere. These values are collected on a meridian surface $\mathbf{V}_y$, which includes the radial vector of the Radon point

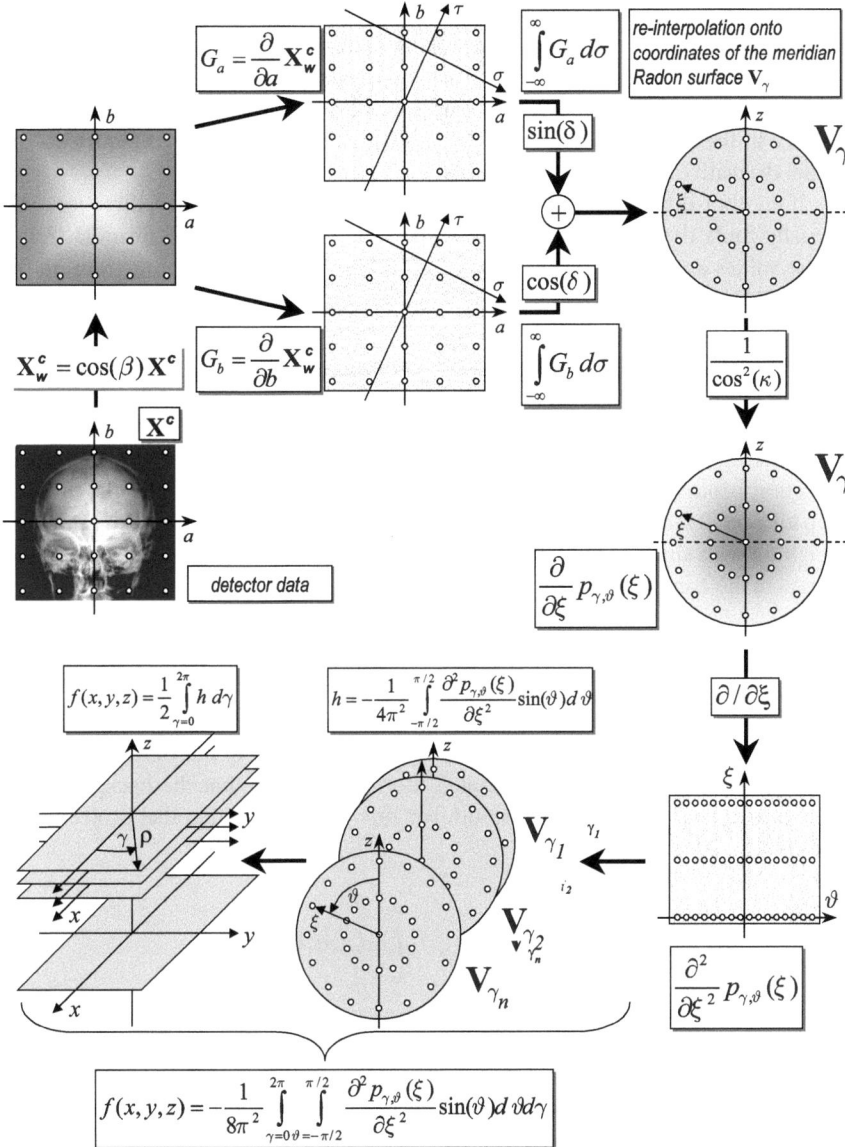

Fig. 8.34. Summary of the three-dimensional reconstruction method proposed by *Pierre Grangeat*

these values must once again be radially differentiated and then azimuthally and polarly backprojected, as required by (8.38).

Figure 8.34 illustrates the individual reconstruction steps schematically. Starting with the cone-beam projection data on the flat-panel detector (center left), the data are pre-weighted in a first step. Next, the horizontal and vertical derivative may be considered separately. Thanks to the linearity of the integration operation in the derivative images, one may now form the line integrals with respect to σ separately.

Both results are added after the sine and cosine weighting step have been carried out. By interpolation, this value can be sorted into a virtual meridian detector surface in the Radon space $\xi = (\xi, \gamma, \vartheta)$. Another weighting step finally yields the radial derivative of the Radon transform.

If another differentiation is carried out once again in a radial ξ direction, one actually finds the integrand that appears in the Radon inversion formula (8.38). These values are schematically shown in the form of a Cartesian sinogram on the lower right side of Fig. 8.34.

The integration with respect to the infinitesimal surface element dS introduced in (8.37) corresponds to two consecutive backprojections: One in a polar direction, ϑ, i.e., inside the interpolated, meridian Radon surfaces, and the other in an azimuthal direction γ, i.e., in parallel horizontal planes around the spatial z-axis.

Filtering of the data prior to backprojection, which is necessary for the reconstruction process, is not readily visible within this method. However, the filtering step takes effect in the penultimate step, because the Radon values, which have been weighted with $\sin(\vartheta)$ and twice radially differentiated, will be backprojected here. This actually corresponds exactly to the required filtering step.

8.5.5
Central Section Theorem and Grangeat's Solution

In this section the second reconstruction phase according to the *Grangeat* method is interpreted in terms of the central section theorem so that the backprojection instruction may better be understood. For this purpose, the parallel projection of the object to be reconstructed

$$X^P\left\{\mu_{\gamma,\vartheta}(\sigma,\xi)\right\} = \int_{-\infty}^{\infty} \mu(\eta,\sigma,\xi)\,d\eta\,, \tag{8.64}$$

must be considered. Figure 8.35 outlines that this parallel projection is measured with vertical, perpendicular detector planes. $\mathbf{n}_\eta$ denotes the detector surface normal vector and, as used in the previous sections, $\mathbf{n}_a$ and $\mathbf{n}_b$ denote the principal detector axes. The detector surface normal vector determines the orientation of the flat-panel detector, which is unequivocally described by the angle γ. The detector surface also represents the coordinates of the planes in the hybrid or generalized Radon space (cf. (8.46)).

With this vertical definition, the unit vectors of the flat-panel detector read

$$\mathbf{n}_\eta = \begin{pmatrix} \cos(\gamma) \\ \sin(\gamma) \\ 0 \end{pmatrix}, \; \mathbf{n}_a = \begin{pmatrix} -\sin(\gamma) \\ \cos(\gamma) \\ 0 \end{pmatrix} \text{ and } \mathbf{n}_b = \begin{pmatrix} 0 \\ 0 \\ 1 \end{pmatrix}. \tag{8.95}$$

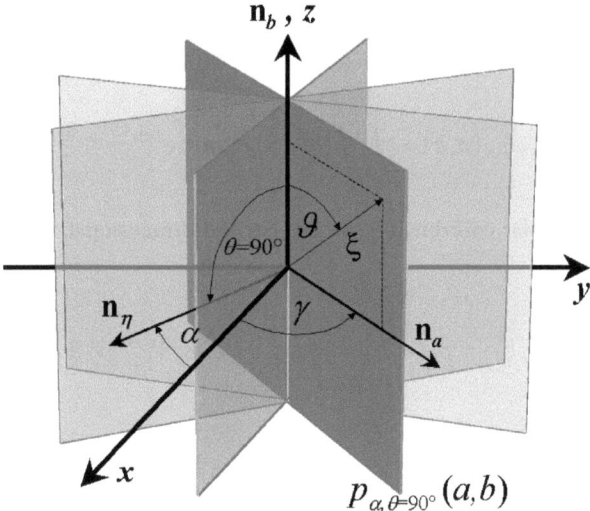

Fig. 8.35. Vertically arranged planar detector surfaces with the unit vectors pointing in direction of the principal detector axes $\mathbf{n}_a$ and $\mathbf{n}_b$. The surface normal of the detector is rotated by the angle $\alpha = 90° - \gamma$ with respect to the x-axis of the fixed patient coordinate system. The X-ray beams vertically pass through the corresponding flat-panel detector surfaces. The measured projection values on the detector represent the values of the hybrid or generalized Radon transform

This constricted detector orientation is obviously the special case, $\theta = 90°$, of the definition of the unit vectors (8.43) through (8.45). The parallel projection $\mathbf{X}^P\{\cdot\}$ thus also obeys the parameter representation of (8.42), i.e.,

$$p_\alpha(a, b) = p_{\alpha, \theta=90°}(a, b) = \mathbf{X}^P\left\{\mu_{\gamma, \vartheta}(\sigma, \xi)\right\} = \int_{-\infty}^{\infty} f(a\mathbf{n}_a + b\mathbf{n}_b + \eta\mathbf{n}_\eta)\,\mathrm{d}\eta\,.$$

$$(8.96)$$

The central section theorem mentioned in ▸ Sect. 8.4.4 ensures that the two-dimensional Fourier transform of the parallel projection is also available on a flat plane in the three-dimensional Fourier space of the object. In this context, the orientation of the detector plane corresponds to the orientation of the flat plane in the frequency domain, since the Fourier transform is rotationally variant. One may therefore write

$$\mathcal{F}_2\left\{p_\alpha(a, b)\right\} = P_\gamma(q, p) = F(q\mathbf{n}_a + p\mathbf{n}_b) = \mathcal{F}_3\left\{f(\mathbf{r})\right\}\big|_{\mathbf{r} = \xi = a\mathbf{n}_a + b\mathbf{n}_b}\,. \quad (8.97)$$

The inverse Fourier transform of $P_\alpha(q, p)$ thus just yields the parallel projection $\mathbf{X}^P\{\cdot\}$ of the object to be reconstructed, i.e.,

$$\mathbf{X}^P\left\{\mu_{y,\vartheta}(\sigma, \xi)\right\} = p_\alpha(a, b) = \int\limits_{-\infty}^{\infty} \int\limits_{-\infty}^{\infty} F(q\mathbf{n}_a + p\mathbf{n}_b)\, e^{2\pi i \mathbf{r}_1^T \cdot (q\mathbf{n}_a + p\mathbf{n}_b)}\, dq\, dp. \quad (8.98)$$

Expressed with polar coordinates, the inverse two-dimensional Fourier transform (8.98) thus reads

$$\mathbf{X}^P\left\{\mu_{y,\vartheta}(\sigma, \xi)\right\} = p_\alpha(a, b) = \int\limits_{-\pi/2}^{\pi/2} \int\limits_{-\infty}^{\infty} F(\mathbf{v})\, e^{2\pi i (\mathbf{r}_1^T \cdot \mathbf{v})}\, |\mathbf{v}|\, d\mathbf{v}\, d\vartheta, \quad (8.99)$$

where

$$\mathbf{q} = q\mathbf{n}_a, \ \mathbf{p} = p\mathbf{n}_b \ \text{and} \ \mathbf{v} = \mathbf{q} + \mathbf{p}. \quad (8.100)$$

If, on the other hand, the Fourier-slice theorem described in ▶ Sect. 8.4.1 is considered here, one may recall that

$$\begin{aligned}
\mathcal{F}_3\left\{f(x, y, z)\right\} = F(u, v, w) &= \int\limits_{-\infty}^{\infty} \int\limits_{-\infty}^{\infty} \int\limits_{-\infty}^{\infty} f(x, y, z)\, e^{-2\pi i q(\mathbf{r}^T \cdot \mathbf{n}_\xi)}\, dx\, dy\, dz \\
&= F_{\text{spherical}}(q, \gamma, \vartheta) \\
&= P_{y,\vartheta}(q) = \int\limits_{-\infty}^{\infty} p_{y,\vartheta}(\xi)\, e^{-2\pi i q\xi}\, d\xi \\
&= \mathcal{F}_1\left\{\mathcal{R}_3\left\{f(x, y, z)\right\}\right\}.
\end{aligned} \quad (8.101)$$

(8.101) means that the one-dimensional radial Fourier transform of the three-dimensional Radon transform is again available as a radial line in the three-dimensional Fourier transform of the object.

(8.97) on the other hand ensures that the radial lines of the three-dimensional Fourier transform of the object located in the Fourier plane of the vertical flat-panel detector correspond to the two-dimensional hybrid Radon transform. One thus finally finds

$$\mathcal{F}_3\left\{f(x, y, z)\right\} = \mathcal{F}_2\left\{p_\alpha(a, b)\right\} = \mathcal{F}_1\left\{\mathcal{R}_3\left\{f(x, y, z)\right\}\right\} \quad (8.102)$$

so that the object to be reconstructed obeys

$$f(x, y, z) = \mathcal{F}_3^{-1}\left\{\mathcal{F}_1\left\{\mathcal{R}_3\left\{f(x, y, z)\right\}\right\}\right\}. \quad (8.103)$$

Figure 8.36 summarizes (8.102) schematically.

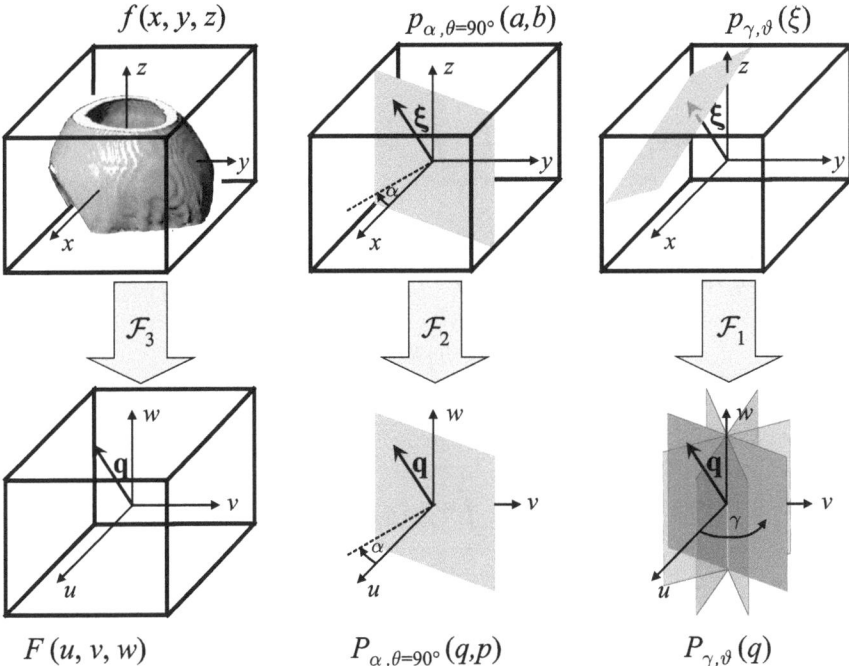

$f(x,y,z)$ $p_{\alpha,\theta=90°}(a,b)$ $p_{\gamma,\vartheta}(\xi)$

$F(u,v,w)$ $P_{\alpha,\theta=90°}(q,p)$ $P_{\gamma,\vartheta}(q)$

Fig. 8.36. Illustration showing the relation worked out in (8.102). The one-dimensional, radial Fourier transform of surface integrals is identical with the radial line with the same orientation of the two-dimensional Fourier transform of the surface of the line integrals. Both frequency points are also available in the three-dimensional Fourier transform of the object

With this identity one may also write (8.99) as follows,

$$\mathbf{X}^P\left\{\mu_{\gamma,\vartheta}(\sigma,\xi)\right\} = p_{\alpha=90°-\gamma}(a,b) = \int_{-\pi/2}^{\pi/2}\int_{-\infty}^{\infty} P_{\gamma,\vartheta}(q)\,e^{2\pi i(\xi^T\cdot\mathbf{q})}\,|\mathbf{q}|\,\mathrm{d}q\,\mathrm{d}\vartheta. \quad (8.104)$$

As (8.104) is consistent with (5.66), it obviously describes the simple two-dimensional filtered backprojection. This means that a parallel projection $\mathbf{X}^P\{\cdot\}$ or the corresponding hybrid Radon transform of the object to be reconstructed respectively may be computed by a two-dimensional filtered backprojection of the radial three-dimensional Radon values in vertical planes.

To understand the benefit of (8.104), one should look once again at the three-dimensional Radon inversion formula

$$f(x,y,z) = -\frac{1}{8\pi^2}\iint_S \frac{\partial^2 p_{\gamma,\vartheta}(\xi)}{\partial\xi^2}\,\mathrm{d}S \quad (8.38)$$

whereby the infinitesimal surface element dS on the unit sphere S is determined by (8.37). Correspondingly, one may substitute the surface element to find

$$f(x, y, z) = -\frac{1}{8\pi^2} \int\limits_{y=0}^{2\pi} \int\limits_{\vartheta=-\pi/2}^{\pi/2} \frac{\partial^2 p_{y,\vartheta}(\xi)}{\partial \xi^2} \sin(\vartheta) \, d\vartheta \, dy \, . \qquad (8.105)$$

The voxel (x, y, z) is reconstructed by the backprojection of the filtered parallel projection. This can be represented by

$$f(x, y, z) = \frac{1}{2} \int\limits_{y=0}^{2\pi} h \left[\mathbf{X}^{\mathbf{P}} \left\{ \mu_{y,\vartheta}(\sigma, \xi) \right\} \right] dy \, , \qquad (8.106)$$

where

$$h \left[\mathbf{X}^{\mathbf{P}} \left\{ \mu_{y,\vartheta}(\sigma, \xi) \right\} \right] = \int\limits_{-\pi/2}^{\pi/2} \int\limits_{-\infty}^{\infty} P_{y,\vartheta}(q) e^{2\pi i \xi q} q^2 \, dq \sin(\vartheta) \, d\vartheta$$

$$= -\frac{1}{4\pi^2} \int\limits_{-\pi/2}^{\pi/2} \frac{\partial^2 p_{y,\vartheta}(\xi)}{\partial \xi^2} \sin(\vartheta) \, d\vartheta \, . \qquad (8.107)$$

The method proposed by *Grangeat* is summarized in Scheme 8.1. Artifacts related to practical implementation are discussed in Lee et al. (2002).

A comparison of (8.107) with (8.104) reveals that the filtered parallel projection differs from the simple parallel projection by a supplementary weighting factor $\sin(\vartheta)$ and that it is filtered with the linear frequency ramp q in the frequency domain. (8.106) and (8.107) correspond exactly to the two final steps in Fig. 8.34. (8.106) can be related to the parallel projection or the hybrid Radon transform $p_\alpha(a, b)$ respectively, by using the central section theorem.

8.5.6
Direct 3D Fourier Reconstruction with the Cone-Beam Geometry

In ▶ Sect. 5.4 the method of direct inverse Radon transform based on the inverse Fourier transform for the two-dimensional case has been discussed. In theory, this approach is straightforward. However, problems related to the radially decreasing density of the sample points in the Fourier space of the Radon transform arise when this approach is implemented in practice. The linogram method (Jacobson 1996), which is described in ▶ Sect. 5.5, was proposed as a countermeasure to these problems. It should be concisely discussed here that it is also possible to reconstruct successfully with a properly adapted linogram method in the case of three-dimensional cone-beam geometry.

Scheme 8.1 Reconstruction with cone-beam geometry according to Grangeat

1. Pre-weighting of the detector data

$$X_w^c \left\{ \mu_{\gamma,\vartheta}(\sigma, \tau(\xi)) \right\} = \cos(\beta) X^c \left\{ \mu_{\gamma,\vartheta}(\sigma, \tau(\xi)) \right\} \qquad (8.92)$$

2. Computation of partial derivatives in the direction of the principal flat-panel detector axes

$$G_a(\sigma, \tau) = \frac{\partial}{\partial a} X_w^c \left\{ \mu_{\gamma,\vartheta}(\sigma, \tau(\xi)) \right\} \text{ and } G_b(\sigma, \tau) = \frac{\partial}{\partial b} X_w^c \left\{ \mu_{\gamma,\vartheta}(\sigma, \tau(\xi)) \right\}$$
$$(8.94)$$

3. Computation of line integrals on the detector and weighting of the results with the factor $1/\cos^2(\kappa)$ yields the radial derivative of the Radon transform, i.e.,

$$\frac{\partial}{\partial \xi} p_{\gamma,\vartheta}(\xi) = \frac{1}{\cos^2(\kappa)} \int_{-\infty}^{\infty} \left\{ G_a(\sigma, \tau) \sin(\delta) + G_b(\sigma, \tau) \cos(\delta) \right\} d\sigma \qquad (8.108)$$

4. Interpolation of the data located on the Radon sphere given by (8.108) into vertical, meridian Radon surfaces.

5. Filtered backprojection within the vertical, meridian Radon surfaces, i.e.,

$$h \left[X^p \left\{ \mu_{\gamma,\vartheta}(\sigma, \xi) \right\} \right] = -\frac{1}{4\pi^2} \int_{-\pi/2}^{\pi/2} \frac{\partial^2 p_{\gamma,\vartheta}(\xi)}{\partial \xi^2} \sin(\vartheta) \, d\vartheta . \qquad (8.107)$$

6. Backprojection within horizontal planes by integration over the plane rotation angle, γ, i.e.,

$$f(x, y, z) = \frac{1}{2} \int_{\gamma=0}^{2\pi} h \left[X^p \left\{ \mu_{\gamma,\vartheta}(\sigma, \xi) \right\} \right] d\gamma . \qquad (8.106)$$

In ▸ Sect. 8.4.1, equation

$$
\begin{aligned}
F(u(q, \gamma, \vartheta), & v(q, \gamma, \vartheta), w(q, \gamma, \vartheta)) \\
&= F(q \cos(\gamma) \sin(\vartheta), q \sin(\gamma) \sin(\vartheta), q \cos(\vartheta)) \\
&= F_{\text{spherical}}(q, \gamma, \vartheta) \\
&= P_{\gamma,\vartheta}(q)
\end{aligned}
\qquad (8.26)
$$

summarized the applicability of the Fourier slice theorem to the three-dimensional reconstruction problem. In this context, it should be recalled here that the Fourier transform of the Radon transform is in line with the Fourier transform of the object on radial lines passing through the origin of the frequency domain. However, the previous sections have shown that direct access to the Radon transform $p_{\gamma,\vartheta}(\xi)$ is not available within cone-beam geometry.

The approach proposed by *Grangeat* only yields the radial derivative of the Radon transform, i.e., $\partial p_{\gamma,\vartheta}(\xi)/\partial\xi$. But *Grangeat* also verified that the derivative is at least half the way to the filtered backprojection. Unfortunately, within the direct Fourier-based reconstruction approach one does not need filtered Radon data, but rather the native Radon transform $p_{\gamma,\vartheta}(\xi)$. The following considerations will show how the expression $\partial p_{\gamma,\vartheta}(\xi)/\partial\xi$ can be integrated into the direct reconstruction approach.

In a first step, the Fourier differentiation property given in (8.33) has to be considered again. Due to this rule, the Fourier transform of the radial derivative of the Radon transform is given by

$$\mathcal{F}_1\left\{\frac{\partial}{\partial\xi}p_{\gamma,\vartheta}(\xi)\right\} = i2\pi q P_{\gamma,\vartheta}(q) . \tag{8.109}$$

The Fourier transform of the Radon transform thus obeys

$$\mathcal{F}_1\left\{p_{\gamma,\vartheta}(\xi)\right\} = \frac{1}{i2\pi q}\mathcal{F}_1\left\{\frac{\partial}{\partial\xi}p_{\gamma,\vartheta}(\xi)\right\} . \tag{8.110}$$

The inverse Fourier transform then yields

$$p_{\gamma,\vartheta}(\xi) = \mathcal{F}_1^{-1}\left\{\frac{1}{i2\pi q}\mathcal{F}_1\left\{\frac{\partial}{\partial\xi}p_{\gamma,\vartheta}(\xi)\right\}\right\} . \tag{8.111}$$

If one takes into account the Fourier result of the signum function mentioned in ▸ Sect. 4.10, i.e.,

$$\text{sign}(\xi) \;\circ\!\!-\!\!-\!\!\bullet\; \frac{1}{i\pi q} , \tag{8.112}$$

the convolution theorem (cf. ▸ Sect. 4.11)

$$p_{\gamma,\vartheta}(\xi) = \mathcal{F}_1^{-1}\left\{\frac{1}{i2\pi q}\mathcal{F}_1\left\{\frac{\partial}{\partial\xi}p_{\gamma,\vartheta}(\xi)\right\}\right\} = \mathcal{F}_1^{-1}\left\{\frac{1}{i2\pi q}\right\} * \mathcal{F}_1^{-1}\left\{\mathcal{F}_1\left\{\frac{\partial}{\partial\xi}p_{\gamma,\vartheta}(\xi)\right\}\right\} \tag{8.113}$$

can be applied to find the normal Radon transform by a convolution between the signum function and the radial derivative of the Radon transform, i.e.,

$$p_{\gamma,\vartheta}(\xi) = \frac{1}{2}\text{sign}(\xi) * \frac{\partial}{\partial\xi}p_{\gamma,\vartheta}(\xi) . \tag{8.114}$$

With (8.103), it is found that

$$\begin{aligned} f(x,y,z) &= \mathcal{F}_3^{-1}\left\{\mathcal{F}_1\left\{\mathcal{R}_3\left\{f(x,y,z)\right\}\right\}\right\} \\ &= \mathcal{F}_3^{-1}\left\{\mathcal{F}_1\left\{p_{\gamma,\vartheta}(\xi)\right\}\right\} \\ &= \mathcal{F}_3^{-1}\left\{\mathcal{F}_1\left\{\frac{1}{2}\text{sign}(\xi) * \frac{\partial}{\partial\xi}p_{\gamma,\vartheta}(\xi)\right\}\right\} . \end{aligned} \tag{8.115}$$

However – similar to the two-dimensional case – the problem that the sampling density decreases with an increasing distance to the origin of the frequency domain still has to be solved. In Jacobson (1996), it has been shown that the linogram sampling method is also helpful here.

8.5.7
Exact Reconstruction using Filtered Backprojection

Kudo and Saito (1994) as well as Defrise and Clack (Defrise and Clack 1994; Clack and Defrise 1994) proposed an initial formulation for a filtered backprojection based on the ideas proposed by *Grangeat* that have been described in the sections above. In order to illustrate the strategy within this approach the following section, similar to Yang and Horn (2002), is based on the general three-dimensional inverse Radon transform

$$f(x, y, z) = -\frac{1}{8\pi^2} \iint_S \frac{\partial^2 p_{y,\vartheta}(\xi)}{\partial \xi^2} \, d\mathbf{n}_\xi . \tag{8.38}$$

Since a backprojection method should be established, a defined, fixed source location does not have to be assumed. Rather, this section deals with an X-ray source trajectory $\mathbf{S}(\lambda)$, parameterized by the continuous variable λ. This is necessary because – similar to all backprojection methods – each backprojection is carried out immediately after the projection has been acquired. Therefore, is not necessary to wait for the complete set of data of the Radon space before the reconstruction can take place. This principle is essentially the reason for the efficiency of backprojection methods.

Thus, it holds that

$$\xi = \mathbf{S}(\lambda)^{\mathrm{T}} \cdot \mathbf{n}_\xi , \tag{8.116}$$

since this expression describes the great circle on the Radon sphere for the source position. Figure 8.37 illustrates the geometrical situation. To be able to reconstruct the point $\mathbf{r}$, the contribution of the detector point $\mathbf{P}$ in the backprojection has to be found.

Taking into account the sifting property of the δ-distribution mentioned in ▶ Sect. 4.6, one may write (8.38) as a convolution

$$f(\mathbf{r}) = -\frac{1}{8\pi^2} \iint_S \left[\int_{-\infty}^{\infty} \frac{\partial^2 p_{y,\vartheta}(\xi)}{\partial \xi^2} \delta(\xi - \mathbf{r}^{\mathrm{T}} \cdot \mathbf{n}_\xi) \, d\xi \right] d\mathbf{n}_\xi , \tag{8.117}$$

which even better indicates that the differentiation takes place at all points $\mathbf{r}$, which obey

$$\xi = \mathbf{r}^{\mathrm{T}} \cdot \mathbf{n}_\xi . \tag{8.118}$$

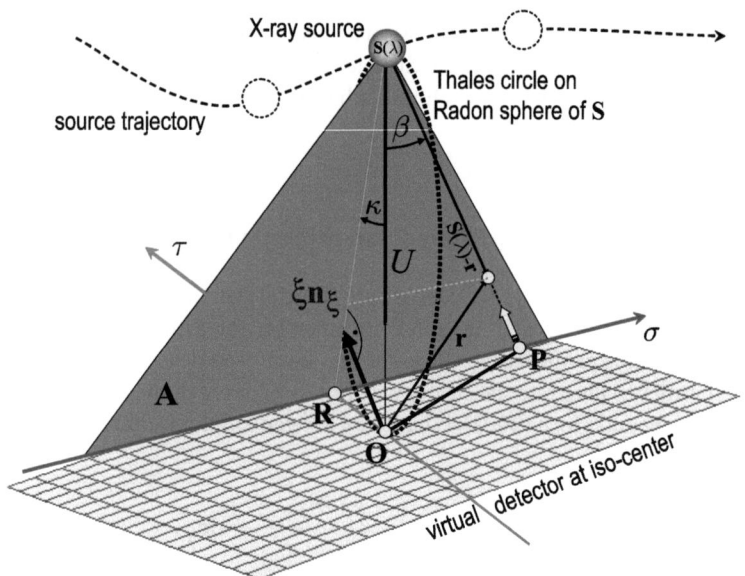

Fig. 8.37. Geometric situation of the cone-beam backprojection. The vector $\mathbf{S}(\lambda)$ has the same direction as the detector surface normal vector and the length of the distance between the detector and source $\|\mathbf{S} - \mathbf{O}\|$. The contribution of the detector value at the point $\mathbf{P}$ must be found in order to reconstruct the point $\mathbf{r}$. $\mathbf{A}$ is the integration surface of the Radon transform. This plane is unequivocally determined by $\boldsymbol{\xi} = \xi \mathbf{n}_\xi$

(8.118) determines the integration surface $\mathbf{A}$ of the Radon transform shown in Fig. 8.37. As the X-ray source must also be located in this plane, (8.116) must obviously be considered as a special case of (8.118).

Here, one actually does not want to consider a movement in the Radon space, but rather a movement of the X-ray source in the spatial domain for the backprojection. Therefore, it must be analyzed how the radial distance, ξ, in the Radon space is changed, in case the motion parameter λ of the X-ray source is varied. The amount of this change is determined by the variation of λ in (8.116), i.e.,

$$\frac{d\xi}{d\lambda} = \left| \mathbf{S}'(\lambda)^{\mathrm{T}} \cdot \mathbf{n}_\xi \right| , \tag{8.119}$$

where $\mathbf{S}'(\lambda)$ is a tangential vector of the source trajectory at the orbit point $\mathbf{S}(\lambda)$. If one writes (8.119) as

$$d\xi = \left| \mathbf{S}'(\lambda)^{\mathrm{T}} \cdot \mathbf{n}_\xi \right| d\lambda , \tag{8.120}$$

it is revealed that the radial sampling density along $\boldsymbol{\xi}$ depends on the trajectory direction. The factor can be understood as a compensation for the generally appearing irregular sampling of the trajectory.

To replace ξ and $d\xi$, (8.116) and (8.120) are substituted into (8.117), which means the trajectory parameter λ of the X-ray source replaces the radial Radon variable.

One finds

$$
f(\mathbf{r}) = -\frac{1}{8\pi^2} \iint_S \left[\int_\Lambda \left. \frac{\partial^2 p_{\gamma,\vartheta}(\xi,\lambda)}{\partial \xi^2} \right|_{\xi = \mathbf{S}(\lambda)^{\mathrm{T}} \cdot \mathbf{n}_\xi} \cdots \right.
$$

$$
\left. \cdots \delta(\mathbf{S}(\lambda)^{\mathrm{T}} \cdot \mathbf{n}_\xi - \mathbf{r}^{\mathrm{T}} \cdot \mathbf{n}_\xi) M(\lambda, \mathbf{n}_\xi) \left| \mathbf{S}'(\lambda)^{\mathrm{T}} \cdot \mathbf{n}_\xi \right| \, d\lambda \right] d\mathbf{n}_\xi . \tag{8.121}
$$

In addition, the correction function

$$
M(\lambda, \mathbf{n}_\xi) = \frac{1}{n(\lambda, \mathbf{n}_\xi)} \tag{8.122}
$$

has been introduced, which acts as a compensation function for redundant measurements of points in the Radon space. This is necessary because (8.118) may have several solutions for arbitrary source trajectories. The function $n(\lambda, \mathbf{n}_\xi)$ describes the number of points of intersection of the source trajectory $\mathbf{S}(\lambda)$ with the integration surface $\mathbf{A}$ of the Radon transform. That is, a particular Radon point $p_{\gamma,\vartheta}(\xi,\lambda)$ is measured for an arbitrary source trajectory generally at $n(\lambda, \mathbf{n}_\xi)$ source positions, or, in other words, the integration surface $\mathbf{A}$ is illuminated at $n(\lambda, \mathbf{n}_\xi)$ positions. As the compensation function $M(\lambda, \mathbf{n}_\xi)$ obviously varies for different points in the Radon space, it is used as a scaling function to prevent some Radon points from excessively contributing to the reconstruction of the object. As long as the sufficiency condition is fulfilled during data acquisition (cf. ▸ Sects. 8.4.5 and 8.6.1), $n(\lambda, \mathbf{n}_\xi)$ does not vanish.

An important issue in (8.122) is the integration in the domain of Λ, i.e., the set of source trajectory points. In this way, the integration over the radial Radon component is substituted here by a more convenient integration direction, required for the backprojection procedure. The order of integration may be changed such that

$$
f(\mathbf{r}) = -\frac{1}{8\pi^2} \int_\Lambda \left[\iint_S \left. \frac{\partial^2 p_{\gamma,\vartheta}(\xi,\lambda)}{\partial \xi^2} \right|_{\xi = \mathbf{S}(\lambda)^{\mathrm{T}} \cdot \mathbf{n}_\xi} \cdots \right.
$$

$$
\left. \cdots \delta\left((\mathbf{S}(\lambda) - \mathbf{r})^{\mathrm{T}} \cdot \mathbf{n}_\xi \right) M(\lambda, \mathbf{n}_\xi) \left| \mathbf{S}'(\lambda)^{\mathrm{T}} \cdot \mathbf{n}_\xi \right| \, d\mathbf{n}_\xi \right] d\lambda \tag{8.123}
$$

is obtained to compute the backprojection instantaneously for each projection measurement.

According to *Grangeat* the first-order derivative of the Radon transform is achieved by a differentiation of the flat-panel detector data. However, in (8.123) the second-order derivative is required. Therefore, it should be analyzed whether one of the two differentiations may be replaced by an appropriate convolution. The required convolution kernel can be obtained by means of the Fourier representation of the first derivative (8.33), which has frequently been used in the previous sections.

The inverse formulation of (8.109) actually reads

$$\frac{\partial}{\partial\xi}p_{\gamma,\vartheta}(\xi,\lambda) = \mathcal{F}_1^{-1}\left\{i2\pi q P_{\gamma,\vartheta}(q,\lambda)\right\}, \tag{8.124}$$

so that the convolution theorem (cf. ▸ Sect. 4.11) yields

$$\frac{\partial}{\partial\xi}p_{\gamma,\vartheta}(\xi,\lambda) = \mathcal{F}_1^{-1}\left\{i2\pi q\right\} * \mathcal{F}_1^{-1}\left\{P_{\gamma,\vartheta}(q,\lambda)\right\}$$
$$= g(\xi) * p_{\gamma,\vartheta}(\xi,\lambda), \tag{8.125}$$

where the convolution kernel obviously reads

$$g(\xi) = \int_{-\infty}^{\infty} i2\pi q\, e^{i2\pi q\xi}\, dq. \tag{8.126}$$

Substituting (8.126) in (8.123) one finds

$$f(\mathbf{r}) = -\frac{1}{8\pi^2}\int_{\Lambda}\left[\iint_S \int_{\xi'=-\infty}^{+\infty} g(\xi'-\xi)\frac{\partial p_{\gamma,\vartheta}(\xi',\lambda)}{\partial\xi'}\delta\left((\mathbf{S}(\lambda)-\mathbf{r})^{\mathrm{T}}\cdot\mathbf{n}_\xi\right)\cdots \right.$$
$$\left. \cdots M(\lambda,\mathbf{n}_\xi)\left|\mathbf{S}'(\lambda)^{\mathrm{T}}\cdot\mathbf{n}_\xi\right|d\xi'\,d\mathbf{n}_\xi\right]d\lambda. \tag{8.127}$$

Since ξ is the radial Radon coordinate, the convolution in (8.127) means that numerous parallel Radon integration surfaces (ξ',γ,ϑ) are needed from different source positions. This, however, jeopardizes the fundamental idea of backprojection, the advantage of which – compared with the Fourier method – is in fact the exclusive and self-contained treatment of each X-ray source position. For a particular λ, the integration surfaces described by ξ and ξ' in the convolution in (8.127) cannot both contain the X-ray source.

On the other hand, for a particular λ on the Radon sphere of the X-ray source position $\mathbf{S}(\lambda)$, the theorem of Thales implies that (8.116) also applies to all other integration surface vectors that describe a great circle on this Radon sphere. For instance, one may also write

$$\xi_1' = \mathbf{S}(\lambda)^{\mathrm{T}}\cdot\mathbf{n}_{\xi_1}. \tag{8.128}$$

(8.127) thus reads

$$f(\mathbf{r}) = -\frac{1}{8\pi^2}\int_{\Lambda}\left[\iint_S \int_{\xi_1'=-\infty}^{+\infty} g(\xi_1'-\xi_1)\frac{\partial p_{\gamma,\vartheta}(\xi_1',\lambda)}{\partial\xi_1'}\delta\left((\mathbf{S}(\lambda)-\mathbf{r})^{\mathrm{T}}\cdot\mathbf{n}_{\xi_1}\right)\cdots \right.$$
$$\left. \cdots M(\lambda,\mathbf{n}_{\xi_1})\left|\mathbf{S}'(\lambda)^{\mathrm{T}}\cdot\mathbf{n}_{\xi_1}\right|d\xi_1'\,d\mathbf{n}_{\xi_1}\right]d\lambda. \tag{8.129}$$

Figure 8.38 illustrates the geometrical situation of the new integration surface $\mathbf{A}'$. Now, the integration surface actually runs through the source as desired, but, unfortunately, the point $\mathbf{r}$ to be reconstructed is no longer included in the surface $\mathbf{A}'$.

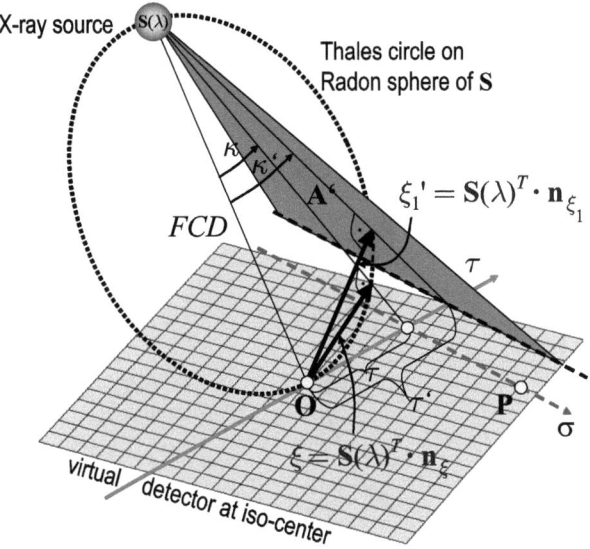

Fig. 8.38. Geometrical situation of the cone-beam backprojection. If one varies the integration surface $\mathbf{A}$ by a variation of the integration surface vector $\boldsymbol{\xi} \to \boldsymbol{\xi}_1'$ on the great circle of the Radon sphere of $\mathbf{S}(\lambda)$, then the new integration surface $\mathbf{A}'$ in fact still includes the X-ray source, but, unfortunately, does no longer include the point to be reconstructed

However, the convolution in (8.129) is a parallel displacement of the surface $\mathbf{A}'$. Fortunately, it can be shown that the undesired shift of the Radon variable ξ_1 can be transformed into a shift of the radial detector variable, τ.

This transformation of the Radon variable into a detector variable is illustrated in Fig. 8.39. From Fig. 8.39a it can be deduced that

$$\xi_1' - \xi_1 = (\tau' - \tau)\cos(\kappa')\frac{U}{FCD} \, . \tag{8.130}$$

The infinitesimal variation of the Radon variable must also be converted into an infinitesimal variation of the detector variable. For this purpose, (8.130) has to be differentiated, i.e.,

$$\frac{d\xi_1'}{d\tau'} = \cos(\kappa')\frac{U}{FCD} \, . \tag{8.131}$$

Analyzing Fig. 8.39b by means of the intercept theorem, the ratio

$$\frac{U}{FCD} = \frac{\tau'\cos^2(\kappa')}{\tau'} \tag{8.132}$$

can be obtained so that the substitution reads

$$d\xi_1' = \cos^3(\kappa')\,d\tau' \, . \tag{8.133}$$

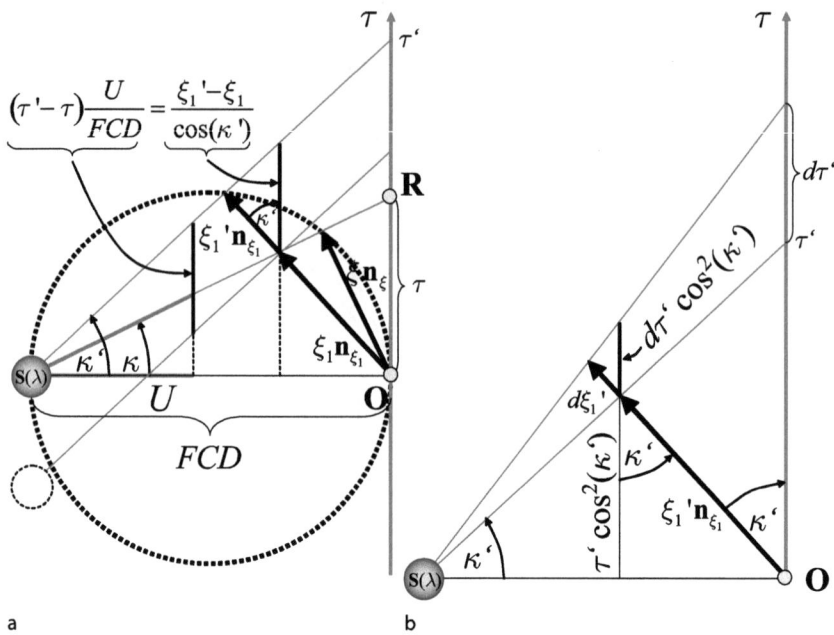

Fig. 8.39. a Intercept theorems in the geometry of the cone-beam backprojection for the two cases $\xi_1' = \mathbf{S}(\lambda)^T \cdot \mathbf{n}_{\xi_1}$ and $\xi = \mathbf{S}(\lambda)^T \cdot \mathbf{n}_{\xi}$, as well as **b** for the infinitesimal change of the radial Radon component ξ_1'

(8.131) and (8.133) are substituted into (8.130) so that one finds

$$f(\mathbf{r}) = -\frac{1}{8\pi^2} \int_\Lambda \left[\begin{array}{c} \displaystyle\iint_S \int_{\tau'=-\infty}^{+\infty} g\left((\tau'-\tau)\cos(\kappa')\frac{U}{FCD}\right)\cdots \\[2mm] \cdots \frac{\partial p_{y,\vartheta}(\xi_1',\lambda)}{\partial \xi_1'} \delta\left((\mathbf{S}(\lambda)-\mathbf{r})^T\cdot\mathbf{n}_{\xi_1}\right)\cdots \\[2mm] \cdots M(\lambda,\mathbf{n}_{\xi_1})\left|\mathbf{S}'(\lambda)^T\cdot\mathbf{n}_{\xi_1}\right|\cos^3(\kappa')\,d\tau'\,d\mathbf{n}_{\xi_1} \end{array} \right] d\lambda . \qquad (8.134)$$

The scaling property of the convolution kernel (8.127) with the argument of (8.131),

$$g\left(\tau\cos(\kappa')\frac{U}{FCD}\right) = \int_{-\infty}^{\infty} i2\pi q\, e^{i2\pi q\left(\tau\cos(\kappa')\frac{U}{FCD}\right)}\,dq , \qquad (8.135)$$

can be found by the substitution

$$q' = q\cos(\kappa')\frac{U}{FCD} . \qquad (8.136)$$

The following can be obtained:

$$g\left(\tau\cos(\kappa')\frac{U}{FCD}\right) = \frac{1}{\left(\cos(\kappa')\frac{U}{FCD}\right)^2}\int_{-\infty}^{\infty} i2\pi q'\, e^{i2\pi q'\tau}\, dq'$$

$$= \frac{FCD^2}{U^2\cos^2(\kappa')}g(\tau)$$

(8.137)

and thus finally for (8.134)

$$f(\mathbf{r}) = -\frac{1}{8\pi^2}\int_\Lambda \begin{bmatrix} \iint_S \int_{\tau'=-\infty}^{+\infty} \dfrac{FCD^2}{U^2}g(\tau'-\tau)\dfrac{\partial p_{\gamma,\vartheta}(\xi_1',\lambda)}{\partial\xi_1'}\cdots \\[2mm] \cdots\delta\left((\mathbf{S}(\lambda)-\mathbf{r})^{\mathrm{T}}\cdot\mathbf{n}_{\xi_1}\right)\cdots \\[2mm] \cdots M(\lambda,\mathbf{n}_{\xi_1})\left|\mathbf{S}'(\lambda)^{\mathrm{T}}\cdot\mathbf{n}_{\xi_1}\right|\cos(\kappa')\,d\tau'\,d\mathbf{n}_{\xi_1} \end{bmatrix} d\lambda .$$

(8.138)

For a particular source position, (8.138) applies not only to the point (8.128), but also to all points on the great circle of the Radon sphere of $\mathbf{S}(\lambda)$; thus, one may substitute back $\xi_1' \to \xi$.

The shift

$$\xi_1' = \mathbf{S}(\lambda)^{\mathrm{T}}\cdot\mathbf{n}_{\xi_1} \;\to\; \xi = \mathbf{S}(\lambda)^{\mathrm{T}}\cdot\mathbf{n}_\xi$$

(8.139)

yields

$$f(\mathbf{r}) = -\frac{1}{8\pi^2}\int_\Lambda\frac{FCD^2}{U^2}\begin{bmatrix} \iint_S \dfrac{\partial}{\partial\tau}\dfrac{\partial p_{\gamma,\vartheta}(\xi,\lambda)}{\partial\xi}\delta\left((\mathbf{S}(\lambda)-\mathbf{r})^{\mathrm{T}}\cdot\mathbf{n}_\xi\right)\cdots \\[2mm] \cdots M(\lambda,\mathbf{n}_\xi)\left|\mathbf{S}'(\lambda)^{\mathrm{T}}\cdot\mathbf{n}_\xi\right|\cos(\kappa)\,d\mathbf{n}_\xi \end{bmatrix} d\lambda ,$$

(8.140)

where the convolution of the radial detector variable τ yields the differentiation operator in analogy to the rules (8.125) through (8.127). The sift property of the δ-distribution may be replaced by changing the integration interval so that (8.140) becomes

$$f(\mathbf{r}) = -\frac{1}{8\pi^2}\int_\Lambda\frac{FCD^2}{U^2}\begin{bmatrix} \iint_{(\mathbf{S}(\lambda)-\mathbf{r})\perp\mathbf{n}_\xi} \dfrac{\partial}{\partial\tau}\dfrac{\partial p_{\gamma,\vartheta}(\xi,\lambda)}{\partial\xi}\dfrac{M(\lambda,\mathbf{n}_\xi)}{\left|\mathbf{S}'(\lambda)^{\mathrm{T}}\cdot\mathbf{n}_\xi\right|\cos(\kappa)}\,d\mathbf{n}_\xi \end{bmatrix} d\lambda .$$

(8.141)

To understand along which directions it must be integrated in the inner integral on the surface of the Radon sphere, Fig. 8.40a also illustrates that the points obeying $(\mathbf{S}(\lambda) - \mathbf{r})^{\mathrm{T}} \cdot \mathbf{n}_\xi = 0$ are located on a circle on this surface.

All points of the circle contribute to the reconstruction of the object point $\mathbf{r}$ in the backprojections, since these are all fan surfaces inside the cone beam, which include the beam $(\mathbf{S}(\lambda) - \mathbf{r})$. All directions $\mathbf{n}_\xi$ perpendicular to this beam, define

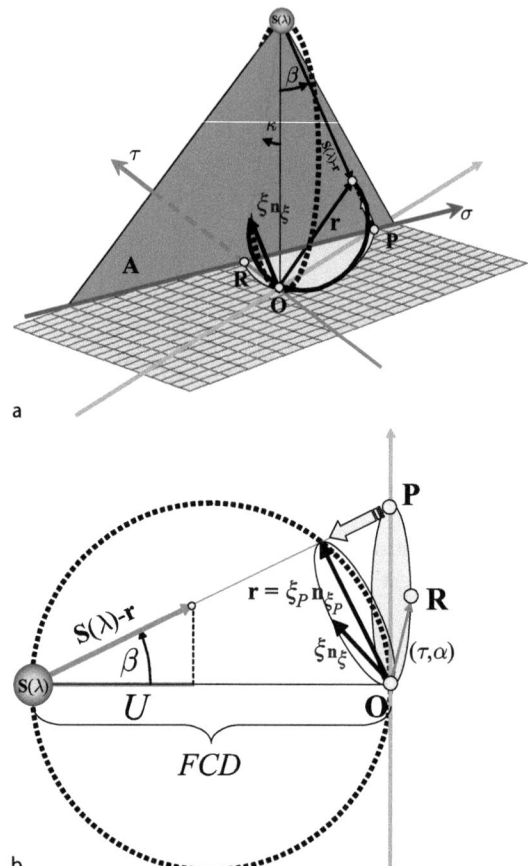

Fig. 8.40. The contributions to the detector backprojection point **P** are collected on a circle within the Radon sphere of $\mathbf{S}(\lambda)$. The δ-distribution in equation (8.140) indicates that it must be integrated along the directional contributions on the Radon sphere, which obey $(\mathbf{S}(\lambda) - \mathbf{r})^\mathrm{T} \cdot \mathbf{n}_\xi = 0$. These are all points $(\mathbf{S}(\lambda) - \mathbf{r}) \perp \mathbf{n}_\xi$ located on a circle of the Radon sphere as illustrated in **a**. **b** shows that the integration over the directions $\mathbf{n}_\xi$ can be mapped on the detector by projection. The data at the position τ will be integrated on the detector over all angles α

a plane passing through the origin **O** and have the normal $(\mathbf{S}(\lambda) - \mathbf{r})$. The section of this plane with the Radon sphere of $\mathbf{S}(\lambda)$ yields a circle with a radius

$$r = |\mathbf{r}| = \left| \xi_P \mathbf{n}_{\xi_P} \right| .\tag{8.142}$$

That means, only the Radon values on this very circle contribute to the reconstruction. This applies to the reconstruction of all object points located on the beam passing through $\mathbf{r}$. Moreover, this is in fact the characteristic feature of the backprojection. The filtered projection values are "smeared" back in the direction of the original projection.

Scheme 8.2 Exact reconstruction by backprojection within cone-beam geometry

1. Computation of the radial derivative of the Radon transform according to Scheme 8.1

$$\mathcal{R}'(f(\tau, \alpha, \lambda)) = \frac{\partial}{\partial \xi} p_{\gamma, \vartheta}(\xi, \lambda)$$

$$= \frac{1}{\cos^2(\kappa)} \frac{\partial}{\partial \tau} \int_{-\infty}^{\infty} \cos(\beta) \mathbf{X}^c \{\mu_{\gamma, \vartheta}(\sigma, \tau(\xi)), \lambda\} \, d\sigma \quad (8.148)$$

2. Weighting of the derivative of the Radon transform

$$\mathcal{R}'_w(f(\tau, \alpha, \lambda)) = \frac{\partial p_{\gamma, \vartheta}(\xi, \lambda)}{\partial \xi} M(\lambda, \mathbf{n}_\xi) |\mathbf{S}'(\lambda)^\mathrm{T} \cdot \mathbf{n}_\xi| \cos(\kappa) \quad (8.143)$$

3. Differentiation in the direction of the radial detector variable τ

$$\mathcal{R}''_w(f(\tau, \alpha, \lambda)) = \frac{\partial}{\partial \tau} \mathcal{R}'_w(f(\tau, \alpha, \lambda)) \quad (8.144)$$

4. Two-dimensional filtered backprojection within the detector plane

$$h_\lambda(a, b) = \int_{\alpha=0}^{2\pi} \mathcal{R}''_w(f(\tau, \alpha, \lambda)) \, d\alpha \quad (8.145)$$

5. Three-dimensional weighted backprojection

$$f(\mathbf{r}) = -\frac{1}{8\pi^2} \int_\Lambda \frac{FCD^2}{U^2} h_\lambda(a, b) \, d\lambda \quad (8.146)$$

In a last step, the appropriate filter operator still has to be found. In this context, Fig. 8.40b illustrates that the projection of the circle located on the Radon sphere onto the detector yields an ellipse, the maximum diameter of which is $\|\mathbf{P} - \mathbf{O}\|$. An arbitrary point $\xi \mathbf{n}_\xi$ on the Radon sphere is projected onto the point τ located on the detector at the angle α described in the coordinate system of the detector. Therefore, in (8.141), the integration over the circle on the Radon sphere can be replaced with an integration over the corresponding ellipse on the detector.

Doing so, the expression

$$f(\mathbf{r}) = -\frac{1}{8\pi^2} \int_\Lambda \frac{FCD^2}{U^2} \left[\int_{\alpha=0}^{2\pi} \frac{\partial}{\partial \tau} \frac{\partial p_{\gamma, \vartheta}(\xi, \lambda)}{\partial \xi} M(\lambda, \mathbf{n}_\xi) \cdot \ldots \right.$$

$$\left. \ldots |\mathbf{S}'(\lambda)^\mathrm{T} \cdot \mathbf{n}_\xi| \cos(\kappa) \, d\alpha \right] d\lambda \quad (8.147)$$

can be found, where the first derivative of the Radon transform must be replaced with the substitution term

$$\frac{\partial}{\partial \xi} p_{\gamma,\vartheta}(\xi,\lambda) = \frac{1}{\cos^2(\kappa)} \frac{\partial}{\partial \tau} \int\limits_{-\infty}^{\infty} \cos(\beta) \mathbf{X}^c \left\{ \mu_{\gamma,\vartheta}(\sigma,\tau(\xi)),\lambda \right\} d\sigma \qquad (8.148)$$

found by *Grangeat*.

The inner integral in (8.147) represents a two-dimensional filtered backprojection in the detector plane. The outer integral is a three-dimensional backprojection of the values from the inner integral, which have been weighted with the squared distance between the X-ray source and the point to be reconstructed. Scheme 8.2 summarizes the exact backprojection algorithm proposed by Defrise and Clack (1994).

8.6
Approximate 3D Reconstructions in Cone-Beam Geometry

The exact methods described in the previous section are based on a complete set of Radon data. However, as already mentioned in ▸ Sect. 8.4.5, complete Radon transforms can only be acquired if certain requirements with regard to the X-ray source trajectory are met. In the frequently used circular X-ray source orbit, one major problem is related to the fact that there are so-called *shadow zones* in which no Radon data are available.

8.6.1
Missing Data in the 3D Radon Space

In principle, the object can be reconstructed in the three-dimensional space. For this purpose, the methods used for the two-dimensional reconstruction must just be extended properly. In ▸ Sect. 7.7 it has already been verified that an exact reconstruction can be achieved with fan-beam geometry, if the coordinates are transformed accordingly. In ▸ Sect. 8.5.2 the method proposed by *Grangeat* has been discussed, which contributes considerably to the solution of the inverse problem related to cone-beam geometry. However, in practice, a complete set of points in the Radon space is difficult to measure. Therefore, the corresponding problems of the frequently used circular acquisition protocol will be described in this section.

Figure 8.41 schematically illustrates the cone-beam geometry in an x–y section. For a circular trajectory of the X-ray source, located within this (x, y) plane, all object points of the plane can be exactly reconstructed in the spatial domain. If one considers a polar representation, the complete set of points in the Radon space belonging to a single point in the spatial domain is located on circles. The diameter of these circles connects the corresponding point in the spatial domain to the origin of

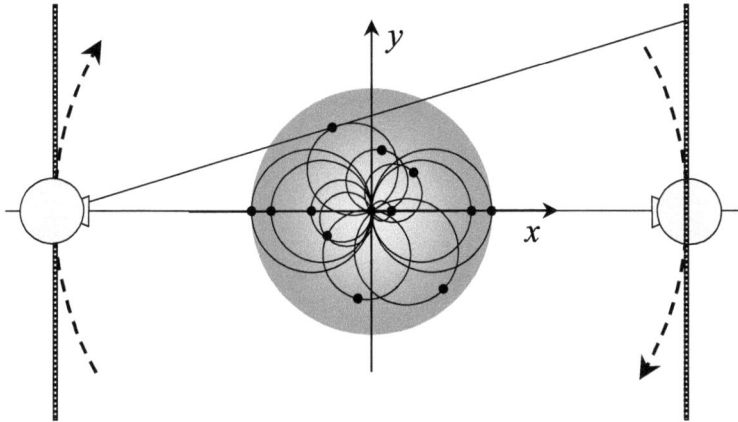

Fig. 8.41. The circular trajectory of the X-ray source (*dashed line*) is located in the (x, y) plane. Thus, the sections of the integration surfaces in space just yield the integration lines known from the two-dimensional reconstruction problem

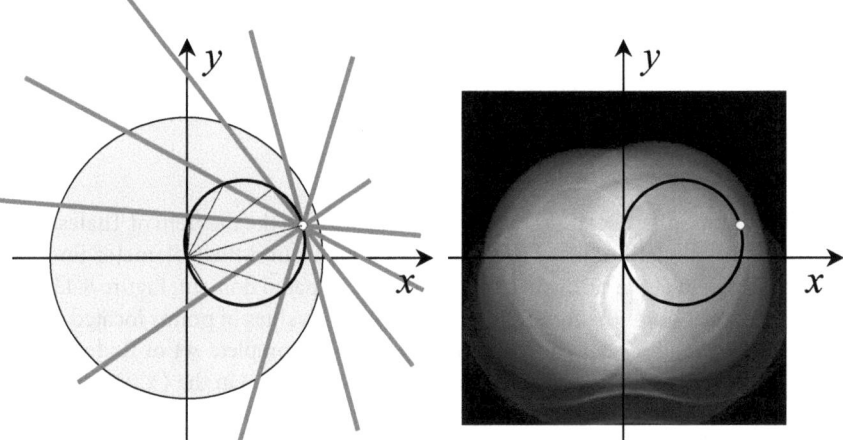

Fig. 8.42. *Left*: As any point must have been illuminated from all directions (from $0°$ up to $180°$) in the reconstruction plane, *circles* in the Radon space according to the theorem of Thales can be found. *Right*: The Radon space is shown with its polar coordinates for a tomogram of the abdomen

the (x, y) plane. For a compact object in space, the Radon space is thus completely filled with those circles.

Figure 8.42 (left) illustrates that the circles in the polar representation of the Radon space can be geometrically derived with the theorem of Thales. Figure 8.42 (right) exemplarily shows the polar Radon space for an abdomen tomogram slice frequently used in the previous chapters. The corresponding circles are clearly visible.

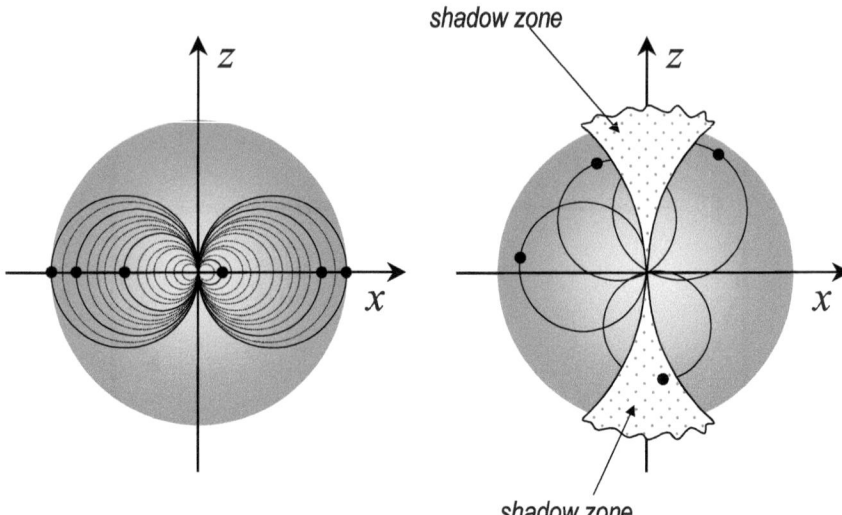

Fig. 8.43. Location of the practically available Radon values in the (x, z) plane. *Left*: As the set of Radon space values is complete for points in the (x, y) plane and the corresponding Radon values for each individual point are located on a spherical surface, the sections again result in *circles*. *Right*: Data are missing in the three-dimensional Radon space for points above and below the (x, y) plane. The zone in which data are missing is called the *shadow zone*

The Radon spheres can also be derived according to the theorem of Thales. They appear as rotational bodies of the circles rotating around their diameter lines between the origin at the reconstruction point in the spatial domain. Figure 8.43 (left) shows the location of the three-dimensional Radon values of points located on the x-axis that can be measured. The availability of the complete set of Radon values again refers to a single cyclic trajectory of the X-ray source in the (x, y) plane.

If there are points located outside the (x, y) plane, it will no longer be possible to measure all the points in the three-dimensional Radon space with this simple source trajectory. However, the following requirement must be met to exactly reconstruct the point. The Radon values of all points whose corresponding integration surfaces **A** intersect the object must be known for an exact reconstruction. According to the so-called *Tuy–Smith* sufficiency condition (Tuy 1983) an exact reconstruction is possible if all surfaces intersecting the object intersect the trajectory of the X-ray source at least once.

The circles, however, have already been the result of the two-dimensional Radon space (cf. ▸ Sect. 5.2). Here, the three-dimensional reconstruction problem shall be solved. If, therefore, integration surfaces instead of integration lines are considered, the circles in the (x, y) plane can be found as sections of corresponding spherical surfaces in the three-dimensional Radon space.

Since the X-ray source must be located in the surface **A** in order to measure its integral, this also makes sense from an intuitive point of view. Points in the Radon space, which cannot be measured with the planar cyclic source trajectory, are located in the so-called shadow zone illustrated in Fig. 8.43 (right).

Figure 8.44 once again shows points that comply with the *Tuy–Smith* condition. Such points are of course located in the (x, y) plane. The conical X-ray beam illuminates a spherical object. If only those points located within the (x, y) plane have a complete set of Radon data, then the sphere cannot be exactly reconstructed with this measurement. Figure 8.44 shows some outermost points with a complete set of Radon values on the front hemisphere of the spherical object.

Figure 8.45 schematically shows the formation of a shadow zone. If points $\mathbf{r} = (x, y, z)^{\mathrm{T}}$ are located outside the (x, y) plane, it will no longer be possible to measure the projection $p_{\gamma, \vartheta}(\xi)$ for all projection angles γ, ϑ and distances from the origin ξ due to obvious geometrical restrictions.

Within the shadow zone the X-ray source is in fact located underneath the horizon of the integration surface **A**. This is exactly the *Tuy–Smith* sufficiency condition. Those points that can, in principle, be measured with geometrical parameters result within a torus. Figure 8.46 shows the filling of the Radon space for the cyclic source orbit in the plane. The resulting torus includes all points that can theoretically be measured in the Radon space. But, in practice, it is not possible to measure all the points of the Radon space inside this torus as a detector has a finite size.

Source orbits meeting the requirements stipulated by *Tuy–Smith* are shown in Fig. 8.47. Figure 8.47a provides a comparison with the cyclic, planar trajectory of the X-ray source, which is responsible for the above-mentioned shadow zone in the Radon space. Figure 8.47b shows one of the possibilities of filling up the shadow zone with Radon data. This is possible by using two circular X-ray source trajec-

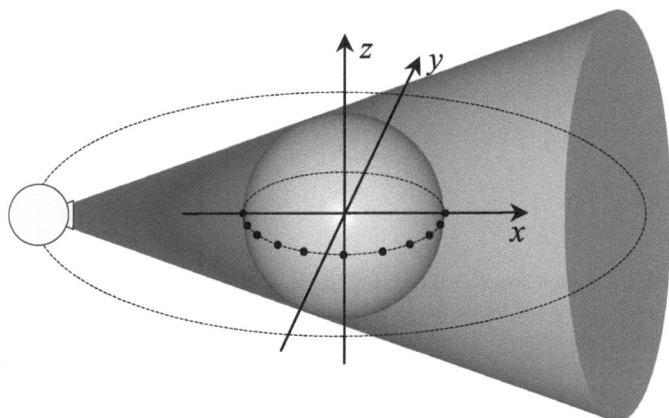

Fig. 8.44. In cone-beam geometry those object points for which the Radon space is complete are located within the plane defined by the X-ray source trajectory. For a spherical object this applies to all those points within the circle that results from a section of the trajectory plane through the sphere. Some outermost points with a complete set of Radon values are exemplarily shown on the front hemisphere of the spherical object

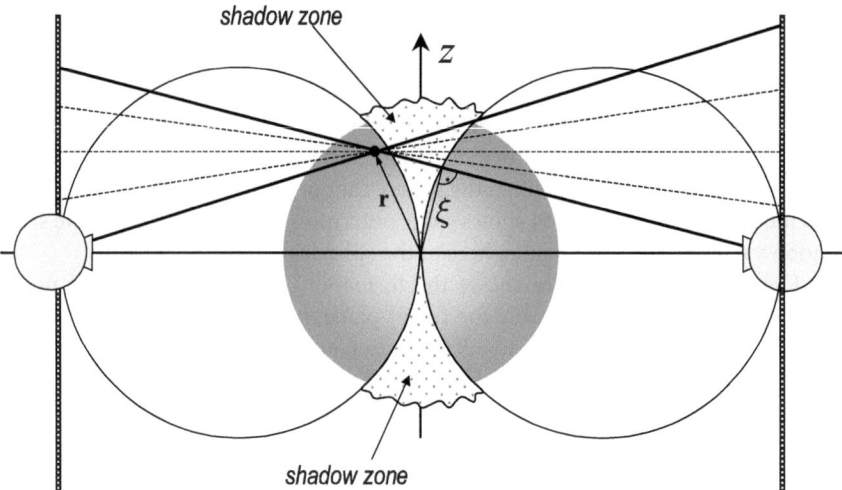

Fig. 8.45. There are points in the Radon space, of the cyclic source orbit in a plane, for which the necessary integration over the surface cannot be carried out. The points assigned in the spatial domain may then not be correctly reconstructed. The Radon values for which the integration surface does not intersect the X-ray source orbit are missing. The missing zone in the Radon space is again the *shadow zone*

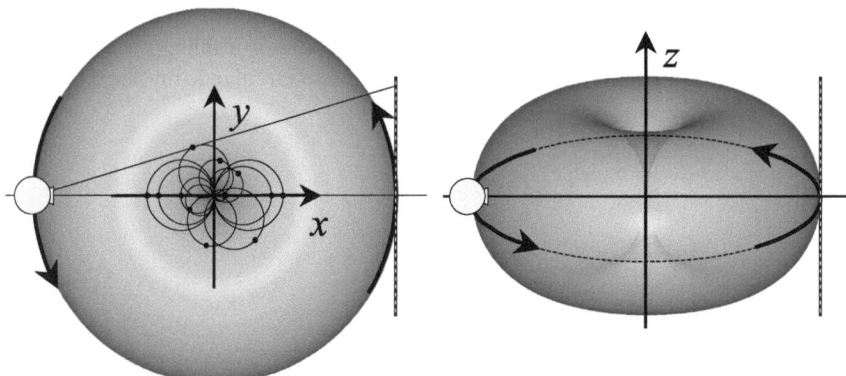

Fig. 8.46. In the Radon space of the cyclic source orbit those points that can theoretically be measured are located within a torus. Nevertheless, it is in fact not possible to measure all theoretically available points of the Radon space with a detector of finite size

tories located in planes that are parallel to each other and connected with a linear trajectory[13]. Figure 8.47c shows two source trajectories that are perpendicular to each other and also circular. The requirements stipulated by *Tuy–Smith* are also met here. However, it can readily be understood that this option is excluded in medical

---

[13] The linear movement is performed anyway, because a planning overview is always measured prior to the tomographic data acquisition (cf. ▶ Sect. 10.2).

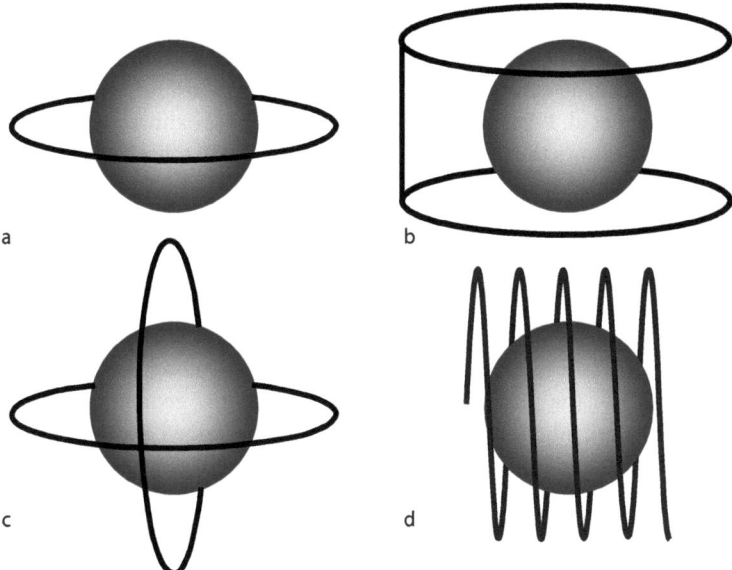

Fig. 8.47a–d. Orbits of the cone-beam X-ray source sampling the Radon space. Apart from the simple cyclic path shown in **a**, all other trajectories meet the requirements stipulated in the sufficiency condition of *Tuy–Smith*. The cone-beam helical path is used in modern medical CT scanners today

applications, because the patient is located in the source path. Practical applications typically use cone-beam helical paths scanning the Radon space with a continuous movement. This path is outlined in Fig. 8.47d. The trajectories shown in this figure are not the only ones being tested at present. Another approach with a central circular orbit and a single supplementary line is proposed by Lin (1999). All these approaches are attempts to fill up the shadow zone of the Radon space as effectively as possible.

8.6.2
FDK Cone-Beam Reconstruction for Planar Detectors

The methods of three-dimensional reconstruction described in ▸ Sect. 8.5 are based on the assumption that a complete set of Radon data is available. However, as discussed in the previous section, the requirements stipulated in the *Tuy–Smith* sufficiency condition are not met for any source trajectory. This is particularly the case for the popular circular X-ray trajectory in cone-beam geometry used in many technical applications, where a complete set of Radon data is therefore not available. However, approximation methods that can also deal with incomplete Radon data of this type are available.

The most frequently used method is the reconstruction approach proposed by Feldkamp, Davis, and Kress (FDK) in 1984 (Feldkamp et al. 1984), which is a derivative-free method. The so-called *FDK cone-beam reconstruction* is an approximation of the exact reconstruction problem developed for technical investigations[14]. As planar detectors are used for these applications, Feldkamp, Davis, and Kress (Feldkamp et al. 1984) initially presented their approximation for this geometry. Figure 8.48 shows the geometry of the sampling unit schematically.

The cone-beam geometry with the planar detector arrangement has a cone angle κ_a at the fan angle, ψ, which is defined by the expression

$$\kappa_a = \arctan\left(\frac{b}{\sqrt{a^2 + FCD^2}}\right). \tag{8.149}$$

With this definition, it is clear that the cone angle varies along the individual rows (i.e. b = const.) of the planar detector. Despite this, the geometry of the planar detector field is well adapted to the backprojection described above. This is due to the fact that the integration over the surfaces can be realized by integrations along straight lines on the detector.

Figure 8.49 shows a tilted fan beam of the X-ray cone beam, which intersects the detector on a line with constant b. This geometry will have to apply to all fan beams inside the cone beam. The contribution of the point (a, b) to the backprojection has to be computed.

The FDK reconstruction is based on the assumption that every fan-shaped surface in the cone beam defined by a detector line b and the X-ray source location must be treated independently, as if dealing with a two-dimensional fan projection as described in ▸ Sect. 7.7.4. Figure 8.49 shows such a surface. Just because a straight line on the detector is assigned to an integration surface (red-shaded fan), with planar detector geometry, one may access the detector independently line by line for the FDK reconstruction.

If the cone-beam projection system is considered under a fixed projection angle θ, then the mathematical treatment of the individual projection fans within the X-ray cone does not differ from the treatment of the fans in ▸ Sect. 7.7.4. Due to the angulation κ_a of the fan off the (x, y) plane, only distances must be adapted appropriately. It is therefore obvious that (7.151) has to be slightly extended to take into account the new distances within the tilted fan. The filtered backprojection may then be carried out exactly with reference to a single, fixed angle under which the projection is considered. Similar to ▸ Sect. 7.7.4 a virtual detector is placed in the iso-center of the system. The geometrical projection situation can then be more easily formulated mathematically.

The results can be applied to the real detector surface by means of the intercept theorem. Figure 8.50 illustrates that the projection of a point $\mathbf{r} = (x, y, z)^T$ in the X-ray cone onto the detector position (a, b) depends on the fan angle ψ via

$$a = FCD \tan(\psi) \tag{8.150}$$

---

[14] *Feldkamp, Davis,* and *Kress* developed this method in the research laboratories of the Ford Motors for material testing.

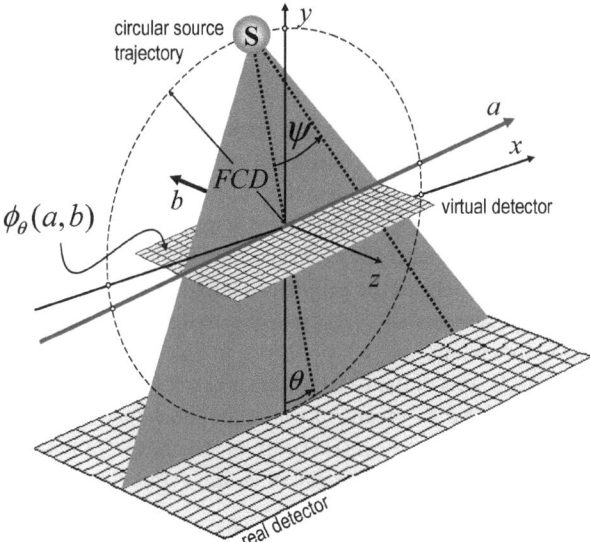

Fig. 8.48. Cone-beam illumination of a flat-panel detector. The X-ray source rotates on a circle with radius FCD (focus center distance). A virtual detector is positioned in the iso-center of the rotation of the sampling unit. $\phi_\theta(a, b)$ are the measured (virtual) projection values. The angle θ between the y-axis and the central beam of the cone – or the x-axis and the a-axis – defines the projection angle of the system

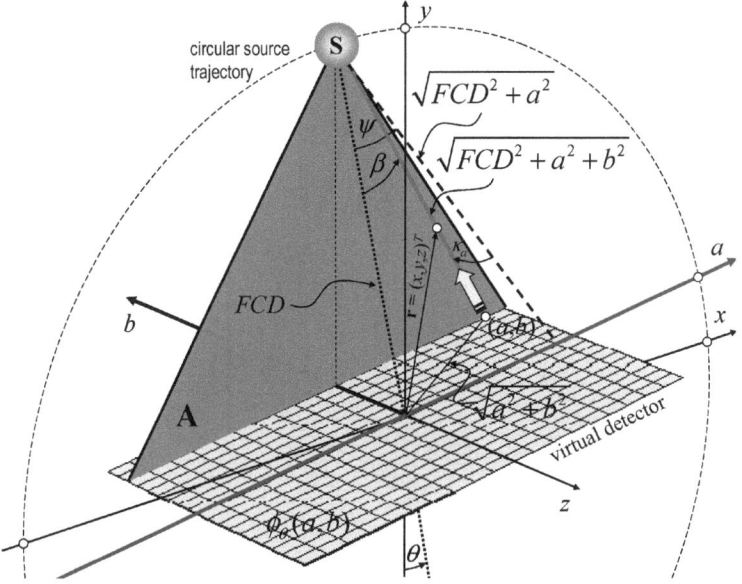

Fig. 8.49. Cone-beam illumination of a flat-panel detector. Only fans are considered within the cone that intersect the detector on a straight line with constant b. The contribution of the detector value at the detector point (a, b) to the backprojection will have to be determined

and the cone angle κ_a via

$$b = \sqrt{FCD^2 + a^2}\,\tan(\kappa_a)\,. \tag{8.151}$$

In this system, the (a, b) detector surface rotates about the z-axis describing the axial direction of the fixed patient coordinate system.

The small principal axis of the detector, b, has been assigned to the negative z direction to ensure that the detector system remains a right-handed system. The angle θ between the y-axis and the central beam of the cone – or the x-axis and the a-axis respectively – defines the projection angle of the system. FCD once again stands for the focus center distance. It is possible to define a new coordinate system, where the unit vectors $\mathbf{n}_\sigma$, $\mathbf{n}_\eta$, and $\mathbf{n}_\xi$ are shown in Fig. 8.50. Here, $\mathbf{n}_\sigma$ points into the direction of the long a-axis of the detector. $\mathbf{n}_\sigma$ is rotated by the angle θ with respect to the x-axis. This is done by means of the orthogonal matrix $\boldsymbol{\Sigma}$. $\mathbf{n}_\eta$ points in the direction of the source along the central beam of the angulated fan and $\mathbf{n}_\xi = \mathbf{n}_\sigma \times \mathbf{n}_\eta$ in the direction of the fan surface normal of $\mathbf{A}$. $\mathbf{n}_\xi$ is thus rotated about the angle

$$\kappa_0 = \arctan\left(\frac{b}{FCD}\right) \tag{8.152}$$

with respect to the z-axis, which means about the $\mathbf{n}_\sigma$-axis. This corresponds to a multiplication with the orthogonal matrix $\boldsymbol{\Xi}$.

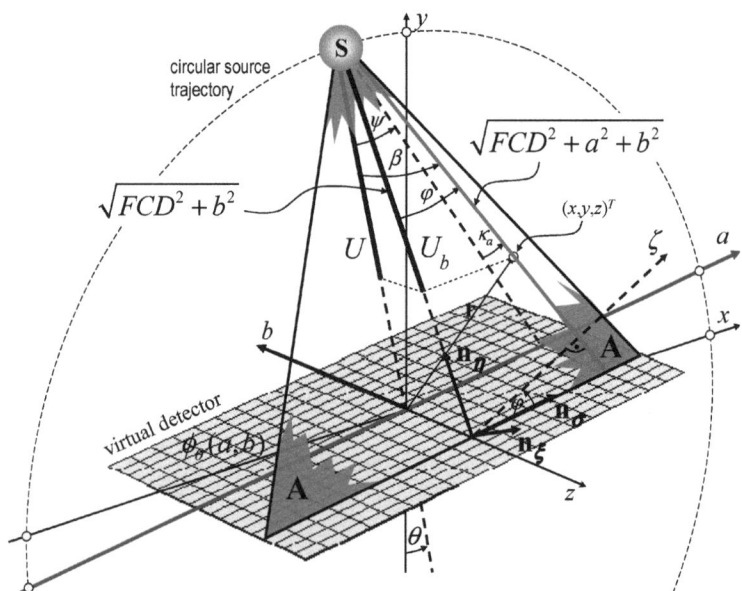

Fig. 8.50. Geometrical situation using a virtual detector in the iso-center of the sampling unit. As a straight line on the detector is assigned to an integration surface (transparent plane) for flat-panel detector geometry, the detector values can be accessed line by line in the case of the Feldkamp, Davis, and Kress (FDK) reconstruction. The unit vectors $\mathbf{n}_\sigma$, $\mathbf{n}_\eta$, and $\mathbf{n}_\xi$ define the fan surface considered, where $\mathbf{n}_\xi = \mathbf{n}_\sigma \times \mathbf{n}_\eta$ is the surface normal of $\mathbf{A}$

Finally, the transformation

$$\begin{pmatrix} r \\ s \\ t \end{pmatrix} = \boldsymbol{\rho} = \Xi\,(\Sigma\mathbf{r} - \mathbf{b}) = \Xi\left(\Sigma\begin{pmatrix} x \\ y \\ z \end{pmatrix} + \begin{pmatrix} 0 \\ 0 \\ b \end{pmatrix}\right) \tag{8.153}$$

is obtained. The point to be reconstructed may thus be expressed with the new co-ordinates as

$$\begin{pmatrix} r \\ s \\ t \end{pmatrix} = \begin{pmatrix} 1 & 0 & 0 \\ 0 & \cos(\kappa_0) & \sin(\kappa_0) \\ 0 & -\sin(\kappa_0) & \cos(\kappa_0) \end{pmatrix}\left(\begin{pmatrix} \cos(\theta) & \sin(\theta) & 0 \\ -\sin(\theta) & \cos(\theta) & 0 \\ 0 & 0 & 1 \end{pmatrix}\begin{pmatrix} x \\ y \\ z \end{pmatrix} + \begin{pmatrix} 0 \\ 0 \\ b \end{pmatrix}\right)$$

$$= \begin{pmatrix} 1 & 0 & 0 \\ 0 & \cos(\kappa_0) & \sin(\kappa_0) \\ 0 & -\sin(\kappa_0) & \cos(\kappa_0) \end{pmatrix}\begin{pmatrix} x\cos(\theta) + y\sin(\theta) \\ -x\sin(\theta) + y\cos(\theta) \\ z + b \end{pmatrix} \tag{8.154}$$

$$= \begin{pmatrix} x\cos(\theta) + y\sin(\theta) \\ -x\sin(\theta)\cos(\kappa_0) + y\cos(\theta)\cos(\kappa_0) + (z+b)\sin(\kappa_0) \\ x\sin(\theta)\sin(\kappa_0) - y\cos(\theta)\sin(\kappa_0) + (z+b)\cos(\kappa_0) \end{pmatrix}.$$

On the other hand, one finds points $\boldsymbol{\rho}$ in the original coordinate system by

$$\mathbf{r} = \Sigma^{\mathrm{T}}\left(\Xi^{\mathrm{T}}\boldsymbol{\rho} - \mathbf{b}\right), \tag{8.155}$$

which means

$$\begin{pmatrix} x \\ y \\ z \end{pmatrix} = \begin{pmatrix} \cos(\theta) & -\sin(\theta) & 0 \\ \sin(\theta) & \cos(\theta) & 0 \\ 0 & 0 & 1 \end{pmatrix}\left(\begin{pmatrix} 1 & 0 & 0 \\ 0 & \cos(\kappa_0) & -\sin(\kappa_0) \\ 0 & \sin(\kappa_0) & \cos(\kappa_0) \end{pmatrix}\begin{pmatrix} r \\ s \\ t \end{pmatrix} - \begin{pmatrix} 0 \\ 0 \\ b \end{pmatrix}\right) \tag{8.156a}$$

$$= \begin{pmatrix} \cos(\theta) & -\sin(\theta) & 0 \\ \sin(\theta) & \cos(\theta) & 0 \\ 0 & 0 & 1 \end{pmatrix}\begin{pmatrix} r \\ s\cos(\kappa_0) - t\sin(\kappa_0) \\ s\sin(\kappa_0) + t\cos(\kappa_0) - b \end{pmatrix} \tag{8.156b}$$

$$= \begin{pmatrix} r\cos(\theta) - \sin(\theta)\,(s\cos(\kappa_0) - t\sin(\kappa_0)) \\ r\sin(\theta) + \cos(\theta)\,(s\cos(\kappa_0) - t\sin(\kappa_0)) \\ s\sin(\kappa_0) + t\cos(\kappa_0) - b \end{pmatrix}. \tag{8.156c}$$

As only those points inside the integration surface **A** are of interest, which are spanned by $\mathbf{n}_\sigma$ and $\mathbf{n}_\eta$, $t = 0$ might be assumed here. Thus, only the fan beam plane has to be dealt with in which

$$
\begin{pmatrix} x \\ y \\ z \end{pmatrix} = \begin{pmatrix} r\cos(\theta) - s\sin(\theta)\cos(\kappa_0) \\ r\sin(\theta) + s\cos(\theta)\cos(\kappa_0) \\ s\sin(\kappa_0) - b \end{pmatrix}
$$

$$
= \begin{pmatrix} r\cos(\theta) - s\sin(\theta)\dfrac{FCD}{\sqrt{FCD^2+b^2}} \\ r\sin(\theta) + s\cos(\theta)\dfrac{FCD}{\sqrt{FCD^2+b^2}} \\ s\dfrac{b}{\sqrt{FCD^2+b^2}} - b \end{pmatrix} \tag{8.157}
$$

assigns to each point $(r,s)^{\mathrm T}$ the corresponding point $(x,y,z)^{\mathrm T}$ patient coordinate system.

In full analogy with the approach described in ▸ Sect. 7.7.4, an appropriate transformation to parallel-beam geometry has to be found to reconstruct the point ρ lying in the fan angulated about the angle κ_0. In this way, the virtual parallel projection $p_\gamma(\zeta)$ under the projection angle γ at the detector position ζ can be calculated within this fan of the X-ray cone beam. The virtual ζ coordinate is shown in Figs. 8.50 and 8.51 by a dashed line. It runs through the origin, which has been shifted by b with respect to the detector coordinate system within the fan, and forms a right angle with the X-ray beam through point ρ.

The relation with the fan-beam coordinates can be – in analogy to (7.69) and (7.71) – expressed by

$$
\zeta = a\cos(\varphi) = a\frac{\sqrt{FCD^2+b^2}}{\sqrt{FCD^2+a^2+b^2}} \tag{8.158}
$$

and

$$
\gamma = \theta_b + \varphi = \theta_b + \arctan\left(\frac{a}{\sqrt{FCD^2+b^2}}\right). \tag{8.159}
$$

The new coordinates are illustrated in Fig. 8.51. Thus, the three-dimensional cone-beam reconstruction problem is reduced to a problem of two-dimensional fan-beam geometry (cf. Fig. 7.23). One may therefore use the equations of the filtered backprojection in the fan-beam plane that have already been introduced in ▸ Sect. 7.7.4. With the new coordinate notations (7.90) now reads

$$
f(r,s) = \frac{1}{2}\int_0^{2\pi}\left(\int_{-\infty}^{+\infty} p_\gamma(\zeta)g(\rho^{\mathrm T}\cdot\mathbf{n}_\zeta - \zeta)\,d\zeta\right)d\gamma, \tag{8.160}
$$

whereby the relation with the x, y, and z components of $\mathbf{r}$ must be realized later by (8.157). The same strategy that has been followed in the previous chapter applies here. The fan-beam coordinates have to be substituted by parallel-beam coordin-

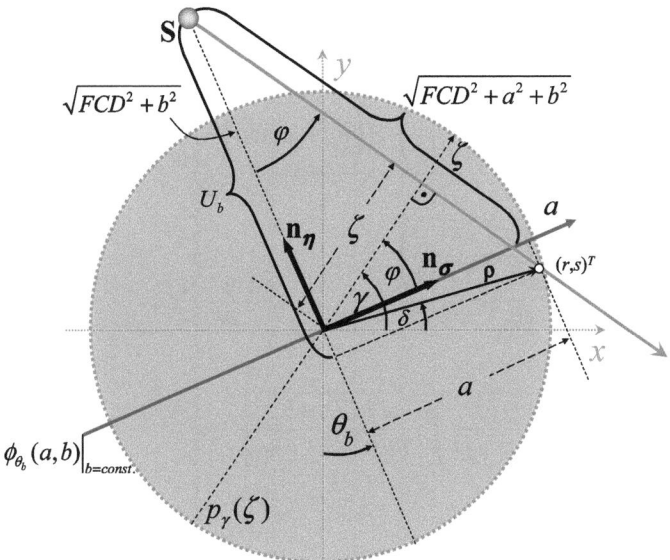

Fig. 8.51. Geometry within the angulated fan-beam plane illustrated in Fig. 8.50. The projection of the fixed patient coordinate system is used as a reference system. This is an approximation only, due to the projective geometry. Nevertheless, this system can be used, because just a fixed reference system is needed in which the sampling unit rotates about the $\mathbf{n}_\xi$ unit vector. As the backprojection is actually carried out over the full angle interval $\theta_b = [0, 2\pi]$ the result does not depend on the precise starting angle. The angles φ, δ, and γ are only required as relative angles with respect to the angle θ_b and do not refer to the actual (x, y, z) coordinate system. The relation between θ_b and θ will be explained later

ates. To do so, the coordinate transformation

$$(\zeta, \gamma) \rightarrow (a, \theta_b) \tag{8.161}$$

has to be carried out. As already shown in several other cases, the new infinitesimal integration element must be scaled by the *Jacobian*. This means that the area element $d\zeta \, d\gamma$ is determined in the new coordinates by $J \, da \, d\theta_b$, where J is given by

$$J \equiv \det \left(\frac{\partial(\zeta, \gamma)}{\partial(a, \theta_b)} \right) = \begin{vmatrix} \dfrac{\partial \zeta}{\partial a} & \dfrac{\partial \gamma}{\partial a} \\[2mm] \dfrac{\partial \zeta}{\partial \theta_b} & \dfrac{\partial \gamma}{\partial \theta_b} \end{vmatrix} . \tag{8.162}$$

If the transformation equations (8.158) and (8.159) are directly substituted in (8.162) one finds

$$
J = \begin{vmatrix}
\dfrac{\partial\left(a\dfrac{\sqrt{FCD^2+b^2}}{\sqrt{FCD^2+a^2+b^2}}\right)}{\partial a} & \dfrac{\partial\left(\theta_b+\arctan\left(\dfrac{a}{\sqrt{FCD^2+b^2}}\right)\right)}{\partial a} \\[4ex]
\dfrac{\partial\left(a\dfrac{\sqrt{FCD^2+b^2}}{\sqrt{FCD^2+a^2+b^2}}\right)}{\partial\theta_b} & \dfrac{\partial\left(\theta_b+\varphi\right)}{\partial\theta_b}
\end{vmatrix}
$$

(8.163)

$$
= \begin{vmatrix}
\left(\dfrac{\sqrt{FCD^2+b^2}}{\sqrt{FCD^2+a^2+b^2}}\right)^3 & \dfrac{\sqrt{FCD^2+b^2}}{FCD^2+a^2+b^2} \\[3ex]
0 & 1
\end{vmatrix}
$$

$$
= \left(\dfrac{\sqrt{FCD^2+b^2}}{\sqrt{FCD^2+a^2+b^2}}\right)^3 = \cos(\varphi)^3 \, .
$$

This means that the infinitesimal area element must be replaced by

$$
d\zeta\,d\gamma \to \left(\dfrac{\sqrt{FCD^2+b^2}}{\sqrt{FCD^2+a^2+b^2}}\right)^3 da\,d\theta_b \, .
$$

(8.164)

If – in analogy to (7.99) – the expression

$$
\boldsymbol{\rho}^{\mathrm{T}}\cdot\mathbf{n}_\zeta = \rho\cos(\gamma-\delta)
$$

(8.165)

is used for the inner product between the point to be reconstructed in the X-ray fan surface and the unit vector in the direction of the virtual axis of an imaginary parallel-beam detector, then the expression

$$
f(r,s) = \frac{1}{2}\int\limits_{-\arctan\left(\frac{a}{\sqrt{FCD^2+b^2}}\right)}^{2\pi-\arctan\left(\frac{a}{\sqrt{FCD^2+b^2}}\right)}\left\{\int\limits_{-a_{min}}^{+a_{max}} p_{\theta_b+\psi}\left(a\dfrac{\sqrt{FCD^2+b^2}}{\sqrt{FCD^2+a^2+b^2}}\right)\cdot\ldots\right.
$$

$$
\ldots\cdot g\left[\rho\cos\left(\theta_b+\arctan\left(\dfrac{a}{\sqrt{FCD^2+b^2}}\right)-\delta\right)-\dfrac{a\sqrt{FCD^2+b^2}}{\sqrt{FCD^2+a^2+b^2}}\right]\ldots
$$

$$
\ldots\cdot\left.\left(\dfrac{\sqrt{FCD^2+b^2}}{\sqrt{FCD^2+a^2+b^2}}\right)^3 da\right\}d\theta_b
$$

(8.166)

can be found for the filtered backprojection with the new coordinates.

In (8.166), the following simplifications can be implemented immediately. The integration limits of the new angle coordinate θ_b have to be changed accordingly as a direct consequence of the substitution. However, since the integration over the full angle 2π is invariant to a constant phase shift, it is also possible to integrate in the new variable from 0 up to 2π. Furthermore, the projection value $p_y(\zeta)$ can be much more easily expressed within the new detector variables, since the coordinate transformation yields

$$
p_y(\zeta) \rightarrow p_{\theta_b+\varphi}\left(a\frac{\sqrt{FCD^2+b^2}}{\sqrt{FCD^2+a^2+b^2}}\right) = \phi_{\theta_b}(a,b)\big|_{b=\text{const.}} . \tag{8.167}
$$

(8.166) is thus simplified to read

$$
f(r,s) = \frac{1}{2}\int_0^{2\pi}\left\{\int_{-a_{\min}}^{+a_{\max}}\phi_{\theta_b}(a,b)g\left[\rho\cos\left(\theta_b+\arctan\left(\frac{a}{\sqrt{FCD^2+b^2}}\right)-\delta\right)\ldots\right.\right.
$$
$$
\left.\left.\ldots-\frac{a\sqrt{FCD^2+b^2}}{\sqrt{FCD^2+a^2+b^2}}\right]\cdot\left(\frac{\sqrt{FCD^2+b^2}}{\sqrt{FCD^2+a^2+b^2}}\right)^3 da\right\}d\theta_b . \tag{8.168}
$$

Using the easily perceivable fact from Fig. 8.51 that

$$
\varphi = \arctan\left(\frac{a}{\sqrt{FCD^2+b^2}}\right) \tag{8.169}
$$

for the subsequent transformation of the argument of the function g[.] in (8.168), one may apply the addition law (7.98) to calculate the argument of $g[.]$ as

$$
\rho\cos(\theta_b+\varphi-\delta)-\frac{a\sqrt{FCD^2+b^2}}{\sqrt{FCD^2+a^2+b^2}} =
$$
$$
\rho\cos(\theta_b-\delta)\cos(\varphi)-\rho\sin(\theta_b-\delta)\sin(\varphi)-\frac{a\sqrt{FCD^2+b^2}}{\sqrt{FCD^2+a^2+b^2}} . \tag{8.170}
$$

With

$$
\cos(\varphi) = \frac{\sqrt{FCD^2+b^2}}{\sqrt{FCD^2+a^2+b^2}} \tag{8.171}
$$

and

$$
\sin(\varphi) = \frac{a}{\sqrt{FCD^2+a^2+b^2}} \tag{8.172}
$$

(8.170) reads

$$
\rho\cos(\theta_b + \varphi - \delta) - \frac{a\sqrt{FCD^2 + b^2}}{\sqrt{FCD^2 + a^2 + b^2}} = \rho\cos(\theta_b - \delta)\frac{\sqrt{FCD^2 + b^2}}{\sqrt{FCD^2 + a^2 + b^2}} + \dots
$$
$$
\dots - \rho\sin(\theta_b - \delta)\frac{a}{\sqrt{FCD^2 + a^2 + b^2}} - \frac{a\sqrt{FCD^2 + b^2}}{\sqrt{FCD^2 + a^2 + b^2}}.
$$
(8.173)

The particular position determined by the point (ρ, δ) and the projection angle θ_b shall now be considered in a similar manner to (7.108) on the basis of the detector variable a. This detector position is denoted a', so that

$$
\tan(\varphi') = \left(\frac{a'}{\sqrt{FCD^2 + b^2}}\right) = \frac{\rho\cos(\theta_b - \delta)}{\left(\sqrt{FCD^2 + b^2} + \rho\sin(\theta_b - \delta)\right)}
$$
(8.174)

and furthermore

$$
a' = \rho\cos(\theta_b - \delta)\frac{\sqrt{FCD^2 + b^2}}{\left(\sqrt{FCD^2 + b^2} + \rho\sin(\theta_b - \delta)\right)}.
$$
(8.175)

Substituting (8.175) into the argument of $g[.]$ in (8.173) yields

$$
\rho\cos(\theta_b + \varphi - \delta) - \frac{a\sqrt{FCD^2 + b^2}}{\sqrt{FCD^2 + a^2 + b^2}} =
$$
$$
= \left(\sqrt{FCD^2 + b^2} + \rho\sin(\theta_b - \delta)\right)\frac{a'}{\sqrt{FCD^2 + a^2 + b^2}} + \dots
$$
$$
\dots - \rho\sin(\theta_b - \delta)\frac{a}{\sqrt{FCD^2 + a^2 + b^2}} - \frac{a\sqrt{FCD^2 + b^2}}{\sqrt{FCD^2 + a^2 + b^2}}
$$
$$
= \left(\sqrt{FCD^2 + b^2} + \rho\sin(\theta_b - \delta)\right)\frac{a'}{\sqrt{FCD^2 + a^2 + b^2}} + \dots
$$
$$
\dots - \left(\sqrt{FCD^2 + b^2} + \rho\sin(\theta_b - \delta)\right)\frac{a}{\sqrt{FCD^2 + a^2 + b^2}}.
$$
(8.176)

For convenience, one may define

$$
U_b = \sqrt{FCD^2 + b^2} + \rho\sin(\theta_b - \delta),
$$
(8.177)

where U_b is the projection of ρ onto the central beam in the angulated $\mathbf{n}_\sigma \times \mathbf{n}_\eta$-plane. Hence, one finds

$$\rho \cos(\theta_b + \varphi - \delta) - \frac{a\sqrt{FCD^2 + b^2}}{\sqrt{FCD^2 + a^2 + b^2}}$$

$$= U_b \frac{a'}{\sqrt{FCD^2 + a^2 + b^2}} - U_b \frac{a}{\sqrt{FCD^2 + a^2 + b^2}} \qquad (8.178)$$

$$= (a' - a) \frac{U_b}{\sqrt{FCD^2 + a^2 + b^2}} ,$$

so that (8.168) may be arranged more concisely to read

$$f(r,s) = \frac{1}{2} \int_0^{2\pi} \left\{ \int_{-a_{min}}^{+a_{max}} \phi_{\theta_b}(a,b) g\left[(a' - a) \frac{U_b}{\sqrt{FCD^2 + a^2 + b^2}} \right] \cdots \right.$$

$$\left. \cdots \left(\frac{\sqrt{FCD^2 + b^2}}{\sqrt{FCD^2 + a^2 + b^2}} \right)^3 da \right\} d\theta_b . \qquad (8.179)$$

The inner integral again represents a convolution. It should be recalled that the function g is still the spatial convolution kernel of the linear frequency ramp, i.e.,

$$g(\xi) = \int_{-\infty}^{+\infty} |q| \, e^{2\pi i q \xi} \, dq . \qquad (8.180)$$

Therefore, if one substitutes here the arguments found for fan-beam geometry with a linear detector (cf. ▸ Sect. 7.7.4), i.e.,

$$g\left((a' - a) \frac{U_b}{\sqrt{FCD^2 + a^2 + b^2}} \right) = \int_{-\infty}^{+\infty} |q| \, e^{2\pi i q \left((a'-a) \frac{U_b}{\sqrt{FCD^2+a^2+b^2}} \right)} \, dq \qquad (8.181)$$

and further uses the substitution

$$q' = q \frac{U_b}{\sqrt{FCD^2 + a^2 + b^2}} , \qquad (8.182)$$

one obtains the scaling behavior of the convolution kernel

$$g\left((a' - a) \frac{U_b}{\sqrt{FCD^2 + a^2 + b^2}} \right) = \frac{FCD^2 + a^2 + b^2}{U_b^2} \int_{-\infty}^{+\infty} |q'| \, e^{2\pi i q'(a'-a)} \, dq'$$

$$= \frac{FCD^2 + a^2 + b^2}{U_b^2} g(a' - a) . \qquad (8.183)$$

Correspondingly, (8.168) finally reads

$$
f(r,s) = \frac{1}{2} \int\limits_0^{2\pi} \left\{ \int\limits_{-a_{\min}}^{+a_{\max}} \phi_{\theta_b}(a,b) \frac{FCD^2 + a^2 + b^2}{U_b^2} g(a' - a) \cdot \ldots \right.
$$

$$
\left. \ldots \left(\frac{\sqrt{FCD^2 + b^2}}{\sqrt{FCD^2 + a^2 + b^2}} \right)^3 da \right\} d\theta_b
$$

$$
= \frac{1}{2} \int\limits_0^{2\pi} \frac{1}{U_b^2} \left\{ \int\limits_{-a_{\min}}^{+a_{\max}} \phi_{\theta_b}(a,b) g(a' - a) \left(\frac{\sqrt{FCD^2 + b^2}}{\sqrt{FCD^2 + a^2 + b^2}} \right)^{-3} da \right\} d\theta_b
$$

$$
= \frac{1}{2} \int\limits_0^{2\pi} \frac{FCD^2 + b^2}{U_b^2} \left\{ \int\limits_{-a_{\min}}^{+a_{\max}} \phi_{\theta_b}(a,b) \left(\frac{\sqrt{FCD^2 + b^2}}{\sqrt{FCD^2 + a^2 + b^2}} \right) g(a' - a) da \right\} d\theta_b
$$

$$
= \frac{1}{2} \int\limits_0^{2\pi} \frac{FCD^2 + b^2}{U_b^2} \left\{ \left(\phi_{\theta_b}(a,b) \frac{\sqrt{FCD^2 + b^2}}{\sqrt{FCD^2 + a^2 + b^2}} \right) * g(a) \right\} d\theta_b . \quad (8.184)
$$

As a last step, the relationship between the rotation angle θ_b around the unit vector $\mathbf{n}_\xi$ in the direction of the normal of the fan surface vector and the actual rotation angle θ of the sampling unit in the (x, y) plane must be determined. Figure 8.52 may be helpful to illustrate what happens to an angular change $\Delta\theta$ if the fan is angulated off the (x, y) plane.

In the central fan beam plane of the X-ray cone beam one finds the lengths

$$
\overline{SO} = \overline{YO} = FCD \tag{8.185}
$$

and in the angulated plane the lengths

$$
\overline{SB} = \overline{YB} = \sqrt{FCD^2 + b^2} . \tag{8.186}
$$

Obviously, the distance on the flat-panel detector is

$$
\overline{OB} = b . \tag{8.187}
$$

For small angular changes, it holds sufficiently true that

$$
\overline{SY} = \Delta\theta \, FCD \approx \Delta\theta_b \sqrt{FCD^2 + b^2} \tag{8.188}
$$

is the arc length of the X-ray source orbit. In this way, the desired variation of the angular change $\Delta\theta_b$ due to an angulation $\Delta\theta$ of the fan can be estimated by

$$
\Delta\theta_b \approx \frac{FCD}{\sqrt{FCD^2 + b^2}} \Delta\theta . \tag{8.189}
$$

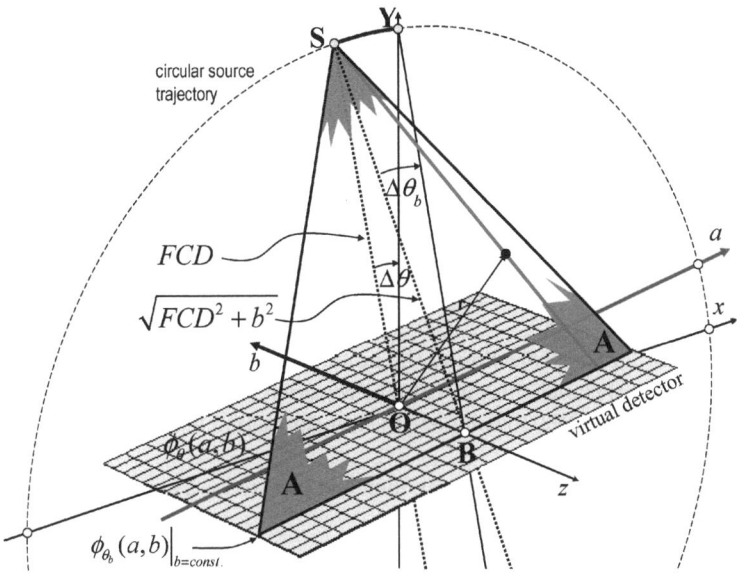

Fig. 8.52. Variation of the rotation angle $\Delta\theta_b$ if the X-ray fan is angulated off the (x, y) plane

However, as mentioned above, (8.188) and (8.189) are only approximations. Just for small angular changes, $\Delta\theta$, $\Delta\theta_b$ is located sufficiently within the angulated fan-beam plane. Figure 8.52 illustrates that the distance

$$\overline{YB} = \sqrt{FCD^2 + b^2} \tag{8.190}$$

is obviously not located within the angulated X-ray fan-beam plane for a large $\Delta\theta$.

Fortunately, one is only interested in infinitesimal angular changes $d\theta_b$ of the sampling unit in (8.184), so that one finds the exact relation

$$d\theta_b = \frac{FCD}{\sqrt{FCD^2 + b^2}} d\theta . \tag{8.191}$$

If (8.191) is substituted into (8.184)

$$f(r, s) = \frac{1}{2} \int_0^{2\pi} \frac{FCD^2 + b^2}{U_b^2} \left\{ \left(\phi_\theta(a, b) \frac{\sqrt{FCD^2 + b^2}}{\sqrt{FCD^2 + a^2 + b^2}} \right) * g(a) \right\} \cdots$$

$$\cdots \cdot \frac{FCD}{\sqrt{FCD^2 + b^2}} d\theta \tag{8.192}$$

$$= \frac{1}{2} \int_0^{2\pi} \frac{FCD^2 + b^2}{U_b^2} \left\{ \left(\phi_\theta(a, b) \frac{FCD}{\sqrt{FCD^2 + a^2 + b^2}} \right) * g(a) \right\} d\theta$$

is obtained. Furthermore, one may use the intercept theorem to read from Fig. 8.50 that

$$\frac{FCD^2 + b^2}{U_b^2} = \frac{FCD^2}{U^2} , \qquad (8.193)$$

whereby, for convenience, the variable U that has been defined in the previous chapter,

$$U = FCD - x\sin(\theta) + y\cos(\theta) , \qquad (7.140)$$

might be used for the projection of the point to be reconstructed onto the central beam in the (x, y) plane. This must also be considered and (8.193) can consequently be substituted into (8.192) to complete the transformation of the rotational variables $\theta_b \to \theta$.

The point to be reconstructed can then be expressed as a function of the x and y coordinates. (8.184) finally reads

$$f(x, y) = \frac{1}{2}\int_0^{2\pi} \frac{FCD^2}{U^2}\left\{\left(\phi_\theta(a, b)\frac{FCD}{\sqrt{FCD^2 + a^2 + b^2}}\right) * g(a)\right\}d\theta . \qquad (8.194)$$

In analogy to (7.151) one may now write the short expression

$$f(x, y) = \int_0^{2\pi} \frac{FCD^2}{U^2}h_\theta(a, b)\,d\theta \qquad (8.195)$$

where

$$h_\theta(a, b) = \frac{1}{2}\left(\phi_\theta(a, b)\frac{FCD}{\sqrt{FCD^2 + a^2 + b^2}}\right) * g(a) . \qquad (8.196)$$

In this context, it is interesting that the weighting step in (8.196) allows the geometrical interpretation

$$\cos(\beta) = \frac{FCD}{\sqrt{FCD^2 + a^2 + b^2}}$$
$$= \frac{FCD}{\sqrt{FCD^2 + a^2}}\frac{\sqrt{FCD^2 + a^2}}{\sqrt{FCD^2 + a^2 + b^2}} = \cos(\psi)\cos(\kappa_a) , \qquad (8.197)$$

cf. Fig. 8.49.

However, at this point of the derivation of the instruction for the reconstruction, only the x and y components of the point to be reconstructed have been obtained. The two-dimensional backprojection result in (8.194) inside the angulated fan must now still be sorted into the fixed three-dimensional (x, y, z) coordinate system. Obviously, the required z component is missing. According to (8.157), this component obeys

$$z = s\frac{b}{\sqrt{FCD^2 + b^2}} - b \,.\tag{8.198}$$

It is, by the way, not surprising that the z component, contrary to the x and y components in (8.157), does not depend on r since the X-ray fan-beam surface **A** has been angulated around the $\mathbf{n}_\sigma$-axis, i.e., around the long detector axis a.

In Fig. 8.53 the situation of the backprojection geometry for the z coordinate is illustrated to assure that (8.198) does indeed yield the correct value for z. However, b is needed as a function of the coordinates $(x, y, z)^T$ in (8.194). (8.198) is somewhat awkward to invert, but, fortunately, Fig. 8.52 illustrates the expression

$$\tan(\kappa_0) = \frac{b}{FCD} = \frac{z}{U} \,.\tag{8.199}$$

This expression, in effect, immediately yields b as a function of the coordinates $(x, y, z)^T$, i.e.,

$$b(x, y, z, \theta) = z\frac{FCD}{FCD - x\sin(\theta) + y\cos(\theta)} \,.\tag{8.200}$$

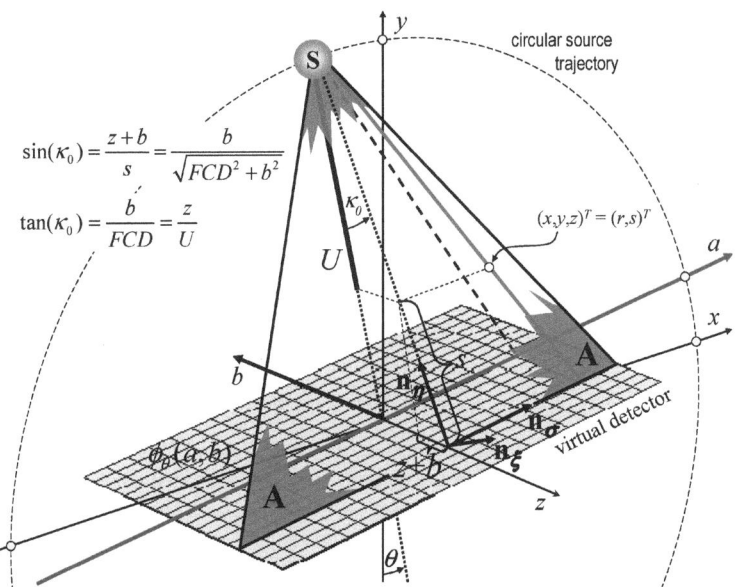

Fig. 8.53. Geometry for the backprojection within the angulated X-ray fan-beam plane. The z component does not depend on r since the fan-beam surface **A** has been angulated about the $\mathbf{n}_\sigma$-axis, i.e., the long detector axis a

Together with (8.200), finally, the three-dimensional reconstruction formula

$$f(x, y, z) = \frac{1}{2} \int_0^{2\pi} \frac{FCD^2}{U^2} \left\{ \left(\phi_\theta(a, b) \frac{FCD}{\sqrt{FCD^2 + a^2 + b^2}} \right) * g(a) \right\} d\theta \qquad (8.201)$$

is obtained. All the instructions for the reconstruction may again be split into three important steps, which are summarized in Scheme 8.3.

Scheme 8.3   Feldkamp, Davis, and Kress (FDK) reconstruction for the planar detector array

1. Coordinate transformation for the ramp filter: The required ramp filter

$$g \left((a' - a) \frac{U_b}{\sqrt{FCD^2 + a^2 + b^2}} \right) = \frac{FCD^2 + a^2 + b^2}{U_b^2} g(a' - a) \qquad (8.202)$$

   must be precisely adapted to the form of the filter for the parallel projection by coordinate transformation.
2. Filtering of the projection signal: The high-pass filtered backprojection signal results from the convolution

$$h_\theta(a, b) = \frac{1}{2} \left(\phi_\theta(a, b) \frac{FCD}{\sqrt{FCD^2 + a^2 + b^2}} \right) * g(a) \qquad (8.203)$$

   of the projection signal weighted with

$$\cos(\beta) = \frac{FCD}{\sqrt{FCD^2 + a^2 + b^2}} = \cos(\psi) \cos(\kappa)$$

   and the filter kernel $g(a)$ in the spatial domain of the flat-panel detector.
3. The filtered backprojection over the complete angle interval of 2π is given by

$$f(x, y, z) = \int_0^{2\pi} \frac{FCD^2}{U(x, y, \theta)^2} h_\theta \left(a(x, y, z, \theta), b(x, y, z, \theta) \right) d\theta \qquad (8.204)$$

   where

$$b(x, y, z, \theta) = z \frac{FCD}{FCD - x \sin(\theta) + y \cos(\theta)} . \qquad (8.205)$$

   This means, within the backprojection, the weighted and filtered projection signals are traced back through the volume to be reconstructed converging toward the X-ray source. Therefore, the values are weighted with the squared reciprocal distance to the source. This makes the process look similar to the light cone of a pocket lamp. The distance U is defined as the projection of the distance between the X-ray source and the current point to be reconstructed onto the central beam of the non-angulated fan.

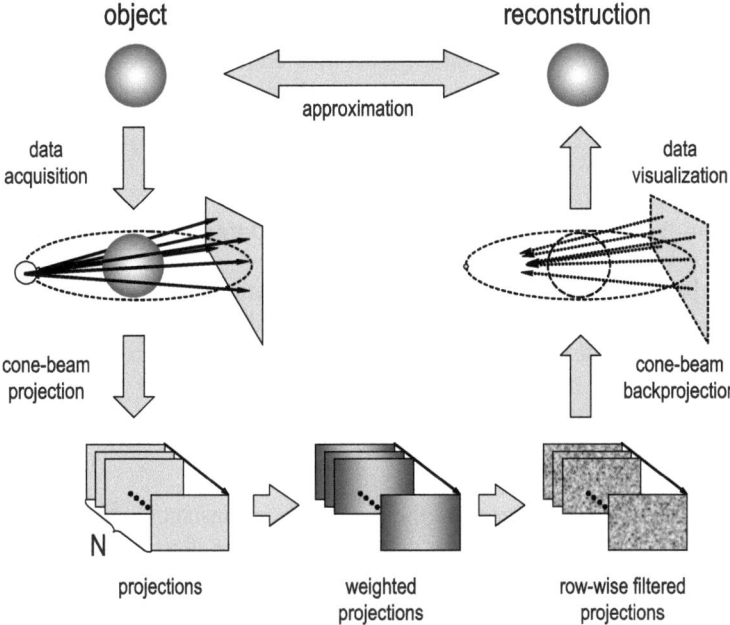

Fig. 8.54. Schematic representation of the FDK method. It is an approximate X-ray cone-beam reconstruction algorithm proposed by *Feldkamp*, *Davis*, and *Kress*

The FDK reconstruction method provides an exact backprojection instruction for the values measured in a plane including the circular X-ray source trajectory. Generally, the FDK reconstruction is of course an approximate method, since the Radon space is incomplete. It has already been pointed out in ▸ Sect. 8.6.1 that the *Tuy–Smith* sufficiency condition is not fulfilled for this trajectory.

For a point $\mathbf{r} = (x, y, z)^{\mathrm{T}}$, which is located outside the (x, y) source plane to be reconstructed, all other projections related to a particular, arbitrarily selected projection fan-beam plane **A** are missing. Therefore, the objects located within this plane **A** cannot be exactly reconstructed. Some Radon data are always inaccessible due to the shadow zone illustrated in Fig. 8.45 for any point outside the (x, y) plane.

However, many studies have revealed that the deficiency of the FDK method due to the incomplete Radon data set of the planar X-ray source trajectory is only a minor deficiency, if the aperture angle of the cone beam is small. Figure 8.54 schematically shows a summary of the FDK method.

The exact three-dimensional X-ray cone-beam backprojection method proposed by *Defrise* and *Clack*, which was described in ▸ Sect. 8.5.7, was not formulated for a special source trajectory. Consequently, if (8.146) and (8.204) are compared, a certain relationship between the two different methods can clearly be seen. In both cases, the backprojection is defined as an integral over the trajectory parameters, whereby a filtered projection term h is used as integrand, weighted with the reciprocal, normalized squared distance $(FCD/U)^2$ of the projection of the distance between the point to be reconstructed and the X-ray source onto the central X-ray beam. In fact, it can be shown that the exact method proposed by *Defrise* and *Clack* is indeed reduced to an approximate FDK method, if a single circle is assumed to be the source trajectory $\mathbf{S}(\lambda)$ (Jacobson 1996).

8.6.3
FDK Cone-Beam Reconstruction for Cylindrical Detectors

Most of the detectors used today for medical applications are cylindrical detectors; therefore, the FDK approximation should be described for this geometry as well. Figure 8.55 shows the cone-beam geometry in the cylindrical detector arrangement with the fan angle ψ and the cone angle κ, which is defined by

$$\kappa = \arctan\left(\frac{\varepsilon}{FCD}\right) . \tag{8.206}$$

With this definition, it is clear that the cone angle is constant along the individual rows ($\varepsilon = $ const.) of the cylindrical detector. Figure 8.55 illustrates that $\varepsilon = b$ only applies to the fan angle $\psi = 0$. In general, one finds the expression

$$b = \frac{\varepsilon}{\cos(\psi)} \tag{8.207}$$

between the planar and the cylindrical detector arrangement.

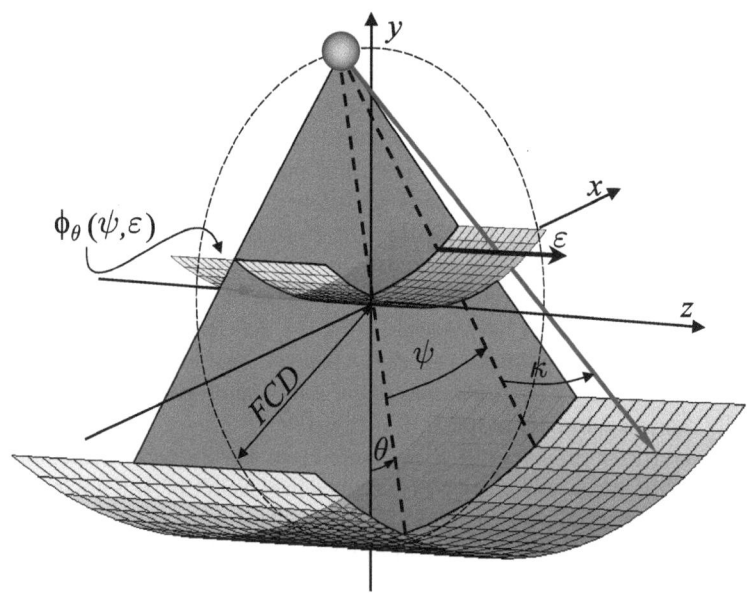

Fig. 8.55. Cone-beam illumination of a cylindrical detector. The X-ray source rotates on a radius *FCD*. A virtual detector is placed in the iso-center of the rotation of the sampling unit

The individual lines of the detector are, of course, not located on a planar surface containing the X-ray source. Planar surfaces, which were used in the previous section, rather produce slightly curved intersections with the cylindrical detector. However, it is possible to re-interpolate the projection data so that they are located on a planar Cartesian grid.

The reconstruction method with cylinder geometry is split into three principal processing steps in Scheme 8.4.

Finally, a few important properties of the FDK reconstruction method should be summarized.

- Exact reconstruction is ensured by the FDK method in the central radiation plane. This is, of course, the minimum requirement to be met by the method. (8.204) or (8.211) reduce to (7.151) or (7.122) respectively, for the corresponding

---

Scheme 8.4 FDK reconstruction with cylinder geometry

1. Pre-weighting of the cylinder detector projection values

$$\tilde{\phi}_\theta(\psi, \varepsilon) = \phi_\theta(\psi, \varepsilon) \frac{FCD}{\sqrt{FCD^2 + \varepsilon^2}} \cos(\psi) \tag{8.208}$$

$$= \phi_\theta(\psi, \varepsilon) \cos(\kappa) \cos(\psi)$$

2. Filtering of the projection signal. The ramp filter

$$\tilde{g}(\psi, \varepsilon) = \frac{1}{2} \left(\frac{\psi}{\sin(\psi)} \right)^2 g(\psi, \varepsilon) \tag{8.209}$$

is a modified version of the filter for the parallel projection and is given in full analogy to (7.121). Combined with the pre-weighting process described in the first step of this scheme, the high-pass filtered signal for backprojection results from the convolution in the angular domain

$$h_\theta(\psi, \varepsilon) = \tilde{\phi}_\theta(\psi, \varepsilon) * \tilde{g}(\psi, \varepsilon) . \tag{8.210}$$

3. Again, in analogy to the equations in ▸ Sect. 7.7.3, the filtered backprojection over the full angle of 2π is given by

$$f_{\mathrm{FDK}}(x, y, z) = \int_0^{2\pi} \frac{1}{L^2} h_\theta(\psi, \varepsilon) \, d\theta , \tag{8.211}$$

where

$$L = \sqrt{(FCD - x\sin(\theta) + y\cos(\theta))^2 + (x\cos(\theta) + y\sin(\theta))^2} \tag{8.212}$$

is defined as the distance between the X-ray source and the actual point to be reconstructed in the (x, y) plane.

two-dimensional X-ray fan-beam reconstruction for the cone angle $\kappa = 0$ (i.e., $b = 0$ for the planar detector or $\varepsilon = 0$ for the cylindrical detector).

- The larger the cone angle κ, the poorer the quality of the reconstructed image. This is due to the fact that an increasing quantity of Radon data is located in the shadow zone (cf. ▸ Sect. 8.6.1).

- The FDK method is exact for objects that are homogeneous in the z direction, i.e., for objects obeying $f(x, y, z) = f(x, y)$. This result is achieved by pre-weighting with $\cos(\kappa)$ in (8.203) for the planar detector or (8.208) for the cylindrical detector. The longer path along which the X-ray beam is attenuated with a larger cone angle κ is weighted by the factor $\cos(\kappa)$ in such a way that the attenuation does not depend on the cone angle and corresponds to the attenuation in the central plane. The pre-weighted lines of the detector are identical for such objects, i.e., the method reduces itself to the fan-beam method in the plane.

Considering the planar detector, it should finally be pointed out that:

- The integral

$$p_z = \int f(x, y, z)\, dz \tag{8.213}$$

is exactly preserved in spite of the approximate character of the FDK method. This is due to the fact that all surface integrals that are perpendicular to the central plane can be measured by a single planar, circular X-ray source trajectory, i.e., they are located within the measurable torus in the Radon space.

8.6.4
Variations of the FDK Cone-Beam Reconstruction

Weighting with the U^{-2}-factor in (8.204) is one of the problems related to the actual implementation of fast three-dimensional reconstructions. As very fast implementations for reconstruction algorithms in parallel-beam geometry are generally available, it seems to be useful to convert the cone-beam geometry into parallel-beam geometry by rebinning the beams in analogy to ▸ Sect. 7.7.1. As data are missing in the Radon space with the cyclic, planar source trajectory, this can be achieved successfully in the fan direction ψ, but not in the cone direction κ.

8.6.4.1
Rebinning with the Planar Detector

For the planar detector, the rebinned projection signal (without derivation, see Turbell [1999]) obeys

$$p_\gamma(\xi, b) = \phi_{\gamma - \psi}(\psi, b) = \phi_{\gamma - \arcsin\left(\frac{\xi}{FCD}\right)}\left(\frac{\xi\, FCD}{\sqrt{FCD^2 - \xi^2}}, b\right). \tag{8.214}$$

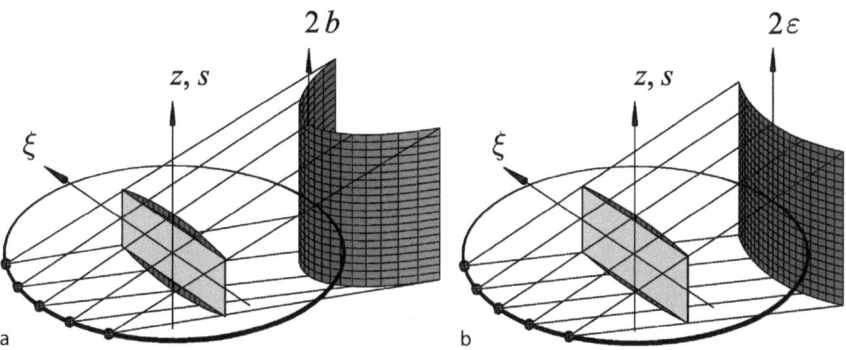

Fig. 8.56. Geometry of the projections after the X-ray cone beams have been rebinned, for **a** the planar and **b** the cylindrical detector. In the iso-center of the rotation the virtual planar detector is illustrated (courtesy of Turbell [2001])

The coordinate of the detector line b remains unchanged here. If one considers this situation from the direction of the z coordinate, parallel-beam geometry has been achieved after completion of the rebinning process. The beams of a particular detector line b match a vertical planar detector (ξ, s) in the iso-center of the rotation along the curves

$$s(b, \xi) = b\left(1 - \frac{\xi^2}{FCD^2}\right). \tag{8.215}$$

Additionally, Fig. 8.56a illustrates what a detector (which is also a virtual detector) would look like, if all beams of a particular detector line b had the same z value.

8.6.4.2
Rebinning with the Cylindrical Detector

For the cylindrical detector the rebinned projection signal – in analogy to (7.75) – obeys

$$p_\gamma(\xi, \varepsilon) = \phi_{\gamma - \psi}(\psi, \varepsilon) = \phi_{\gamma - \arcsin\left(\frac{\xi}{FCD}\right)}\left(\arcsin\left(\frac{\xi}{FCD}\right), \varepsilon\right). \tag{8.216}$$

The coordinate of the detector line ε remains unchanged here. If one again considers this situation from the direction of the z coordinate, parallel-beam geometry has been achieved after completion of the rebinning process. The beams of a particular detector line ε match a virtual planar detector (ξ, s) in the iso-center of the rotation along the curves

$$s(\varepsilon, \xi) = \varepsilon\sqrt{1 - \frac{\xi^2}{FCD^2}}. \tag{8.217}$$

The intersection curves on the virtual, cylindrical detector are obviously not as curved as they are for the planar detector, as described in (8.215).

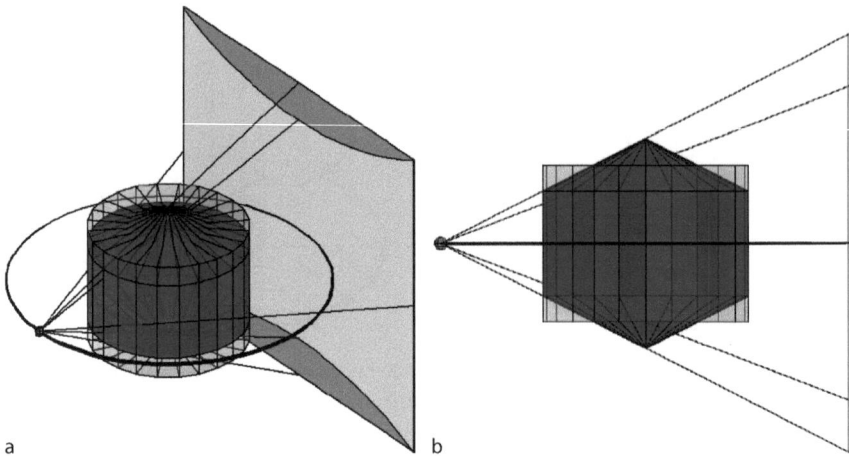

a   b

Fig. 8.57a,b. Once the data have been rebinned, the geometry of the illuminated volume is similar to a tent with a rectangular base area (courtesy of Turbell [2001])

8.6.4.3
Tent-FDK Method

Although one is able to accelerate the backprojection process by rebinning the data as described above, the artifacts are still the same (Turbell 2001). However, this changes if one takes into account the approach of Grass et al. (2000). They proposed to re-interpolate the data $p_y(\xi, b)$ from (8.214) or the data $p_y(\xi, \varepsilon)$ from (8.216), on the virtual (ξ, s) detector in the iso-center of the rotation in such a way that they lie on horizontal lines.

This method can only be applied reliably, i.e., a ramp filter can only be applied along horizontal lines, if the data along these lines on the detector are complete. Figure 8.57a shows that the data in the shaded area may therefore not be used for the so-called T-FDK method[15]. In Fig. 8.57b, it can be seen that the cross-sectional area produced with the T-FDK method in the three-dimensional space is shaped like a tent with a rectangular base area. In practical applications, it is thus possible to reduce the radiation dose by suitable collimation of the required data. The T-FDK method is able to improve the quality of the reconstructed image.

8.6.4.4
Sequential FDK Method

Köhler et al. (2000) suggested sampling the patient projection data not on a helical path but in discrete distances ΔZ on circular paths. If the real detector is high enough so that a virtual detector in the iso-center of the rotation has just the

---

[15] T-FDK is the abbreviation of Tent-FDK, because the data form a tent after completion of the rebinning process.

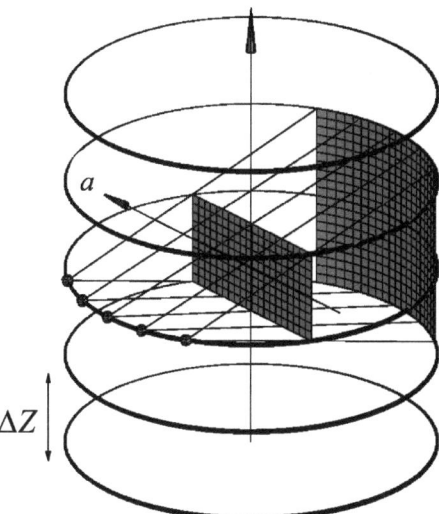

Fig. 8.58. The beam geometry of the sequential FDK method (S-FDK) is a straightforward extension of the FDK method in the axial direction (courtesy of Turbell [2001])

height ΔZ, then all voxels receive exactly one backprojection contribution from each projection angle. Figure 8.58 is displaying this arrangement.

The voxels inside the double cone shown in Fig. 8.57b receive their backprojection contribution from the source trajectory with the smallest distance to this voxel. All other points to be reconstructed receive contributions either from this trajectory or from the following circular trajectory. A point is illuminated from the consecutive circular trajectory at a projection angle $\theta + \pi$ only, if it is not illuminated at a projection angle θ from the nearest neighboring circular trajectory. According to this sequential FDK (S-FDK) principle, a complete set of measured projection values is made available for all points to be reconstructed.

8.6.4.5
FDK-SLANT Method

Within the original FDK method, a horizontal line on the (planar) detector at a particular projection angle θ is considered for a point to be reconstructed. This line is related to a projection plane **A**, which also contains the X-ray source and the point itself. The projection values on the detector line can be pre-weighted, filtered, and then backprojected in the direction of the X-ray source in the plane, independent of all other projection angles and all other angulations, κ. If one switches over to another projection angle θ', one has to leave this plane and search for a new plane, which starts from the X-ray source and then runs through the point to be reconstructed at another angulation angle κ'.

Turbell (2001) proposed to vary the FDK method in such a way that the plane is held when the projection angle changes. Figure 8.59a shows a fan consisting of

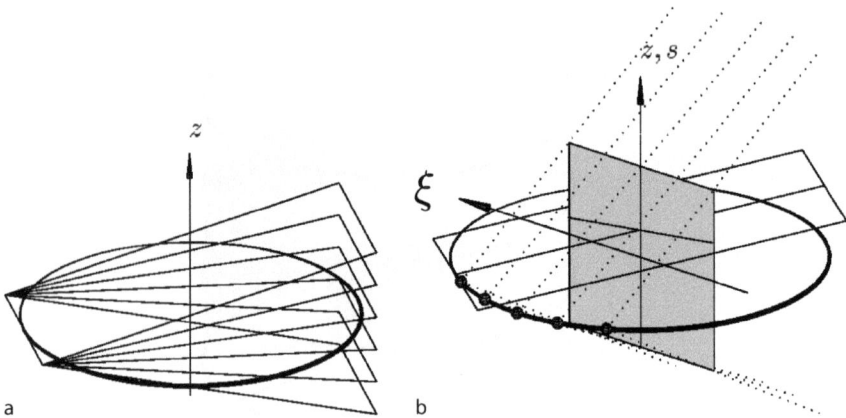

Fig. 8.59a,b. The beams of the FDK-SLANT method have a geometrical pattern that is similar to the pages of a book. The intersection line between a book page and the detector plane is non-horizontal. This line on the (ξ,s) detector corresponds to the filtering direction (courtesy of Turbell [2001])

held planes with horizontal intersection lines on the planar detector and touching the X-ray source trajectory at the projection angle θ_i at the point

$$\mathbf{S}_i = (-FCD\sin(\theta_i), FCD\cos(\theta_i), 0). \tag{8.218}$$

This geometrical situation is similar to a half-opened book, the pages of which are stitched together at the point $\mathbf{S}_i$ in (8.218). *Turbell* then uses a number N of such books with the anchor points $\mathbf{S}_i$, with $i = 1,\ldots,N$ being uniformly distributed on the cyclic source trajectory. The first five anchor points are shown in Fig. 8.59b.

The filtering direction of a filtered backprojection at the projection angle θ is given by the intersection line of the detector plane and the book page with the anchor point $\mathbf{S}_i$. Figure 8.59b clearly illustrates that the intersection lines on the virtual detector are slanted straight lines. The key idea of this method is that the backprojection is to remain as far as possible an interaction between the points within a single plane.

8.7
Helical Cone-Beam Reconstruction Methods

In the previous sections of this chapter, object reconstruction with cone-beam geometry based on initially arbitrary source trajectories (exact methods proposed by *Grangeat*, ▸ Sect. 8.5.2, and *Defrise and Clack*, ▸ Sect. 8.5.7) and then special circular, cyclic X-ray source trajectories (approximate methods proposed by *Feldkamp, Davis, and Kress* and their extensions, ▸ Sect. 8.6) have been discussed. The derivative-free reconstruction method proposed by Feldkamp et al. (1984) has been frequently used in technical applications. However, it is an approximate method, the use of which in medical applications is not unproblematic due to the required image quality. As long as the cone angle aperture of the X-ray beam is small, this method yields acceptable results. However, today, the development of new detector generations focuses on even larger flat-panel detectors; thus, the reconstruction algorithms must be adapted accordingly. Also, inherently three-dimensional methods must be implemented because the three-dimensional adaptations of the two-dimensional approaches reach their limits.

In ▸ Sect. 8.6.1 it has been explained that a single circular, cyclic source trajectory cannot completely fill the Radon space. As Kalender et al. (1989) has successfully demonstrated with the spiral method for the one-dimensional detector array (cf. ▸ Sect. 8.3), here it can be shown that the methods developed for cone-beam geometry can be adapted to spiral or helical trajectories

$$\mathbf{S}(\lambda) = (-FCD\sin(\lambda), FCD\cos(\lambda), h\lambda) \qquad (8.219)$$

of the X-ray source. In Fig. 8.47 it has already been illustrated that the helix represents an X-ray source orbit, which, in principle, may completely fill the Radon space. In this case, $\lambda \in \Lambda \subset \mathbb{R}$ is again the parameterization of the trajectory. The parameter h denotes the rise of the helix. The parameter *pitch* introduced in ▸ Sect. 8.3, explaining the simple spiral CT method, has no direct meaning in an X-ray cone. However, the pitch parameter can be adapted to the new situation. This can be done for instance by defining $P = 2\pi h$, meaning the table feed in z direction per full rotation of the sampling unit on the gantry (Proksa et al. 2000). In this way, the patient

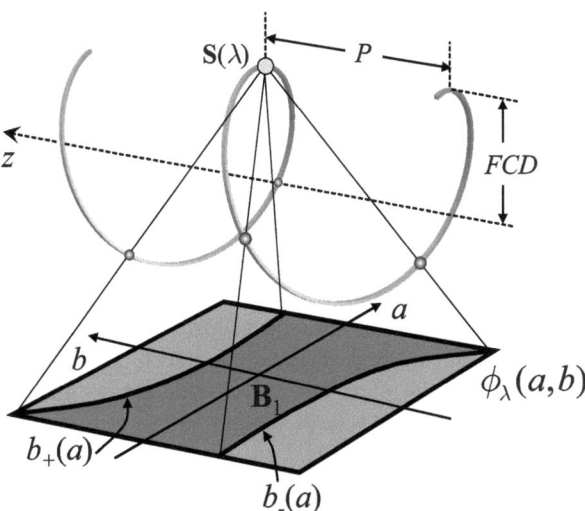

Fig. 8.60. *Tam–Danielsson* window $\mathbf{B}_1$ on the flat-panel detector array $\phi_\lambda(a, b)$, which is exceptionally located outside the source trajectory area here. The data window $\mathbf{B}_1$ on the detector used for the reconstruction process is defined by two successive arcs of the spiral. It is called a PI window

table feed in the axial direction can be expressed by

$$z(\lambda) = \lambda \frac{P}{2\pi} + z_0 \, . \tag{8.220}$$

Figure 8.60 illustrates the geometrical situation. The detector is exceptionally not placed in the iso-center of the helical rotation, although the mathematics of the object reconstruction would be more concise if a virtual detector were considered in the iso-center.

The data measured by a rectangular flat-panel detector must be windowed to prevent projection values being overrepresented in the reconstruction algorithm. For this purpose, one only considers the detector field in the area **B**. The so-called *Tam–Danielsson* window (Tam 1995; Turbell 2001) is defined by restricting the X-ray beams to the area between two successive spiral arcs opposite the source position **S**(λ). On the planar (a, b) detector with the projection values $\phi_\lambda(a, b)$, a projection yields the restriction of the data to be used between the curves

$$
\begin{aligned}
b_+(a) &= P\left(1 + \frac{a^2}{FCD^2}\right)\left(\frac{\arctan\left(a/FCD\right)}{2\pi} + \frac{1}{4}\right) \\
&= h\left(1 + \frac{a^2}{FCD^2}\right)\left(\arctan\left(a/FCD\right) + \frac{\pi}{2}\right)
\end{aligned}
\tag{8.221}
$$

and

$$
\begin{aligned}
b_-(a) &= P\left(1 + \frac{a^2}{FCD^2}\right)\left(\frac{\arctan\left(a/FCD\right)}{2\pi} - \frac{1}{4}\right) \\
&= h\left(1 + \frac{a^2}{FCD^2}\right)\left(\arctan\left(a/FCD\right) - \frac{\pi}{2}\right) \, .
\end{aligned}
\tag{8.222}
$$

Due to this restriction, this method is classified as a so-called PI method.

The object may now be reconstructed according to the approach proposed by *Pierre Grangeat*, which has been described in ▶ Sect. 8.5.2. For this purpose, the derivative of the three-dimensional Radon transform, i.e., the differentiation of the integration over the Radon surfaces through the object, has to be computed. Figure 8.61 shows the orientation of an arbitrary Radon surface through the object. As the aperture of the X-ray cone is limited, the Radon surface must be assembled from several segments.

The points of intersection of the helical X-ray source orbit with the Radon surface in consideration (cf. Fig. 8.61a) are the nodes of a triangulation of the Radon surface. In Fig. 8.61b, visualization along the surface normal of the Radon surface onto the resulting triangulation is given. The *Tam–Danielsson* window restricts the measured values in such a way that each triangle is illuminated exactly once. Thus,

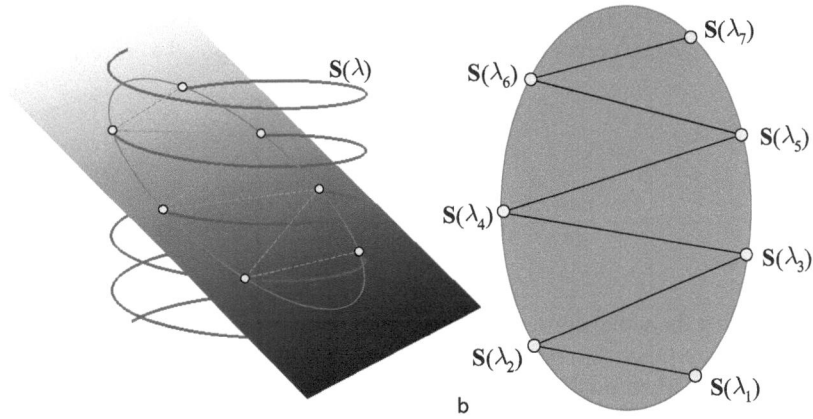

Fig. 8.61. a The intersection points of the helix are basis points by means of which any Radon surface may be triangulated. **b** If one looks perpendicularly onto the Radon surface, one sees the segments of the Radon surface, which is thus completely illuminated, but not in an overlapping manner

if the source has reached the position $S(\lambda_4)$, then the data made available are restricted by the points $S(\lambda_3)$ and $S(\lambda_5)$ in accordance with (8.221) and (8.222), so that the corresponding triangle is exclusively illuminated by the X-ray source at position $S(\lambda_4)$. If one assembles all these segments, an angulated ellipse can be found within the helix (cf. Fig. 8.61a).

The key idea of the exact reconstruction for incomplete projections is the combination of the individual contributions of the segments shown in Fig. 8.61. The intermediate result of *Grangeat* was

$$\frac{\partial}{\partial \xi} p_{\gamma,\vartheta}(\xi) = \frac{1}{\cos^2(\kappa)} \int_{-\infty}^{\infty} \frac{\partial}{\partial \tau} X_w^c \left\{ \mu_{\gamma,\vartheta}(\sigma, \tau(\xi)) \right\} d\sigma , \tag{8.91}$$

i.e., replacing the derivative in the radial direction of the Radon transform with the weighted radial detector derivative and line integration.

The surface can be decomposed into triangular segments

$$\frac{\partial}{\partial \xi} p_{\gamma,\vartheta}(\xi, \lambda_i) \approx \frac{1}{\cos^2(\kappa)} \int_{\sigma_1}^{\sigma_2} \frac{\partial}{\partial \tau} X_w^c \left\{ \mu_{\gamma,\vartheta}(\sigma, \tau(\xi), \lambda_i) \right\} d\sigma , \tag{8.223}$$

whereby the sum over all N_Δ triangular elements

$$\frac{\partial}{\partial \xi} p_{\gamma,\vartheta}(\xi) = \sum_{i=1}^{N_\Delta} \frac{\partial}{\partial \xi} p_{\gamma,\vartheta}(\xi, \lambda_i) \tag{8.224}$$

only yields the desired derivative of the Radon transform.

Kudo et al. (1998) have shown that the single addition in (8.224) unfortunately does not provide an exact reconstruction result. They proposed a correction term

$$
\frac{\partial}{\partial \xi} p_{\gamma,\vartheta}(\xi,\lambda_i) = \frac{1}{\cos^2(\kappa)} \int_{\sigma_1}^{\sigma_2} \frac{\partial}{\partial \tau} \mathbf{X}_w^c \left\{ \mu_{\gamma,\vartheta}(\sigma,\tau(\xi)),\lambda_i \right\} d\sigma + \ldots
$$

$$
\ldots - \frac{\tan(\kappa)}{FCD} \left(\sigma_1 \mathbf{X}_w^c \left\{ \mu_{\gamma,\vartheta}(\sigma_1,\tau(\xi)) \right\} - \sigma_2 \mathbf{X}_w^c \left\{ \mu_{\gamma,\vartheta}(\sigma_2,\tau(\xi)),\lambda_i \right\} \right)
$$

$$(8.225)$$

considering the actual borders of the integration intervals. Turbell (2001) interprets this correction term as a compensation of a minor change in the triangular surface, which arises when the Radon surface moves along the surface normal, while the derivation process is carried out. On the basis of this piece of Radon data, the filtered backprojection may, for instance, now be implemented according to the approach proposed by Defrise and Clack (1994).

Proksa et al. (2000) have shown that this PI method can of course be extended to a n-PI method, if larger flat-panel detector areas are used. If one defines the detector window more generally as

$$
\begin{aligned}
b_n(a) &= P\left(1 + \frac{a^2}{FCD^2}\right)\left(\frac{n}{4} - \frac{\arctan(a/FCD)}{2\pi}\right) \\
&= h\left(1 + \frac{a^2}{FCD^2}\right)\left(\frac{n\pi}{2} - \arctan(a/FCD)\right)
\end{aligned}
$$

$$(8.226)$$

where

$$
\mathbf{B}_n = \{(a,b)| - b_n(-a) \le b \le b_n(a)\} ,
$$

$$(8.227)$$

the data used cover a multiple of the helix pitch. This extension is called the n-PI window[16].

The ellipse resulting from a section of the Radon surface under consideration with the helix contains an odd number n_o of piercing points, $\mathbf{S}(\lambda_i)$, of the source trajectory with the Radon surface. For the n-PI detector Proksa et al. (2000) found that all Radon surfaces with n_o piercing points and $n_o < n$ have an n_o-fold illumination or over-representation. If $n_o \ge n$, the Radon surfaces have an n-fold over-representation. The corresponding over-representation is compensated for by means of a function M, which has already been mentioned in ▸ Sect. 8.5.7.

---

[16] All positive integer odd numbers are in principle acceptable in this context. But in practice, only $n = 1, 3, 5$, and 7 are used.

In the case here, the M function reads

$$M_n(\lambda, \mathbf{n}_\xi) = \begin{cases} \dfrac{1}{n_o(\lambda, \mathbf{n}_\xi)} & \text{for } n_o(\lambda, \mathbf{n}_\xi) < n \\ \dfrac{1}{n} & \text{otherwise} \end{cases}, \qquad (8.228)$$

so that one may finally formulate the reconstruction shown in Scheme 8.5.

Today, a multitude of algorithms or implementation strategies are available for three-dimensional helix-based X-ray cone-beam geometries.

Finally, the *rebinning* strategy, which has already been formulated for a tilted gantry (Kachelrieß et al. 2000, 2001), should be addressed. Compared with the n-PI methods, the essential difference is due to the possibility of replacing the computationally expensive, three-dimensional backprojection step with a two-dimensional backprojection step on appropriately defined planes. The appropriate slice is defined such that its inclination matches the helical X-ray source trajectory locally.

Scheme 8.5 Exact cone-beam helical reconstruction for n-PI detector geometry

1. Determination of the number of intersection points n_o with the helix for all Radon surfaces.
2. Computation of the positions $\mathbf{S}(\lambda_i)$ for all intersection points from step 1.
3. Determination of the integration intervals $[\sigma_1, \sigma_2]$ belonging to the individual positions.
4. Computation of the derivative of the individual partial Radon surfaces

$$\frac{\partial}{\partial \xi} p_{\gamma,\vartheta}(\xi, \lambda_i) = \frac{1}{\cos^2(\kappa)} \int_{\sigma_1}^{\sigma_2} \frac{\partial}{\partial \tau} \mathbf{X}_w^c \{\mu_{\gamma,\vartheta}(\sigma, \tau(\xi)), \lambda_i\} \, d\sigma + \dots$$

$$\dots - \frac{\tan(\kappa)}{FCD} \left(\sigma_1 \mathbf{X}_w^c \{\mu_{\gamma,\vartheta}(\sigma_1, \tau(\xi)), \lambda_i\} + \dots \right.$$

$$\dots \left. - \sigma_2 \mathbf{X}_w^c \{\mu_{\gamma,\vartheta}(\sigma_2, \tau(\xi)), \lambda_i\} \right) . \qquad (8.229)$$

5. Computation of the compensation function

$$M_n(\lambda, \mathbf{n}_\xi) = \frac{1}{\min\{n_o(\lambda, \mathbf{n}_\xi), n\}} . \qquad (8.230)$$

6. Summation of the individual Radon segments

$$\frac{\partial}{\partial \xi} p_{\gamma,\vartheta}(\xi) = M \sum_{i=1}^{N_\Delta} \frac{\partial}{\partial \xi} p_{\gamma,\vartheta}(\xi, \lambda_i) . \qquad (8.231)$$

7. Computation of the object with the Radon inversion formula

$$f(x, y, z) = -\frac{1}{8\pi^2} \iint_S \frac{\partial^2 p_{\gamma,\vartheta}(\xi)}{\partial \xi^2} \, dS . \qquad (8.38)$$

As illustrated in Fig. 8.62 the two-dimensional reconstruction plane nutates about the patient z-axis like a gyroscope with an inclination angle, α. The nature of the different reconstruction artifacts is described by Köhler et al. (2002).

As a last remark regarding helix-based cone-beam geometry, it should be mentioned that Katsevich (2001) proposed an exact backprojection method that also works for long objects in such a way that one does not have to include data acquired in areas far away from each other.

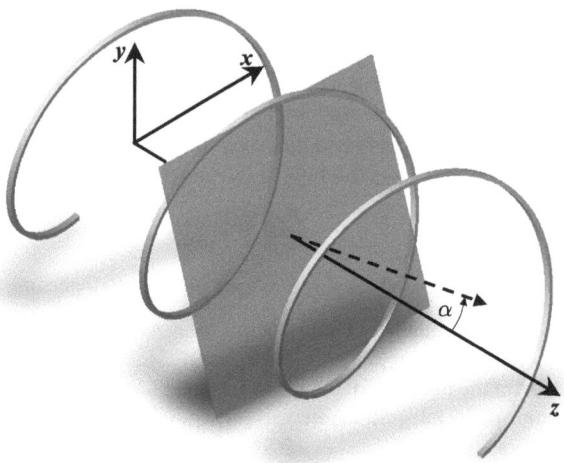

Fig. 8.62. Nutation of the two-dimensional reconstruction plane around the axial direction. The surface normal is tilted by the angle α with respect to the axial direction

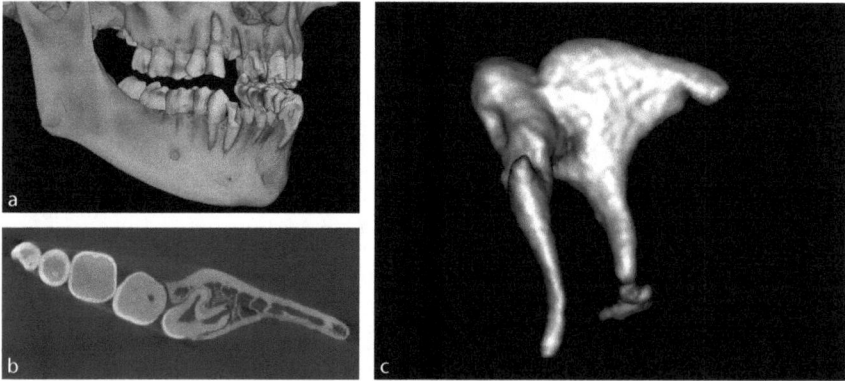

Fig. 8.63. Performance of modern cone-beam CT scanners with flat-panel detectors. **a** and **b** show the three-dimensional visualization of a human jaw and a corresponding section through a row of teeth. **c** shows the smallest human bones, i.e., the hammer, anvil, and stirrup in the middle ear. The pictures have an isotropic spatial resolution of 140 μm (courtesy of General Electric Medical Systems: Pfoh [2002])

Figure 8.63 demonstrates the performance of a cone-beam CT prototype scanner. The scanner is equipped with a flat-panel detector measuring 41 cm × 41 cm with 2.048 × 2.048 detector elements (cf. ▸ Sect. 2.5.3). Figure 8.63a shows the volume CT image of a human jaw. Figure 8.63b shows the corresponding section through a row of teeth, revealing extremely small details with a resolution of 140 µm. Even the smallest bones in the human body, i.e., the hammer, anvil, and stirrup in the middle ear, can be excellently imaged with an isotropic spatial resolution of 140 µm. Figure 8.63c shows a volume CT image of the three bones (Pfoh 2002).

9 Image Quality and Artifacts

Contents

9.1	Introduction	403
9.2	Modulation Transfer Function of the Imaging Process	404
9.3	Modulation Transfer Function and Point Spread Function	410
9.4	Modulation Transfer Function in Computed Tomography	412
9.5	SNR, DQE, and ROC	421
9.6	2D Artifacts	423
9.7	3D Artifacts	445
9.8	Noise in Reconstructed Images	462

9.1
Introduction

The physical attenuation values $f(x, y)$ of the human tissue within an axial slice are distributed continuously in value and space. In the previous chapter it has become evident that in practice it is indispensable for computed tomography (CT) to digitize the acquired data in order to carry out the mathematical object reconstruction. This means that the attenuation of the radiation must be measured, digitalized, and stored. Within this processing chain, the continuous physical signal is discretized in the spatial domain and the values have to be quantized. Figure 9.1 shows that this process can be modeled as a signal transmission chain. One can thus evaluate the changes – in general the deterioration – to which the signal $f(x, y)$ is subjected during acquisition and discretization into a series of numbers $g(n, m)$ and further on until the tomogram $c(n, m)$ is displayed and presented to the clinician.

The signal path is to be modeled with five layers. The first layer, which is called the physical layer or X-ray imaging layer, describes the beam characteristic[1]. This layer must be considered analogously to the lens system of a camera because the physical quality of the X-ray optical system is described by the focus spot size in

---

[1] The physical parameters influencing the image quality can of course be evaluated retrospectively only, i.e., after the image has been reconstructed. This is due to the fact that an image $f(x, y)$ is not available at the focal spot of the X-ray tube.

the X-ray tube and by the slice collimation of the X-ray fan. The transfer functions (H_1 and H_2) describe the change of the spatially continuous and value-continuous input signal $f(x, y)$. The physical nature of the signal initially remains unchanged. However, the interaction between the imaging radiation and the object generally also changes the physical nature of the radiation regarding hardening of the spectrum of polychromatic X-ray. This may be compared with the chromatic aberration of a camera lens system. The image, which has been degenerated by beam hardening (H_3), is denoted by $b(x, y)$.

The sensor or detector layer (analogous to the film plane of a conventional optical camera) is the second layer. Here, the signal is subdivided by an anti-scatter grid beam collimation (H_4) and then physically transformed. That is, the X-ray photons are, for example, detected with a scintillation crystal and then converted into an electrical signal $e(x, y)$ with an attached photomultiplier or photodiode. Here, among others, crystal properties such as afterglow (H_5) have to be modeled.

In practice, spatial discretization will already occur in the sensor or detector layer. However, for a better separation of each cause of signal deterioration, this effect is modeled in the third layer – the electronic or digitalization layer – in which the electrical signal of the photomultiplier or the photodiode (cf. Fig. 2.19) is spatially discretized (H_6). Furthermore, the values are quantized (H_7) by an acquisition with an analog–digital converter (ADC).

In the fourth layer – the reconstruction or algorithm layer – the size of the image matrix to be reconstructed together with the corresponding detector interpolation (H_8) as well as the special properties of the filter kernels used for the filtered back-projection (H_9) are modeled.

The last layer – the image processing and display layer – models the type of image representation. An example is, for instance, the mapping of the physical attenuation values onto a gray-value interval (H_{10}), which is appropriate for human perception as well as the quality of the visualization medium, for example, the monitor (H_{11}).

In the following sections of this chapter, it will be shown how the individual factors of the imaging chain that influence each part can be quantified. In this context, the so-called *modulation transfer function* (*MTF*) plays an important role.

9.2
Modulation Transfer Function of the Imaging Process

With the transfer functions H_1 through H_{11} in the chart shown in Fig. 9.1, the imaging, or more precisely, the alteration of the representation of an object $f(x, y)$ can be described mathematically (cf. ▸ Sect. 4.9). These functions are generally available as amplitude and phase curves as a function of the spatial frequency, i.e., as $H(u, v)$. From a technical point of view, it is interesting how the spatial resolution is changed by each individual element of the system chain during the imaging process. The commonly used unit of the spatial resolution is [*lp/mm*], i.e., line pairs per

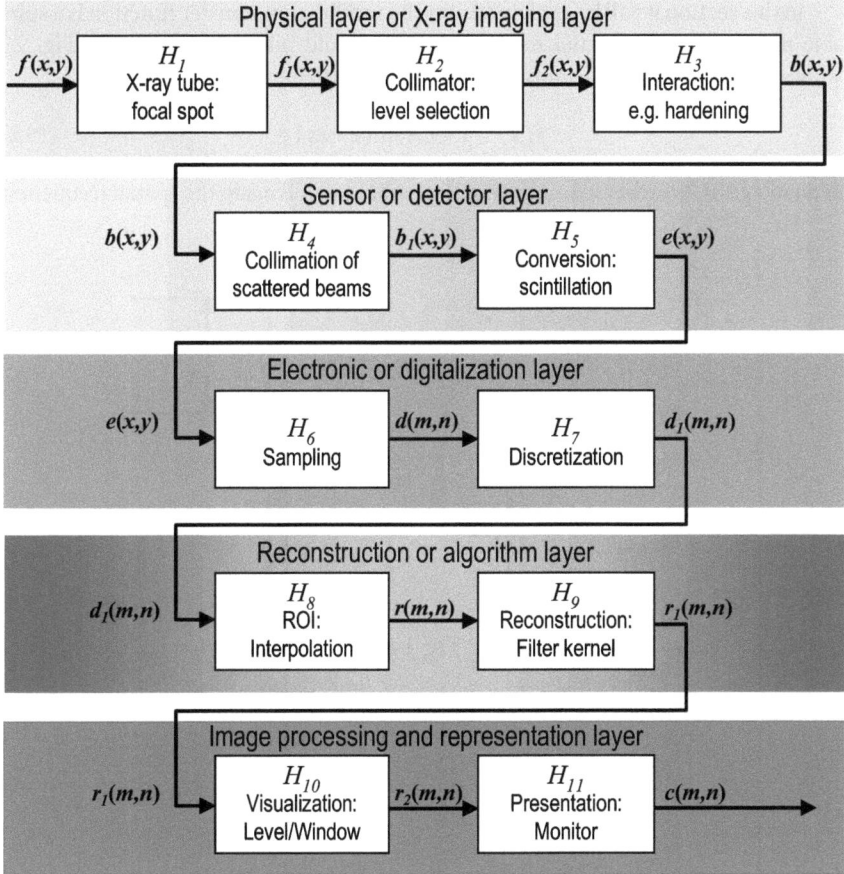

Fig. 9.1. The image acquisition process can be modeled in signal transmission layers from the X-ray tube to the visualization of the image data on a display

millimeter[2]. The spatial resolution thus indicates how close two neighboring lines can get to each other before they can no longer be distinguished due to the vanishing modulation of the image values, i.e., the variation of the gray values between the lines. An important parameter related to the evaluation of the spatial resolution is the contrast[3]

$$C = \frac{\max\{f(x,y)\} - \min\{f(x,y)\}}{\max\{f(x,y)\} + \min\{f(x,y)\}} \tag{9.1}$$

of the image, i.e., the variation interval of the image values around the mean value, normalized by the mean value.

---

[2] This is because resolution cannot be defined by the visibility of a small single object. It is rather defined by the ability to sufficiently separate two neighboring objects.

[3] The contrast is also called *modulation*.

In this section it will be explained why the modulation transfer function is a suitable measure for the spatial resolution. One should initially assume that Fig. 9.2 shows a one-dimensional sinusoidal gray-value distribution

$$f(x) = 1 + C_0 \sin(2\pi u x) \tag{9.2}$$

of an image. It is further assumed that the parameter u is again the spatial frequency.

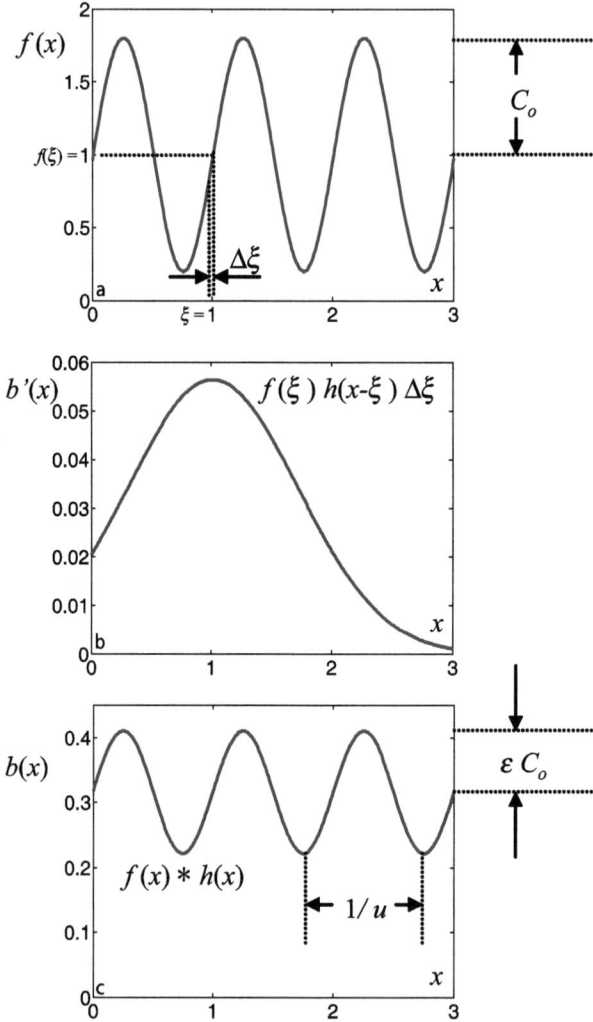

Fig. 9.2. Motivation of the modulation transfer function (*MTF*). *Top*: A sinusoidal signal is blurred by a Gaussian impulse response. *Center*: Transfer of a narrow strip with a width of $\Delta\xi$ via $f(\xi)h(x - \xi)\Delta\xi$. *Bottom*: The entire image results from the convolution of the original signal with the impulse response $f(x) * h(x)$

With the impulse response

$$h(x) = \frac{1}{\sqrt{\pi}} e^{-x^2} ,$$ (9.3)

which causes a Gaussian blurring, the following mapping of a narrow image strip $\Delta\xi$ at the point ξ can be found

$$b'(x) = f(\xi) h(x - \xi)\Delta\xi = \frac{1}{\sqrt{\pi}} f(\xi) e^{-(x-\xi)^2} \Delta\xi .$$ (9.4)

(9.4) represents the blurring caused by the imaging or transfer system. The entire image obviously results from the convolution $b(x) = f(x) * h(x)$ of the original signal $f(x)$ with the impulse response $h(x)$, i.e., from

$$b(x) = \int_{-\infty}^{\infty} f(\xi) h(x - \xi)\,d\xi = \frac{1}{\sqrt{\pi}} \int_{-\infty}^{\infty} f(\xi) e^{-(x-\xi)^2}\,d\xi .$$ (9.5)

Since the convolution is symmetric, one may also write

$$b(x) = \int_{-\infty}^{\infty} f(x - \xi) h(\xi)\,d\xi = \frac{1}{\sqrt{\pi}} \int_{-\infty}^{\infty} f(x - \xi) e^{-\xi^2}\,d\xi .$$ (9.6)

If the gray-value distribution (9.2) is substituted into (9.6), one finds

$$b(x) = \int_{-\infty}^{\infty} (1 + C_0 \sin(2\pi u(x - \xi)))\, h(\xi)\,d\xi .$$ (9.7)

Taking into account the addition law for the sine

$$\sin(\alpha - \beta) = \sin(\alpha)\cos(\beta) - \cos(\alpha)\sin(\beta) ,$$ (9.8)

(9.7) can be converted into

$$b(x) = \int_{-\infty}^{\infty} h(\xi)\,d\xi + \int_{-\infty}^{\infty} C_0 \sin(2\pi u x)\cos(2\pi u \xi) h(\xi)\,d\xi + \dots$$

$$\dots - \int_{-\infty}^{\infty} C_0 \cos(2\pi u x)\sin(2\pi u \xi) h(\xi)\,d\xi$$

(9.9)

$$= \int_{-\infty}^{\infty} h(\xi)\,d\xi + C_0 \sin(2\pi u x) \int_{-\infty}^{\infty} \cos(2\pi u \xi) h(\xi)\,d\xi + \dots$$

$$\dots - C_0 \cos(2\pi u x) \int_{-\infty}^{\infty} \sin(2\pi u \xi) h(\xi)\,d\xi .$$

As the two last integral terms in (9.9) look like components of a vector, the expression

$$\varepsilon = \sqrt{\left(\int_{-\infty}^{\infty} \cos(2\pi u x) h(x)\, dx\right)^2 + \left(\int_{-\infty}^{\infty} \sin(2\pi u x) h(x)\, dx\right)^2} \tag{9.10}$$

may be interpreted as the length of a vector. From this point of view, the phase can be expressed via

$$\cos(\varphi) = \frac{1}{\varepsilon} \int_{-\infty}^{\infty} \cos(2\pi u x) h(x)\, dx \tag{9.11}$$

or by

$$\sin(\varphi) = \frac{1}{\varepsilon} \int_{-\infty}^{\infty} \sin(2\pi u x) h(x)\, dx\,. \tag{9.12}$$

Since the impulse response is normalized to 1, i.e.,

$$\int_{-\infty}^{\infty} h(x)\, dx = 1\,, \tag{9.13}$$

the image can be expressed as

$$b(x) = 1 + \varepsilon C_0 \sin(2\pi u x) \cos(\varphi) - \varepsilon C_0 \cos(2\pi u x) \sin(\varphi) \tag{9.14}$$

or, again using the addition law (9.8), as

$$b(x) = 1 + \varepsilon C_0 \sin(2\pi u x - \varphi)\,. \tag{9.15}$$

(9.15) is thus the mapping of the object, represented by (9.2), via the transfer system.

To be able to interpret the value ε, one still has to verify that ε always ranges between 0 and 1. This is directly due to the normalization (9.13) of the impulse response, since

$$1 = \int_{-\infty}^{\infty} h(x)\, dx = \left(\int_{-\infty}^{\infty} h(x)\, dx\right)^2 \geq \left|\int_{-\infty}^{\infty} h(x) e^{-i2\pi u x}\, dx\right|^2$$

$$= \left|\int_{-\infty}^{\infty} h(x) \cos(2\pi u x)\, dx - i \int_{-\infty}^{\infty} h(x) \sin(2\pi u x)\, dx\right|^2 \tag{9.16}$$

$$= \left(\int_{-\infty}^{\infty} h(x) \cos(2\pi u x)\, dx\right)^2 + \left(\int_{-\infty}^{\infty} h(x) \sin(2\pi u x)\, dx\right)^2 = \varepsilon^2\,.$$

Thus, ε may be interpreted as the *MTF*, since this variable obviously controls the modulation of the image during the transfer through a system with the impulse response $h(x)$.

Again, looking at (9.10) reveals that ε depends on the spatial frequency such that one may write

$$\varepsilon = MTF(u) = \left| \int_{-\infty}^{\infty} h(x) e^{-i2\pi ux} \, dx \right| , \qquad (9.17)$$

where the normalization $MTF(0) = 1$ shall apply.

The relation with the contrast function (9.1) can now be interpreted as follows. The object contrast – for simplicity described for the one-dimensional case – is defined as

$$C_0 = \frac{\max\{f(x)\} - \min\{f(x)\}}{\max\{f(x)\} + \min\{f(x)\}} . \qquad (9.18)$$

In analogy, the contrast of the mapped pattern is defined as

$$C_b = \varepsilon C_0 = \frac{\max\{b(x)\} - \min\{b(x)\}}{\max\{b(x)\} + \min\{b(x)\}} . \qquad (9.19)$$

This means that the $MTF(u)$ reads

$$\varepsilon = MTF(u) = \frac{C_b}{C_0} = \frac{\left(\dfrac{\max\{b(x)\} - \min\{b(x)\}}{\max\{b(x)\} + \min\{b(x)\}} \right)}{\left(\dfrac{\max\{f(x)\} - \min\{f(x)\}}{\max\{f(x)\} + \min\{f(x)\}} \right)} \qquad (9.20)$$

and is thus the frequency-dependent ratio of the image contrast to the object contrast. Figure 9.3 shows a typical profile of the *MTF*. The profile must be interpreted such that the constant component of the image, i.e., $MTF(0)$, is always perfectly

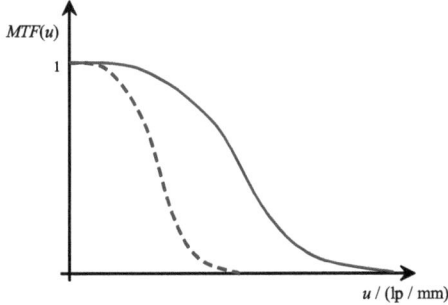

Fig. 9.3. Profile of the frequency-dependent $MTF(u)$. Steeply decreasing curves characterize poor image transmission quality (*dashed curve*)

transmitted. With increasing spatial frequency of the structures in the object to be imaged, the contrast in the image decreases. The faster the *MTF* decreases, the poorer the quality of the transmitted image becomes.

The complete *MTF* of a system consisting of several components in the imaging chain is the product of the corresponding *MTF*s of the individual components. Regarding Fig. 9.1, this means, for example,

$$MTF_{\text{system}}(u) = MTF_{\text{X-ray layer}}(u) \cdot MTF_{\text{detector layer}}(u) \cdot MTF_{\text{electronics layer}}(u) \cdot \ldots$$

$$\ldots \cdot MTF_{\text{algorithm layer}}(u) \cdot MTF_{\text{display layer}}(u)$$

$$(9.21)$$

with

$$MTF_{\text{X-ray layer}}(u) = MTF_{\text{X-ray source}}(u) \cdot MTF_{\text{collimator}}(u) \cdot \ldots$$

$$\ldots \cdot MTF_{\text{photon-matter interaction}}(u)$$

$$(9.22)$$

and further layers in analogy to the above equations. (9.21) indicates that the weakest transfer element in the chain dominates the *MTF* of the entire system and thus the imaging quality.

9.3
Modulation Transfer Function and Point Spread Function

The *MTF* can be easily set in relation to the Fourier transform of the *point spread function (PSF)*. Therefore, one has to assume that a δ spike is located in the Cartesian x–y image plane. As shown in ▸ Sect. 5.6, the imaging system generates a special image $g(x, y)$ of that point – the point image. With a δ spike as an input signal, i.e. a point being the object to be imaged, the resulting image is actually identical to the *PSF* of the imaging system.

Taking into account the conditions of linearity and shift invariance of the imaging system one can use the Fourier transform to find the expression

$$OTF(u, v) = \int_{-\infty}^{\infty} \int_{-\infty}^{\infty} h(x, y) e^{-2\pi i(xu + yv)} \, dx \, dy \,, \qquad (9.23)$$

called the *optical transfer function (OTF)* (Lehmann et al. 1997). The absolute value of the optical transfer function is then the *MTF*

$$MTF(u, v) = |OTF(u, v)| \,. \qquad (9.24)$$

In general, the *OTF* is a complex expression so that it can be divided into the absolute value and phase

$$OTF(u, v) = MTF(u, v) e^{iPTF(u,v)} \,. \qquad (9.25)$$

The *PTF* is called the *phase transfer function*. In summary, (9.23) may thus be described by

$$PSF \; \circ\!\!-\!\!\bullet \; OTF(u,v) = MTF(u,v)\,e^{iPTF(u,v)} \; . \tag{9.26}$$

If the radial frequency is defined as

$$q = \sqrt{u^2 + v^2} \tag{9.27}$$

and if it is initially assumed that $v = 0$, to use the example described in the previous section, (9.23) reads

$$\begin{aligned}
OTF(q,0) &= \int\limits_{-\infty}^{\infty} \int\limits_{-\infty}^{\infty} h(x,y)\,e^{-2\pi i x q}\,dx\,dy \\
&= \int\limits_{-\infty}^{\infty} \left(\int\limits_{-\infty}^{\infty} h(x,y)\,dy \right) e^{-2\pi i x q}\,dx \; .
\end{aligned} \tag{9.28}$$

The bracketed expression is the *line spread function* $L(x)$. One therefore writes

$$MTF(q) = |OTF(q,0)| = \left| \int\limits_{-\infty}^{\infty} L(x)\,e^{-2\pi i x q}\,dx \right| \; . \tag{9.29}$$

According to Krestel (1990) the *MTF* is thus the absolute value of the one-dimensional Fourier transform of the line spread function. In the one-dimensional example, described in the previous section, the line spread function can be identified by the *PSF*. For this purpose, one may compare the (9.29) with (9.17).

In practice, it may happen that the phase changes during the signal transfer through the system. If, for example, the image grid $b(x)$ is shifted by π with respect to the object grid $g(x)$, then a wave trough of $b(x)$ just falls onto a wave crest of $g(x)$. In doing so, dark structures will be displayed bright and bright structures dark. This effect is called *pseudo-sharpness*. Figure 9.4 illustrates this effect on the *MTF*. In the frequency range $q_1 < q < q_2$, the image is reproduced with inverted phases (Morneburg 1995).

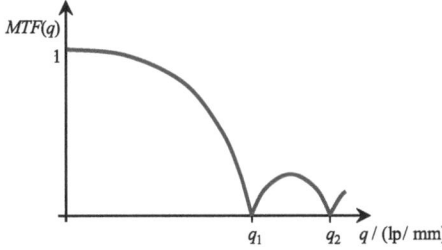

Fig. 9.4. Profile of the frequency-dependent $MTF(u)$ with pseudo-sharpness

9.4
Modulation Transfer Function in Computed Tomography

For quantitative and thus objective evaluation of the imaging quality in CT, the concept of the *MTF* has indeed proved to be useful. The corresponding approach is described by means of (9.21), where the system is decomposed into its components. The first step consists of separating the technical image generation process from the image reconstruction algorithm, thus

$$MTF_{CT}(q) = MTF_{\text{imaging hardware}}(q) \cdot MTF_{\text{imaging software}}(q) . \tag{9.30}$$

The first factor represents layers 1 through 3 of Fig. 9.1 and the second factor layer 4. Layer 5 will be not be dealt with here.

Concerning the hardware part that is related to the image generation process, two components, namely the X-ray source and the detector, play a particularly important role. In the following, the focal spot size of the X-ray tube and the spatial discretization of the detector array will be discussed in detail. These two issues are represented in Fig. 9.1 by H_1 in layer 1 and by H_6 in layer 3. As suggested by (9.22), one may also consider these two components separately and then multiply the individual *MTFs* of the components, i.e.,

$$MTF_{\text{imaging hardware}}(q) = MTF_{\text{X-ray source}}(q) \cdot MTF_{\text{sampling}}(q) . \tag{9.31}$$

Starting with the X-ray source, the size of the target point of the electrons on the anode has to be interpreted as the spatial dimension of the X-ray source. In fact, the size of the focal spot on the rotating anode depends on several parameters, such as on the angle under which the detector "sees" the spot. Furthermore, the tube power plays an important role, too. Typical focal spots in clinical applications have a diameter of approximately 1 mm (cf. ▸ Sect. 2.2.4).

Figure 9.5 shows the situation for an X-ray source with a focus diameter F. The detector, which has a distance *FDD* from the source, is initially assumed to be ideally punctiform because one actually considers the components separately. Therefore, the detector does not contribute to image deterioration in this first step.

A point located in the center of the measurement field (the distance between the focus and the center will be denoted the focus center distance, *FCD*, and the focus detector distance, *FDD*, will be assumed to be 2 *FCD*) generates on the punctiform detector – according to the intercept theorem – an inverse rectangular function of the length

$$b_{\text{F}} = F\frac{FDD - FCD}{FDD} = \frac{F}{2} . \tag{9.32}$$

This rectangular signal is the *PSF*, i.e., the impulse response of the imaging system, since it is in fact the image of a δ object in the center of the measurement field. If

$$I_{\xi'}(\xi) = I_0 \left(1 - \frac{1}{b_{\text{F}}} \text{rect}\left(\frac{\xi}{b_{\text{F}}}\right)\right) \tag{9.33}$$

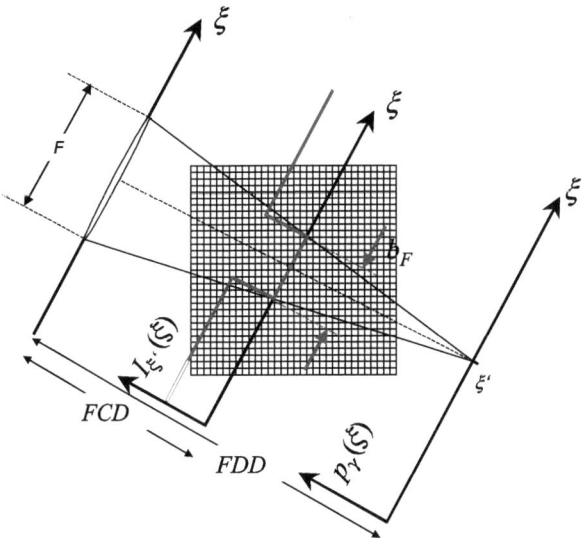

Fig. 9.5. Deconstruction of the imaging system into its components to determine the individual *MTFs* for each element. Illustration of a real X-ray source with a focal spot diameter *F* and an ideal punctiform detector

is the inverse rectangle with a length b_F – where $\tilde{b}_F$ is a dimensionless normalization length $b_F/(1\,\mathrm{m})$ – then the modulation transfer function may be easily determined on the basis of the Fourier transform of the impulse response, i.e., by

$$MTF_{\text{X-ray source}}(q) = \left| \frac{\sin(\pi \tilde{b}_F q)}{\pi \tilde{b}_F q} \right| \tag{9.34}$$

(cf. ▶ Sect. 3.10).

If the detector is considered a second component of the imaging chain, then one readily sees that a finite aperture must always be taken into account for real systems. In analogy to the first step, one also starts here by idealizing all the other components. The focus of the X-ray source will thus be assumed to have a zero diameter in this second step.

The object is scanned with a rectangular sensitivity profile here, i.e. the detector is integrated over its length $\Delta\xi$. For a δ object in the center of the measurement field this means that a signal is acquired, which – similar to (9.32) and also according to the intercept theorem – has the length

$$b_D = \Delta\xi \frac{FCD}{FDD} = \frac{\Delta\xi}{2} . \tag{9.35}$$

Figure 9.6 readily illustrates this fact. This sensitivity profile represents the required impulse response

$$I_j(\xi) = I_0 \left(1 - \frac{1}{\tilde{b}_D} \mathrm{rect}\left(\frac{\xi}{b_D} \right) \right) . \tag{9.36}$$

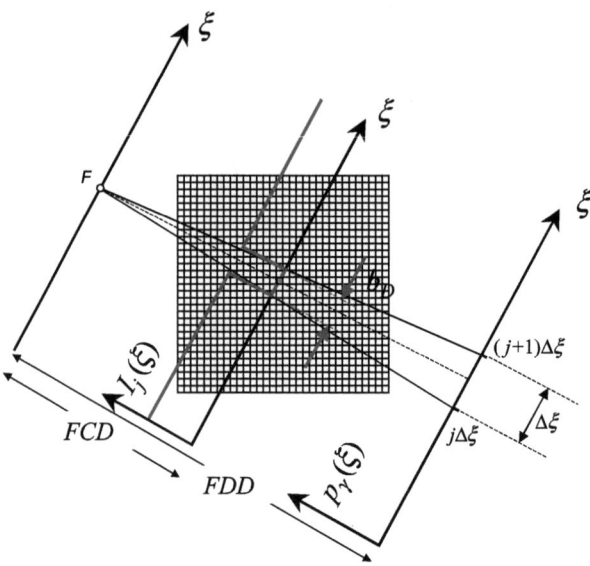

Fig. 9.6. Deconstruction of the imaging system into its components to determine the individual *MTFs*. Illustration of an ideal punctiform X-ray source and a real detector consisting of elements of size $\Delta\xi$

The *MTF* can again be determined by means of the Fourier transform from this inverse rectangle with a length b_D (where $\tilde{b}_\mathrm{D}$ is the dimensionless normalization length $b_\mathrm{D}/(1\mathrm{m})$). The result reads

$$MTF_{\text{sampling}}(q) = \left| \frac{\sin(\pi b_\mathrm{D} q)}{\pi b_\mathrm{D} q} \right| . \tag{9.37}$$

For the major hardware parts of the imaging system chain one now finds the *MTF*

$$
\begin{aligned}
MTF_{\text{imaging hardware}}(q) &= MTF_{\text{X-ray source}}(q) \cdot MTF_{\text{sampling}}(q) \\
&= \left| \frac{\sin(\pi b_\mathrm{F} q)}{\pi b_\mathrm{F} q} \right| \left| \frac{\sin(\pi b_\mathrm{D} q)}{\pi b_\mathrm{D} q} \right| .
\end{aligned}
\tag{9.38}
$$

Figure 9.7 shows the behavior of the total *MTF*. Indeed it is obvious that the weakest component of the imaging system chain impairs the final result. The first root of the *MTF* is called the *cut-off* frequency. This is the maximum frequency of spatial structures that can be resolved by the system.

The magnitudes of the *cut-off* frequencies given in Fig. 9.7 are simply estimated values that have to be adapted to the corresponding real situation. Depending on the design of the tube, which is frequently designed as a flying focal spot system, the terms in (9.38) must be computed accordingly.

In order to provide another example for the assessment of the image quality, the *MTF* of the reconstruction algorithm will be estimated according to (9.30). Regarding the filtered backprojection, the *MTF* shown in layer 3 of Fig. 9.1 consists of the

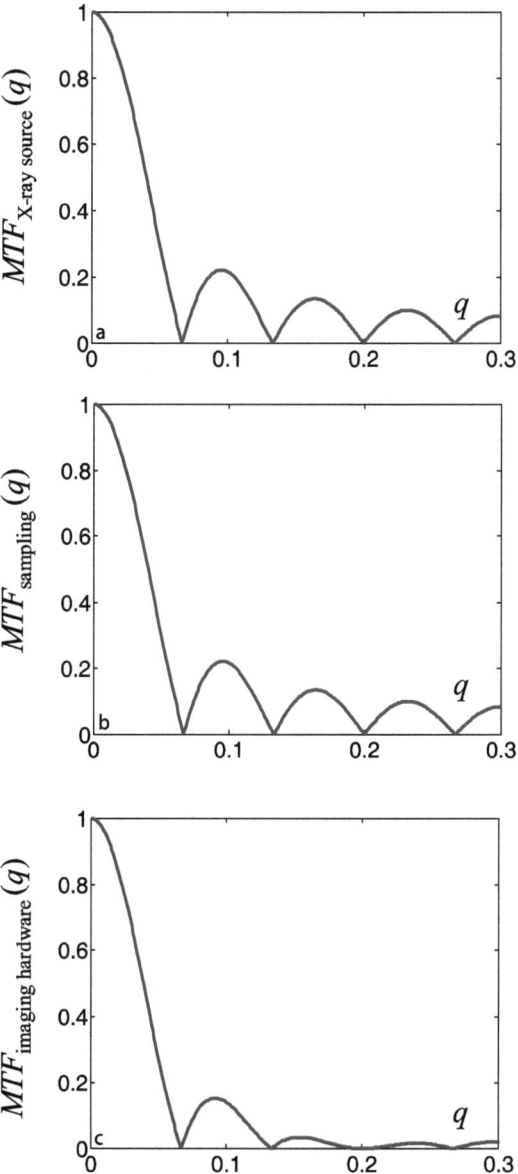

Fig. 9.7. *MTF*s for the X-ray source, for the sampling process with a finite detector aperture, and for the major hardware parts of the imaging system chain (*from top to bottom*)

following components

$$MTF_{\text{algorithm}}(q) = MTF_{\text{interpolation}}(q) \cdot MTF_{\text{filter kernel}}(q) . \qquad (9.39)$$

The first term in (9.39) describes the image transfer behavior through interpolation. The projection values must be interpolated because the projection profile is scanned with detector elements with a width $\Delta\xi$. As already described in ▸ Sect. 7.3.2, the required point of the image mask is usually not available on an exact sampling point of the projection or the filtered projection. Figure 9.8 illustrates this problem with a very coarse sampling grid, $\Delta\xi$, of the filtered projection.

In order to illustrate the *MTF* for the interpolation process, one has to discuss the interpolation itself. If a required value is not found on the grid of the available values, then this value must be interpolated from the neighboring points.

This, formally, is the convolution

$$h_\gamma(\xi') = \sum_j h_\gamma(j\Delta\xi)l(\xi' - j\Delta\xi) . \tag{9.40}$$

In this context, the ideal convolution kernel $l(\xi)$ results from the sampling theorem (cf. ▸ Sect. 4.18). As sampled signals have a periodic spectrum, an appropriate window has to be chosen within the inverse Fourier transform when the signal is reconstructed from the spectrum. Under ideal conditions, a multiplication of the spectrum with a rectangular window is used for this purpose. In the spatial domain, the rectangle is represented by a *sinc* function, i.e., the ideal convolution kernel for the interpolation process is the function

$$l_{\text{ideal}}(\xi) = \frac{\sin\left(\pi\dfrac{1}{\Delta\xi}\xi\right)}{\pi\dfrac{1}{\Delta\xi}\xi} . \tag{9.41}$$

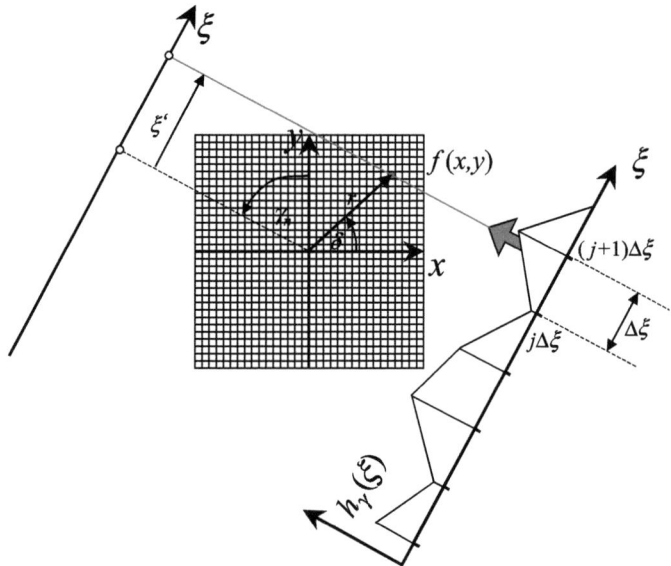

Fig. 9.8. Interpolation of the sampling values for the filtered projection signal. The grid value $(x, y)^T$ usually does not yield an integer sampling value of $h_\gamma(j\Delta\xi)$

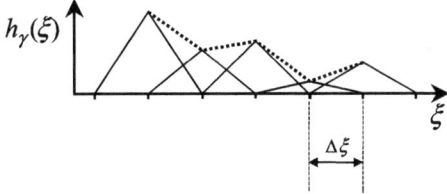

Fig. 9.9. With the linear interpolation based on the superposition of triangle functions, one can approximate sufficiently well the ideal *sinc* function for many practical cases. The triangle function is the required impulse response

Since the impact of the real spatial signal is limited to the detector length, the idealized kernel (9.41) cannot be used. Therefore, an approximation of the convolution kernel $l_{ideal}(\xi)$ must be found.

An appropriate approximation is given by the linear interpolation. As outlined in Fig. 9.9, the linear interpolation yields the triangle function[4], i.e.,

$$l_{\text{linear interpol.}}(\xi) = \begin{cases} 1 - \dfrac{|\xi|}{\Delta\xi} & \text{for } |\xi| \leq 1 \\ 0 & \text{otherwise} \end{cases} . \tag{9.42}$$

The function (9.42) is the exact impulse response of the linear interpolation, since a single sampled δ impulse is represented simply by a triangle in this case. This is the reason why the *MTF* of the linear interpolation is the absolute value of the Fourier transform of the triangle function.

For the frequency representation of the triangular convolution kernel, one may go back to the derivation in (7.32) and carry out the following changes,

$$L_{\text{lin. interp.}}(q) = \int_{-\Delta\xi}^{\Delta\xi} \left(1 - \frac{|\xi|}{\Delta\xi}\right) e^{-i2\pi\xi q} \, d\xi$$

$$= \int_{-\Delta\xi}^{\Delta\xi} e^{-i2\pi\xi q} \, d\xi + \dots$$

$$\dots - \frac{1}{\Delta\xi}\left(\int_{-\Delta\xi}^{0} (-\xi) e^{-i2\pi\xi q} \, d\xi + \int_{0}^{\Delta\xi} \xi e^{-i2\pi\xi q} \, d\xi\right) \tag{9.43}$$

$$= \int_{-\Delta\xi}^{\Delta\xi} e^{-i2\pi\xi q} \, d\xi + \dots$$

$$\dots + \frac{1}{\Delta\xi}\int_{-\Delta\xi}^{0} \xi e^{-i2\pi\xi q} \, d\xi - \frac{1}{\Delta\xi}\int_{0}^{\Delta\xi} \xi e^{-i2\pi\xi q} \, d\xi .$$

---

[4] Limited to one detector element in this definition.

Integration by parts yields

$$L_{\text{lin. interpol.}}(q) = \left[-\frac{1}{2\pi iq}e^{-2\pi iq\xi}\right]_{-\Delta\xi}^{\Delta\xi} + \frac{1}{\Delta\xi}\left[\left(\frac{-2\pi iq\xi-1}{(2\pi iq)^2}\right)e^{-2\pi iq\xi}\right]_{-\Delta\xi}^{0} + \dots$$

$$\dots - \frac{1}{\Delta\xi}\left[\left(\frac{-2\pi iq\xi-1}{(2\pi iq)^2}\right)e^{-2\pi iq\xi}\right]_{0}^{\Delta\xi}.$$

$$\text{(9.44)}$$

After applying the limits, the above reads

$$L_{\text{lin. interpol.}}(q) = -\frac{1}{2\pi iq}e^{-2\pi iq\Delta\xi} + \frac{1}{2\pi iq}e^{2\pi iq\Delta\xi} + \dots$$

$$\dots - \frac{1}{\Delta\xi}\left[+\frac{1}{(2\pi iq)^2} + \left(\frac{2\pi iq\Delta\xi-1}{(2\pi iq)^2}\right)e^{2\pi iq\Delta\xi}\right] + \dots \quad \text{(9.45)}$$

$$\dots + \frac{1}{\Delta\xi}\left[\left(\frac{2\pi iq\Delta\xi+1}{(2\pi iq)^2}\right)e^{-2\pi iq\Delta\xi} - \frac{1}{(2\pi iq)^2}\right].$$

If the bracketed expressions in (9.45) are evaluated

$$L_{\text{lin. interpol.}}(q) = -\frac{1}{2\pi iq}e^{-2\pi iq\Delta\xi} + \frac{1}{2\pi iq}e^{2\pi iq\Delta\xi} + \dots$$

$$\dots + \frac{1}{\Delta\xi}\left[-\frac{1}{(2\pi iq)^2} - \frac{2\pi iq\Delta\xi}{(2\pi iq)^2}e^{2\pi iq\Delta\xi} + \frac{1}{(2\pi iq)^2}e^{2\pi iq\Delta\xi}\right] + \dots$$

$$\dots + \frac{1}{\Delta\xi}\left[\frac{2\pi iq\Delta\xi}{(2\pi iq)^2}e^{-2\pi iq\Delta\xi} + \frac{1}{(2\pi iq)^2}e^{-2\pi iq\Delta\xi} - \frac{1}{(2\pi iq)^2}\right].$$

$$\text{(9.46)}$$

If one rearranges the terms of the second and third line of (9.46), one finds the complex expression of the *sine*

$$L_{\text{lin. interpol.}}(q) = \frac{1}{2\pi iq}e^{2\pi iq\Delta\xi} - \frac{1}{2\pi iq}e^{-2\pi iq\Delta\xi} + \dots$$

$$\dots + \frac{1}{\Delta\xi}\left[\frac{1}{(2\pi iq)^2}e^{2\pi iq\Delta\xi} - \frac{2}{(2\pi iq)^2} + \frac{1}{(2\pi iq)^2}e^{-2\pi iq\Delta\xi}\right] + \dots$$

$$\dots + \frac{1}{\Delta\xi}\left[\frac{\Delta\xi}{2\pi iq}e^{-2\pi iq\Delta\xi} - \frac{\Delta\xi}{2\pi iq}e^{2\pi iq\Delta\xi}\right]$$

$$\text{(9.47)}$$

so that

$$L_{\text{lin. interpol.}}(q) = \frac{1}{\pi q}\left(\frac{e^{2\pi i q \Delta\xi} - e^{-2\pi i q \Delta\xi}}{2i}\right) + \dots$$

$$\dots + \frac{1}{\Delta\xi}\left(\frac{1}{\pi q}\right)^2\left(\frac{e^{\pi i q \Delta\xi} - e^{-\pi i q \Delta\xi}}{2i}\right)^2 - \frac{1}{\pi q}\left(\frac{e^{2\pi i q \Delta\xi} - e^{-2\pi i q \Delta\xi}}{2i}\right)$$

$$= \Delta\xi\frac{\sin(2\pi q\Delta\xi)}{2\pi q\Delta\xi} + \Delta\xi\left(\frac{\sin(\pi q\Delta\xi)}{\pi q\Delta\xi}\right)^2 - 2\Delta\xi\frac{\sin(2\pi q\Delta\xi)}{2\pi q\Delta\xi}$$

$$= \Delta\xi\left(\frac{\sin(\pi q\Delta\xi)}{\pi q\Delta\xi}\right)^2 .$$

$$(9.48)$$

This is in fact the *MTF* of the linear interpolation

$$MTF_{\text{lin. interpol.}}(q) = \left|\frac{\sin(\pi\Delta\xi q)}{\pi\Delta\xi q}\right|^2 . \qquad (9.49)$$

The second term in (9.39) consists of

$$MTF_{\text{filter kernel}}(q) = \frac{|G(q)|}{|q|}, \qquad (9.50)$$

where $G(q)$ is the Fourier transform of the kernel $g(\xi)$. Corresponding examples have already been listed in ▸ Sect. 7.2.

Deviating from the ideal reconstruction filter $|q|$ of the filtered backprojection, special band limitations were introduced in ▸ Sect. 7.2.1, which – on the one hand – are necessary because the spectrum of the sampled projection signal is a periodic spectrum. On the other hand, special window functions are selected for band limitation to attenuate higher band noise. With respect to the windowing, it is obvious that the exact window function will influence the resolution of the image to be reconstructed, since the higher bands contain spatial structures of high frequency that are no longer resolvable. This means that windowing is actually identical to shifting the *cut-off* frequency to lower values.

Regarding (9.50), it immediately becomes apparent that one obviously has to normalize the *MTF* of the filter kernel with the ramp filter $|q|$. This may be explained by the fact that $G(q)$ already includes the idealized ramp. However, the idealized (infinitely long and continuous) ramp in fact just produces an exact reconstruction of the object with a theoretical *MTF* for which $MTF(q) \equiv 1$ holds. Deviations from this may only be achieved due to windowing since

$$G(q) = |q|W(q) . \qquad (9.51)$$

Therefore, $W(q)$ denotes the band-limiting window function. Strictly speaking, (9.50) should more clearly arranged and actually read

$$MTF_{\text{filter kernel}}(q) = \frac{|G(q)|}{|q|} = \frac{\||q|\,W(q)\|}{|q|} = \frac{|q|\,|W(q)|}{|q|} = |W(q)| \, . \tag{9.52}$$

If the frequently used standard *Shepp–Logan* window is used – (cf. ▸ Sect. 7.2, (7.17)) one obtains

$$MTF_{\text{filter kernel}}(q) = \left| \frac{\sin(\pi \Delta \xi q)}{\pi \Delta \xi q} \right| \, . \tag{9.53}$$

Finally, one finds the special case of the *MTF* for the reconstruction algorithm with *Shepp–Logan* windowing and linear interpolation

$$MTF_{\text{algorithm}}(q) = MTF_{\text{interpolation}}(q) \cdot MTF_{\text{filter kernel}}(q)$$
$$= \left| \frac{\sin(\pi q \Delta \xi)}{\pi q \Delta \xi} \right| \cdot \left| \frac{\sin(\pi q \Delta \xi)}{\pi q \Delta \xi} \right|^2 = \left| \frac{\sin(\pi q \Delta \xi)}{\pi q \Delta \xi} \right|^3 \, . \tag{9.54}$$

At this point, the major terms of the overall *MTF* of CT imaging can be summarized. According to (9.30), one thus finds

$$MTF_{CT}(q) = MTF_{\text{imaging hardware}}(q) \cdot MTF_{\text{imaging software}}(q)$$
$$= MTF_{\text{X-ray source}}(q) \cdot MTF_{\text{sampling}}(q) \cdot \ldots \tag{9.55}$$
$$\ldots \cdot MTF_{\text{interpolation}}(q) \cdot MTF_{\text{filter kernel}}(q)$$

i.e.,

$$MTF_{CT}(q) = \left| \frac{\sin(\pi b_F q)}{\pi b_F q} \right| \left| \frac{\sin(\pi b_D q)}{\pi b_D q} \right| \left| \frac{\sin(\pi q \Delta \xi)}{\pi q \Delta \xi} \right| \left| \frac{\sin(\pi q \Delta \xi)}{\pi q \Delta \xi} \right|^2 \, . \tag{9.56}$$

As illustrated in Fig. 9.7, for the imaging hardware layer, the first root of $MTF(q)$ determines the total transfer behavior of the system. In summary, one may say that any one of focus spot size, detector aperture, reconstruction kernel, and interpolation strategy can impair the image resolution on its own. Inversely, all components must contribute positively to a good imaging system. In addition to the above parameters, there are others that influence the image quality that are not part of the *front-end* scanner unit. For example, the viewing station – which is the so-called *back-end* unit – also influences the image quality.

Due to technical progress in the field of detector technology with the latest generation of two-dimensional flat-panel detectors with sensor sizes of less than 200 μm, which have already been tested as prototypes, in combination with the continuous improvement of X-ray tube technology, the resolution limits of clinical CT scanners are continuously decreased. Figure 9.10 shows the performance of these prototypes. The reconstructed images of the mouse have almost micro-CT-like image quality. In fact, the clinical prototypes of CT scanners with flat-panel detectors are getting closer to a resolution to the order of micro-CT (Pfoh 2002).

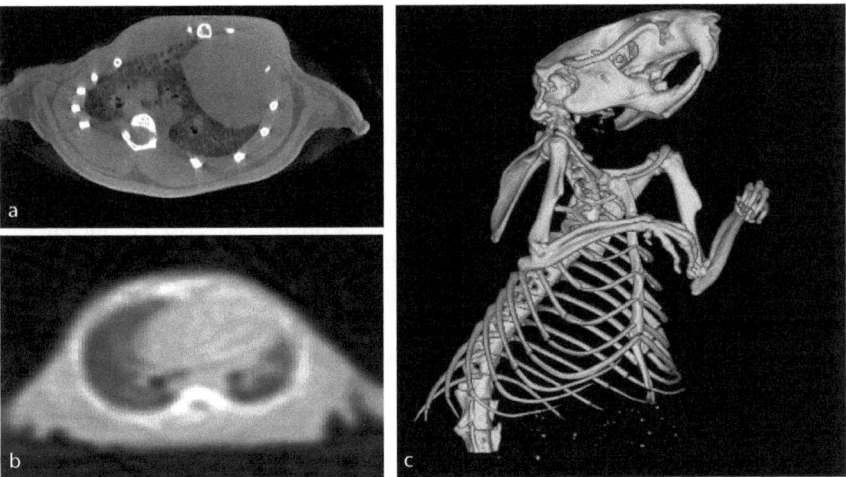

Fig. 9.10. Illustration of the performance of modern cone-beam CT scanners with two-dimensional flat-panel detectors. **a** and **b** compare axial slices of a mouse thorax. **a** shows the result of a cone-beam system with the resolution of 0.05 mm and **b** the result of a clinical CT system with a slice thickness of 1.25 mm. **c** shows the variety of details of a three-dimensional reconstruction of the mouse with an isotropic resolution of 0.14 mm (courtesy of General Electric Medical Systems: Pfoh [2002])

9.5
SNR, DQE, and ROC

Noise impairs the detectability of signals. It therefore seems to be useful to define the quotient between the signal level and the fluctuation range as a quality characteristic. In ▸ Sect. 2.6.4 it has been already mentioned that the so-called *signal-to-noise ratio* (*SNR*)

$$SNR = \frac{\text{signal level}}{\text{noise level}} = \frac{\mu}{\sigma} \tag{9.57}$$

is an important parameter in evaluating image quality. In the above equation, μ and σ denote the mean value and the standard deviation respectively. Poisson-distributed X-ray photons obey

$$SNR = \frac{\langle n \rangle}{\sqrt{\langle n \rangle}} = \sqrt{\langle n \rangle} \tag{9.58}$$

i.e., the signal-to-noise ratio is increased with the root of the mean value[5] of the quantum number. In principle, the image quality might thus be arbitrarily improved. However, the number of X-ray quanta is proportional to the dose and can thus not be arbitrarily increased.

---

[5] The mean value $\langle n \rangle$ is the estimation of the expectation value n^*.

We therefore have to take into account that

$$SNR \propto \sqrt{\text{dose}} . \qquad (9.59)$$

At small dose values the quantum noise is thus the dominant factor influencing the image quality (Kamm 1998). It is well known that an SNR of at least 5 is necessary to reliably detect a detail in a homogeneous noise background (Neitzel 1998).

Not only the inherent fluctuation of the X-ray quanta, but also the efficiency of the detectors, i.e., the degree according to which X-ray is converted into visual light and finally to electrical signals, plays an important role with respect to image-quality assessment. This factor is quantified by the so-called *detective quantum efficiency* (*DQE*), which is related to the signal-to-noise ratio via

$$DQE = \frac{\left(SNR_{\text{detector output}}\right)^2}{\left(SNR_{\text{detector input}}\right)^2} . \qquad (9.60)$$

Taking into account the detective quantum efficiency, it is possible to evaluate to what extent the detector further deteriorates the signals already impaired by the quantum noise. According to (9.58) one obviously must take into account that

$$DQE = \frac{\langle n \rangle_{\text{detector output}}}{\langle n \rangle_{\text{detector input}}}, \qquad (9.61)$$

so that the detective quantum efficiency only reaches the maximum value 1 with an ideal detector.

If one wants to indicate the quanta that have actually been detected, one has to take into account the *noise equivalent quanta* (*NEQ*)

$$NEQ = DQE\langle n \rangle_{\text{detector input}} = \langle n \rangle_{\text{detector output}} . \qquad (9.62)$$

The *MTFs* defined in the previous sections and the detective quantum efficiency (*DQE*) are evaluation parameters, which cannot be optimized simultaneously and independently (Dössel 2001).

Regardless of these technical parameters, in practice it is, frankly speaking, only interesting whether a physician is able to recognize a diagnostically relevant structure or not. The answer to this question is of course rather subjective, since physiological perception processes are involved, which are naturally subject to a certain variability. All experiments must therefore be precisely planned to measure the so-called *receiver operating characteristic* (*ROC*).

With this method, the test structures are submitted to a group of skilled observers who have to make a decision. The imaging quality is then evaluated by means of a statistical assessment of the test objects, which have been correctly or incorrectly detected (Morneburg 1995). In this context, the terms "sensitivity" and "specificity" play a key role. The first one describes the number of correct positive decisions compared with the total number of positive cases, while the second one is the number of correct negative decisions compared with the total number of negative cases. The sensitivity is proportional to the success probability p and the specificity is proportional to $1 - q$, where q is the so-called *false alarm probability*. The *ROC* curve plots p versus q to evaluate the detection performance.

9.6
2D Artifacts

Artifacts are image errors that may emerge due to a variety of reasons. Artifacts can originate from a simplification of the reconstruction method – to date usually the filtered backprojection – which assumes monochromatic radiation or continuous representation of the projection signal. Artifacts may also stem from the use of special sampling technologies and detector arrangements, or simply from defective detector elements. Corrective actions may only be taken if the causes of such artifacts are known. Such counter-measures are in fact very important, since the filtered backprojection has the disadvantage that artifacts are projected back over the entire image so that the overall diagnostic value of the image is reduced or completely destroyed.

9.6.1
Partial Volume Artifacts

If a detail of an object consists of a sharply contrasted boundary, the limited resolution of a detector system of course becomes particularly noticeable. The boundary will usually not be located directly at the edge from one detector element to another. Therefore, the intensity of X-ray on the corresponding element that has to image this boundary, will be linearly averaged[6] over the detector width, $\Delta\xi$. Due to this averaging step the object is blurred.

In the following, problems, which are due to the partial overlap of object structures on the detector elements, will be illustrated. This is done by taking an axial abdomen slice with a strong absorbing vertebra as an example. For this purpose, Fig. 9.11a shows schematically that blurring of the details at the object boundary from the vertebral body to the surrounding tissue is more and more reduced the smaller the detector width, $\Delta\xi$.

A change in the projection signal due to the object structure within the detector width $\Delta\xi$ is of course problematic for the reconstruction process with regard to a violation of the sampling theorem. However, the corresponding reconstruction problem in the field of CT, which is referred to as "partial volume artifact," is due to the logarithm of the intensity values, i.e.,

$$I(\xi) = I_0(\xi)\,e^{-\int \mu(\xi,\eta)\,d\eta} \Leftrightarrow p(\xi) = \int \mu(\xi,\eta)\,d\eta \,. \tag{9.63}$$

---

[6] In this context, the detector element is assumed to have a rectangular sensitivity profile.

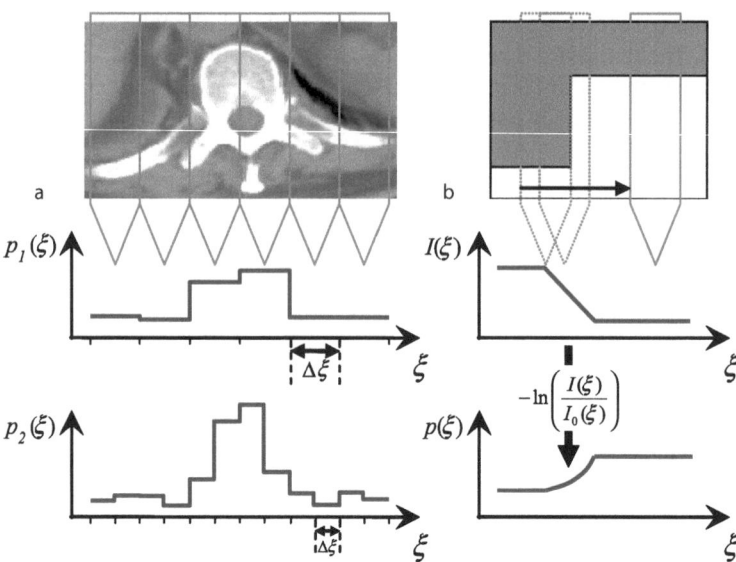

Fig. 9.11. a The detectors of the array have a finite width, $\Delta\xi$. Sharp anatomical object boundaries that are represented here as high-contrast, gray-value edges located within a beam with the width $\Delta\xi$ are thus not imaged faultlessly. The smaller the detector width, the more details are revealed in the projection signal. In **b** an artificial edge image schematically illustrates that the partial volume error is a non-linear effect

Figure 9.11b shows graphs representing (9.63). The projection values, $p(\xi)$, which are the basis of all the reconstruction methods described in the previous chapters, are the result of the negative logarithm of the ratio between the intensities in front of and behind the attenuating object. As the reconstruction process assumes linearity of the projection integral, sharp object boundaries encounter the problem that

$$\ln\left(\alpha I(\xi_1) + (1-\alpha)I(\xi_2)\right) \neq \ln\left(\alpha I(\xi_1)\right) + \ln\left((1-\alpha)I(\xi_2)\right) . \tag{9.64}$$

(9.64) states that the logarithm is a non-linear function. Expressing (9.64) in words, the logarithm of the linearly averaged intensities $I(\xi_1)$ and $I(\xi_2)$ does not correspond to the sum of the logarithms of the individual partial intensities. In this context, the factor $\alpha \in (0,1)$ describes to what extent the projection of the high-contrast object boundary effectively covers the corresponding detector element with respect to the detector width $\Delta\xi$. As a result, one does not only obtain a smoothed edge, as illustrated in Fig. 9.11b, but also a mean attenuation, which does not match the mean intensities. Consequently, the estimated effective attenuation coefficient is too small (Morneburg 1995).

Due to the superposition of filtered backprojections from all directions, this inconsistency leads to artifacts within the reconstructed image, which are visible as streaks from the origin of the inconsistency along the backprojection path. Partial volume artifacts are thus observed, for instance, as ghost lines that extend particularly straight object boundaries. This is due to the fact that the backprojections from the other directions are not able to consistently correct an erroneously detected value, which has been projected back over the entire image.

9.6.2
Beam-Hardening Artifacts

X-ray produced in electron-impact sources, where fast electrons are entering a solid metal anode, cannot be monoenergetic or monochromatic. In ▸ Sect. 2.2.2 the different spectral X-ray components, such as the continuous spectrum of the bremsstrahlung (fast electrons are decelerated by the Coulomb fields of the atoms in the anode material) and the characteristic emission lines (originating from the direct interaction of the fast electrons with the inner shell electrons of the anode material), have already been discussed in detail. The latter effect represents a fingerprint of the anode material.

Colloquially, X-ray is known to have the property of effectively penetrating material. If this phenomenon is considered from a physical point of view in detail, it can be seen that the radiation attenuation does not only depend on the path length, but is also a function of the specific, wavelength-dependent interaction between X-ray and the material concerned. The physical photon–matter processes involved in this interaction have already been discussed in detail in ▸ Sect. 2.3.2.

In order to describe the basic mathematical reconstruction procedures (as already indicated in ▸ Sect. 2.3.1), the attenuation is modeled in a first step by Lambert–Beer's law

$$I(s) = I(0)\, e^{-\int_0^s \mu(\eta)\,d\eta} \tag{9.65}$$

for the intensity of X-ray having passed through a material along the path s. A crucial issue in this context is the fact that attenuation coefficients, μ, which only depend on the spatial coordinate, η, are summed along the X-ray path. This indeed is a simplification. If, in addition, the energy dependence of the attenuation coefficient $\mu = \mu(\xi, \eta, E)$ is taken into account, one finds

$$I(s) = \int_0^{E_{max}} I_0(E)\, e^{-\int_0^s \mu(\xi,\eta,E)\,d\eta}\, dE \,, \tag{9.66}$$

where $I_0(E)$ is the X-ray source spectrum. The reconstruction is in this case – similar to the partial volume artifact described in the previous section – impaired by the non-linearities, which occur here. That is, the simple negative logarithmic ratio between the incident and the output intensity does not sufficiently take into account the energy dependence of the attenuation. If one describes the incident intensity with

$$I_0 = \int_0^{E_{max}} I_0(E)\, dE \,, \tag{9.67}$$

the projection integral must be extended to

$$p(\xi) = -\ln\left(\frac{1}{I_0}\int_0^{E_{max}} I_0(E)\, e^{-\int_0^s \mu(\xi,\eta,E)\,d\eta}\, dE\right). \tag{9.68}$$

The so-called beam-hardening artifact is caused by the non-linear relation between the attenuation values, μ, and the measured values of the projection, p. If an X-ray beam with a broad-band energy spectrum passes through an object, the spectrum changes along the path. This is due to the fact that different bands of the frequency spectrum are differently attenuated, depending on the specific attenuation coefficients $\mu = \mu(\xi,\eta,E)$ of the material being radiographed. In general, the low-energy, i.e., soft, X-ray beams are more strongly absorbed than the high-energy, hard X-ray beams. This is the reason why this effect is named hardening of the X-ray spectrum and the corresponding image error is named beam-hardening artifact.

In Fig. 9.12, the mass attenuation coefficients are given for water, bone, amalgam, and gold. If the mass attenuation coefficients are considered as a function of the incident radiation energy, a complicated structure is revealed. Water and bone, for instance, show quite well the λ^3 dependence of the photoelectric absorption; however, amalgam and gold have their characteristic absorption edges within the

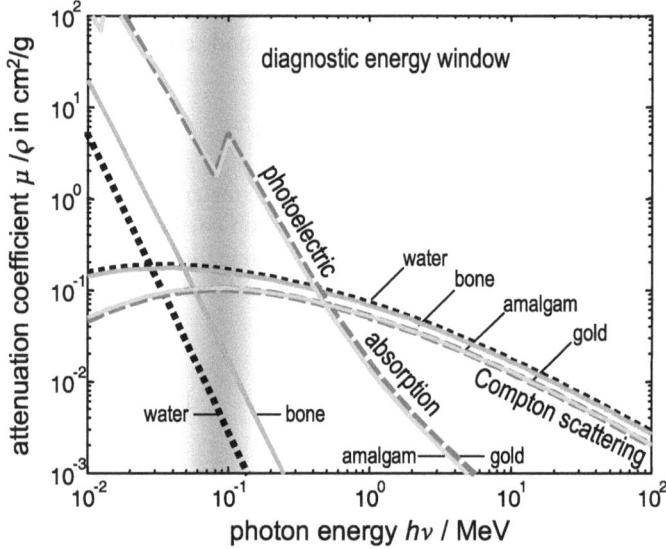

Fig. 9.12. Mass attenuation coefficient, μ/ρ, versus incident photon energy for water, bone, amalgam, and gold. For the diagnostic energy window of CT, $E = [50\,keV\text{–}140\,keV]$, photoelectric absorption is dominant for amalgam and gold. However, Compton scattering is dominant for water. For bone, Compton scattering and photoelectric absorption show comparable contributions to the total attenuation (compiled with data from the web database XCOM [Berger et al. 2004])

diagnostic window. Additionally, the contribution of the Compton effect in relation to the photoelectric absorption is different for each material. The variety of materials inside the human body makes it therefore impracticable to simultaneously calibrate and compensate for the beam hardening caused by any material.

Similar to the partial volume artifact, the beam-hardening artifact can be explained by the inconsistency of the individual projection values from different directions, which cannot complement each other correctly within the filtered backprojection method. The individual detectors in practice only measure the integral intensity over all wavelengths, i.e., they cannot differentiate distinct energies. Figure 9.13a illustrates this situation schematically. A section of an axial abdomen image with a vertebral body is to be reconstructed. Having a look at the image area above the vertebral body it can be seen that the soft, low-energy X-ray quanta of the horizontal, polychromatic X-ray beam are attenuated by the tissue, while the hard, high-energy X-ray quanta pass through the tissue almost unattenuated. The incident and output spectra are shown schematically.

Along the vertical line the X-ray beam, which is required for the reconstruction of the marked field of view, passes through the vertebral body, which – contrary to soft tissue – attenuates high-energy radiation significantly. The schematic output spectrum illustrates that here – contrary to the horizontal X-ray beam – the curve of the intensity versus the energy is considerably lowered for all wavelengths (compare the K_β lines of the spectra in Fig. 9.13a).

Overall, the mean energy of the radiation, i.e., the first moment of the distribution, is shifted to higher energies due to the beam-hardening effect. The mean intensities measured for the individual projections in each detector element are therefore not consistent. This is the reason why streaks arise in the filtered backprojection, which often spread along the backprojection directions over the entire image. Certain anatomic regions are particularly sensitive to these beam-hardening effects. These image errors are for instance disturbing in the area of the cerebellum, where both beam-hardening and partial volume artifacts occur.

One corrective method applied in virtually all CT scanners consists of filtering the soft radiation next to the source, i.e., before the radiation reaches the tissue. This may, for example, be done with thin aluminum or copper foils. In Fig. 2.11 the influence of an aluminum and a copper filter on the X-ray spectrum can be seen.

It should be noted that for a single material with known properties, it is of course possible to correct for beam hardening computationally. In general, for simplicity, the material properties of water are introduced into this computational correction step, since the properties of the soft tissues of importance in this context just differ slightly from those of water.

Figure 9.14 illustrates the so-called *cupping artifact* induced by the non-linear characteristics of beam hardening. Due to the fact that the actual measured projection value is always below the "true" (expected) value (cf. Fig. 9.13b), a simple calibration method can be carried out by elevating any measured projection to its ideal curve in a pre-processing step. Since this calibration step is usually performed with a water phantom, this pre-processing is named *water correction*.

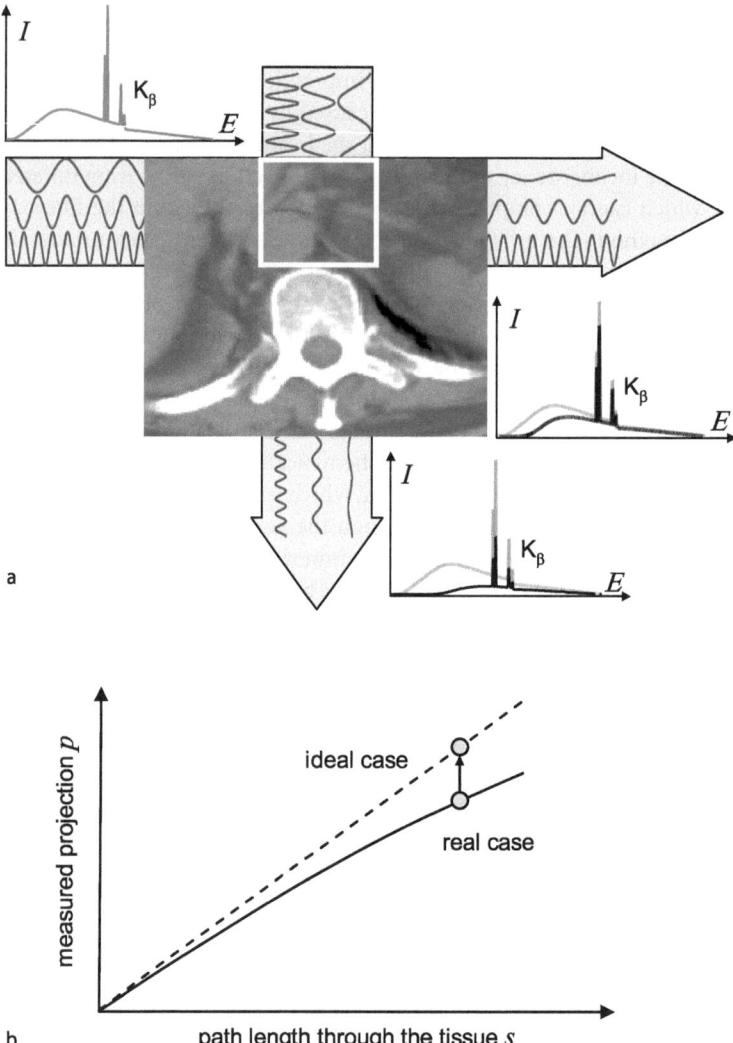

Fig. 9.13. The attenuation of X-ray passing through the tissue depends on the length of the path and the energy-specific attenuation coefficient $\mu = \mu(\xi, \eta, E)$. **a** The dependency of the attenuation coefficient upon the radiation energy results in inconsistencies during the reconstruction process. The different energies are illustrated here by the different wavelengths within the radiation field. The marked anatomical area located above the vertebra shall be reconstructed. The logarithmic ratio of the intensities is not the same for the different energies within one projection direction. Since the detector integrates all incoming X-ray quanta in the same manner, the change in the spectral shapes results in effective projection values that do not match each other and thus cause image errors called beam-hardening artifacts. **b** The projection values actually measured, p, should ideally be proportional to the path length s of the beam through the object. However, in the real situation p is always smaller than the ideal value. The real curve can be corrected in a calibration step indicated by the *arrow*

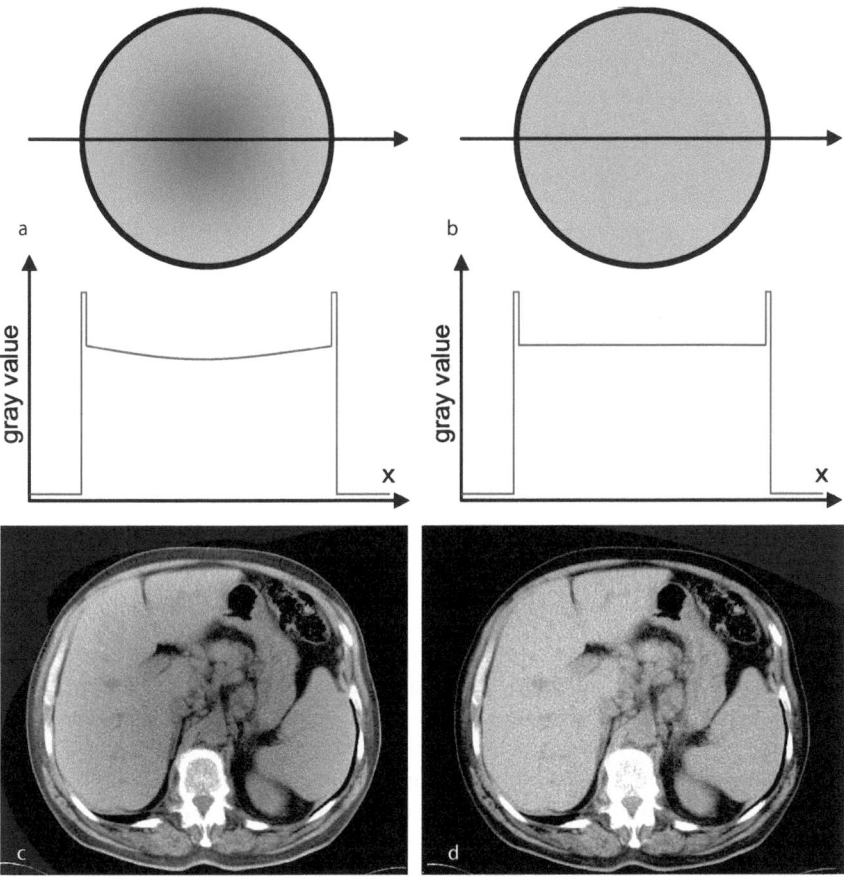

Fig. 9.14. a, b Cupping artifact due to beam hardening and the result of water correction for a skull phantom and **c, d** for an axial abdominal tomogram (**c, d**: Courtesy of Siemens Medical Solutions)

More precisely, Fig. 9.14a illustrates the consequences of an uncalibrated reconstruction. The cupping artifact is schematically reproduced for a simple head phantom, which consists of a high absorbing skull and a water-like X-ray absorbing soft tissue. The center gray-value profile shows a sagged line, which may indicate an area being affected by pathology. After water correction has been carried out, the center gray-value profile should be a straight horizontal line as illustrated in Fig. 9.14b. In Fig. 9.14c and d the real reconstruction of an abdominal slice without and with water correction respectively can be seen.

However, this a priori correction cannot be performed for any potentially unknown object with arbitrary attenuation spectra, which differ considerably from those of water. Even if water correction has been carried out, beam-hardening artifacts might therefore occur in the area of thicker bony structures. This is illustrated

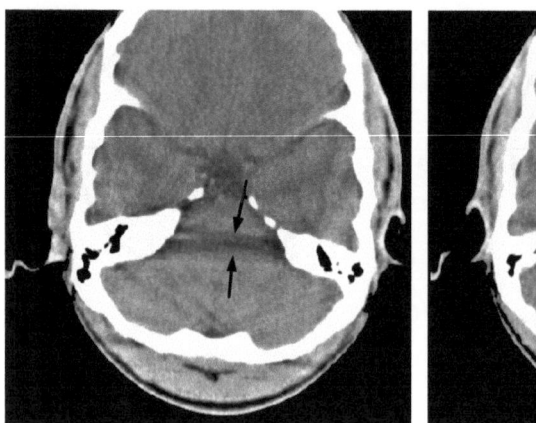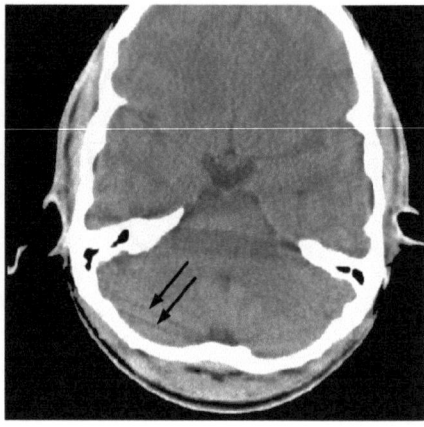

Fig. 9.15. Streak artifacts in the base of the skull caused by beam hardening. The *arrows* mark the most peculiar image errors, which are due to the change in the X-ray spectrum while the beams pass through thicker bony structures. The shadow between the petrous bones is called the Hounsfield bar. Since filtered backprojection was used as a reconstruction algorithm, image errors spread from the structure, causing this error across the entire image

in Fig. 9.15, which provides two CT reconstructions of the skull base, where so-called Hounsfield bars, sometimes also referred to as *interpetrous lucency artifacts*, emerge between the petrous bones. These image errors are marked with arrows in Fig. 9.15.

However, it would be possible to reconstruct images without beam-hardening artifacts with any wavelength or energy of radiation. Due to the specific attenuation property of the different tissues, a slightly different image would be produced for each individual energy. Due to this variety of information provided by images obtained with the above method, their combination might be useful in diagnostics. However, CT scanners do not produce monochromatic radiation. Nevertheless, the energy-dependent attenuation properties of the tissue are used in practice by scanning the patient with different accelerating voltages[7]. With these images it is for instance possible to estimate the mineral content of bones and to compute images that look as acquired with monochromatic radiation. This allows an elegant beam-hardening correction. In the *dual-energy* approach, the energy dependence of the attenuation coefficient must be modeled as

$$\mu(\xi, \eta, E) = k_{\text{absorption}}(\xi, \eta) \cdot \alpha_{\text{absorption}}(E) + k_{\text{scattering}}(\xi, \eta) \cdot \mu_{\text{scattering}}(E) .$$
$$(9.69)$$

Here, the spatial and energy dependence of the attenuation is factorized for the two main photon–matter interactions in the diagnostic energy window, i.e., photoelectric absorption and Compton scattering (cf. ▸ Sect. 2.3.2). The energy dependence

---

[7] In conventional fluoroscopy, this method is also known as the *dual-energy* method.

of the photoelectric absorption is given by (2.32)

$$\alpha_{\text{absorption}} \propto \frac{1}{(h\nu)^3} \, . \tag{9.70}$$

The energy dependence of the Compton scattering is given by the Klein–Nishina equation (2.36), i.e.,

$$\mu_{\text{Compton}} = n \cdot 2\pi r_e^2 \left[\left(\frac{1 + \mathcal{E}}{\mathcal{E}^2} \right) \left(2 \frac{(1 + \mathcal{E})}{1 + 2\mathcal{E}^2} - \frac{\ln(1 + 2\mathcal{E})}{\mathcal{E}} \right) + \frac{\ln(1 + 2\mathcal{E})}{2\mathcal{E}} - \frac{1 + 3\mathcal{E}}{(1 + 2\mathcal{E})^2} \right], \tag{9.71}$$

where $\mathcal{E} = h\nu/(m_e c^2)$ is the reduced energy of the incoming photon and n is the number of target atoms per unit volume. If (9.69) is substituted into (9.68), one obtains

$$p_\gamma(\xi) =$$

$$= -\ln \left(\frac{1}{I_0} \int_0^{E_{\max}} I_0(E) e^{-\int_0^s (k_{\text{absorption}}(\xi,\eta) \cdot \alpha_{\text{absorption}}(E) + k_{\text{scattering}}(\xi,\eta) \cdot \mu_{\text{scattering}}(E)) \, d\eta} \, dE \right)$$

$$= -\ln \left(\frac{1}{I_0} \int_0^{E_{\max}} I_0(E) e^{-\alpha_{\text{absorption}}(E) \int_0^s k_{\text{absorption}}(\xi,\eta) \, d\eta - \mu_{\text{scattering}}(E) \int_0^s k_{\text{scattering}}(\xi,\eta) \, d\eta} \, dE \right)$$

$$= -\ln \left(\frac{1}{I_0} \int_0^{E_{\max}} I_0(E) e^{-\alpha_{\text{absorption}}(E) K_{\text{absorption}}(\xi) - \mu_{\text{scattering}}(E) K_{\text{scattering}}(\xi)} \, dE \right) \tag{9.72}$$

where

$$K_{\text{absorption}}(\xi) = \int_0^s k_{\text{absorption}}(\xi, \eta) \, d\eta \tag{9.73}$$

and

$$K_{\text{scattering}}(\xi) = \int_0^s k_{\text{scattering}}(\xi, \eta) \, d\eta \, . \tag{9.74}$$

The integrals $K_{\text{absorption}}(\xi)$ and $K_{\text{scattering}}(\xi)$ can now be approximated by dual-energy measurements. That means, for instance, two different scans may be carried out at two different X-ray tube voltages[8]. This leads to

$$p_{1,y}(\xi) = -\ln\left(\frac{1}{I_1}\int_0^{E_{max}} I_1(E)\,e^{-\alpha_{absorption}(E)K_{absorption}(\xi)-\mu_{scattering}(E)K_{scattering}(\xi)}\,dE\right)$$

(9.75)

and

$$p_{2,y}(\xi) = -\ln\left(\frac{1}{I_2}\int_0^{E_{max}} I_2(E)\,e^{-\alpha_{absorption}(E)K_{absorption}(\xi)-\mu_{scattering}(E)K_{scattering}(\xi)}\,dE\right),$$

(9.76)

where

$$I_1 = \int_0^{E_{max}} I_1(E)\,dE \text{ and } I_2 = \int_0^{E_{max}} I_2(E)\,dE. \tag{9.77}$$

Obviously, two projection integrals, $p_{1,y}(\xi)$ and $p_{2,y}(\xi)$, are measured for two unknown variables $K_{absorption}(\xi)$ and $K_{scattering}(\xi)$. In this way, the integrals $K_{absorption}$ and $K_{scattering}$ can be estimated and, further, the distributions $k_{absorption}(\xi,\eta)$ and $k_{scattering}(\xi,\eta)$ can be obtained by inverting the integrals (9.73) and (9.74) using the reconstruction methods from ▸ Chap. 5. Knowing the distributions $k_{absorption}(\xi,\eta)$ and $k_{scattering}(\xi,\eta)$ in (9.69), artifact-free images can be reconstructed in an energy range that is correctly modeled with (9.70) and (9.71).

9.6.3
Motion Artifacts

So far it has been assumed that the morphology in the slice to be reconstructed is not changed during data acquisition. But if one also has to take into account the temporal variation of the attenuation coefficient, $\mu = \mu(\xi,\eta,E,t)$, one faces the problem of image reconstruction with a changing data basis. That is, the data measured during the rotation are inconsistent.

One example of the occurrence of system-inherent or deliberately accepted inconsistencies is the spiral CT method, which has already been described in ▸ Sect. 8.3. Due to the continuous table feed, none of the measured projection profiles consistently matches the other ones. However, as one knows the speed of the change, one is able to complete the spiral CT data by interpolation – within certain limits. These limits refer to the *pitch* factor. How image reconstruction is impaired if the selected *pitch* factor is too high will be described later in this chapter.

---

[8] Split detector and sandwich principles or alternating X-ray tube voltages during rotation of the sampling unit have been proposed as well (Kak and Slaney 1988).

Another example of deliberately induced changes is the administration of contrast medium. If the absorption of contrast medium is to be observed dynamically, then the spatial distribution of the attenuation coefficient with respect to time also changes – however, this is exactly the physiological information one is interested in. But with regard to the imaging quality, there are also unwanted motions such as colon peristalsis, respiration, and the beating heart, which result in inconsistent projection data.

Figure 9.16 shows the simulation of the administration of contrast medium. A Perspex bar is introduced during a 360° rotation into the central hole of a Perspex cylindrical phantom with five holes. In order to ensure that this change occurs in the projection angle interval ranging from 180° to 225°, a robot is used to introduce the Perspex bar into the plane that is to be reconstructed. The slice thickness is restricted to 3 mm by means of collimators. Figure 9.16b shows the sinogram with the detector data given as a function of the projection angle. The different holes of the phantom can be readily distinguished by the corresponding sine curves in the diagram – they are numbered accordingly. The correspondence is illustrated in Fig. 9.16c, which provides the reconstructed image free from any motion artifacts. In order to be able to compare the different reconstructions, Fig. 9.16e shows the image with the completely inserted central bar. This image does not show any motion artifacts either. The reconstruction corresponding to the sinogram of Fig. 9.16b is shown in Fig. 9.16d. The artifacts, spread over large image areas in a way that is characteristic of the backprojection method, are readily visible.

Recently, a data-driven method has been proposed that learns the motion parameters from the raw data (Schumacher and Fischer 2007). It combines the reconstruction and the motion correction in a single step. Such an augmented reconstruction can be formulated as a regularized optimization problem

$$(\mathbf{f}, \omega)^*_{\min} = \min_{\mathbf{f} \in \Omega} \left\{ \frac{1}{2} \sum_{i=1}^{K} \|\mathbf{A}_i \mathbf{T}(\omega_i)\mathbf{f} - \mathbf{p}_i\|^2 + \alpha R(\mathbf{f}) \right\} . \tag{9.78}$$

The number K denotes the number of projection sets where the object is at rest. The system matrix $\mathbf{A}_i$, which models the projection geometry and the projection $\mathbf{p}_i$, is partitioned into K sets. Compared with the equations in ▸ Sect. 6.5, the novel step within (9.78) is the introduction of the motion parameter ω describing the transformation $\mathbf{T}$. Since (9.78) represents an over-determined system of equations, the image $\mathbf{f}$ and the motion parameters ω can be estimated by means of a least-squares minimization.

However, one fundamental goal for engineers developing new CT scanner generations is the acceleration of the data acquisition process, particularly with respect to the time constants related to anatomical and physiological motions. The presently used scanners are multi-slice sub-second CTs, which, however, are not able to display perfect radiographs of beating hearts without ECG triggering.

The acquisition speed cannot be arbitrarily increased at present due to technical restrictions. On the one hand, the mechanical equipment reaches the load limit to

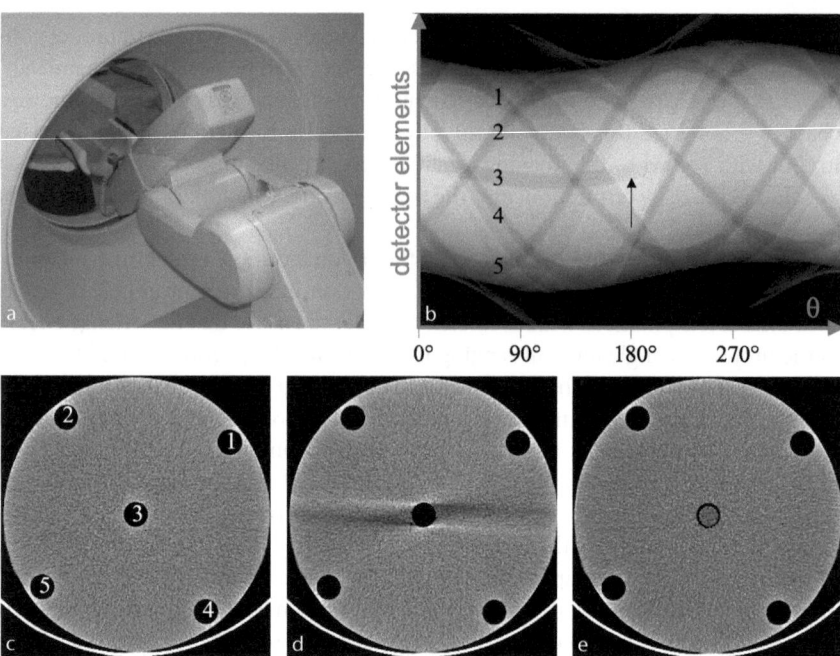

Fig. 9.16. Simulation of the administration of contrast medium. During a 360° rotation of the sampling unit, a matching Perspex bar is introduced into the central hole of a cylindrical Perspex phantom with five holes. **a** The bar is introduced by means of a robot pushing the bar into the hole while the sampling unit rotates by 45°, starting with a projection angle $\theta = 180°$. **b** The sinogram shows the raw data of this measurement for a rotation about 360°. On the first half of the sinogram the five holes can be readily distinguished and are numbered accordingly. Beyond an angle of $\theta = 180°$ the central hole fades more and more over an angular interval of 45° (this corresponds to the partial volume effect over a slice thickness of 3 mm) and then disappears. The tomogram corresponding to this sinogram is shown in **d**. Due to the inconsistencies related to the backprojection, this artifact is not limited to hole "3" into which the bar has been introduced, but extends over wide image areas. To be able to compare the different models, the tomograms without artifact – **c** without bar and **e** with completely introduced bar – are also shown. **c** also shows the correspondence between the five holes to the sine curves in the sinogram given in **b**

be taken into account with the high angular velocities and the corresponding angular moments of a 1,000-kg gantry. On the other hand, data transmission from the rotating sampling unit to the fixed gantry is a bottle neck due to which the data rate cannot arbitrarily be increased as well. Electron beam computed tomography (EBCT) described in ▸ Sect. 3.7 is therefore still the technology providing the shortest acquisition times. However, recently, it has been claimed that a dual-source CT system, which will briefly be discussed in ▸ Sect. 11.4, actually achieves comparable acquisition times.

9.6.4
Sampling Artifacts

In ▸ Sect. 7.8 it has already been pointed out that, as in general in any signal processing task, Shannon's sampling theorem must not be violated in CT. This applies to both the reconstruction of an axial slice and the subsequent reconstruction of 3D data presentations by slice stacking. Sub-sampling of a signal also results in the typical aliasing artifacts. The inherent sampling problem discussed in ▸ Sect. 7.8 arises particularly for the detector array with a rectangular sensitivity profile of the single detector elements. That is, the individual elements would have to be arranged at half the distance of their own width. As this requirement cannot be met due to obvious technical reasons (cf. ▸ Sect. 7.8) one makes use of an elegant mechanical trick. The corrective action, which is nowadays used to prevent aliasing, is either the quarter detector shift or the so-called *flying focus* of the X-ray tube.

9.6.5
Electronic Artifacts

There are several electronic defects deteriorating the image and in most cases destroying it. The most famous, or better, most notorious, electronic defect is the failure of a detector channel. In third-generation CT scanners such a detector defect will result in prominent so-called ring artifacts.

Figure 9.17 outlines the occurrence of the ring artifacts. As the X-ray source and the detector array are tightly joined at the sampling unit, the failure of an individual detector element or the corresponding processing channel respectively be-

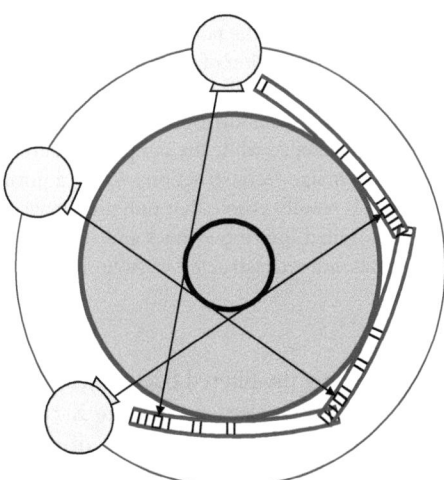

Fig. 9.17. In third-generation CT scanners, each detector element is responsible for the data collection on a circle in the object space. If one element fails, the data may only be acceptably reconstructed within the corresponding circle

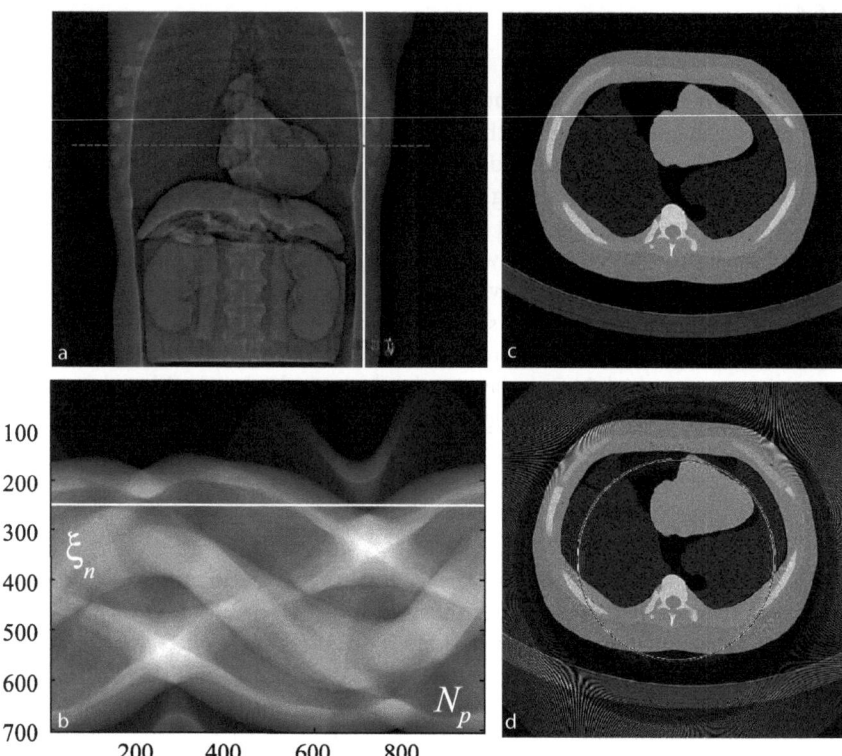

Fig. 9.18. Example of a detector channel failure. **a** The planning overview of an anthropomorphic torso phantom shows a defective detector as a straight vertical line resulting from the fixed position of the X-ray tube while the table moves forward along the feed direction. The *dashed horizontal line* marks the location of the axial tomogram to be measured. **b** The sinogram also contains a straight line since the corresponding detector position ξ_n is shown as a function of the projection angle. The horizontal detector position indicates the defective channel. This position is not changed while the sampling unit rotates. The figures on the right side exemplarily show the image reconstructions with **c** a normal detector array and **d** a defective detector array. It is readily visible that only the image data located inside the corresponding circle, which is called the ring artifact and results from the tangents of the backprojected erroneous signals, are reconstructed without major errors

comes specifically visible. During the filtered backprojection the virtual lines connecting the corresponding detector element and the X-ray source, which sometimes are called defective beams, form the tangents of a circle. This means that all values outside the circle are seriously concerned by this artifact. Inconsistencies with the measured values of the corresponding other projection directions in fact arise for each point of each line. Due to the backprojection, all image areas are again affected by the artifact. However, the area inside the circle encounters this problem less frequently (cf. Fig. 9.18). It is in fact possible to reconstruct the image almost without artifacts inside the circle formed by the defective beam tangents.

Figure 9.18 provides an example of the failure of a detector channel while an anthropomorphic torso phantom is scanned. In Fig. 9.18a the planning overview of the phantom can be seen. The axial slice is to be reconstructed along the gray dashed line through the heart. In the planning overview, it can be already seen that one of the measuring channels is defective. Due to the fixed anterior-posterior orientation of the X-ray tube/detector array unit, the defective channel results in a straight line along the table feed direction.

Figure 9.18b shows the Radon space as a sinogram. Although the sampling unit rotates over $360°$ with approximately $N_p = 1,000$ angular steps, the defective channel still stands out as a straight line. This can also be easily understood as the horizontal lines $\xi_n = $ const. always correspond to a fixed line connecting the radiation source and the detector array elements. Figure 9.18c first shows the artifact-free axial tomogram of the axial slice indicated with a dashed line in Fig. 9.18a. Figure 9.18d illustrates the consequences of the detector failure on the reconstruction process. The circle or ring is formed by accumulation of the circle tangents, which are backprojected over the entire image over $360°$. The generation principle of this ring artifact had already been illustrated schematically in Fig. 9.17. Outside the circle the image is defective. Different Moiré patterns, which arise during reconstruction according to the number of the measured projection angles, N_p, can be observed.

As already mentioned above, the result of the reconstruction process inside the ring is almost artifact-free. However, smaller waves are visible in the vicinity of the ring artifact inside the circle, which are also reconstruction artifacts. These waves are due to the convolution of the projection signal before it is backprojected. As a result of the convolution, the error in the filtered projection is finally broader (cf. Fig. 7.4) than in the raw projection. However, the data error quickly decays with increasing distance from the circle toward the circle center.

9.6.6
Detector Afterglow

Images can also deteriorate due to too long a glow time of the X-ray detectors. The fluorescence time of the detector material after conversion of an X-ray quantum into a visible photon should be as short as possible. Otherwise, afterglow artifacts appear in the reconstruction, which manifest as smeared object boundaries in the image. Although afterglow is a short-term effect, it becomes significant at the high rotation speed of the sampling unit. Afterglow is usually modeled as a linear effect on the intensities, independent on the detector element position. It decays exponentially and, if the temporal signature is known, the artifact can be suppressed by means of deconvolution.

9.6.7
Metal Artifacts

One known problem in CT is the appearance of metal artifacts in reconstructed CT images. Low-energy X-rays are attenuated more strongly than high-energy X-rays (cf. ▸ Sect. 9.6.2). Recall that the absorption is given by

$$\alpha \propto Z^4 \lambda^3 . \tag{2.33}$$

Due to the Z^4 dependence, this beam-hardening effect is prominent for metals that are introduced into the human body, such as dental fillings or hip prostheses, and leads to inconsistencies in the Radon or projection space. These inconsistencies observed in the integral attenuation values are due to the polychromatic X-ray spectrum produced by the X-ray tube. Additionally, without applying the dual-energy principle, the total attenuation of the X-ray intensity is an a priori unknown combination of the photoelectric effect and the Compton effect. This often leads in the reconstructed images to artifacts in the form of dark stripes between metal objects with light, pin-striped lines covering the surrounding tissue. Besides beam hardening, another origin of the metal artifacts is a higher ratio of scattered radiation to primary radiation, causing a low SNR in the metal shadow. This effect will be discussed in the subsection below. Additionally, the partial volume effect (▸ Sect. 9.6.1) is a source of metal artifacts in transmission CT images. Especially in those cases in which the radiation is completely absorbed[9] due to the thickness of the materials, very bright strips are found radially around the object; thus, the complete image loses its diagnostic value.

   If there are materials with high attenuation coefficients in the object to be examined, then strong streak artifacts arise, which are spread across the whole image. In practice, the system detects an infinitely high attenuation coefficient of the object. In the reconstructed image, one obtains on the backprojection lines through the object extremely high numerical values, which are spread across the entire image, as already stated above, and more importantly may not be compensated for by any other projection direction. But, as mentioned above, even if the radiation is not completely absorbed, the arising beam-hardening artifacts and – due to the usually extended sharp-edged metal implant objects – partial volume artifacts are so strong that the diagnostic value is considerably reduced.

   Figure 9.19 shows the occurrence of metal artifacts in the tomographic reconstructions of the dental area. Figure 9.19a again shows the planning overview scan on which the position of the slices to be measured is selected. Figure 9.19b–d shows three axial reconstructions. Major artifacts are not visible in Fig. 9.19b; however, on Fig. 9.19c and d, it can be seen that bright strike artifacts arise with an increasing quantity of highly absorbing substance (in the tooth area this is frequently amalgam) through which the radiation must obviously penetrate inside the slice to be

---

[9] From a physical point of view, due to (9.65) the radiation is of course never completely absorbed. But in practice the intensity may drop below the sensitivity of the detector or the noise level may be higher than the signal level.

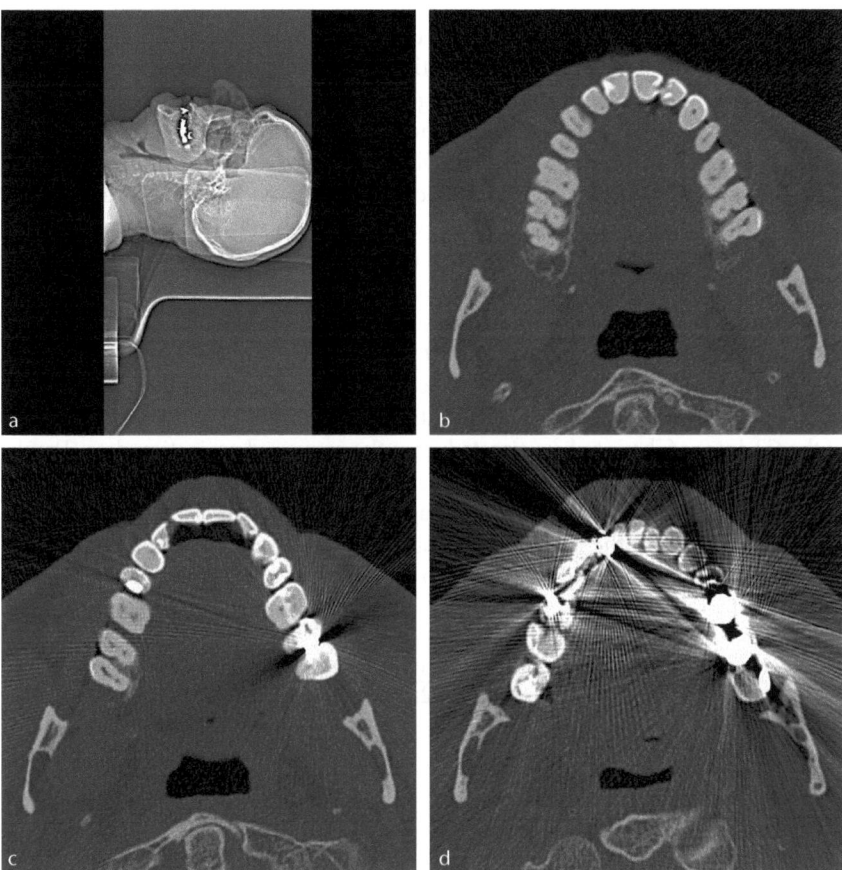

Fig. 9.19. Metal artifacts in CT images of a jaw. **a** presents the overview scan used to plan the axial slices. **b–d** then show the axial reconstructions at different slice positions throughout the tooth area. It is readily visible that the metal strike artifacts are considerably increased with the solid amalgam tooth fillings

reconstructed. Along the lines connecting different fillings, the artifacts are particularly striking, because the radiation must then pass through two metal volumes. The fact that the artifacts occur star-like around the metal object is a result of the principle of the filtered backprojection. Since the physical origin of the artifact is always located in the respective X-ray path of the backprojection under the diverse angles, the streaks appear radially distributed around the origin.

Although the mechanism is clear, standard filtered backprojection does not cope with these inconsistencies in the integral attenuation values. Thus, artifacts emerge within filtered backprojection, because the Radon space values are inherently considered as a complete set of consistent absorption integrals of a monochromatic process.

On the other hand, statistical reconstruction techniques (cf. ▸ Chap. 6) are especially capable of dealing with the metal artifact reduction problem. The core of these methods consists of an a priori knowledge of the statistical distribution of photon counts and a system matrix that models the physical X-ray absorption process. Each row of the system matrix represents a single X-ray beam running through the measurement volume. Contrary to filtered backprojection, the influence of each single beam on the image reconstruction can be weighted separately. In this way, beams through the metal objects can be treated appropriately.

Within the maximum likelihood (MLEM) algorithm discussed in ▸ Chap. 6, the image **f** is estimated that best fits the measured projections **p** under consideration of the statistical photon-count variations. It is based on the realistic assumption that the photon counts follow a Poisson process. The parameters of the joint Poisson distribution, which resemble the expectations of the linear attenuation coefficients of the CT image **f**, have to be estimated from the measured projection data. Thanks to the exponential Lambert–Beer attenuation law, the photon count is coupled to the attenuation values that form the image **f**. Therefore, the MLEM algorithm is generally superior to the filtered backprojection with respect to streaking artifacts in the reconstructed image. This is due to the fact that, generally, MLEM for transmission CT

$$f_r^{*(n+1)} = f_r^{*(n)} \frac{\sum\limits_{i=1}^{M} a_{ir} e^{-\sum\limits_{j=1}^{N} a_{ij} f_j^{*(n)}}}{\sum\limits_{i=1}^{M} a_{ir} e^{-p_i}} \qquad (6.105)$$

has an inherent weighting. Any X-ray that intersects a metal object will naturally have a low intensity value. So, such rays will automatically get less weight, because the projection integral

$$p_i^* = \sum\limits_{j=1}^{N} a_{ij} f_j^* \qquad (9.79)$$

of the estimated forward projection and the actual measured projection value p_i weight the influence of the corresponding terms in the sums of the nominator and denominator respectively in (6.105). This is due to comparably high values of p_i^* and p_i.

However, inconsistent projection data can also be repaired with surrogate data created by interpolation or the inconsistent data can be treated as missing data in order to ignore the inconsistent data in the statistical approach[10]. The problem with the interpolated surrogate data is that these always include residual inconsistencies due to missing information during interpolation. On the other hand, within the missing data approach, a brute force method is applied to eliminate inconsistencies. However, the problem with this strategy is that the reconstruction actually suffers

---

[10] This is not possible for a filtered backprojection based reconstruction.

from voids in the projection data. Recently, it has been shown (Oehler and Buzug 2007) that both strategies may be combined. A weighted MLEM approach can be used to reduce the influence of the residual inconsistencies from interpolation in such a way that optimal imaging quality is obtained by optimizing the compromise between residual inconsistencies and void data.

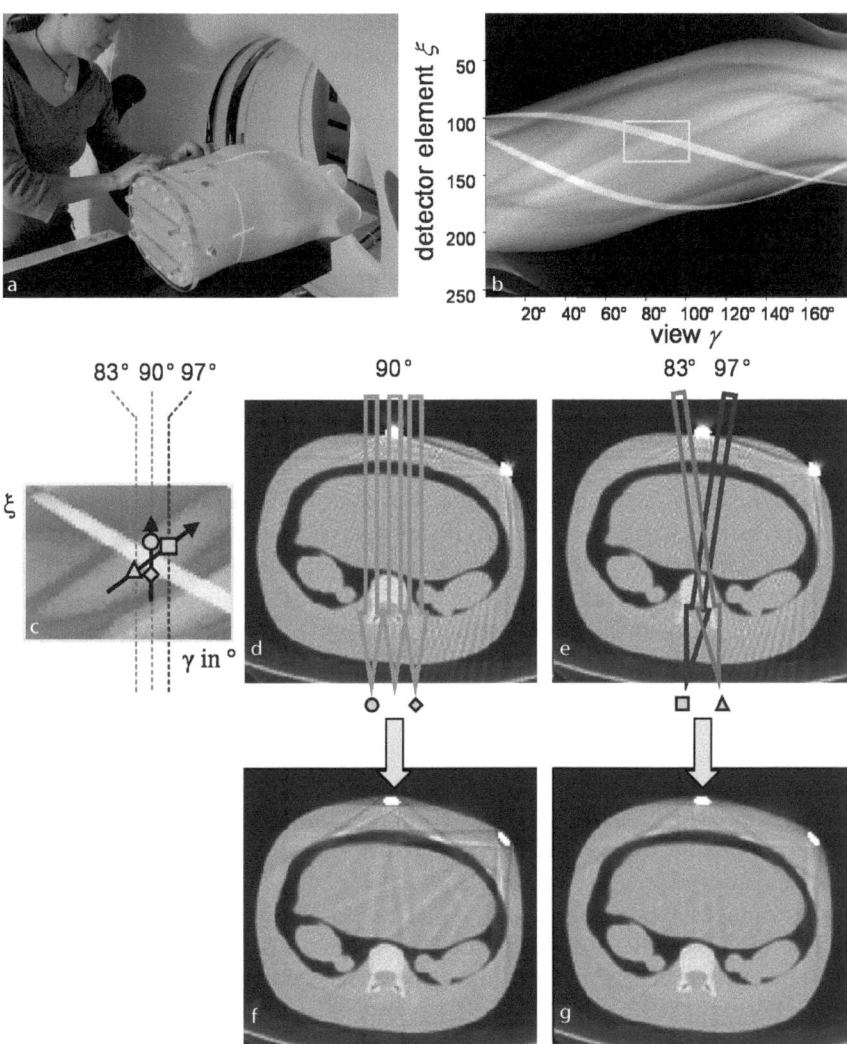

Fig. 9.20. a Philips CT scanner with CIRS torso phantom marked with steel markers. **b** Sinogram of the torso phantom marked with two steel markers. **c** Region of interest from the sinogram (**b**). Different interpolation methods displayed in the region of interest. **d** Interpolation inside one projection under one angle; **e** directional interpolation. **f,g** Reconstruction of torso phantom data. **f** λ-MLEM reconstruction with linear interpolation and **g** λ-MLEM reconstruction with directional interpolation (Oehler and Buzug 2007)

Raw data of an anthropomorphic torso phantom marked with two metal markers in a certain slice were acquired (cf. Fig. 9.20a). Flat metal cylinders with a diameter of 15 mm and a height of 3 mm – similar to those used as fiducial markers in image-guided surgery – are used as markers in the experiments. The markers are made of steel of high atomic number. This leads to an amplification of the beam-hardening artifact, introducing inconsistent raw data. The markers are placed outside the phantom, because this configuration ensures that the sinusoidal metal trace inside the Radon space has a maximum amplitude crossing all other sinogram structures of the phantom. This is the most complicated case for evaluating the interpolation. In Fig. 9.20b the sinogram data of the torso phantom equipped with the steel markers are shown.

In a first step, inconsistent projection data are bridged by interpolation. For this, the metal object is labeled in a preliminary filtered backprojection reconstruction using a simple threshold. Subsequently, a forward projection of the metal-marker image is calculated, resulting in metal-trace projection data. This is used to label the projections running through a metal object, i.e., the metal traces shown in Fig. 9.20b.

To demonstrate that different interpolation methods lead to different results in terms of metal artifact reduction in the reconstructed images, the inconsistent data are bridged by two different interpolation methods. The first is the standard method (Glover and Pelc 1981): A linear interpolation inside one projection under one fixed view is used (Fig. 9.20c,d: Interpolation at an angle of 90° between $\diamond$ and $\bigcirc$). The second method, which considers the arrangement of the underling raw data, is a directional interpolation filling the gaps of inconsistent data following the flow of the sinogram data outside the metal traces. The principle of directional interpolation is shown in Fig. 9.20c, e (interpolation between $\triangle$ and $\square$). This method leads to superior results compared with the first interpolation schemes. Due to interpolation, the projection values of the soft tissue lying behind the metal markers are estimated.

Unfortunately, any interpolation-based repair scheme of the Radon space is based on a weak underlying physical model and hence it cannot be expected that the estimated projection data will perfectly fit the projection data measured without metal markers. Therefore, projections p_i running through a metal object, as well as the corresponding rows of the system matrix $\mathbf{A} = \{a_{ij}\}$, have to be weighted by an appropriate confidence parameter $0 \leq \lambda \leq 1$. Recall that $j \in \{1, \ldots, N\}$ corresponds to the number of image pixels and $i \in \{1, \ldots, M\}$ to the number of projections. A careful choice of the confidence parameter λ optimizes the compromise between residual inconsistencies and void data. The corresponding fixed point iteration is given by

$$f_r^{*(n+1)} = f_r^{*(n)} \frac{\sum\limits_{i=1}^{M} \lambda_i a_{ir} \, e^{-\sum\limits_{j=1}^{N} \lambda_i a_{ij} f_j^{*(n)}}}{\sum\limits_{i=1}^{M} \lambda_i a_{ir} \, e^{-\lambda_i p_i}}. \tag{9.80}$$

This modified MLEM algorithm is called λ-MLEM (Oehler and Buzug 2007). Figure 9.20f and g display the results of the different interpolation methods – linear interpolation (Fig. 9.20f) inside one projection under one single view and directional interpolation (Fig. 9.20g) – as a pre-processing step with consecutive λ-MLEM reconstruction. An appropriate weighting λ depends on the interpolation scheme and the size of the metal object. Linear interpolation is a frequently used approach to bridge metal data. It can be seen that this strategy works for the main artifact between the two markers.

However, strong new artifacts are introduced that stem from a sharp non-differentiable connection between the measured and interpolated Radon space data. A directional interpolation follows the flow of the Radon space (cf. Fig. 9.20c). Contrary to the vertical interpolation schemes, the flow consists of sinusoidal traces that connect projection values of actual objects inside the image. Consequently, better results are achieved with the λ-MLEM algorithm in combination with directional interpolation of the inconsistent data. As a practical rule of thumb, it has been recommended to set the parameter of the approach to $\lambda = 0.5$. This always improves image quality compared with pure interpolation or missing data concept.

9.6.8
Scattered Radiation Artifacts

In ▸ Sect. 2.3.2 the photon–matter interaction mechanisms of X-rays have already been discussed in detail. Whereas for the detector element located in the unscattered, direct beam path it does in principle not make a difference which physical mechanism of the interaction actually reduces the intensity, as long as

$$p(s) = -\ln\left(\frac{I(s)}{I(0)}\right) = \int_0^s \mu(\eta)\,\mathrm{d}\eta \qquad (9.81)$$

holds (cf. (5.10)). Other detector elements located outside the direct line of sight may in fact be impaired by certain interactions. Particularly in the area of strong attenuating anatomical objects such as the shoulder, abdomen, and pelvis (Morneburg 1995), the measured values may be distorted due to scattered radiation. These scattered X-rays may become a considerable part of the overall signal.

Whereas the scattered radiation is almost the same for all projection angles, it is very different for the wanted signal. In projection directions in which highly absorbing objects are located one after another, the wanted signal may become extremely weak so that the scattered radiation dominates the signal (cf. Fig. 9.21a). Within the filtered backprojection, inconsistencies then arise from this projection direction, which result in streak artifacts.

With regard to the interfering radiation caused by scattering reaching the detector, the third-generation CT scanners are superior to the fourth-generation CT scanners. Figure 9.21b and c outlines the scanners of both generations. Figure 9.21b shows the third-generation scanner. The detector array of this scanner is designed

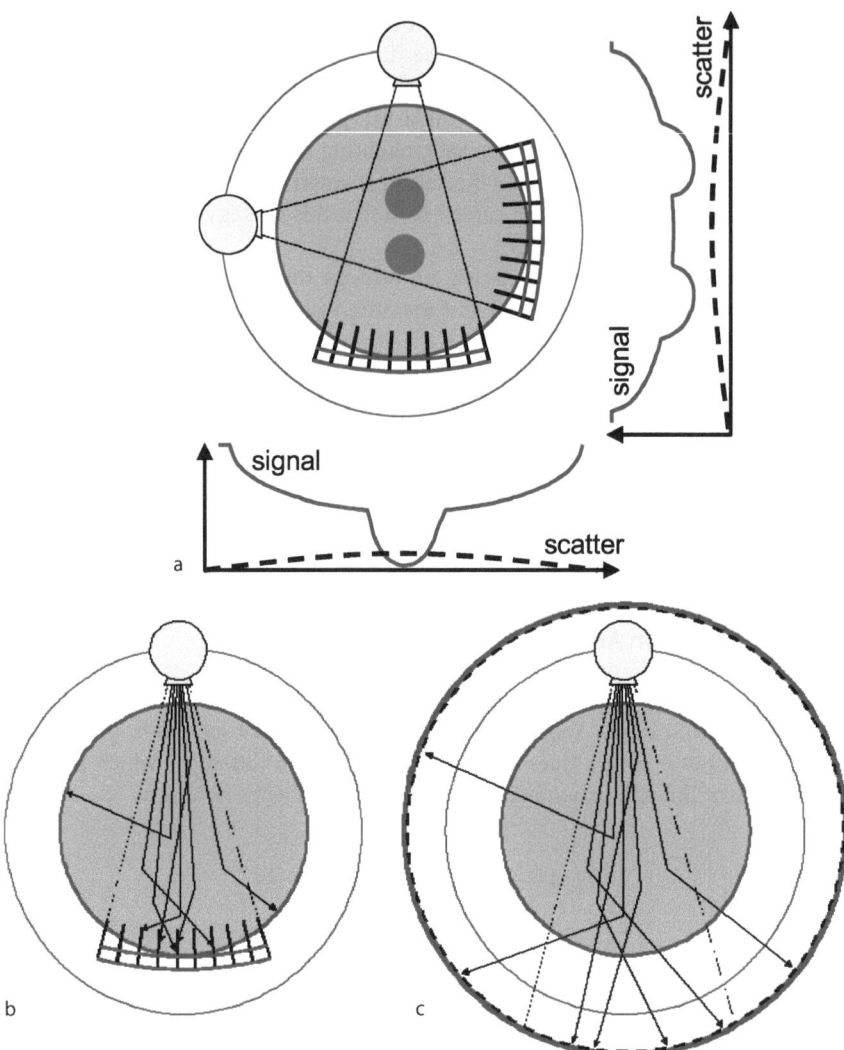

Fig. 9.21. a The background scatter signal may cover the wanted signal. Comparison of **b** the third and **c** the fourth generation of CT scanners with regard to their radial geometry. As the detector array of the third generation represents a circle segment, the center of which is located in the X-ray tube, it is possible to design an anti-scatter grid with septa pointing toward the X-ray focus

such that the row of detector elements is arranged on a circular segment whereby the X-ray source is located in the circle center. It is therefore possible to collimate the radiation by septa so that scattered radiation – with an incident angle into a "false" detector element, which is larger than a threshold angle – is effectively shielded. The critical angle is determined by the length and the spacing of the detector segments.

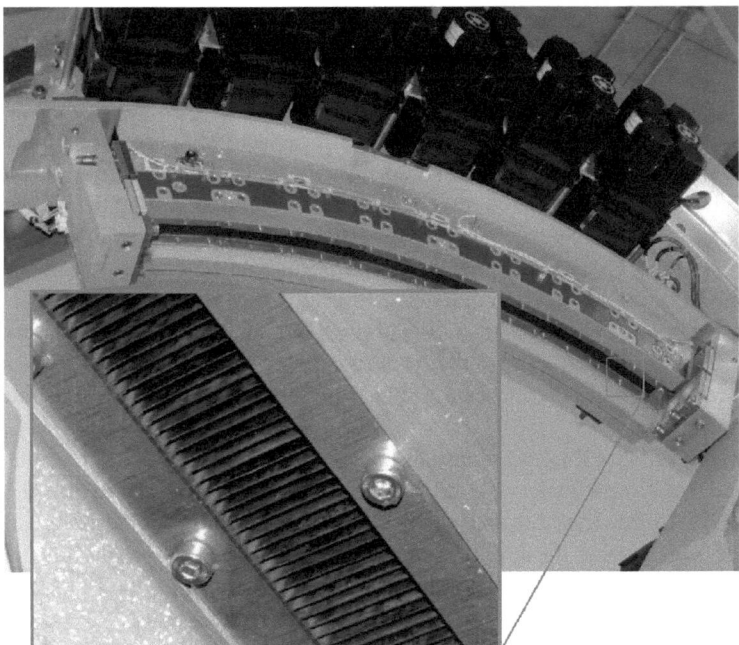

Fig. 9.22. The anti-scatter grid at the detector array; here the mobile Philips Tomoscan M – a third-generation scanner – is shown. The septa of the grid are readily visible in the close-up

In the fourth-generation scanners (c.f. Fig. 9.21c) the detectors are located on a circle, the center of which is the iso-center of the measurement field. An X-ray source, to which a detector collimation might be focused on successfully, obviously cannot be located there. Figure 9.22 shows the segmented scattered radiation grid at the detector array of the mobile CT scanner Philips Tomoscan M – a third-generation scanner.

9.7
3D Artifacts

In connection with the so-called secondary reconstruction of three-dimensional images based on CT image stacks[11], but also with the implementation of 3D acquisition techniques such as spiral CT, cone-beam reconstruction, and their combinations – the spiral or helical cone-beam reconstruction – a variety of new artifacts may arise, which will only briefly be discussed in this section.

---

[11] In the form of multiplanar reformatting (MPR) or surface rendering.

9.7.1
Partial Volume Artifacts

Partial volume artifacts due to the finite detector width arise inside an axial slice. These artifacts, which have been discussed in ▸ Sect. 9.6.1, are called transversal partial volume artifacts. In addition, artifacts must also be expected in the axial or z direction, i.e., the table feed direction, due to the finitely collimated slices. These artifacts are called longitudinal or axial partial volume artifacts (Morneburg 1995). Figure 9.23 shows that averaging plays an important part, not only within one slice, but also in the axial direction due to the restricted slice thickness Δz.

In regions with a multitude of details rich in contrast, the partial volume artifacts may considerably disturb the reconstruction process. The slices in these anatomic regions should therefore be as thin as possible. Figure 9.24 demonstrates this effect.

Even if the slice thickness is usually given by a single scalar value, one must bear in mind that X-rays are leaving the focal spot of the X-ray tube in all directions (cf. ▸Chap. 2). In this way, a source-sided collimation does not actually define a "slice thickness," but rather the aperture angle of the fan beam in the longitudinal or axial direction. Figure 9.24a shows the beam geometry for a thick slice and Fig. 9.24b

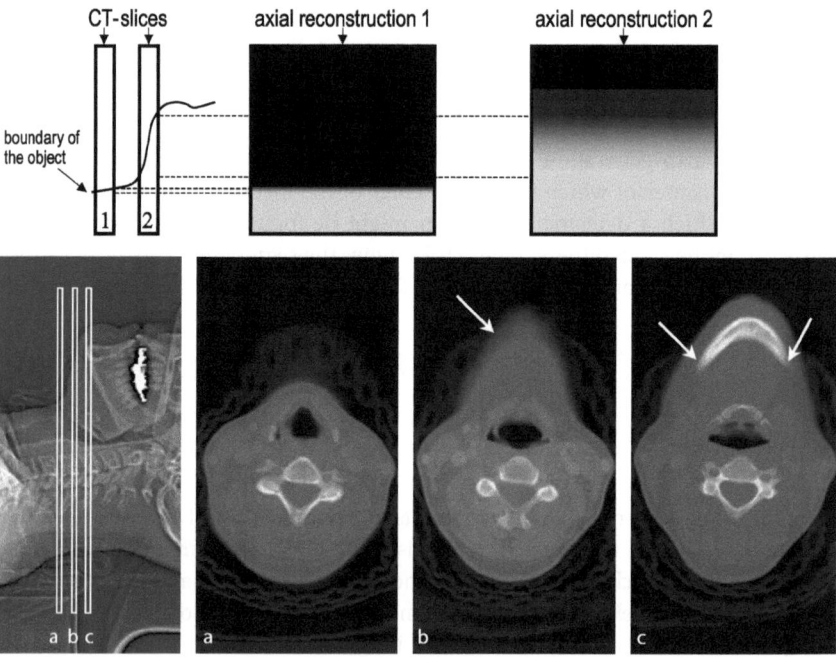

Fig. 9.23. Object boundaries passing through an axial slice with a high slope are blurred (schematically on top of the figure). **a** This effect is readily visible in the area of the jaw. The *arrows* in the slices **b** and **c** indicate the blurred image areas

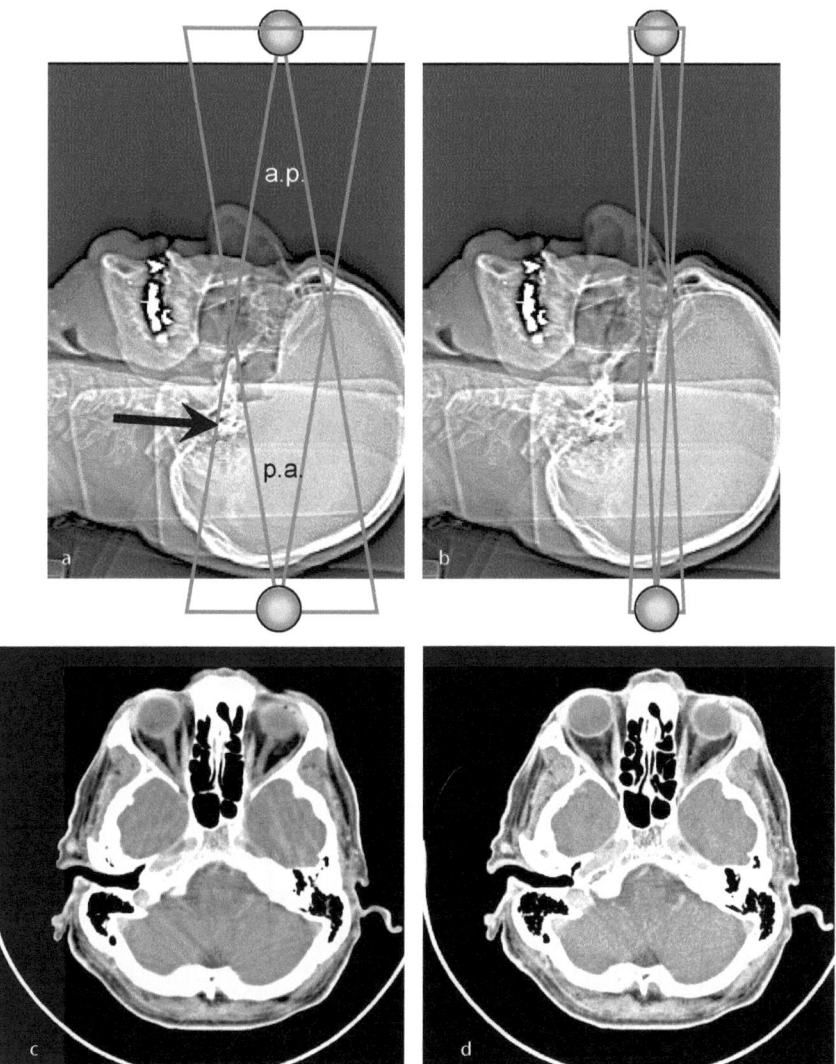

Fig. 9.24. Longitudinal or axial partial volume artifact. **a** If the axial aperture of the X-ray fan beam is comparably large, different information (indicated by an *arrow*) is measured from the anterior–posterior (a.p.) and the posterior–anterior (p.a.) direction. **b** This leads to inconsistency artifacts in the reconstruction. **c** The amount of inconsistency can be reduced by a finely collimated X-ray beam. **d** In this way, the artifacts in the reconstruction are reduced

that for a finely collimated slice. Due to this reason, the projection measurements always carry inconsistencies to some extent.

However, these inconsistencies are particularly disturbing if the axial aperture angle is large and the object information is of high contrast. Bony structures, for instance, indicated with an arrow in Fig. 9.24a, show highly contrasted structures that are included in the measurement only in the anterior–posterior (a.p.) direction. As a consequence, the typical streak artifacts are visible in the reconstruction displayed in Fig. 9.24c. As a counter-measure, the beam may be collimated finely (cf. Fig. 9.24b) and the corresponding artifacts will be reduced (cf. Fig. 9.24d).

9.7.2
Staircasing in Slice Stacks

If fast changes in object boundaries rich in contrast along the table feed direction arise in the object to be imaged, then steps can be encountered side by side in the three-dimensional reconstruction, leading to a stair-like appearance in the object visualization. These steps are artifacts produced by averaging the object boundary within the sensitivity profile of the axial slice. Figure 9.25 shows this effect schematically. At the skull surface visualization (cf. Fig. 9.25d; the slice stack is also shown

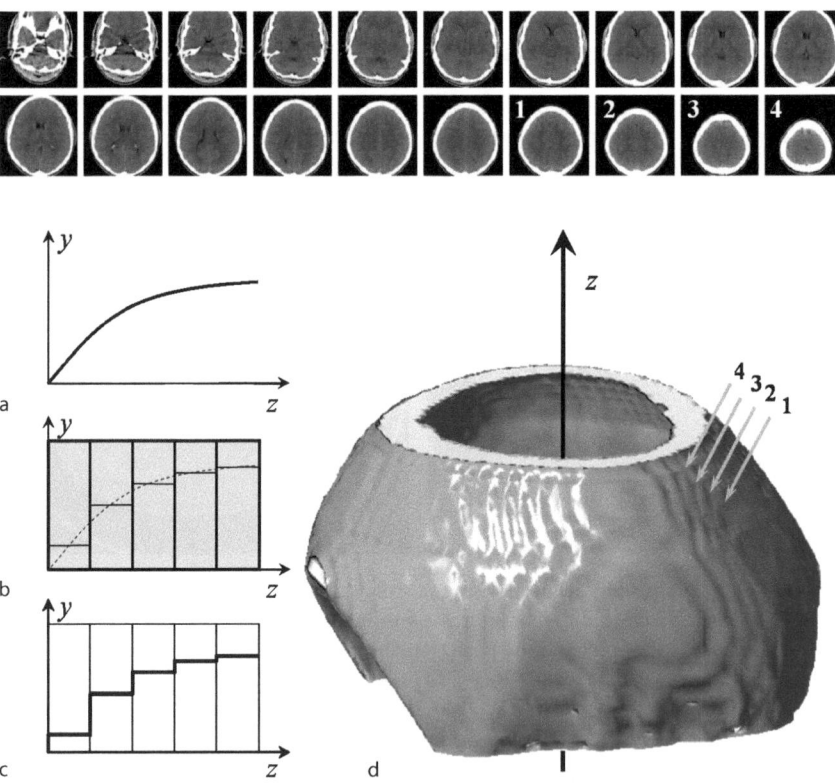

Fig. 9.25. If the object in the CT scanner is defined by boundaries that steeply pass through the individual slices (**a**), then averaging within one slice (**b**) results in staircasing of the three-dimensional reconstruction of the slice stack (**c** and **d**). *Upper rows*: Slice stack of the skull. The last four slices are assigned to the three-dimensional reconstruction with numbers

in the two top rows of this figure), it can be seen that the steps in the area in which the surface is subjected to considerably fast changes with respect to the axial (in this figure vertical) z direction are particularly pronounced.

This image error may be reduced by arranging very small collimated slices that potentially overlap. However, the more elegant method is the spiral CT approach, by means of which this artifact type may be effectively reduced. In principle, arbitrarily dense slice stacks may be computed retrospectively with the method described in ▸ Sect. 8.3. In the case of fast changes in structures that are rich in contrast along the axial direction, a *pitch* factor p (cf. Eq. 8.9) smaller than 1 is recommended (Blanck 1998).

Figure 9.26 shows the staircasing artifact on the images of a spine phantom. The individual slices scanned with conventional CT acquisition technology are 6 mm thick here. This figure should also be compared with Figs. 8.7 and 8.16 in the previous chapter. The staircasing effect, which expresses the anisotropy of the underlying data set, is particularly problematic if very small details must be analyzed.

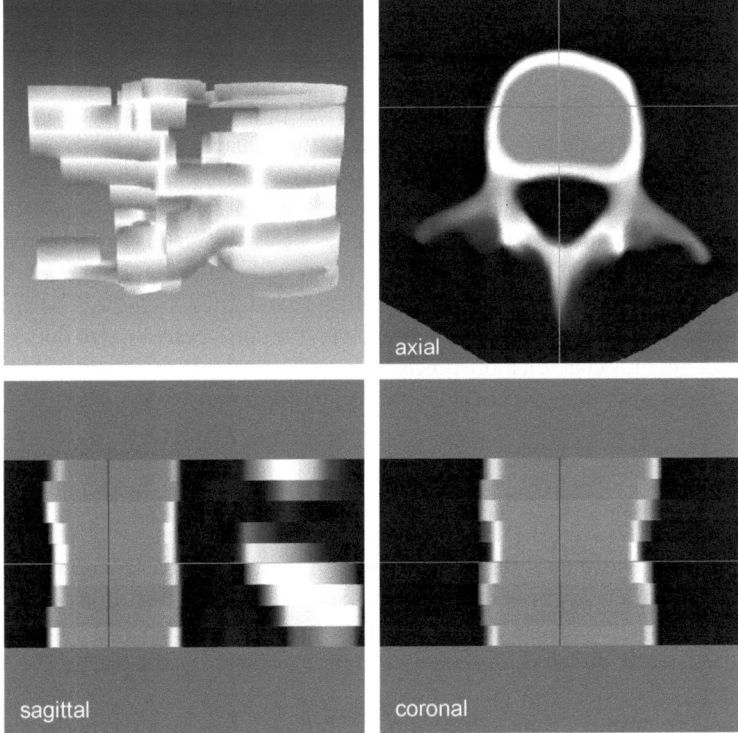

Fig. 9.26. Three-dimensional reconstruction of a vertebral phantom. Typical orthogonal reformatting and volume rendering can be seen. The slice thickness selected for the measurement is 6 mm. The staircase pattern of the three-dimensional rendering and of the coronal and sagittal slices can be clearly seen

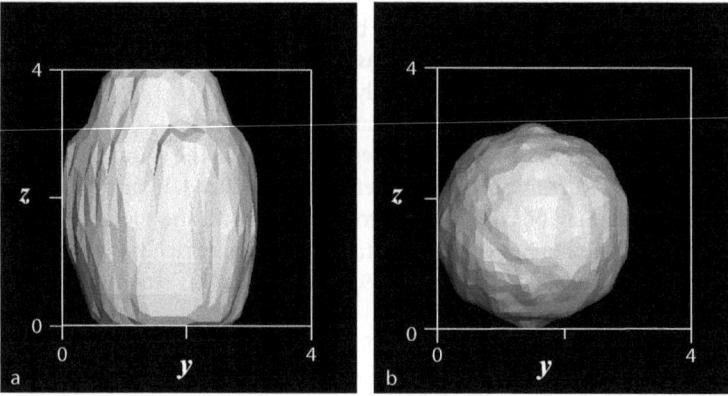

Fig. 9.27. Comparison of the reconstruction quality of a spherical tumor phantom with a radius of 2 mm. **a** shows the typical result of a conventional clinical CT system. The length extension of the sphere in the z direction is due to averaging within the slice. **b** shows the reconstruction result of a prototypical volume CT equipped with a flat-panel X-ray detector with 0.14 mm isotropic spatial resolution (courtesy of General Electric Medical Systems: Pfoh [2002])

Figure 9.27 shows the image of a spherical tumor phantom with a radius of 2 mm. Figure 9.27a demonstrates the typical length extension of the object due to averaging within the slice measuring 1.25 mm thick. Inside the slice the resolution is of course better and amounts to approximately 0.5 mm in this example. In the isotropic presentation acquired with a volume or cone-beam CT with a spatial resolution of 140 µm, in Fig. 9.27b the spherical shape of the phantom is revealed. Therefore, errors with regard to the volume estimation in tumor studies can be reduced. The considerably higher spatial resolution with the cone-beam system does in consequence not only enable a more precise quantitative volume measurement, but also the potential early detection of small tumors in oncology (Pfoh 2002).

9.7.3
Motion Artifacts

Motion artifacts in a two-dimensional image, i.e., in the reconstruction of a single axial slice, have already been described in ▸ Sect. 9.6.3. In the two-dimensional case, image errors such as streaks, blurring, and ghost images found in the backprojection were particularly due to inconsistencies in the raw data, i.e., the data of the Radon space.

Another problem now arises with the three-dimensional, secondary reconstruction, in which the size and position of objects to be imaged by several axial slices in a stack no longer match with regard to the original object in consecutive slices. Figure 9.28 shows one example in which the data set is blurred by respiration. On the upper left of the figure, in the three-dimensional visualization of the

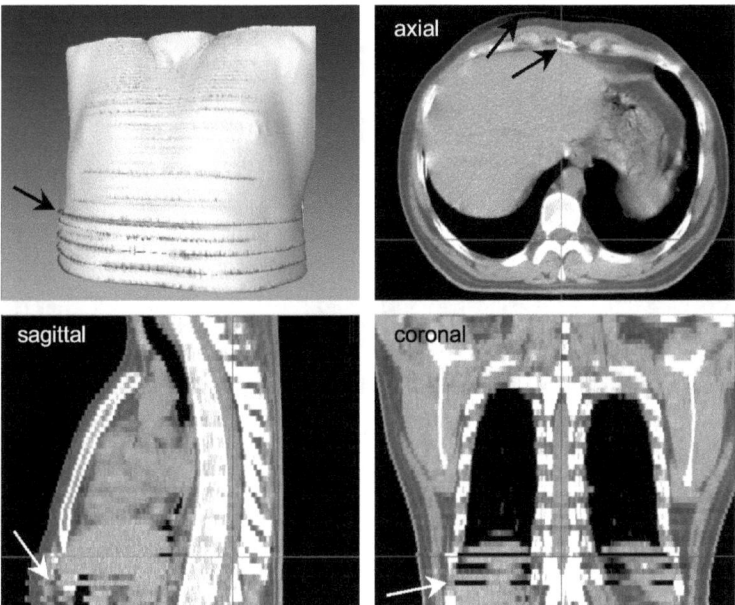

Fig. 9.28. Motion artifacts in the three-dimensional reconstruction. After approximately three-quarters of all planned axial slices from the shoulder to the abdomen had been measured, the patient could no longer hold his breath. In the last quarter of the data set this resulted in offsets and blurring marked by *arrows* in the corresponding images (courtesy of Ruhlmann, Medical Center Bonn)

skin surface, it can be seen that the patient data in the lower quarter of the slice stack (see arrow) have body contours that do not match each other.

In this example, the patient was not able to hold his breath during the acquisition time. If one selects an axial view of such a slice blurred by respiration, this blurring effect and the resulting ghost structures (marked by arrows) can readily be seen. The disturbances of course continue in the corresponding reformatting, i.e., in the calculated sagittal and coronal slices. The correspondence among the axial, sagittal, and coronal slice is indicated by vertical and horizontal lines in the images.

9.7.4
Shearing in Slice Stacks Due to Gantry Tilt

The spatial normal vectors of the individual slices need not necessarily match with the z-axis, i.e., the table feed axis of the CT system. Under certain anatomic conditions, the gantry, i.e., the frame of the rotating sampling unit with the X-ray tube and the detector array, must be inclined so that the relevant structures are located within one slice and need not be assembled from a slice stack. Figure 9.29a and b shows the planning overview for a skull tomogram with two possible slice sequences. In Fig. 9.29a, the normal slice stack is planned. This corresponds to the

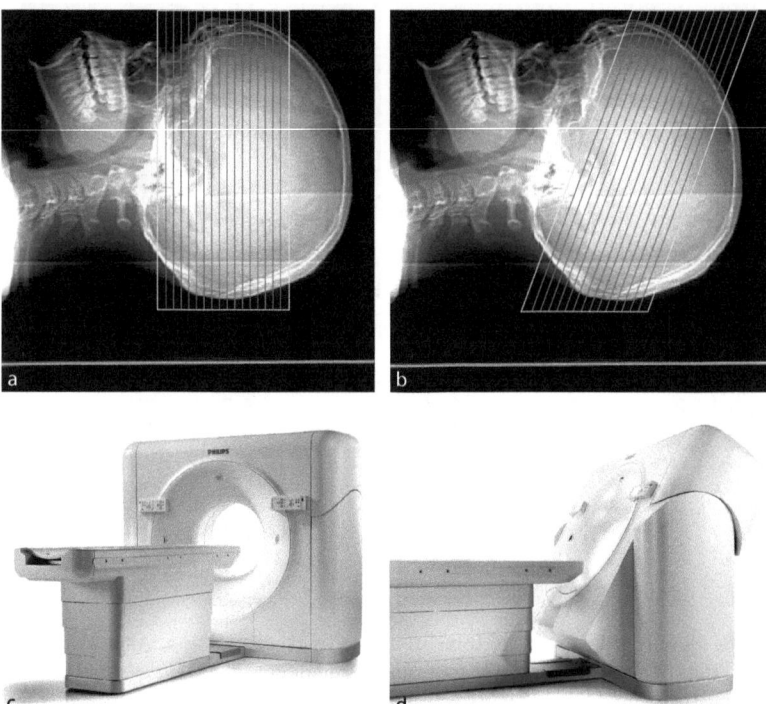

Fig. 9.29. Planning overview of a cranial tomogram. **a**, **c** Normal slice sequence; **b**, **d** angulated slice sequence. Due to the anatomical orientation of the brain inside the skull, the angulated slice sequences are more appropriate. The eyes are, for example, not exposed to direct radiation. The measured slice stack is shown in Fig. 9.25

upright position of the gantry (cf. Fig. 9.29c). In Fig. 9.29b, an angulated slice sequence is planned. This corresponds to the tilted gantry shown in Fig. 9.29d. Due to the anatomical orientation of the cortex inside the skull, an angulated sequence of images is frequently used to prevent unnecessarily high doses of direct radiation on the sensitive eyes.

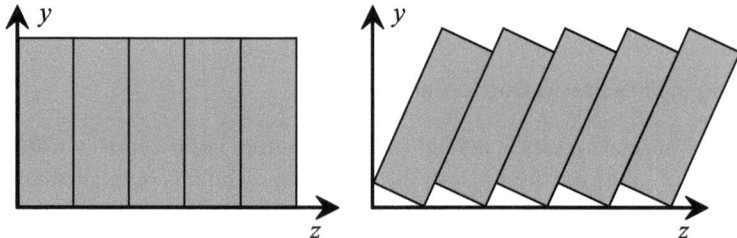

Fig. 9.30. Schematic sketch of the slice sequence of conventional data acquisition, not taking into account the sampling theorem. *Left*: Slice acquisition vertical to the z-axis. *Right*: Tilted slice acquisition

Figure 9.30 once again shows schematically the orientation of the individual slices with the normal and angulated form. If one wants to compose a slice stack measured in the angulated orientation of the gantry to provide a three-dimensional presentation, then the orientation of the data set must be corrected slice by slice. Figure 9.31 illustrates the correction process schematically and Fig. 9.32 shows the result on the skull tomograms presented in Fig. 9.25 as a slice sequence. Shearing does not only occur with an inclined gantry. When the table moves into the meas-

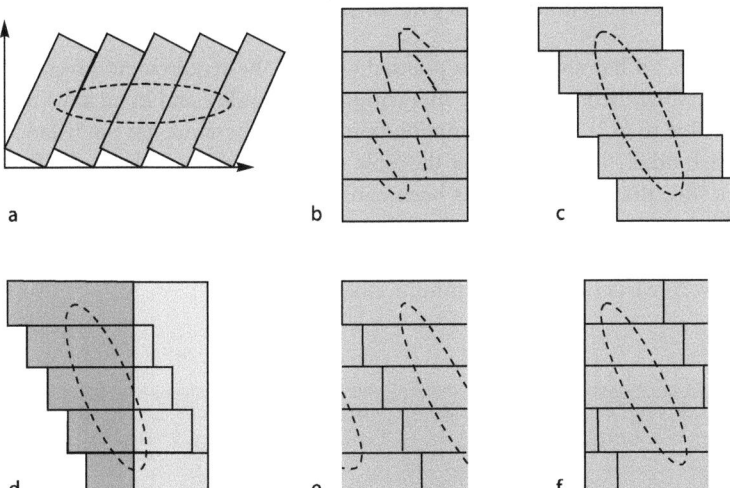

Fig. 9.31. a If one assembles the data measured from an angulated slice sequence without any correction into a slice stack, **b** the individual slices will be sheared with respect to one other. **c** If one reverses this shearing process the data are available as they were in the original measurement (**a**). However, the image will not have a rectangular format. **d** If one corrects this deficiency by increasing the cyclic movement of the slices, **e** the object seems to be cyclically shifted. **f** This deficiency can be corrected in a last step – by synchronous cyclic shifting of all slices

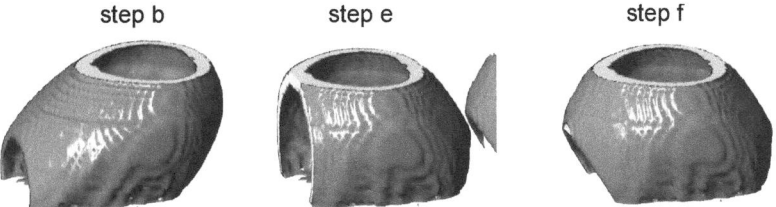

Fig. 9.32. If the skull measured with an angulated slice sequence is assembled by stacking one slice on top of another without any further corrective action, the single slices seem to be sheared with respect to one other. This is readily visible on the *left-hand* picture (cf. **b** in Fig. 9.31). Staircasing indicates the thickness of the slices. Shearing has been corrected on the picture in the *center*, but the object seems to be cyclically shifted (cf. **e** in Fig. 9.31). On the *right-hand* picture, the cyclic shifting has also been corrected (cf. **f** in Fig. 9.31)

uring field, the gravity moment acting on the table increases more and more the longer the path of the table through the gantry. This table deformation, the amount of which depends on the feed length, also results in shearing of the individual slices. However, this deformation is rather small and does not play a role at all with respect to diagnosis. But if the CT data are used for image-guided surgery, then the table deformation must be corrected (Zylka and Wischmann 1996).

9.7.5
Sampling Artifacts in Secondary Reconstruction

In ▶ Sect. 8.2 it has already been pointed out that the requirement given in Shannon's sampling theorem must be met within an axial slice and in an axial direction as well. Due to (8.1) the axial sampling points must therefore have a distance $\Delta z/2$ from each other. This means that the table must not be moved by more than $\Delta z/2$ when a slice thickness of Δz has been chosen. Thus, at least two slices have to be measured per slice thickness. Otherwise, the secondary reconstruction might be subjected to aliasing artifacts. Figure 9.33 demonstrates this aliasing effect in a secondary reconstruction of an aluminum hole phantom.

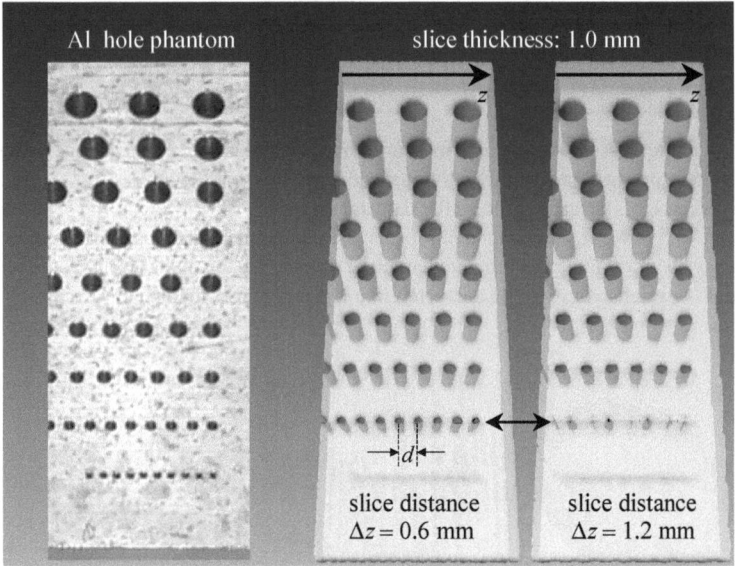

Fig. 9.33. Tomographic survey of a hole phantom (Al phantom on the left-hand side) with a slice thickness of 1 mm each. If a slice is acquired every $\Delta z = 0.6$ mm (*center*) or every $\Delta z = 1.2$ mm (*right side*) respectively, and a secondary three-dimensional reconstruction is carried out, aliasing artifacts in the horizontal hole grid with $d = 1.5$ mm (diameter = distance) can be seen

9.7.6
Metal Artifacts in Slice Stacks

If considerably strong artifacts arise within a slice, then this slice destroys the secondary three-dimensional reconstruction of the volume in this slice. Figure 9.34 shows the effects of metal artifacts in the three-dimensional image. Looking at the

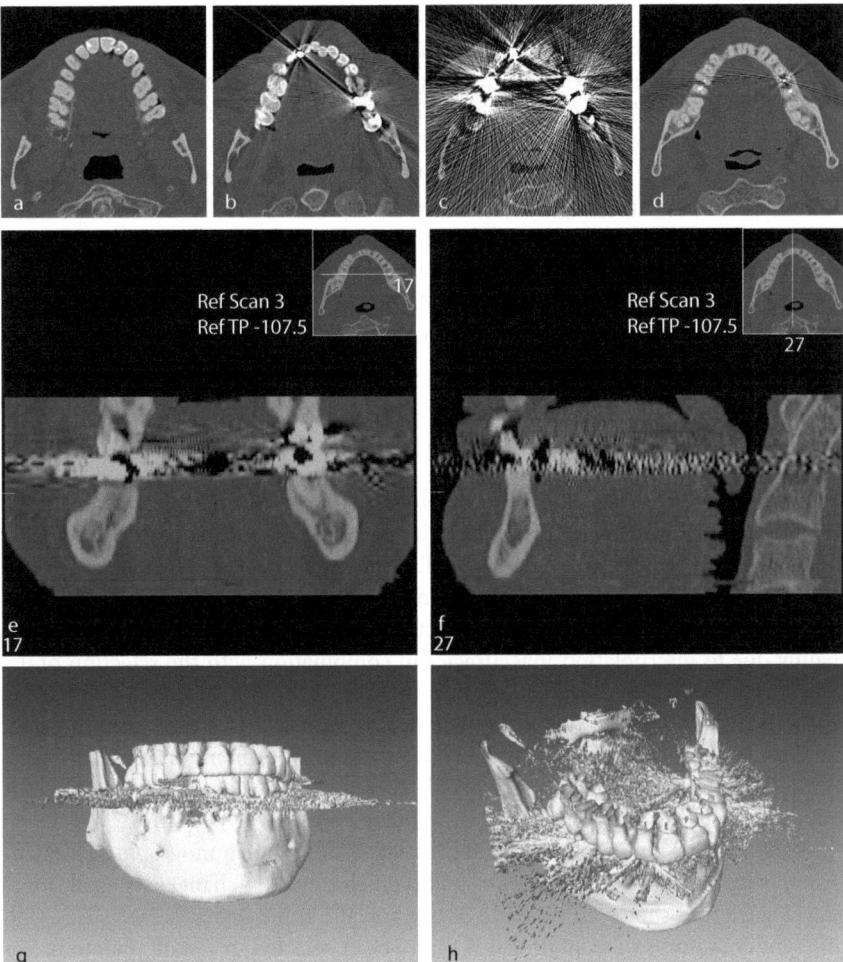

Fig. 9.34. Metal artifacts in the three-dimensional reconstruction. The figures **a** through **d** show axial slices with artifacts of different strengths, depending on the number of tooth fillings in the slice. **e** and **f** show a coronal and sagittal slice reconstruction (or reformatting) from an axial stack of 55 slices respectively. The spatial orientations of this reformatting are always shown on the upper right on **e** and **f**. **g** and **h** show surface renderings based on a simple gray-value threshold in which one can readily see the metal artifacts in the corresponding axial slices (courtesy of Ruhlmann, Medical Center Bonn)

jaw tomography that has already been presented in ▸ Sect. 9.6.7, once again some axial slice reconstructions can be seen in the sequence of Fig. 9.34a–d (anatomically from bottom to top). In Fig. 9.34a one initially does not see any significant artifacts.

In Fig. 9.34b one can readily see that two metal objects located in the beam path produce strong artifacts in the area of their connecting lines. In Fig. 9.34c a multitude of tooth fillings are located in the area of the slice to be reconstructed so that the whole image is completely degraded by streaks due to the inconsistencies of the projections from different directions. Figure 9.34d once again illustrates the radial propagation of the streaks in the area of a single tooth filling.

If one stacks the individual slices one above another, the image of a coronal slice (Fig. 9.34e) and of a sagittal slice (Fig. 9.34f) in orthogonal reformatting can be computed. The location of the corresponding slices is inserted in an artifact-free axial slice on the upper right-hand side. One can clearly see that the complete slices are contaminated with artifacts in the area of the tooth fillings.

Figure 9.34g and h shows three-dimensional visualizations of the jaw. These are surface renderings based on gray-value threshold bone segmentations. Both figures show the same object, each with a different viewing angle. Outside the slices impaired by the tooth fillings, the image quality of the tomogram is very good.

9.7.7
Spiral CT Artifacts

In principle, image errors also arise in spiral CT that are similar to those arising in the direct reconstruction of single axial slices. However, these artifacts are generally not as strong as they are in single-slice CT, since spiral CT reduces considerably, for example, motion artifacts or staircasing in the secondary reconstruction. However, new spiral-related reconstruction artifacts arise with this CT method, which are mainly due to interpolation inaccuracies in combination with the selection of the *pitch* factor. In principle, one finds here as many image error types as interpolation methods. In this section, the interpolation problem cannot be discussed in detail, but the so-called *scalloping* (Blanck 1998) artifact should be mentioned, which is due to the fact that the slice sensitivity profile is increased in spiral CT so that partial volume artifacts also become stronger (Wilting 1999).

Scalloping is a phenomenon arising, for example, in skull tomographies, particularly in slice positions in which the skull diameter quickly changes its axial direction. Figure 9.35 shows an example of this specific spiral artifact. Two slices with different curvatures in axial directions have been selected in a skull phantom. For comparison, the slices with a thickness of 1 mm were measured for both positions conventionally, i.e., without a continuous table feed.

The result is shown in Fig. 9.35a and b. The virtually different thicknesses of the skull in the two slice positions is due to the fact that the angle between the corresponding reconstruction layer and the local skull surface normal vector varies. One can also readily see the partial volume artifacts arising as a thin fringe around the skull structure.

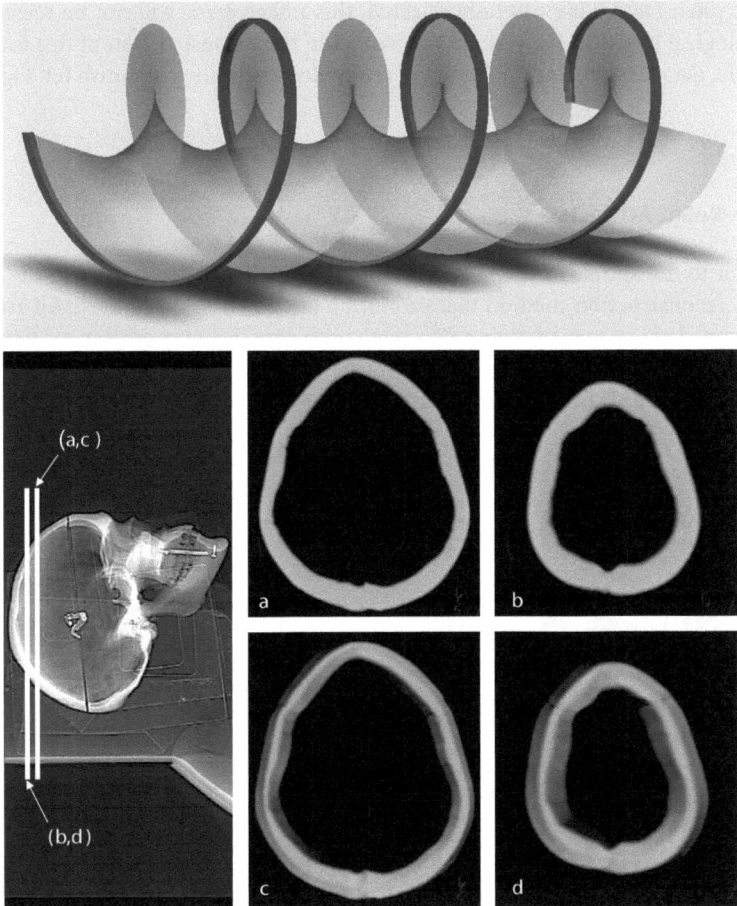

Fig. 9.35. Interpolation artifact in spiral CT. *Top*: The data structure looks like a cut radish. On the *lower left* one sees the planning overview of a skull phantom in which two slices have been selected. **a** and **b** show the corresponding, conventionally acquired reconstructed slices with the partial volume artifacts described in ▸ 9.6.1 and 9.7.1. These existing artifacts are overlaid by an interpolation artifact if an excessively high pitch factor is chosen. The stronger the curvature of the object in the axial direction, the more this error becomes visible in the corresponding parts **c** and **d**

If the spiral acquisition method is used in this geometrical situation, errors may arise with an excessively high *pitch* factor. These are shown in Fig. 9.35c and d. The fringe around the bone structure is considerably enlarged and it is shifted as a function of the angular position of the sampling system. This fringe is due to the fact that data are used for interpolation that are not really close to each other due to the quick axial change of the skull structure. This effect is usually small. To make it nevertheless noticeable, an excessively high pitch factor of $p = 10$ has been chosen.

If the pitch factors are carefully selected, this image error will not be relevant in radiological practice. However, it is revealed in the reconstruction of this example that the data structure of the spiral CT resembles a helically cut radish (cf. Fig. 9.35, top).

9.7.8
Cone-Beam Artifacts

Similar to the spiral CT method, a multitude of artifacts also arise with the cone-beam reconstruction method that cannot be discussed in detail here. All artifacts mentioned above may of course also occur with cone-beam geometry, even though some image errors such as motion artifacts may be reduced due to the increased acquisition speed. But again, new image errors arise with this method, which are

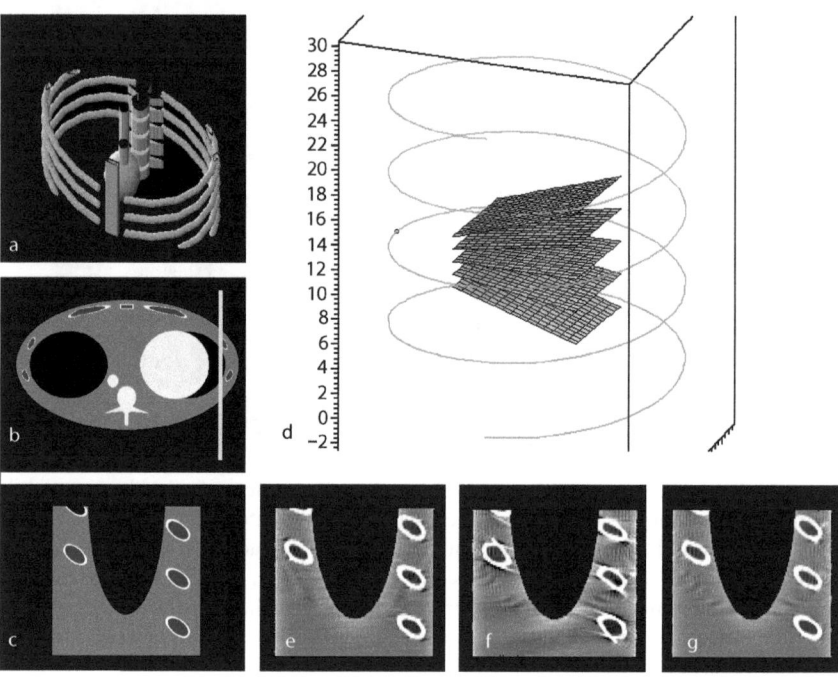

Fig. 9.36. Result of the cone-beam correction with larger aperture angles of the X-ray cone. **a–c** show an anthropomorphic thorax phantom. **c** is the reformatted plane selected by the vertical line in **b**. An important factor influencing the quality of the cone-beam reconstruction is the adaptation of the reconstruction plane to the spiral geometry. **d** shows the corresponding reconstruction planes. The different reconstruction results are shown in **e–g**. **e** shows the standard reconstruction for a 4 × 1-mm slice with a feed of 12 mm/s, **f** the standard reconstruction for a 16 × 1-mm slice with a feed of 48 mm/s, and **g** the reconstruction with the plane correction for a 16 × 1-mm slice with a feed of 48 mm/s (courtesy of Siemens Medical Solutions)

for example due to the shadow zone of the Radon space in cases of incomplete data sets (Wang et al. 1995; Lee et al. 2002).

In practice, the choice of the orientation of the reconstruction plane within the helical cone-beam geometry is particularly important. Figure 9.36a–c shows a simulation of an anthropomorphic thorax phantom.

Figure 9.36d illustrates the adaptation of the inclination of the reconstruction plane to the cone-beam geometry and the Fig. 9.36e–g illustrates the corresponding reconstruction results (Stierstorfer et al. 2001). The result of the reconstruction using the data of a 4-line detector with a slice thickness of 1 mm each and a feed of 12 mm/s is still acceptable for clinical applications with standard reconstruction, although the cone aperture angle has not been taken into account (cf. Fig. 9.36e).

However, the mathematical approximations of the standard reconstruction with regard to the actual inclination of the reconstruction plane in the three-dimensional space are no longer valid with larger cone aperture angles. The standard reconstruction therefore produces considerable artifacts if, for instance, a 16-line detector is used with a slice thickness of 1 mm each and a feed of 48 mm/s (cf. Fig. 9.36f). Figure 9.36g illustrates that the image quality is considerably improved by adapting the reconstruction plane.

9.7.9
Segmentation and Triangulation Inaccuracies

The viewing workstation frequently presents very smooth surfaces to the clinician, which, however, does not allow a conclusion to be drawn regarding the parameters of the scan protocol, such as the slice thickness or the spiral reconstruction index. Figure 9.37 shows the visualization results for the vertebral phantom presented in Fig. 8.7 in the previous chapter. On the left-hand side a volume rendering is shown, in which the slice thickness can be clearly concluded from the figure. On the right-hand side a surface rendering is shown, which does not give any indication of the

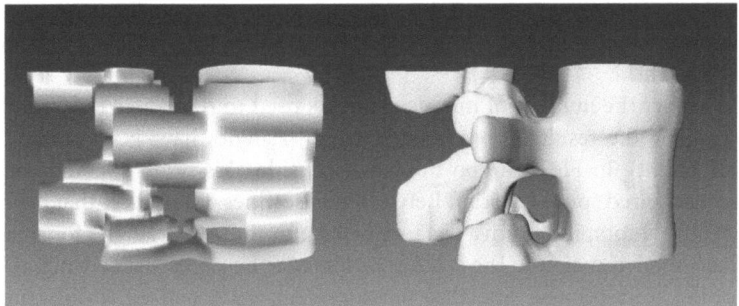

Fig. 9.37. Result of the volume and the surface representation of a vertebral phantom. In the volume representation (*left*) the slice thickness of 6 mm is evident. The surface rendering (*right*) does not give an indication of the slice thickness

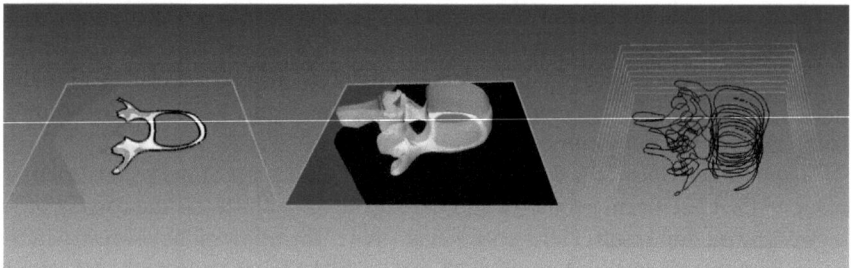

Fig. 9.38. *Left*: An iso-line is selected in the measured slice by an individually chosen gray-value threshold. *Center*: Orientation of the measured slice in the data volume. *Right*: A stack of iso-lines of nine measured slices is the basis for triangulation

slice thickness. However, both representations are based on the same data acquisition protocol.

To understand why the slice thickness is no longer discernible in the surface representation, the visualization process must be briefly discussed. Within the visualization process, the user may decide to show special ranges of gray values and to mask out other ones selectively. If the user selects a constant gray value as a threshold, then all spatial points with this gray-value attribute are displayed as iso-surfaces in space. A single iso-line within the selected slice is shown on the left-hand side of Fig. 9.38. This example illustrates that the type of segmentation may already lead to errors. The spinous processes, for example, which are more faintly imaged due to the partial volume effect, are not enclosed by the iso-line. In the middle of Fig. 9.38 the location of the slice in the three-dimensional data volume is shown to improve the orientation. The right-hand side of Fig. 9.38 illustrates the stack of the nine measured slices with the individual iso-lines.

The stack of the iso-lines is the basis for visualization since the gray-value threshold-based iso-surface is reproduced with triangles on the basis of these lines. Figure 9.39 illustrates the result of the triangulation. The iso-surface is shown as a virtual surface by assigning each of the triangles with a reflectance and then illuminating the mosaic consisting of the triangle tiles with a virtual light source (Fig. 9.39, right).

The higher the number of mosaic elements used to reproduce the surface, the more realistic the result. Approximately 1,600 triangles have been computed in Fig. 9.39 (right). In Fig. 9.40, on the left-hand side, this result is compared with a densely acquired spiral CT scan. Between the iso-lines of 100 slices that have been retrospectively computed with a distance of 0.5 mm, 500,000 triangles form the surface (dark shaded surface). The accuracy difference becomes visible when the surfaces of the conventional nine-slice measurement are visualized within the same volume (bright shaded surface).

Figure 9.40, on the right-hand side, illustrates once again why the result of the coarse measurement must be less accurate. The actual data basis is superimposed onto a transparent surface representation with a sagittal slice. The location of the

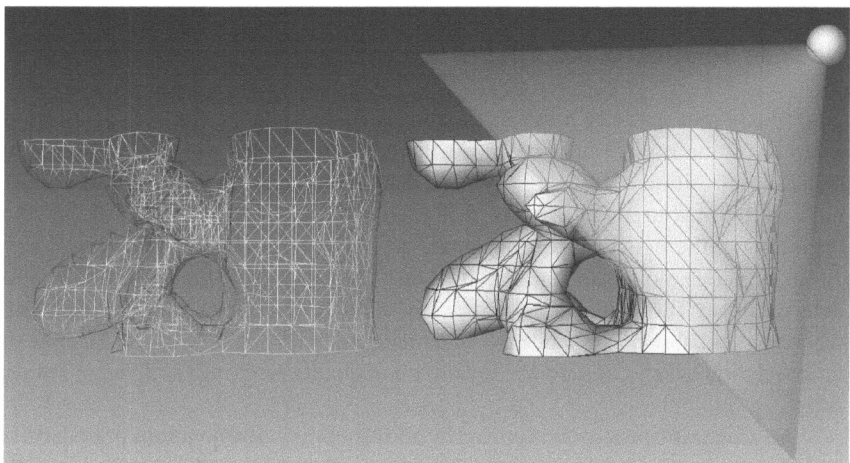

Fig. 9.39. *Left*: Grid model for the triangulation based on the iso-lines. *Right*: Each of the triangles is assigned a reflectance and illuminated with a virtual light source

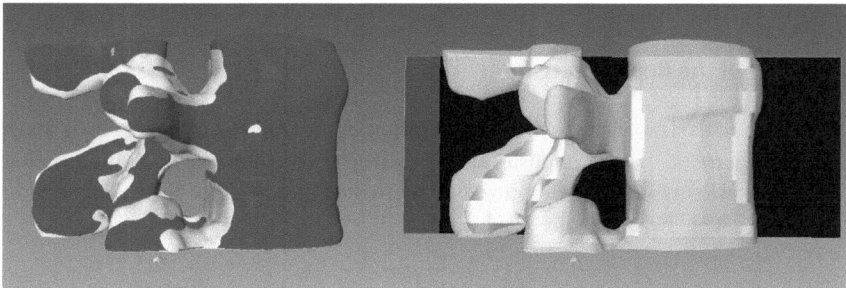

Fig. 9.40. *Left*: The surface representations of the finely acquired spiral CT scan with 100 retrospectively computed slices with a reconstruction distance of 0.5 mm (*dark shaded surface*) and a conventional measurement of nine slices with a slice thickness of 6 mm (*bright shaded surface*) are compared. *Right*: The sagittal slice in the visualization of the transparent surfaces reveals the difference between the coarse data basis and the smoothening triangulation

surface in space is obviously the result of an interpolation. Furthermore, the selection of the gray value threshold for the iso-surface is obviously subject to a certain arbitrariness and, in addition, the location of the surface is influenced by the partial volume artifacts. This must always be taken into account when planning a surgical intervention, for example, when so-called pedicle screws are to be placed through the spinous processes, so that the screws are always placed exactly at the planned location.

9.8
Noise in Reconstructed Images

To evaluate and quantify noise in the reconstructed images, the data acquisition chain has to be studied once again. At the end of the chain the CT images are in fact a superposition of filtered projection data. However, the physical source of the noise is the measurement process of X-ray quanta and this process indicates the beginning of the signal processing chain.

In principle, noise must be classified into the categories of quantum noise and detector noise. The quantum noise is produced by stochastic fluctuations regarding scattering and absorption while the radiation passes through the object. In spite of identical radiographic situations the number of quanta reaching the detector always varies.

This stochastic process is modeled by additive noise. The quantum noise level is linearly increased with the intensity. The quantum noise reveals that the energy is transmitted to the detector in small portions, the X-ray quanta. In contrast, detector noise is a thermal noise produced by the electrons in the detector, does not depend on the X-ray exposure, and above all can also be measured without radiation.

9.8.1
Variance of the Radon Transform

The noise analysis is started with consideration of the number of quanta on a single detector element. As the intensity of the radiation is proportional to the number of quanta measured in the detector, one first takes into account Beer's attenuation law – cf. (2.42) – which reads

$$n_\gamma(\xi) = n_{\gamma,0}(\xi)\,e^{-\int \mu(\eta)\,d\eta}\,,\tag{9.82}$$

where $n_\gamma(\xi)$ is the number of quanta measured under the projection angle γ at the detector element ξ, and $n_{\gamma,0}(\xi)$ is the reference number of quanta without any absorbing objects in the measurement field. Due to the linear relation between the radiation intensity and the number of quanta, the projection integral – cf. (5.10) – can be given by

$$p_\gamma(\xi) = -\int \mu(\xi,\eta)\,d\eta = \ln\left(\frac{n_{\gamma,0}(\xi)}{n_\gamma(\xi)}\right) = \ln\left(n_{\gamma,0}(\xi)\right) - \ln\left(n_\gamma(\xi)\right).\tag{9.83}$$

The values of the projection integral represent the Radon space. As the number of quanta measured during the counting process is subject to Poisson distribution, the average value $\langle n_\gamma(\xi)\rangle$ and the standard deviation $\sigma_{n,\gamma}(\xi)$ obey the relation[12]

$$\langle n_\gamma(\xi)\rangle = \left(\sigma_{n,\gamma}(\xi)\right)^2\tag{9.84}$$

---

[12] Recall that the mean value is an estimate of the expectation value.

(cf. ▶ Sect. 2.6), so that the measurement of the number of quanta on the individual detectors can be expressed as

$$n_\gamma(\xi) = \langle n_\gamma(\xi) \rangle \pm \sigma_{n,\gamma}(\xi) = \langle n_\gamma(\xi) \rangle \pm \sqrt{\langle n_\gamma(\xi) \rangle} \,. \tag{9.85}$$

As the next step in the signal processing chain consists of the logarithmation of the values measured with the detector, one is interested in the standard deviation of the projection integral $p_\gamma(\xi)$.

If one substitutes (9.85) in (9.83), one finds

$$
\begin{aligned}
p_\gamma(\xi) &= \ln\left(n_{\gamma,0}(\xi)\right) - \ln\left(\langle n_\gamma(\xi)\rangle \pm \sqrt{\langle n_\gamma(\xi)\rangle}\right) \\
&= \ln\left(n_{\gamma,0}(\xi)\right) - \ln\left(\langle n_\gamma(\xi)\rangle \left[1 \pm \frac{\sqrt{\langle n_\gamma(\xi)\rangle}}{\langle n_\gamma(\xi)\rangle}\right]\right) \\
&= \ln\left(n_{\gamma,0}(\xi)\right) - \ln\left(\langle n_\gamma(\xi)\rangle \left[1 \pm \frac{1}{\sqrt{\langle n_\gamma(\xi)\rangle}}\right]\right) \\
&= \ln\left(n_{\gamma,0}(\xi)\right) - \ln\left(\langle n_\gamma(\xi)\rangle\right) + \ln\left(1 \pm \frac{1}{\sqrt{\langle n_\gamma(\xi)\rangle}}\right).
\end{aligned}
\tag{9.86}
$$

With the series expansion

$$\ln(1 \pm x) = -\sum_{i=1}^{\infty} \frac{(\mp 1)^i x^i}{i} \quad \text{for } -1 < x < +1, \tag{9.87}$$

it can be concluded for $\langle n_\gamma(\xi)\rangle^{1/2} \gg 1$ and the truncation of the expansion (9.87) after the linear term that

$$p_\gamma(\xi) = \ln\left(n_{\gamma,0}(\xi)\right) - \ln\left(\langle n_\gamma(\xi)\rangle\right) \pm \frac{1}{\sqrt{\langle n_\gamma(\xi)\rangle}}. \tag{9.88}$$

According to (9.83), the first two terms in (9.88) can be considered as the average value of the projection integral such that

$$p_\gamma(\xi) = \langle p_\gamma(\xi)\rangle \pm \frac{1}{\sqrt{\langle n_\gamma(\xi)\rangle}}. \tag{9.89}$$

The standard deviation of the projection integral therefore obeys

$$\sigma_{p,\gamma}(\xi) = \frac{1}{\sqrt{\langle n_\gamma(\xi)\rangle}} \tag{9.90}$$

and the variance is given by

$$\left(\sigma_{p,\gamma}(\xi)\right)^2 = \frac{1}{\langle n_\gamma(\xi)\rangle}. \tag{9.91}$$

9.8.2
Variance of the Reconstruction

The next step in the reconstruction chain considered here is the filtered backprojection, which can be formulated with its discrete form – cf. (7.11) – as

$$f(x, y) = \frac{\pi}{N_p} \sum_{n=1}^{N_p} h_{\gamma_n}(x \cos(\gamma_n) + y \sin(\gamma_n)) . \tag{9.92}$$

If the convolution is carried out in the spatial domain of the detector array to filter the projection signal, one can write, in analogy to (7.51),

$$f(x, y) = \frac{\pi}{N_p} \Delta\xi \sum_{n=1}^{N_p} \sum_{k=-D/2}^{D/2-1} p_{\gamma_n}(k\Delta\xi) g_{\gamma_n}(x \cos(\gamma_n) + y \sin(\gamma_n) - k\Delta\xi) . \tag{9.93}$$

Taking into account (9.89), it can be seen that

$$f(x, y) = \frac{\pi}{N_p} \Delta\xi \sum_{n=1}^{N_p} \sum_{k=-D/2}^{D/2-1} \left(\langle p_{\gamma_n}(k\Delta\xi) \rangle \pm \frac{1}{\sqrt{\langle n_{\gamma_n}(k\Delta\xi) \rangle}} \right) g_{\gamma_n}(x \cos(\gamma_n) + \dots$$
$$\dots + y \sin(\gamma_n) - k\Delta\xi) , \tag{9.94}$$

so that the individual pixel values may compactly read

$$f(x, y) = \langle f(x, y) \rangle \pm \sigma_f(x, y) \tag{9.95}$$

where

$$\langle f(x, y) \rangle = \frac{\pi}{N_p} \Delta\xi \sum_{n=1}^{N_p} \sum_{k=-D/2}^{D/2-1} \langle p_{\gamma_n}(k\Delta\xi) \rangle g_{\gamma_n}(x \cos(\gamma_n) + y \sin(\gamma_n) - k\Delta\xi) \tag{9.96}$$

and

$$\sigma_f(x, y) = \frac{\pi}{N_p} \Delta\xi \sum_{n=1}^{N_p} \sum_{k=-D/2}^{D/2-1} \frac{1}{\sqrt{\langle n_{\gamma_n}(k\Delta\xi) \rangle}} g_{\gamma_n}(x \cos(\gamma_n) + y \sin(\gamma_n) - k\Delta\xi) . \tag{9.97}$$

As the individual projections represent statistically independent fluctuations, the variance can be expressed as

$$\sigma_f^2(x, y) = \left(\frac{\pi}{N_p} \Delta\xi \right)^2 \sum_{n=1}^{N_p} \sum_{k=-D/2}^{D/2-1} \frac{g_{\gamma_n}^2(x \cos(\gamma_n) + y \sin(\gamma_n) - k\Delta\xi)}{\langle n_{\gamma_n}(k\Delta\xi) \rangle} , \tag{9.98}$$

because the mixed terms add up to zero.

The variance in the origin[13] is thus

$$\sigma_f^2(0,0) = \left(\frac{\pi}{N_p}\Delta\xi\right)^2 \sum_{n=1}^{N_p} \sum_{k=-D/2}^{D/2-1} \frac{g_{\gamma_n}^2(k\Delta\xi)}{\langle n_{\gamma_n}(k\Delta\xi)\rangle} , \tag{9.99}$$

since a comparison with ▸ Sect. 7.2.3 shows that g is an even function so that $g(-k\Delta\xi) = g(k\Delta\xi)$. For a cylindrical, homogeneous object, the fluctuation of the center beam is always the same under any projection angle; thus,

$$\sigma_f^2(0,0) = \left(\frac{\pi}{N_p}\Delta\xi\right)^2 \left(\sum_{n=1}^{N_p} \frac{1}{\langle n_{\gamma_n}(0)\rangle}\right) \left(\sum_{k=-D/2}^{D/2-1} g_\gamma^2(k\Delta\xi)\right)$$

$$= \left(\frac{\pi}{N_p}\Delta\xi\right)^2 \frac{N_p}{\langle n(0)\rangle} \left(\sum_{k=-D/2}^{D/2-1} g_\gamma^2(k\Delta\xi)\right) . \tag{9.100}$$

If the general relation for integral discretization, i.e.,

$$\Delta\xi \sum g_\gamma^2(k\Delta\xi) \approx \int g_\gamma^2(\xi)\,d\xi \tag{9.101}$$

and, furthermore, Rayleigh's theorem (cf. ▸ Sect. 4.12), is used, then the real signal $g_\gamma(\xi)$ obeys

$$\int g_\gamma^2(\xi)\,d\xi = \int |G_\gamma(q)|^2\,dq \tag{9.102}$$

and, thus, the variance of the pixel in the origin obeys

$$\sigma_f^2(0,0) = \left(\frac{\pi}{N_p}\Delta\xi\right)^2 \frac{N_p}{\langle n(0)\rangle} \frac{1}{\Delta\xi} \int_{-\frac{1}{2\Delta\xi}}^{+\frac{1}{2\Delta\xi}} |G_\gamma(q)|^2\,dq$$

$$= \frac{\pi^2\Delta\xi}{N_p\langle n(0)\rangle} \int_{-\frac{1}{2\Delta\xi}}^{+\frac{1}{2\Delta\xi}} |G_\gamma(q)|^2\,dq . \tag{9.103}$$

Obviously, the pixel noise of the reconstructed image depends on the selected filter function of the filtered backprojection. The variance $\sigma_f^2(0,0)$ of the pixel $f(0,0)$ is proportional to the area under the quadratic norm of the filter functional in the frequency space. Several filter functions have been presented in ▸ Sect. 7.2. If the functionals (7.14) through (7.28) are combined with (9.103), one clearly sees that the image noise can be reduced by minimizing the area

$$\sigma_f^2(0,0) \propto \int_{-\frac{1}{2\Delta\xi}}^{+\frac{1}{2\Delta\xi}} |G_\gamma(q)|^2\,dq . \tag{9.104}$$

---

[13] Without restriction of generality, the behavior of the center pixel only is discussed.

On the other hand, it must be taken into account that any deviation from ideal filtering

$$G(q) = |q| \text{ for all } q,\tag{9.105}$$

results in a reduction of the spatial resolution because, as mentioned earlier in this chapter, the MTF of the reconstructed image depends on the type of band limitation and windowing of the ideal linear filter function. In ▸ Sect. 9.4 it was explained that the spatial resolution is determined by

$$MTF_{\text{filter kernel}}(q) = \frac{|G(q)|}{|q|}.\tag{9.106}$$

This means that the spatial resolution cannot be maximized without increasing the image noise, too. The selection of the filter functional is an optimization process in which one has to find a trade-off between acceptable image noise and the required spatial resolution, depending upon the anatomical problem. If (9.50) is substituted into (9.103) one finds the expression

$$\sigma_f^2(0,0) = \frac{\pi^2 \Delta \xi}{N_p \langle n(0) \rangle} \int_{-\frac{1}{2\Delta\xi}}^{+\frac{1}{2\Delta\xi}} \left(q MTF_{\text{filter kernel}}(q) \right)^2 dq.\tag{9.107}$$

Minimizing the area under the integral would thus unfortunately mean that the MTF would be minimized.

If a fixed filter functional $G(q)$ is assumed, the normalization term

$$\sigma_f^2(0,0) \propto \frac{\pi^2 \Delta \xi}{N_p \langle n(0) \rangle}\tag{9.108}$$

of (9.103) can be analyzed. Obviously, noise is reduced with an increasing number of projections, N_p.

Furthermore, (9.108) supports the impression that the width of the detector elements $\Delta \xi$ must be kept as small as possible to achieve a small variance in the reconstructed pixels. This is actually a false conclusion since, on the one hand, the average number of quanta $\langle n(0) \rangle$, measured in one single detector element, linearly depends on the detector width. On the other hand, the width of the detector elements appears within the limits of the integral term of (9.107). Therefore, one has to solve this integral to draw a final conclusion on the dependence of variance and detector width. For this purpose, ideal (rectangular) filtering – cf. (7.14) – is assumed; thus, band limitation results in

$$\sigma_f^2(0,0) = \frac{\pi^2 \Delta \xi}{N_p \langle n(0) \rangle} \int_{-\frac{1}{2\Delta\xi}}^{+\frac{1}{2\Delta\xi}} (q)^2 dq = \frac{\pi^2 \Delta \xi}{N_p \langle n(0) \rangle} \frac{1}{3} \left[q^3 \right]_{-\frac{1}{2\Delta\xi}}^{+\frac{1}{2\Delta\xi}}$$

$$= \frac{\pi^2 \Delta \xi}{N_p \langle n(0) \rangle} \frac{1}{12 (\Delta \xi)^3} = \frac{\pi^2}{12 N_p \langle n(0) \rangle (\Delta \xi)^2}.\tag{9.109}$$

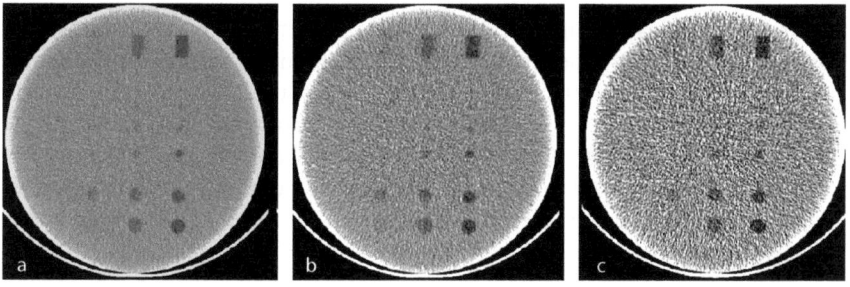

Fig. 9.41. Influence of the radiation intensity and the number of projections on the noise in the reconstructed image. The phantom was scanned with different numbers of projections, N_p, and tube currents, I (**a**: N_p = 1,640 and I = 100 mA; **b**: N_p = 656 and I = 100 mA; **c**: N_p = 656 and I = 50 mA)

In fact, the variance increases inversely proportional to the square of the detector width.

In (9.109), $\langle n(0) \rangle$ is the mean number of transmitted X-ray quanta of the center beam. For a circular disk of radius R and homogeneous attenuation μ, this number can be estimated by

$$\langle n(0) \rangle = \langle n_0(0) \rangle e^{-\int_0^{2R} \mu \, d\eta} = \langle n_0(0) \rangle e^{-2R\mu} . \tag{9.110}$$

Under any angle, the mean value of the quanta number reaching the detector is equal to the number of quanta generated in the X-ray source, multiplied by a constant attenuation factor. As the mean number of quanta $\langle n(0) \rangle$ is inversely proportional to the variance of the image noise, the noise may obviously be reduced by increasing $\langle n_0(0) \rangle$. The number of quanta $\langle n_0(0) \rangle$ depends on the current I of the X-ray tube (cf. 2.55 and 2.56). The higher the tube current, the higher the intensity of the radiation and the more image noise is reduced. Figure 9.41 illustrates this effect on a phantom scanned with different numbers of projections, N_p, and different tube currents, I.

9.8.3
Dose, Contrast, and Variance

The X-ray dose to which the patient is exposed is proportional to the tube current, I. The relation between the dose and the contrast, which is marginally visible, is given by the so-called low-contrast resolution defined by a line of holes in a phantom (Morneburg 1995). According to this, the dose D and the just identifiable contrast K_d are related by the proportionality

$$K_d \propto \frac{1}{\sqrt{D}} , \tag{9.111}$$

where d is the diameter of the individual holes of the hole line.

Strictly speaking, not the current but the product of current and measuring time, the so-called *mAs* product (cf. ▸ Sect. 11.5), determines the number of photons and thus the image quality. If one doubles the *mAs* product, the signal-to-noise ratio is improved by the factor $\sqrt{2}$ (cf. 2.96). This of course improves the contrast. As the dose increases linearly with the *mAs* product, one must therefore make a compromise in practice. Additionally, one has to take into account in practice that heavier patients absorb more quanta than thinner patients. The signal-to-noise ratio is thus reduced with heavier patients if the same *mAs* is used as for thin patients.

The influence of slice thickness on image quality has already been discussed in ▸ Sects. 9.6.1, 9.7.1, and 9.7.2 in the context of partial volume artifacts and staircasing. These averaging problems resulted in a deterioration of the image. However, it must be taken into account with respect to the image noise that, obviously, the number of quanta increases proportionally to the slice thickness. Therefore, if one doubles the slice thickness, the signal-to-noise ratio is improved by the factor $\sqrt{2}$. These two counteracting effects may not be generally compared, because, in practice, the slice thickness to be selected always depends on the anatomic region that is actually scanned.

Intelligent scanner concepts use (9.111) such that the anode current is dynamically adapted to the corresponding anatomical area. Figure 11.7 illustrates this approach schematically. In the area of the lung the tube current and thus the dose may be reduced without having to accept a poorer image quality.

The relation between the signal-to-noise ratio and the dose, as well as the detector size is particularly interesting. Especially from a practical point of view, there is a high demand for small detector elements in order to visualize smaller anatomical details. In this context, the combination of (9.109) and (9.110) directly yields

$$\sigma_f^2(0,0) = \frac{\pi^2}{12 N_p \left(\Delta\xi\right)^2 \langle n_0(0)\rangle e^{-2R\mu}} \,. \tag{9.112}$$

As the mean for the attenuation value to be reconstructed is μ everywhere for a homogeneous disk or cylindrical object, the average value of the original pixel particularly obeys

$$\langle f(0,0)\rangle = \mu \,. \tag{9.113}$$

Taking into account the definition of the signal-to-noise ratio (9.57), we therefore find

$$SNR(0,0) = \frac{\langle f(0,0)\rangle}{\sigma_f(0,0)} = \frac{\mu}{\pi}\sqrt{12 N_p \left(\Delta\xi\right)^2 \langle n_0(0)\rangle e^{-2R\mu}} \tag{9.114}$$

for the pixel in the origin.

The dose applied in the center pixel, $f(0,0)$ with the pixel edge length b of a very thin disk with the radius R during measurement with X-ray, amounts to

$$D \propto \frac{N_p \langle n_0(0)\rangle}{b^2} \,, \tag{9.115}$$

(cf. formula of Brooks and DiChiro [1976] discussed in ▸ Sect. 11.4) and is thus proportional to the quantum number per pixel area[14]. If one assumes that the pixel edge
length b approximately corresponds to the size of the detector element, i.e., $b \approx \Delta\xi$,
then (9.114) yields

$$SNR(0,0) \propto \frac{\mu}{\pi}\sqrt{12\,(\Delta\xi)^4\,D} \qquad (9.116)$$

and one finds the important relation

$$D \propto \frac{(SNR)^2}{(\Delta\xi)^4} \qquad (9.117)$$

among the detector resolution, dose, and signal-to-noise ratio. (9.117) means that,
with a constant SNR, the dose is increased with the fourth power when the detector
size is reduced. Thus, the size of the detector elements can of course not be arbitrarily reduced because the dose cannot be arbitrarily increased. This limits the chase
after smaller detector elements in future detector systems.

---

[14] The exact conversion of the quantum number/mm² into the dose depends on the anode
voltage, the beam-hardening filter, and the tissue type (Dössel 2001).

10 Practical Aspects of Computed Tomography

Contents

10.1   Introduction 471
10.2   Scan Planning 471
10.3   Data Representation 475
10.4   Some Applications in Medicine...................................... 482

10.1
Introduction

In the following sections, some important practical aspects of computed tomography (CT) will be discussed. Among others, these aspects concern scan planning, data processing, and representation – particularly gray value scaling. In particular, scan planning plays an important role in clinical applications of CT since the scans cannot arbitrarily be repeated due to the system-inherent radiation dose to which the patient will be exposed.

Therefore, it is furthermore not only important to properly plan the scan, i.e., to prepare the image acquisition, but also to be informed about how the image data may be represented. That is, how to appropriately post-process the acquired data, in order to maximize the information that is available from the images by means of modern visualization techniques. Both aspects – planning and post-processing – will be discussed in this chapter. The different scan protocols for the corresponding radiological aspects will not be discussed here; they are described, for example, by Romans (1995). However, in the last section some examples of the medical application of modern CT systems are given.

10.2
Scan Planning

The first and most important step in planning a CT scan is the acquisition of an overview scan, which is called a topogram, scanogram, scout view or pilot scan by the different manufacturers. To acquire this overview scan, the rotation of the sampling unit is stopped at a desired angle. In principle, any angulation is possible, but typical positions are *anterior–posterior* (a.p.), i.e., X-ray examination from the patient's front to the patient's back, and *lateral*, i.e., X-ray examination from the side.

During acquisition of the overview scan the patient table is continuously moved through the measuring field. Figure 10.1 exemplarily shows images resulting from an a.p. and a lateral overview scan. The resulting images are quite similar to conventional X-ray images. However, both imaging techniques differ in that a parallax in the axial direction of the patient does not occur for the overview scan because of the minimal divergence of the X-ray beam due to slice collimation.

By means of the a.p. overview scan, it is possible to plan and program a particular slice plane, the slice thickness, and the number of slices or the volume. When using the lateral overview scan it is also possible to program special slice orien-

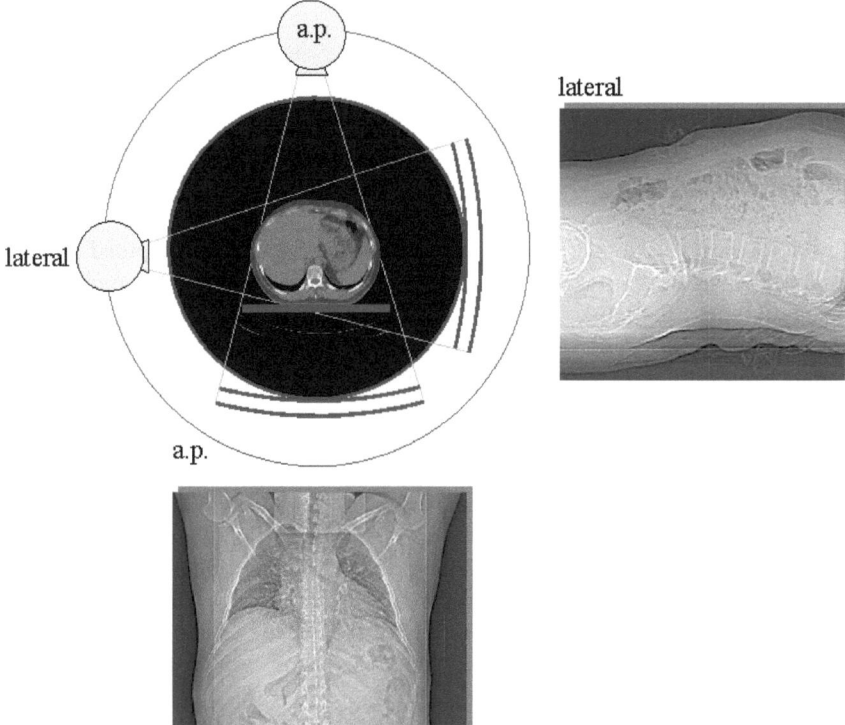

Fig. 10.1. Preparation of an overview scan to plan the slice plane position. With a fixed tube detector position, the table is continuously moved through the *gantry*. This produces projection images similar to X-ray fluoroscopy. Two geometries are typically used: Lateral – The patient is X-rayed from the side; anterior–posterior (*a.p.*) – The patient is X-rayed from the front to the back. It is often also possible to adjust to the opposite direction, i.e., posterior–anterior X-ray examination

tations by appropriately tilting the entire gantry. This is especially useful in skull radiographs and spinograms to exclude sensitive organs (e.g., eyes) from the scan.

Figure 10.2a shows the a.p. overview scan of a thorax. Here, axial slices with a thickness and distance of 8 mm up to 10 mm are typically scanned. If the lumbar vertebrae are to be examined, the gantry must be tilted such that it is adapted to the orientation of the individual vertebral bodies. Figure 10.2b highlights that this is the only possibility of acquiring planar images of, for instance, the intervertebral disks.

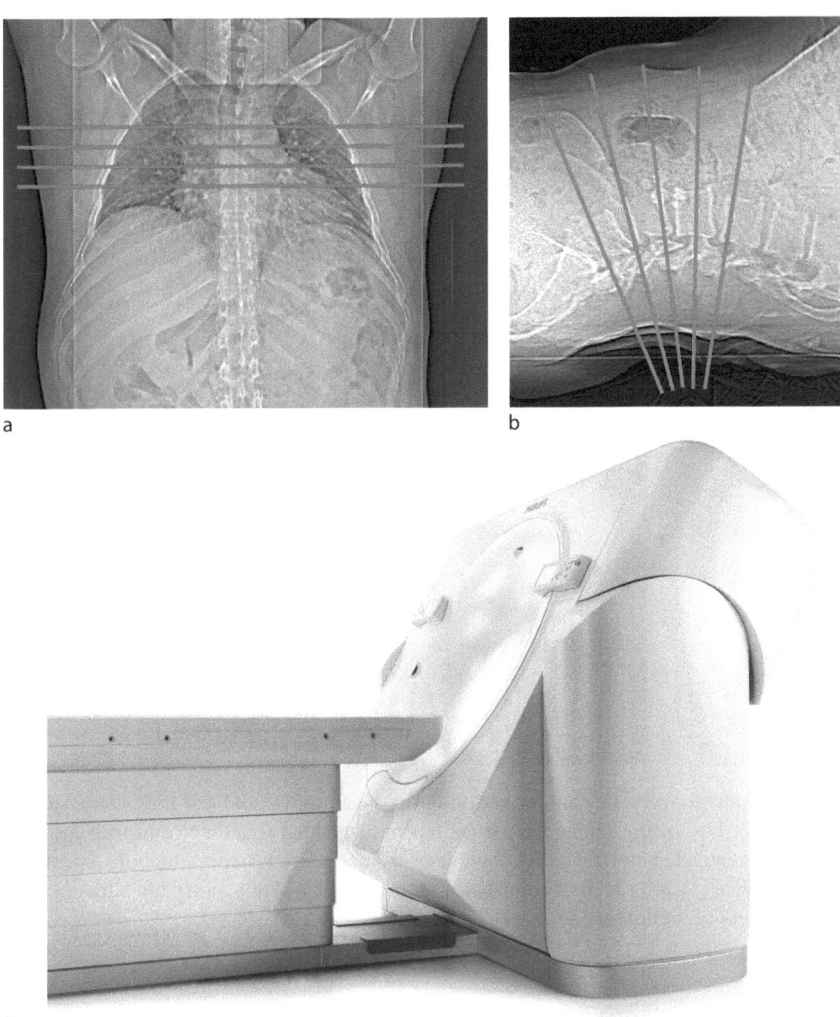

a

b

c

Fig. 10.2. a An a.p. overview scan is used to plan the slice position in thorax radiographs. With this overview scan, it is only possible to plan axial radiographs without tilting the gantry. **b** If different lumbar vertebrae are to be examined, it is recommended to adapt the slice orientation to the anatomical situation by programming the corresponding gantry tilts (**c**)

Figure 10.3 shows two different orientations of the skull. If one aims to acquire coronal skull images, it is necessary to appropriately place the patient on the table. In Fig. 10.3a a lateral overview scan of the patient can be seen. The patient is lying on his abdomen with his head being heavily hyper-extended. This hyper-extension of the head is indeed required due to the limitations in the angle of gantry tilt. Two tomographic slices are shown on the right side in Fig. 10.3a. Such radiographs are frequently used either to diagnose chronic sinusitis or to evaluate bone fractures. In slice 1, artifacts due to dental fillings can clearly be identified. Slice 2 shows a section of the frontal calvarium of the skull with spongiosa. Both slices allow unobstructed paranasal sinuses to be discerned.

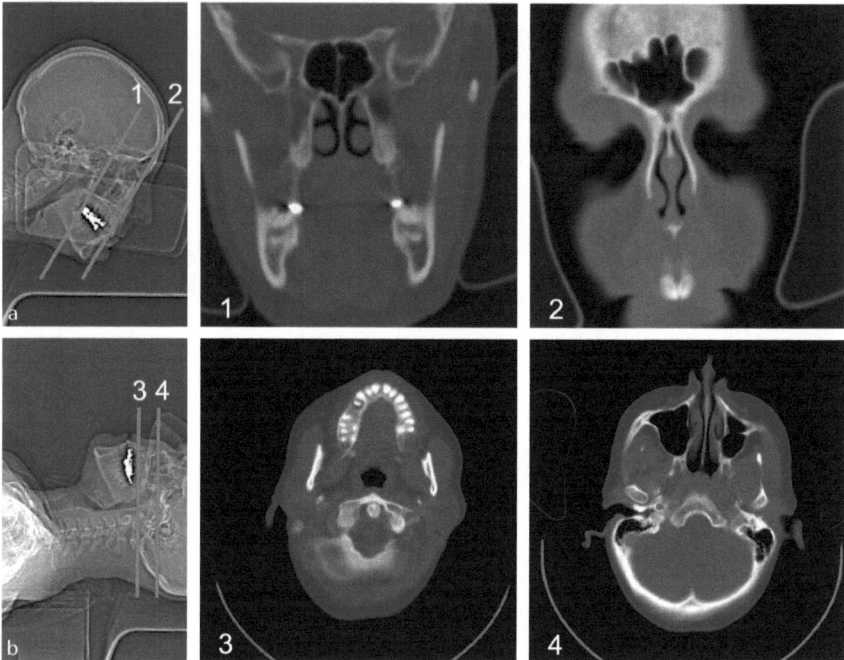

Fig. 10.3. Lateral overview scan for planning the slice position of head tomograms. **a** Due to the limited tilt of the gantry the patient has to be bedded face-down and with his head being hyperextended into the neck to be able to acquire a coronal representation of the facial cranium: Slice (1) and (2). **b** In comparison, axial images without gantry tilt: Slices (3) and (4). Courtesy of Ruhlmann, Medical Center Bonn

10.3
Data Representation

10.3.1
Hounsfield Units

In CT, the attenuation values μ from equation

$$I(\eta) = I_o e^{-\mu\eta} \tag{10.1}$$

are usually represented as gray values. In this context, an approach developed by Hounsfield has proven to be appropriate and is nowadays commonly used. Here, the attenuation values are transformed onto a dimensionless scale and are related to the attenuation value of water. The definition of these CT values reads

$$\text{CT-Value} = \frac{\mu - \mu_{\text{water}}}{\mu_{\text{water}}} 1000 \ . \tag{10.2}$$

In honor of *Hounsfield*, the unit of these values,

$$[\text{CT-Value}] = \text{HU}, \tag{10.3}$$

is called the Hounsfield unit (HU). On this scale, the CT value $-1,000\,\text{HU}$ is assigned to air and the CT value $0\,\text{HU}$ to water. In principle, this is an open-ended scale, but in practice it ends at approximately $3,000\,\text{HU}$. The range of $4,000\,\text{HU}$ overall can be captured quite well by means of 12-bit gray-value images. This scaling is arbitrary, but nevertheless has practical consequences. Since the attenuation values of almost all organs – except the bones – are quite similar to those of water, the difference from the attenuation value of water that is given by (10.2) is of per mill.

Radiologists are accustomed to considering the CT values as absolute values, which can unequivocally be assigned to the organs. Deviations of these CT values for certain organs indicate pathology. ▸ Chapter 2 already pointed out the dependence of the X-ray attenuation on the wavelength of the radiation and ▸ Sect. 9.6.2 discussed the potential artifacts arising from this. This problem, which emerges in the case of all CT scanners used for diagnostic imaging, is a consequence of the usage of polychromatic radiation spectra. While passing through the body, the spectral distribution of the radiation changes such that an unequivocal assignment of attenuation values is actually not possible.

Nevertheless, the view of the radiologists is largely justified, since most organs behave like water with regard to radiation physics. Therefore, it is possible to correct the beam hardening for these objects by means of a calibration measurement performed with a water phantom (cf. ▸ Sect. 9.6.2). Thus, for the CT values of soft tissue, the definition of the Hounsfield value

$$\text{CT-Value} = \frac{\rho - \rho_{\text{water}}}{\rho_{\text{water}}} 1000 \tag{10.4}$$

is directly linked to the tissue's density ρ. In this context, the lower part (b) in Fig. 10.4 is of special interest because different organs and organ changes can be readily distinguished here according to (10.4). However, (10.4) does not hold for bones, contrast media, metal implants, and similar materials, but depends on the X-ray spectrum and the beam-hardening correction (Morneburg 1995).

10.3.2
Window Width and Window Level

It is sensible to divide the whole Hounsfield scale into diagnostically relevant intervals as shown in Fig. 10.4. Figure 10.4a provides the histogram of the relative frequencies of the CT values of an abdominal slice. The accumulations of air, the foam plastic of the patient bed as well as the fat and the organs can be seen. In the representation of the tomographical slice of the parenchymatous organs in Fig. 10.4b, the problem arises that many organs are mapped into overlapping intervals of CT values. Therefore, a sound diagnosis is actually not easy; thus, textures of organs are also important in clinical practice.

The human visual system cannot resolve the complete dynamic range from $-1,000\,HU$ up to $3,000\,HU$ with 4,000 gray-value steps. This is the reason why in practice gray-value discriminations of only 512 steps are resolved on the display devices. Recent studies have shown that the human observer is able to discriminate between 700 and 900 shades of gray for the available luminance range of current medical displays and in optimal conditions (Kimke and Tuytschaever 2006). To be able to detect differences between organs that have rather similar visual representations in their attenuation, it is necessary to appropriately map the respective anatomically sensible Hounsfield interval to the perceptible gray-value range. For this, one uses the piecewise linear function

$$G = 511 \cdot \begin{cases} 0 & \text{for CT values } \leq WL - \dfrac{WW}{2} \\[2mm] WW^{-1}\left(\text{CT value} - WL + \dfrac{WW}{2}\right) \\[2mm] 1 & \text{for CT values } \geq WL + \dfrac{WW}{2} \,. \end{cases} \tag{10.5}$$

Thereby WW denotes the window width and WL the window level.

Figure 10.5 shows the corresponding piecewise linear function for a bone window ($WL = +300\,HU$, $WW = 1,500\,HU$) and a soft tissue window ($WL = +50\,HU$, $WW = 350\,HU$), as well as their effects on the representation of an abdominal tomogram. Density differences in the spinal process are visible only in the bone window, but a differentiation of the soft tissue is hardly possible due to the large width of the window. In the soft tissue window, organs such as the liver and the kidney can be distinguished quite well. However, in this relatively narrow window, all CT values above $+225\,HU$ are displayed undifferentiated as white areas.

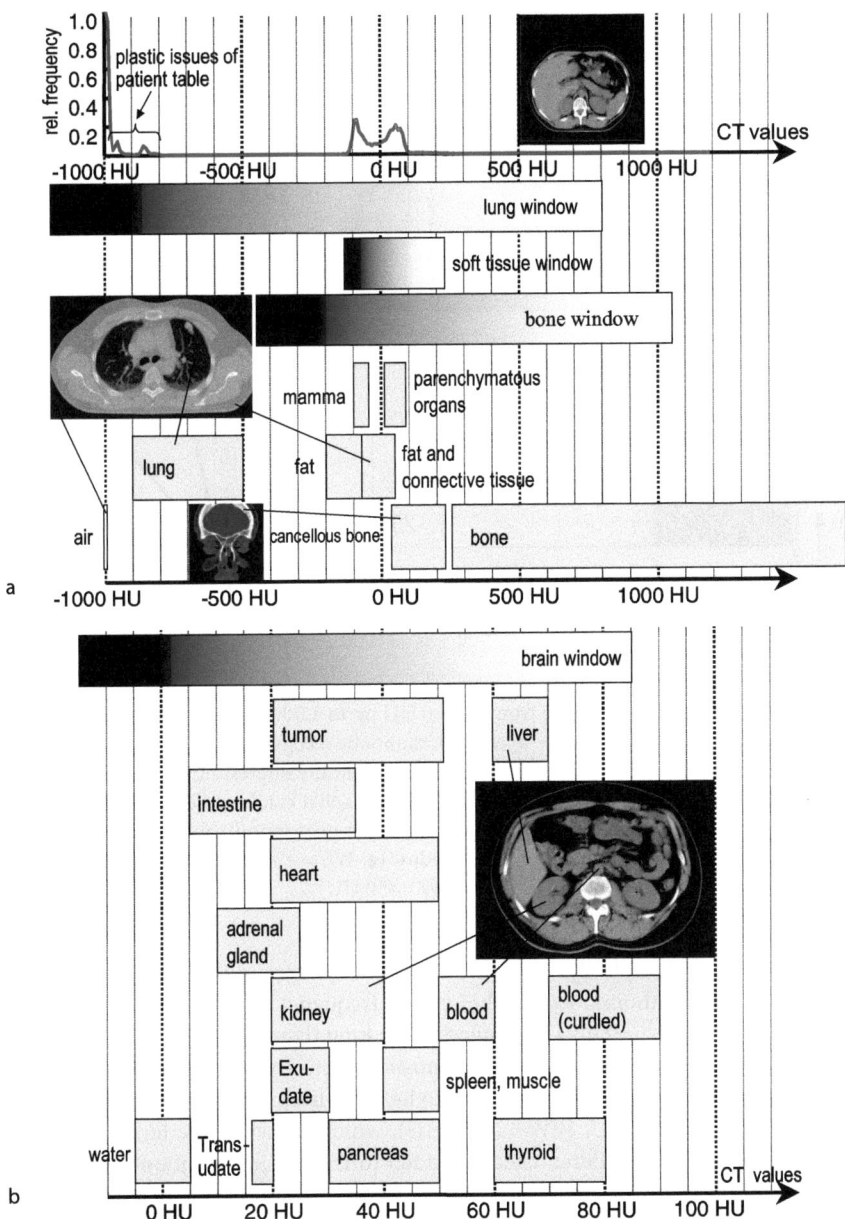

Fig. 10.4. Attenuation values in Hounsfield units **a** of all tissues and **b** of the soft tissue interval only. The data are compiled from the collections in Krestel and Hofer (Krestel 1990; Hofer 2000)

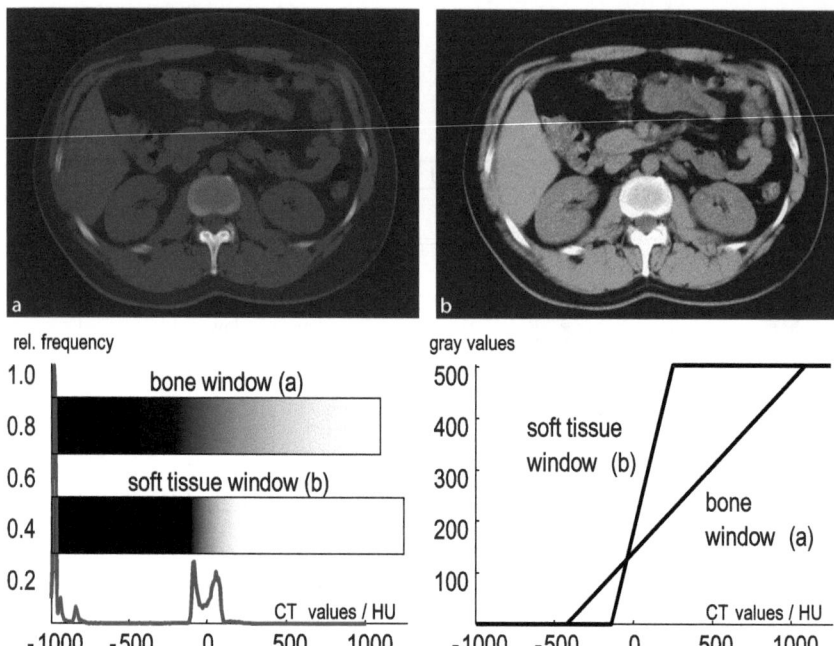

Fig. 10.5. The principle of windowing: The complete interval of practically sensible Hounsfield values (HU), ranging from –1,000 HU up to 3,000 HU (cf. relative frequency of the values for the abdomen area – *lower left*), cannot be recognized and distinguished by the human visual system. Therefore, the different, anatomically interesting Hounsfield intervals have to be mapped to appropriate gray-scale intervals that can be differentiated. The *lower right* image provides the characteristic curves for two anatomically relevant windows. The *upper* images show the result of the bone window (**a**: $WL = +300$ HU, $WW = 1,500$ HU) and the soft tissue window (**b**: $WL = +50$ HU, $WW = 350$ HU)

Figure 10.6 exemplarily shows the results of applying different windows to an image slice of the thorax. Again, the relative frequency of the CT values is given. These images are especially challenging, since lung tissue, soft tissue, and the bone tissue might be interesting from a diagnostic point of view. For the classification of the areas, three windows turned out to be quite useful. Thus, the lung or pleura window ($WL = -200$ HU, $WW = 2,000$ HU), which allows to the lung tissue with its low density to be differentiated, is added to the already mentioned soft tissue window and bone window.

A quite narrow window has to be arranged in order to be able to image the small density differences within the nerve tissue of the brain. This is exemplarily illustrated in Fig. 10.7a–c.

Typically, a window width of $WW = 80$–100 HU is selected for a window level of $WL = +35$ HU (Hofer 2000). Again, Fig. 10.7 also shows the relative frequency distribution of the CT values. Therefore, the cumulation representing the nerve tissue is readily visible. When the brain window is adapted accordingly, it is even

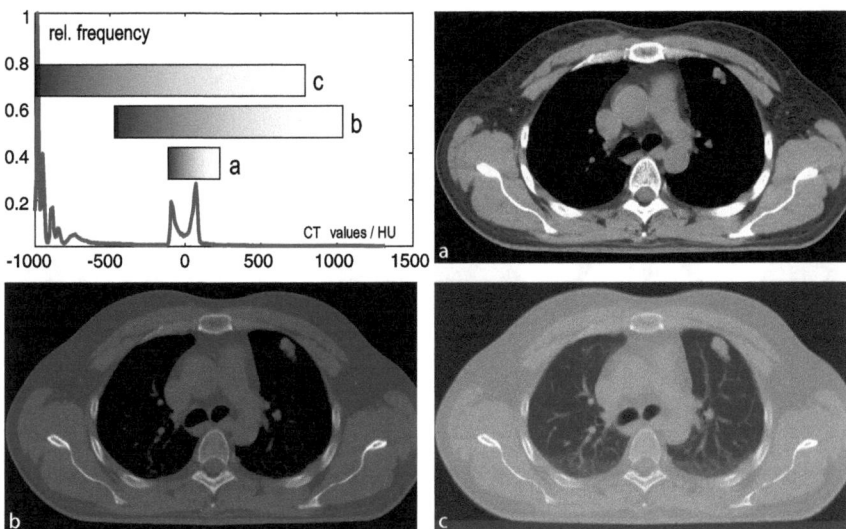

Fig. 10.6. Comparison of the differentiability of tissue when using a soft tissue window (**a**: $WL = +50$ HU, $WW = 350$ HU), a bone window (**b**: $WL = +300$ HU, $WW = 1,500$ HU), and a lung or pleura window (**c**: $WL = -200$ HU, $WW = 2,000$ HU); data courtesy of Ruhlmann

possible to visualize density differences between the white and gray matter of the brain. However, it is not possible to assess the calvarium of the skull at the same time, since all CT values larger than 75–85 HU are again mapped appearing equally white. For this purpose, one has to select the bone window to be able to differentiate the higher CT values representing the cranial bone. The bone window representations in Fig. 10.7b and d even allow the spongiosa structure of the cranial bone to be identified.

10.3.3
Three-Dimensional Representation

Figure 10.8b and c provides images of the volume rendering technique. This method assigns a physical light reflection and scattering value to each voxel (spatial pixel). The computer is then used to light up this *data fog* with a virtual light source and to model the optical impression artificially. To do so a real reflection and scattering of light is simulated.

If bones, organs or contrast media-enhanced vessels are assigned to different optical properties, interesting insights into the pathological status of the patient can be gained (cf. Figs. 10.8b–c and 10.9f). In a post-processing step, certain tissue types can be suppressed. In Fig. 10.8d for example, bone is eliminated by a special bone removal technique for better visualization of the vessel tree.

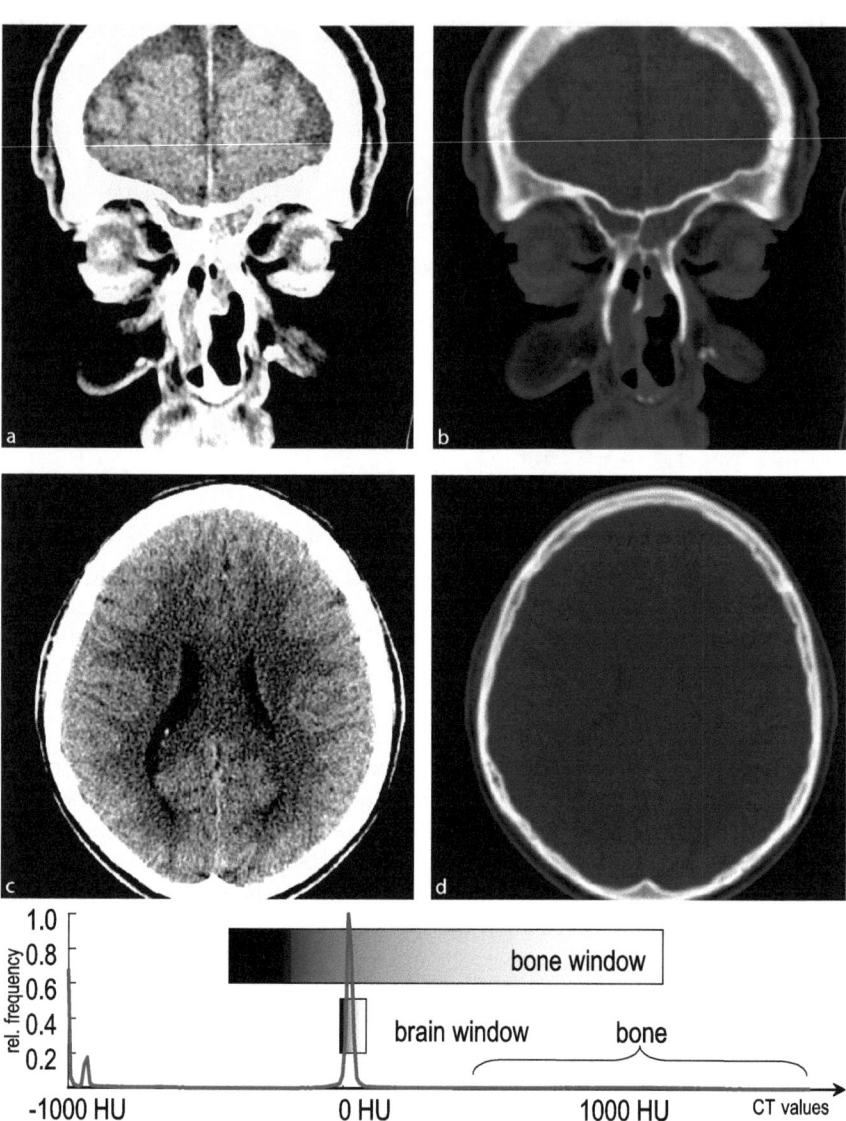

Fig. 10.7. Differentiability of tissue when using a brain window (**a:** $WL = +35$ HU, $WW = 100$ HU) and a bone window (**b:** $WL = +300$ HU, $WW = 1{,}500$ HU). The frequency distribution indicates how the brain window has been selected); data courtesy of Ruhlmann

An alternative to volume rendering is surface rendering, as illustrated in Fig. 10.9. The individual shades of gray of the layers (images) in the data stack represent the degree of physical attenuation of the X-ray beam. In a clinical context, deviations from the normal distribution of these values may indicate pathological changes in the patient.

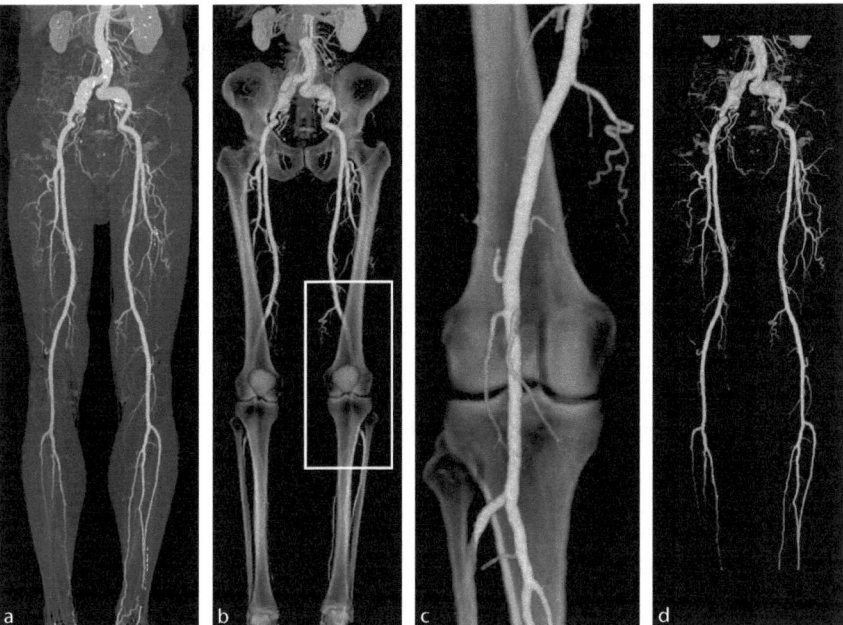

Fig. 10.8. Angiographic image acquisition with CT (CTA): **a** *Maximum-Intensity Projection* (MIP), **b** volume rendering, **c** zoom into the knee and visualization from the back, and **d** virtual bone elimination with the bone removal technique (courtesy of Philips Medical Systems)

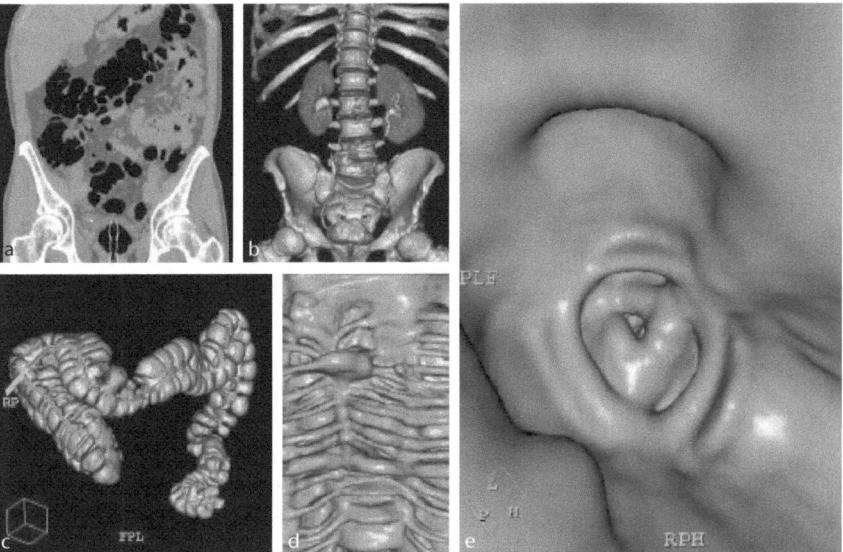

Fig. 10.9. Three-dimensional representation of the abdomen. **a** Coronal reformatting, **b–e** volume and surface rendering, **c–e** virtual endoscopy of the intestine together with the representation of the virtual endoscopy trajectory, and **e** virtually opened intestine in cylinder coordinates (courtesy of Philips Medical Systems)

During visualization, it is possible to visualize certain ranges of values and to selectively suppress others. If the viewer chooses a constant gray value, i.e., a threshold, all spatial points with this value may be displayed in space as an iso-surface. This surface is approximated by triangles using a technique called triangulation. Then the mosaic of triangles is again lit and displayed by the virtual method described above. The larger the chosen number of mosaic pieces for the reconstruction of the surface, the more lifelike the result.

In addition to the three-dimensional representation of organs or bone surfaces, which may be interactively rotated on the computer screen, an interior view of hollow organs and airways of contrast-enhanced vessels can be produced. In this way, virtual rides into the body – for example into the bronchial tubes or the intestine – can be visualized. In Fig. 10.9e, a virtual insight into the human intestine is given. The value of such a three-dimensional diagnostic approach lies in the data reduction otherwise being inherent in the hundreds of single slices.

10.4
Some Applications in Medicine

In the final section of this chapter, the broad variety of clinical applications of modern CT imaging is summarized by means of some examples. In Fig. 10.10, four typical applications are given. Important fields of application are summarized in the following:

- **Brain perfusion:** The blood flow in the brain (cerebral blood flow – CBF), blood volume in the brain (cerebral blood volume – CBV), mean transit time (MTT), and peak time of the bolus maximum (time to peak – TTP) are acquired and displayed as colored overlay onto the relevant CT slices. In this way, colored maps of tissue vitality give indications of an acute or chronic infarct (cf. Fig. 10.10a).
- **Liver perfusion:** Arterial and portal measurements of perfusion in liver studies.
- **Tumor perfusion:** Characterization of known lesions via their perfusion.
- **Lung measurement:** Diagnostics of lung emphysema (cf. Fig. 10.10b); automatic detection of lung nodules.
- **Calcium scoring:** Quantification of coronary calcification.
- **Vessel analysis (CTA):** Visualization of vessel trees (cf. Fig. 10.10c); analysis of stenoses and aneurysms; planning of stents.
- **Cardio CT:** Identification and quantification of stenoses; planning of stents and visualization of implanted stents (cf. Fig. 10.10d).
- **Virtual endoscopy:** Three-dimensional CT data as a basis for anatomical interior views of hollow organs (cf. Fig. 10.9e) and contrast-enhanced vessel trees.
- **Trauma:** Fast imaging of the entire body for diagnostics of accident injuries.
- **Dental planning:** Three-dimensional reconstruction and slices through the maxilla and mandible as a planning basis for the implantation of prostheses for the oral surgeon.

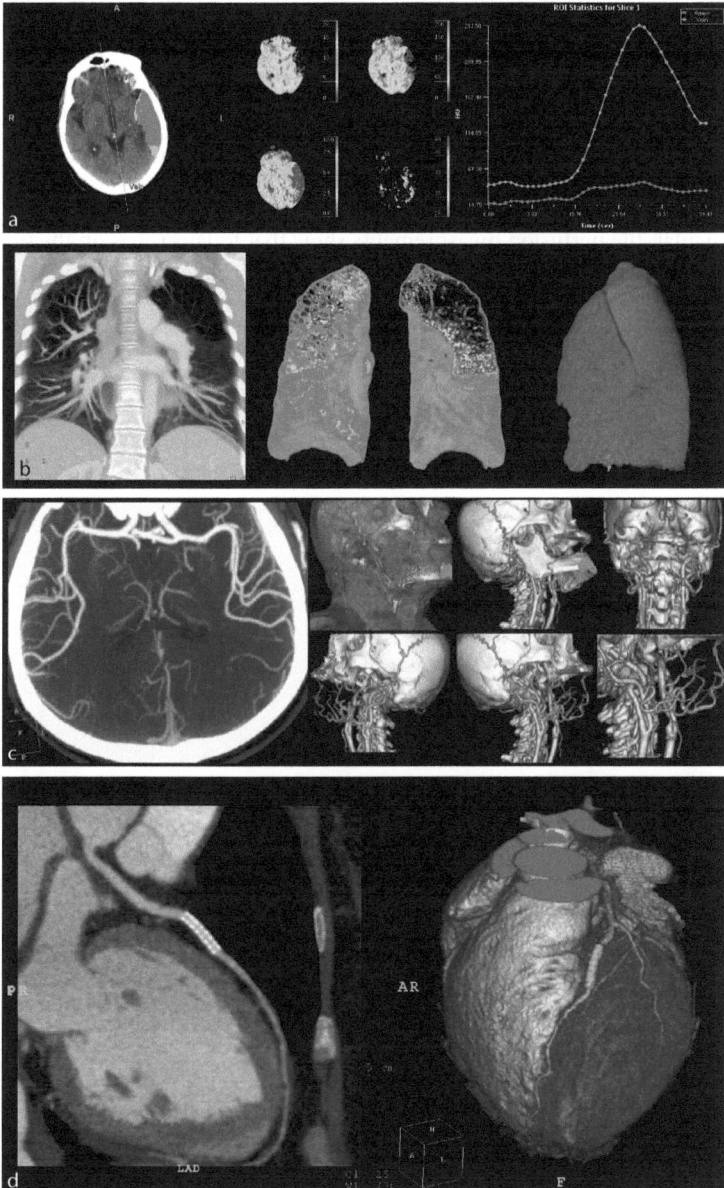

Fig. 10.10. a Perfusion measurement for stroke patients, **b** visualization of lung emphysema, **c** CT angiography of the cranial vessels, and **d** imaging of a coronary stent (courtesy of Philips Medical Systems)

- **Planning of radiotherapy**: Three-dimensional CT reconstruction as a basis for dose planning in the radiation therapy of tumors.
- **Image-guided surgery**: Three-dimensional CT reconstruction as a basis for the planning and navigation of surgical interventions.
- **Interventional imaging**: Visualization of a surgical instrument tip during a biopsy.

In particular, the perfusion measurements show that today's CT is on the upswing toward becoming a functional modality. This technique allows the blood flow to be measured after contrast media injection into different organs.

11 Dose

Contents

11.1	Introduction	485
11.2	Energy Dose, Equivalent Dose, and Effective Dose	486
11.3	Definition of Specific CT Dose Measures	487
11.4	Device-Related Measures for Dose Reduction	493
11.5	User-Related Measures for Dose Reduction	499

11.1
Introduction

The gain in diagnostically valuable information accompanying the advent of computed tomography (CT) was considered to be exceptional. This is one of the main reasons why the applied dose was not considered to be of vital importance during this developmental stage. Based on the number of devices currently installed in Germany and an average of 3,500 examinations per year, an annual total number of several hundreds of millions of slices can be estimated to be acquired each year. This approximately lies within the same range as the total number of standard projection radiographs acquired in clinical practice. Considering the different types of radiological examinations (X-ray examinations), one immediately becomes aware of the following discrepancy. Although CT examinations represent only about 4% of all radiological examinations, their share of the total dose amounts up to approximately 35%. In short, CT accounts for the largest portion of medically related X-ray exposure.

Considering new generations of CT scanners, such as spiral multislice CT, the applied dose indeed is not reduced. In fact, it is more likely that the amount of dose applied during one single imaging session will increase. This is mainly due to the fact that in modern scanners longer sequences and thinner slices can be easily measured (Nagel 2002). Initiated by reports about unnecessarily high X-ray exposure in pediatric CT examinations (Sternberg 2000), which were actually performed on the basis of the same scanning protocols as were used in human adults, a high sensibility concerning the applied dose has emerged among both manufacturers and users. Recent developments such as automatic exposure control, take the problem of dose into account. With this technology, one can easily adjust the amount of dose according to the anatomy of the patient or likewise take into account whether the individual under examination is an adult or an infant.

The second section of this chapter briefly introduces the essential physical interactions between ionizing radiation and biological tissue as well as their physical parameters. The third section provides dose rates that are characteristic of CT. Finally, in the last two sections, different methods of dose reduction are discussed.

11.2
Energy Dose, Equivalent Dose, and Effective Dose

The general definition of energy dose is

$$D = \frac{\text{absorbed energy}}{\text{tissue mass}} . \tag{11.1}$$

That is, the absorbed energy is related to the mass of the exposed tissue. The unit of dose is Gray ($[D] = \text{J/kg} = \text{Gy}$). Since the resulting biological effects of different types of radiation having the same energy dose varies, an additional biological weighting of the energy dose is necessary. This is done using the so-called equivalent dose. In ICRP paper no. 60, 1991 (ICRP 1991) an equivalent dose for a certain organ or tissue is defined as

$$H_T = \sum_R w_R \cdot D_{T,R} , \tag{11.2}$$

where $D_{T,R}$ represents the dose applied to the organ T with respect to the type of radiation R. The equivalent dose H_T is measured in units of Sievert ($[H_T] = \text{J/kg} = \text{Sv}$). Here, w_R represents the so-called radiation weighting factor, which only depends on the type and energy of the applied radiation. Table 11.1 provides the weightings for different types of radiation. Obviously, the weighting factor for X-ray is $w_R = 1$.

In order to take the varying radiosensitivity of different organs and tissues into account, a tissue weighting factor, w_T, has been introduced, resulting in the so-called effective dose E (as opposed to the equivalent dose H_T). Table 11.2 provides a collection of organs, which is subdivided into four distinct classes according to their sensitivity.

The effective dose E is the sum of the weighted equivalent doses in all the tissues and organs. It is given by

$$E = \sum_T w_T \cdot H_T = \sum_T w_T \cdot \sum_R w_R \cdot D_{T,R} . \tag{11.3}$$

The effective dose E is measured in units of Sievert $[E] = \text{J/kg} = \text{Sv}$.

Table 11.1. Weighting factors (w_R) for different radiation types (ICRP 1991)

Radiation type and energy range		Radiation weighting factor w_R
Photons (all energies)		1
Electrons und muons (all energies)		1
Neutrons	$E < 10\,\text{keV}$	5
	$10\,\text{keV} < E \leq 100\,\text{keV}$	10
	$100\,\text{keV} < E \leq 2\,\text{MeV}$	20
	$2\,\text{MeV} < E \leq 20\,\text{MeV}$	10
	$E > 20\,\text{MeV}$	10
Protons (other than recoil protons) $E > 2\,\text{MeV}$		5
α-Particles, fission fragments, heavy nuclei		20

Table 11.2. Tissue weighting factors to be used for calculation of the effective dose (ICRP 1991)

Tissue or organ	Tissue weighting factor (w_T)
Gonads	0.20
Bone marrow (red)	0.12
Colon	0.12
Lung	0.12
Stomach	0.12
Bladder	0.05
Breast	0.05
Liver	0.05
Esophagus	0.05
Thyroid	0.05
Skin	0.01
Bone surface	0.01
Remainder	0.05

11.3
Definition of Specific CT Dose Measures

The spatial dose distribution in CT fundamentally differs from the one observed in traditional projection radiography. Figure 11.1a provides an illustration of the resulting relative iso-dose lines as seen in an anterior posterior projection radiograph of a skull. Since tissue attenuates the penetrating X-ray beam, the dose decreases continuously along the y-axis, i.e., from the tube-adjacent side toward its opposite side. As a result of Beer–Lamberts law the applied dose decreases exponentially. In contrast, Fig. 11.1b provides the iso-dose lines resulting from a CT of the same skull. Here, the object is X-rayed from all sides, so that the resulting dose distri-

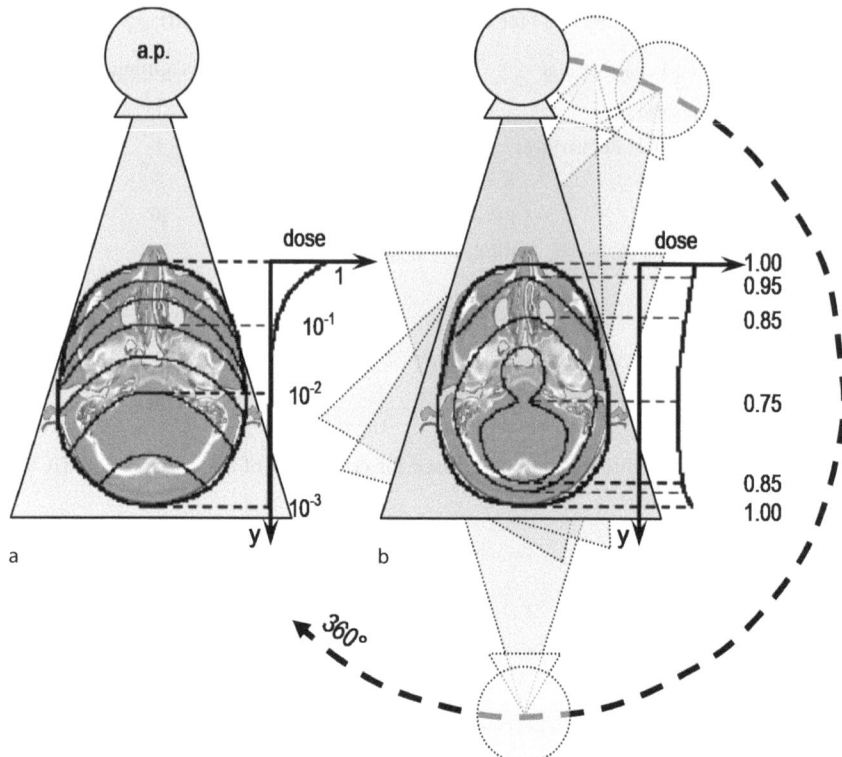

Fig. 11.1. Illustration of the spatial dose distribution. Isodose lines are drawn onto a cranial slice for **a** normal projection radiography and **b** 360° CT

bution is almost homogeneous inside the skull. Of course, every single projection yields exactly the same spatial dose distribution as seen in conventional projection X-ray imaging with respect to the direction of the X-ray radiation. However, due to the accumulation across all angles the spatial exposure in total is almost homogeneous.

Additionally, a difference in the simultaneously irradiated volume can be observed. If the CT acquisition is limited by means of collimation to one single slice, the patient is primarily just irradiated in this respective layer with a thickness of only a few millimeters. Figure 11.2a shows the dose profile for one single slice given a nominal slice thickness of $d = 10$ mm. It immediately becomes apparent that the measured dose profile does not conform to an ideal rectangular function. Therefore, even outside the slice adjusted by means of collimation, the dose is applied to the patient. The main reason for this is the scattered radiation being produced in the respective layer. Typically, the nominal thickness of a layer is within the range of the full width at half maximum (FWHM) of the dose profile. Considering the total amount of dose, the additional dose applied by scattered radiation outside the X-ray fan beam must not be neglected. These circumstances are taken into account

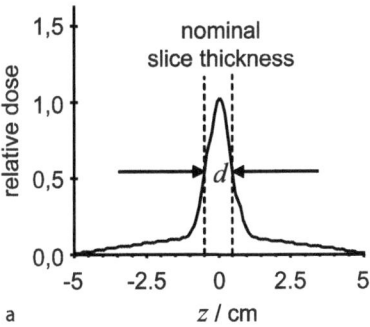 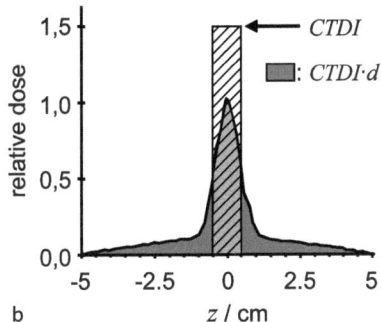

Fig. 11.2. a Illustration of a single-slice dose profile for a nominal slice thickness of $d = 10$ mm. **b** The *Computed Tomography Dose Index* (*CTDI*) is obtained via the area of dose profile for a certain slice thickness d

via *CTDI*, the Computed Tomography Dose Index. *CTDI* represents an important, CT-specific dose quantity, which relates the total amount of a dose to an ideal rectangular dose profile along the z-axis. *CTDI* is calculated as follows

$$CTDI = \frac{1}{d} \int_{-\infty}^{\infty} D(z)\, dz. \tag{11.4}$$

Here, $D(z)$ denotes the function of the dose along the z-axis. Figure 11.2b schematically illustrates the *CTDI* as the height of the hatched rectangular area. Therefore, the *CTDI* represents the equivalent value of the dose inside the nominal slice with an ideal rectangular profile that comprises the total amount of radiation that was actually introduced (gray-shadowed area in the background of Fig. 11.2b).

Since in CT applications a series of adjacent and overlapping slices is commonly acquired, it has to be taken into account that the dose of one single slice increases due to the contribution of the neighboring slices. This is expressed by the *Multiple Scan Average Dose* (*MSAD*)

$$MSAD = \frac{CTDI}{p}. \tag{11.5}$$

Here, p is the pitch factor, whose definition from ▸ Sect. 8.3 also holds for a sequence of single slices. Figure 11.3 shows how an *MSAD* dose plateau develops in the presence of a sufficient number of overlapping slices.

In contrast to (11.4) the summation of the dose contribution is actually limited by the FDA requirement for CT systems to an interval of ±7 times the slice thickness in front of and behind the slice under consideration. That is,

$$CTDI_{FDA} = \frac{1}{d} \int_{-7d}^{+7d} D(z)\, dz. \tag{11.6}$$

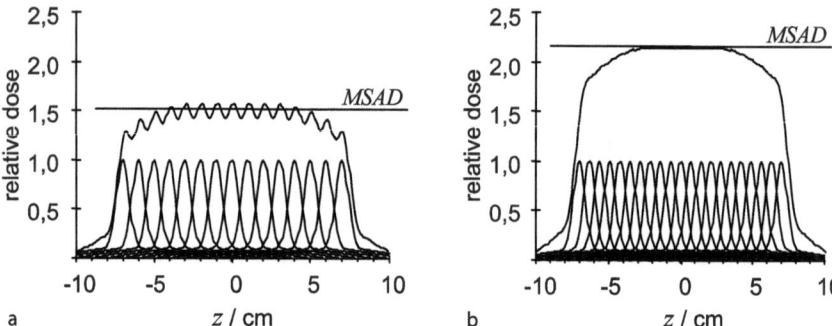

Fig. 11.3. Dose profile of multiple slice scan series with slice thickness of 10 mm. **a** Series of 15 slices with 10-mm table feed (pitch = 1), **b** series of 21 slices with 7-mm table feed (pitch = 0.7). Definition of *MSAD* = *Multiple Scan Average Dose* (courtesy of Nagel [2002])

The FDA regulations obligate each CT manufacturer to specify the $CTDI_{FDA}$. The shortcoming of this definition of dose is obvious. The interval in which the dose contribution of the scattered radiation is to be summed is actually reduced along with the slice thickness. However, the range of the scattered radiation is almost independent of the slice thickness. Therefore, the thinner the slices are, the more incomplete the coverage of the generated dose contribution will be. In order to avoid such underestimations of the dose, the so-called *Practical CTDI* has been introduced. It is defined as

$$CTDI_{100} = \frac{1}{d} \int\limits_{-50\,mm}^{+50\,mm} K_a(z)\,dz \tag{11.7}$$

i.e., fixed to a constant integration interval of 100 mm. Furthermore, instead of the absorbed dose in plexiglass, air kerma[1] K_a is used.

A special significance is given to the so-called weighted *CTDI*. This quantity is denoted by $CTDI_w$. Current dose recommendations as well as dose displays at the scanner backend are based on the $CTDI_w$. The weighted dose index is defined by

$$CTDI_w = \frac{1}{3}CTDI_{100,c} + \frac{2}{3}CTDI_{100,p}\ , \tag{11.8}$$

i.e., resulting from the $CTDI_{100}$, one-third of which is related to the center of the patient and two-thirds to the periphery. In analogy to (11.5), considering the total examination, the dose-increasing and dose-reducing influence of a pitch value $p \neq 1$ has to be taken into account. This is done via the definition of the effective *CTDI*

$$CTDI_{w,eff} = \frac{CTDI_w}{p}\ . \tag{11.9}$$

---

[1] The word Kerma originates from "*kinetic energy released per unit mass*".

Occasionally, *CTDI* values found in technical datasheets of CT devices are given for a certain current–time product

$$Q = I \cdot t \, . \tag{11.10}$$

The quantity

$$_nCTDI_{xyz} = \frac{CTDI_{xyz}}{Q} \tag{11.11}$$

is called the normalized *CTDI* value. This is a device-specific constant for the dose rate, which on its own does not allow any statement about the radiation exposure.

Another interesting dose quantity is the so-called organ dose

$$D_{org} = \frac{\text{absorbed energy}}{\text{organ mass}} \, . \tag{11.12}$$

Here, the absorbed energy is referred to the mass of the organ. The dose D, as well as all of the above-mentioned values derived from it, are measured in units of Gray $[D] = \text{J/kg} = \text{Gy}$. This quantity can be defined by means of tabulated conversion factors from the dose free-in-air on the system axis

$$D_{org} = \frac{1}{p} CTDI_{air} \sum_{z} f(\text{organ}, z) \, . \tag{11.13}$$

Therewith, the dose free-in-air on the system axis $CTDI_{air}$ is defined analogously to (11.7), thus

$$CTDI_{air} = \frac{1}{d} \int\limits_{-50\,mm}^{+50\,mm} K_a(z) \, dz \, . \tag{11.14}$$

The table by Zankl et al. (1991), which is used in Germany, contains conversion factors separated according to gender. The conversion factors are graphically illustrated in Fig. 11.4, as observed in a patient along a patient's axial direction.

The relevance of the organ dose lies in the fact that one can estimate the probability of a radiation-induced cancer of the organ from the multiplication of the organ dose by the corresponding risk factor.

At this point a new value has to be introduced, which describes the entire amount of radiation exposure of a complete CT examination with for instance 20 adjacent slices being acquired. The question regarding the applied dose in the case of 20 slices in relation to the measurement of one single slice is frequently answered incorrectly, because, surprisingly, the dose remains the same. That is to say, the term dose has, according to the definition given in (11.12), the meaning of a density. The mass density does not increase proportionally along with an enlargement of the considered volume. In order to quantify the total effect of the radiation of n slices, the so-called dose length product (*DLP*) is introduced.

$$DLP_w = CTDI_{w,eff} \cdot p \cdot n \cdot d \, . \tag{11.15}$$

Here, d is given in centimeters, so that the unit of the *DLP* is $[DLP_w] = \text{Gy} \cdot \text{cm}$.

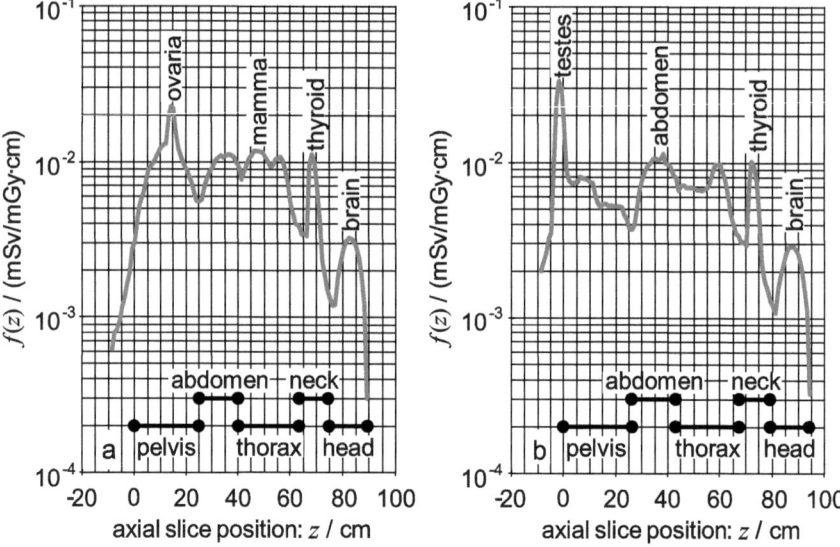

Fig. 11.4. Conversion factor f per cm slice thickness from $CTDI_{air}$ to an effective dose for normal weight adults: **a** female, **b** male. Tube voltage: $U_{ref} = 125\,kV$; 9-mm Al filter equivalent (courtesy of Nagel [2002])

By taking advantage of the DLP the effective dose defined in (11.3) can be estimated in a simplified form by

$$E = DLP_{air} \cdot f_{average} \, . \tag{11.16}$$

Therefore, the DLP free-in-air on the system axis is composed of

$$DLP_{air} = \frac{CTDI_{air}}{CTDI_{w,X}} DLP_w \, . \tag{11.17}$$

Table 11.3. Average conversion factor $f_{average}$ (in mSv/mGy · cm) from DLP based on $CTDI_{air}$ to effective dose for normal weight adults: Female, male. Tube voltage: $U_{ref} = 125\,kV$; 9-mm Al filter equivalent (courtesy of Nagel [2002])

Body region	Female ($f_{average}$)	Male ($f_{average}$)
Head	0.0022	0.0020
Neck	0.0051	0.0047
Thorax	0.0090	0.0068
Upper abdomen	0.0100	0.0091
Pelvis	0.0110	0.0062
Complete abdomen	0.0100	0.0072

The subscript X in (11.17) refers to the employed body phantom (16-cm head phantom or 32-cm torso phantom). The average values of the conversion factors with respect to different anatomical areas, a tube reference voltage of $U_{ref} = 125\,\text{kV}$ and a 9-mm Al filter equivalent (see Fig. 11.10) are summarized in Table 11.3.

11.4
Device-Related Measures for Dose Reduction

In ▸ Sect. 9.8.3 it has already been discussed that dose and image quality are closely related to each other. Thereby, the formula of Brooks and DiChiro (1976) is frequently quoted, according to which this relation is given by

$$ D \propto \frac{B}{\sigma^2 \cdot b^2 \cdot \Delta \xi \cdot d} . \tag{11.18} $$

$B = \exp(L\mu)$ is the object's mean attenuation factor, where L denotes the thickness of the object and σ describes the noise, i.e., the variance, of the image values. b is the sampling distance given at the rotation center, $\Delta \xi$ the width of the detector element, and d the slice thickness. Sophisticated scanners take advantage of this connection in such a way that the anode current and therefore, the dose, are dynamically adapted to the respective anatomical range, such that the resulting total dose can be lowered whilst preserving the signal-to-noise ratio. Thereby, one basically takes advantage of the relation $D \propto B$ (at a constant SNR). The three concepts – automatic exposure control, longitudinal dose modulation, and angular dose modulation, all of which are described in the following – are based on this idea.

Figure 11.5 schematically illustrates that one initially has to differentiate patients according to their physique and between infants and adults respectively. If we adjust the same tube current–time product for infants (small values for L in (11.18)), as used in scan protocols for adults (Fig. 11.5 left), the obtained intensity of the X-ray at the detector and thus the applied dose, with respect to the achievable diagnostic quality (i.e., the $SNR = \mu/\sigma$), is too high. Therefore, in the case of a pediatric examination, it indeed is reasonable to decrease the anode current to a level that allows a comparable diagnostic quality to be obtained, as achieved in the examination of an adult (Fig. 11.5, right). The automatic exposure control (AEC) can be steered by means of the scanogram.

Figure 11.6 provides an illustration of the dynamic adaptation of tube current or dose respectively with respect to the anatomical situation observed along the axial z-axis (longitudinal dose modulation). In anatomical areas in which, as known a priori, a small attenuation of the X-ray intensity is to be expected (as for example within the lung area) the tube current and thus the dose can be lowered, without impairing the image quality. If all areas are operated on with the same tube current, this will either result in dose rates that are too high or noise levels that are too strong.

Figure 11.7 illustrates the dynamic adaptation of the tube current across the projection angle γ to the integral attenuation of the radiated body area (angular dose

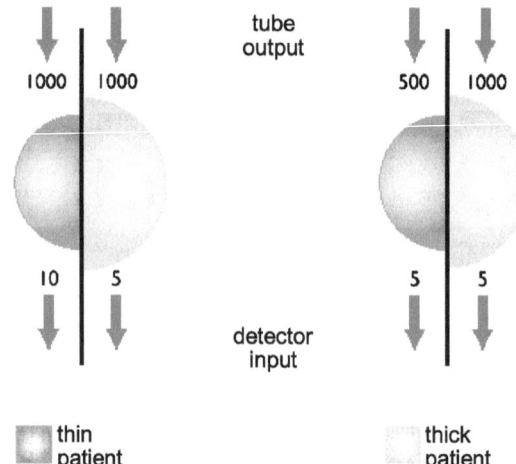

Fig. 11.5. Automatic exposure control (AEC) differentiates between heavier and thinner patients (courtesy of Nagel [2002])

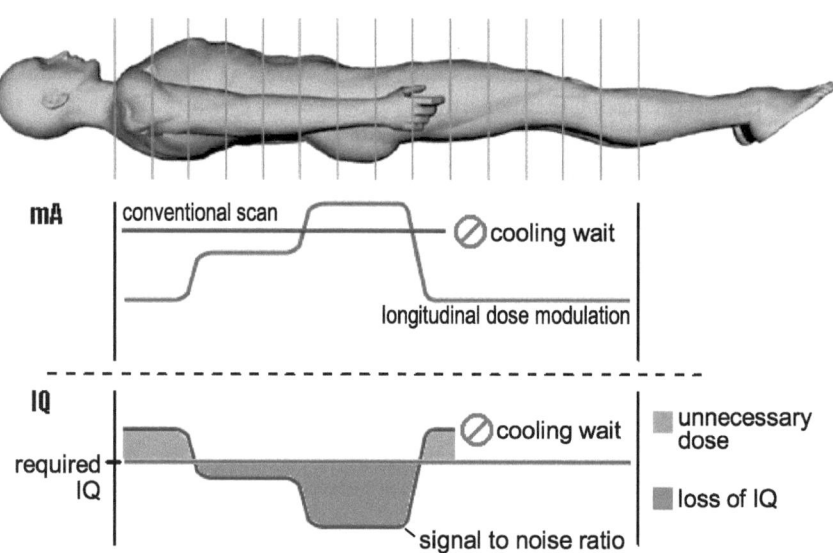

Fig. 11.6. The longitudinal dose modulation is a dynamic adaptation of the tube current or dose to the anatomical situation. In axial locations with low attenuation of X-ray intensity, such as the lung, the tube current can be reduced without loss of image quality. The conventional scan protocol, i.e., constant tube current along the axial scan direction, leads to an unnecessarily high dose or unacceptably high noise level (courtesy of General Electric Medical Systems)

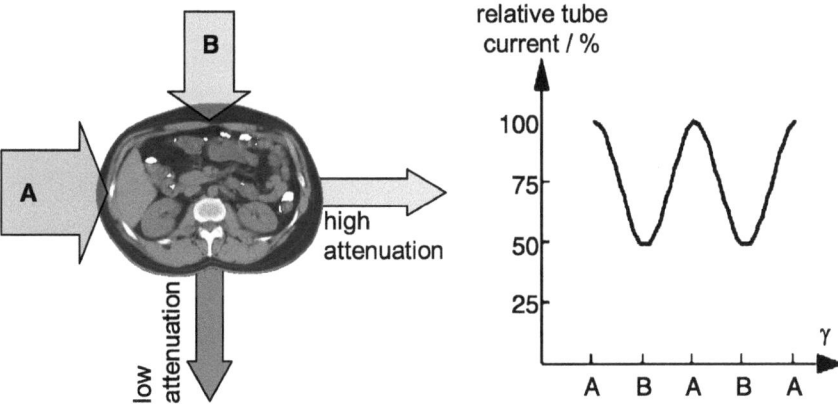

Fig. 11.7. The angular dose modulation is a dynamic adaptation of the tube current or dose to the shape of the axial body section. Especially in neck, thorax or abdomen imaging, the attenuation of X-ray intensity in the anterior–posterior direction is lower than the attenuation in the lateral direction. With an angular modulation of the tube current the signal level at the detector can be held constant

modulation). At all sections through the body, which are oval rather than circular, the angular dose modulation is to be applied. In particular for image acquisitions of the shoulder, a strong dose modulation is to be used.

Figure 11.8 shows the effect of the combination of longitudinal and angular dose modulation on the attenuation of the X-ray beams and on the tube current. At conventional scan protocols with a constant tube current along the patient's axial direction, the dose is usually selected in such a way that all anatomical areas with a good signal-to-noise-ratio are measured. This adjustment of the tube current, however, is oversized for most slice positions. In Fig. 11.8, one easily can recognize that only in the shoulder and in the pelvic areas is it necessary to employ a higher incoming intensity due to the fact that the attenuation is higher in these areas. This is implemented by means of a longitudinal dose modulation. Furthermore, it is possible to decrease the mean dose, provided that the differences in attenuation between frontal and lateral radiation are considered. At additional angular dose modulation, it is even possible to temporarily increase the dose during lateral radiation compared with conventional scan protocols. Overall, this allows an enhancement of the image quality whilst simultaneously reducing the overall dose applied.

Another possibility for dose reduction arises in cardiac imaging. In Fig. 11.9 a trigger sequence of an ECG-triggered image acquisition of the heart is depicted. Since data are only acquired within the resting phase of the beating heart, the overall amount of the dose can be significantly reduced, provided that the tube current is switched off outside the data window. This effective and efficient procedure is called temporal dose modulation.

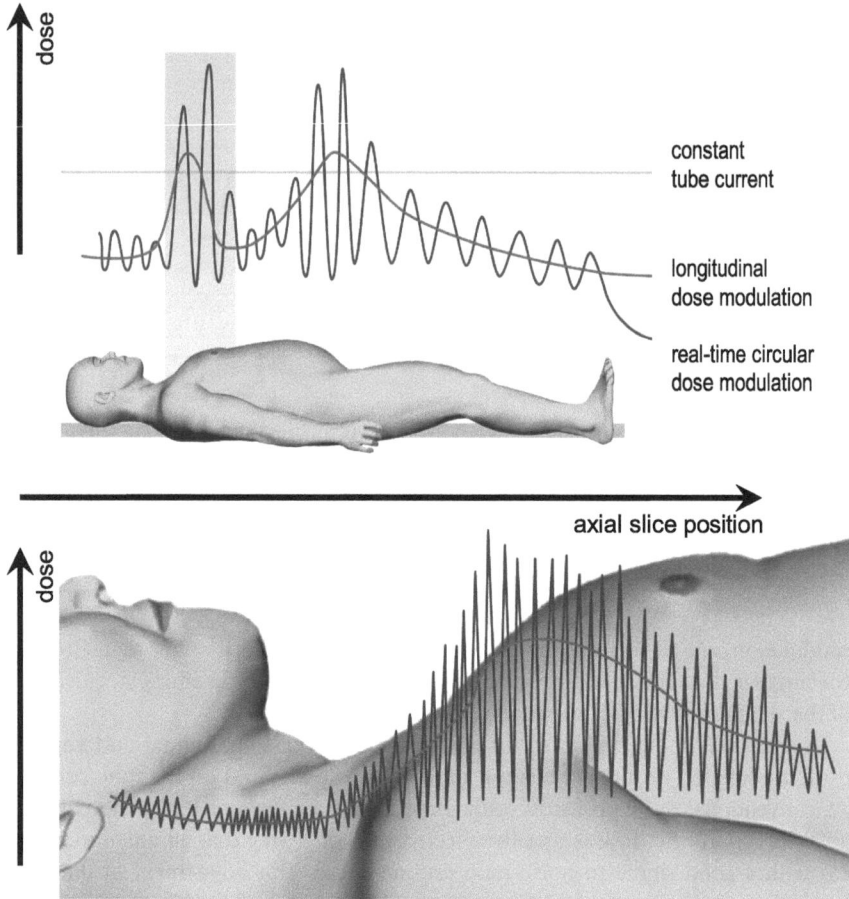

constant
tube current

longitudinal
dose modulation

real-time circular
dose modulation

axial slice position

Fig. 11.8. Combination of longitudinal and angular dose modulation (courtesy of Siemens Medical Solutions)

In Fig. 11.10 two other efficient methods of device-operated dose reduction are provided. First, a flat metal filter, which is installed directly behind the X-ray tube, allows the absorption of low-energy radiation present in the polychromatic X-ray spectrum. Since the soft X-ray beam passing through the patient is attenuated more strongly than the hard X-ray beam, its contribution to the overall dose is relatively high – this is distressing since its contribution to the imaging result is quite small. Due to its low energy, it is absorbed in the patient and therefore does not reach the detector. This part of the X-ray spectrum correspondingly only contributes to an increase in patient dose, without improving the image quality. Therefore, DIN 6815 demands at least a minimal value of the total filtering of 2.5 mm Al for CT. Another advantage of this filter is the reduction of beam-hardening artifacts (cf. ▸ Sect. 9.6.2). However, as a general rule, employing stronger filters will coevally

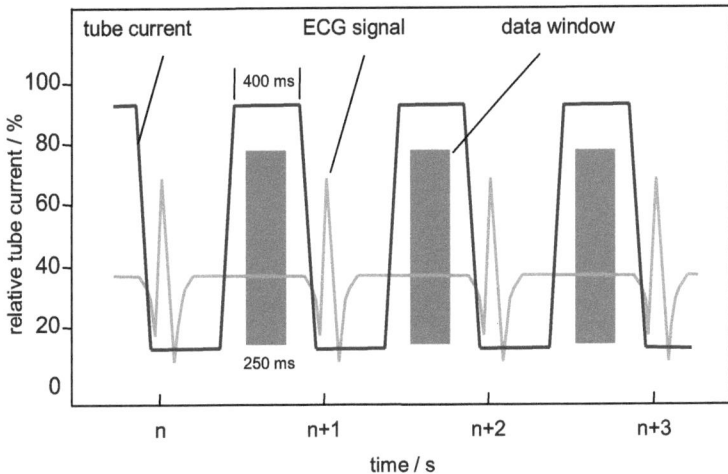

Fig. 11.9. The temporal dose modulation is an ECG triggered switching of either the tube current or the dose. In cardiac imaging, best results are achieved when projection data are acquired in the diastolic phase of the cardiac cycle. The dose can be reduced considerably if the tube is switched off outside the data window (Nagel 2002)

reduce the contrast and, moreover, result in higher noise due to the fact that less radiation reaches the detector. In order to compensate for the loss of contrast-to-noise-ratio, an increase in the mAs product is necessary. This in turn counteracts the positive effects arising from the reduced radiation exposure of the patient by means of filtering (Hidajat 2001). Therefore, it is necessary to cope with a trade-off between dose reduction and contrast-to-noise ratio, both of which are directly related to the effects of filtering, as already mentioned above. Modern devices are frequently equipped with a source filter of 1–3 mm Al and, device-specific, an additional filter of 0.1 mm Cu. The overall filtering therefore amounts to approximately 5–6 mm Al-equivalent (Hidajat 2001).

Using a specially shaped filter, called a *bowtie* filter, it can be considered in a second step that the total distances covered by beams of the X-ray fan passing through the patient's body differ from each other. The white lines in Fig. 11.10 schematically illustrate that the covered distance through the patient at the border of the fan is typically shorter than in the center. This offers the chance to reduce the dose of the rays situated at the border of the fan via a gradual attenuation in so far as the signal-to-noise-ratio is comparable across the entire fan beam. The *bowtie* filter also allows optimization of the requirements for the dynamic range of the detector system. In order to create absorption behavior comparable to that observed in the anatomy, specially shaped filters often consist of plastic or other materials with a lower atomic number Z. Some devices work with differently shaped filters according to the anatomical area to be examined (Hidajat 2001).

Figure 11.11a shows a recent development in the field, the so-called dual-source CT. Two complete sampling units, both of which consist of an X-ray tube and a de-

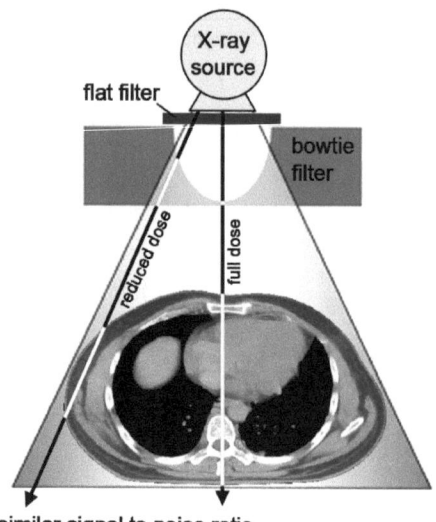

Fig. 11.10. With a flat metal filter a low-energy X-ray is attenuated. In a second step, a so-called bowtie filter leads to a spatially increasing attenuation from the center to the border of the X-ray fan

tector array, are installed perpendicular to each other. Best images of the beating heart are obtained when the acquisition interval is set to the diastolic phase. Since the patients are usually agitated, this resting phase of the heart is often so short that motion artifacts are inevitable.

In Fig. 11.11b it is schematically shown that images have to be acquired across 180°, in order to obtain a complete raw data set within the relatively short resting phase of the heart. Figure 11.11c depicts that by employing a sampling unit taking advantage of the dual-source technology, only a rotation about 90° is necessary. Thereby, the sampling time is halved, such that motion artifacts are significantly reduced. If one wants to abandon the administration of beta blockers, which are used to lower the pulse, an artifact-free image reconstruction by means of a single-source scanner is no longer possible. Figure 11.11d schematically depicts that single-source scanners in this case must employ multi-segment protocols with a high dose. Here the strength of the halved acquisition time (cf. Fig. 11.11e) becomes apparent. Using these modern dual-source subsecond scanners a temporal resolution can be achieved, which only differs by a factor of 2 from the acquisition speed achieved by electron-beam CT (cf. ▸ Chap. 3). Considering the applied dose, the ECG-supported triggering of the acquisition window – a principle that already has been discussed in Fig. 11.9 – is of great importance. Here, compared with the standard CT scan protocol in which the tube current is not modulated, a dose reduction of up to 50% can be achieved.

At this point it must be noted that for the calculation of the effective dose, device-specific correction factors have to be introduced, if one does not accept es-

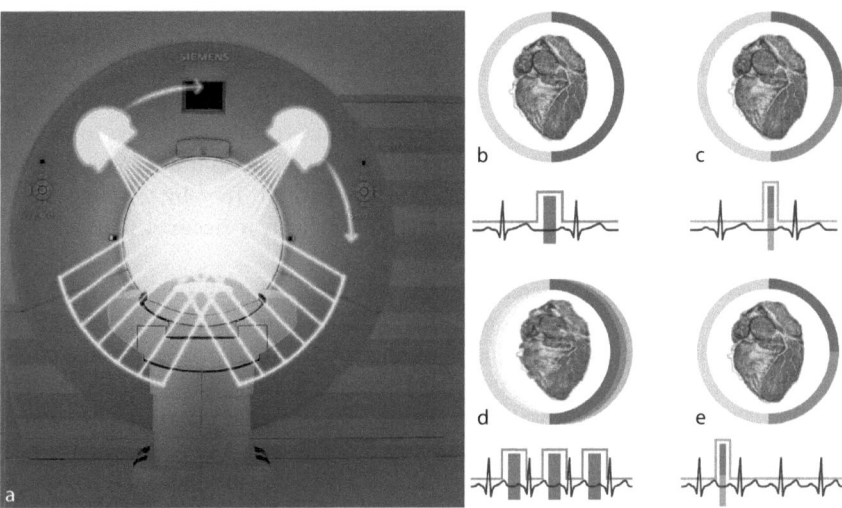

Fig. 11.11. Dual-source CT consists of two independent sampling units. **a** A second X-ray tube and a second detector array has been added perpendicular to the first sampling unit. **b–e** Especially in cardiac imaging, it is of great significance for simultaneous data acquisition to double the temporal resolution (courtesy of Siemens Medical Solutions)

timation errors of up to 150%. The values for the conversion factors as given in Fig. 11.4 refer to devices, that are operated with a reference voltage of $U_{ref} = 125\,kV$ and are employed with a radiation filtering of a 9-mm Al-equivalent. Furthermore, it is assumed that no special shaped filters are installed and that the focus center distances are approximately 70–76 cm. Since these assumptions are by no means representative, (11.16) has to be corrected via

$$E = DLP_{air} \cdot f_{average} \cdot k_{CT} . \tag{11.19}$$

An overview of the device-specific correction factors k_{CT} can be found in Nagel (2002).

11.5
User-Related Measures for Dose Reduction

Apart from device-related measures for the reduction of radiation exposure, for which the manufacturers bear responsibility, there is a set of user-related measures, which also affect the applied dose. The quantities and their connection to the dose, adjustable in the scan protocol, are given in the following. For a detailed discussion on the radiation exposure in CT, one is referred to the book by Nagel (2002).

- **Current–time product (mAs product):** Dose and mAs product, i.e., the product between the X-ray tube current and acquisition time, has a linear dependency. However, the standard deviation, i.e., the image noise, increases inversely proportional to the root of the mAs product.

- **Acquisition time**: At constant X-ray tube current, the dose increases linearly along with the acquisition time. However, the mAs product is always to be considered as a total, so that at a constant dose the acquisition time can be reduced along with simultaneously increasing the tube current. Thereby, the image noise is unaffected; however, motion artifacts are less likely to occur at short acquisition times.
- **Tube voltage**: With an increasing tube voltage, U_B, the efficiency of the X-ray tube as well as the penetration of the radiation is increased. The intensity of the radiation increases with U_B on average by the power of 3.5. With a larger penetration of the radiation, the image contrast obviously decreases. However, this is more than compensated for by the better quantum statistics. Therefore, the image quality generally is improved – obviously at the expense of a higher dose being applied to the patient. If the voltage is for instance increased from 120 kV to 140 kV, in order to reduce the dose whilst maintaining image quality, the mAs product must be reduced by 40%. Thus, a dose reduction of approximately 15% is obtained (Nagel 2002). If one wants to calculate the effective dose at a voltage U that differs from the reference voltage $U_{ref} = 125$ kV on the basis of (11.19), then two adjustments to this equation are required. The normalized *CTDI* free-in-air (cf. (11.11)) typically changes with the 2nd power of the voltage, such that the corresponding correction factor reads

$$k_{U,1} = \left(\frac{U}{U_{ref}} \right)^2 .$$
(11.20)

In addition, the conversion factors change (cf. Fig. 11.4 and Table 11.3) – indirectly related to device-specific factors – on average by approximately 5% if the voltage is increased by 10 kV. This can be considered by a second correction factor

$$k_{U,2} = \left(\frac{U}{U_{ref}} \right)^{0.5} .$$
(11.21)

- **Thickness of the object to be imaged**: For infants and frail patients one has to bear in mind that a smaller mAs product has to be adjusted. Due to lower attenuation, the statistics of the X-ray quanta are still as good as for adults or heavier patients. Correspondingly, the image quality is not affected. For heavier patients, an increase in X-ray tube voltage is preferable to an increase in the mAs product, since the increase in radiation exposure is less intense.
- **Slice thickness**: The slice thickness can typically be adjusted by means of a tube-sided collimator from 1 mm up to 10 mm. Thereby, the slice thickness does not affect the dose if the same body section is to be measured. The advantage of a finer slice sequence is the reduction of partial volume artifacts as well as step artifacts occurring at coronal or sagittal reformatting of the image. The disad-

vantage is that at a finer collimation fewer X-ray quanta will reach the detector such that the image noise will correspondingly increase. If the image quality is to be kept constant, the mAs product and thus the dose will have to increase inversely proportional to the slice thickness.

- **Pitch factor**: A pitch factor of $p = 1$ means that in the case of a rotation of the sampling unit through $360°$, the patient table is linearly moved by a length equal to the adjusted slice thickness. If $p < 1$, the individually measured slices have a larger overlap, such that the image quality is increased. However, this results in an increased dose. Accordingly, if $p > 1$, the dose can be reduced. In theory, an artifact-free image reconstruction should be possible up to a pitch of $p = 2$ (cf. ▸ Sect. 8.3). With this, the scanning length is increased such that in the presence of the same mAs product less X-ray quanta are available for the image formation of a coevally larger body section. Therefore, the image noise increases.

- **Scanning length**: If the imaged body section is enlarged, the dose applied to the patient will accordingly increase. This is expressed by means of the effective dose or the dose-length product. The number of slices always has to be limited to the diagnostically relevant section, which has to be specified in the overview scan.

- **Filter kernel**: The choice of the respective filter kernel at first glance does not directly influence the dose. However, as described in ▸ Sects. 7.2 and 9.8.2, the choice of the high-pass filter of the filtered backprojection does indeed affect the image resulting from the reconstruction, since the choice of the filter kernel represents a trade-off between noise and spatial resolution. If one wants to reduce the noise while maintaining a high spatial resolution, this is only possible at the expense of an increase in dose. Thus, it always depends on the diagnostic question as to whether an appropriate choice of the filter kernel can be used for dose reduction.

- **Window width**: The window width (cf. ▸ Chap. 10) used for adjusting the display of the CT images does not initially have a direct influence on the dose. However, the higher the selected contrast by constricting the window width, the stronger the present noise in the images. Conversely, it is possible to smooth the image by enlarging the window width. If there is a contrast reserve due to the diagnostic question, a smoothing of the visualization can be already incorporated during the planning step. This will decrease the mAs product and thus decrease the radiation exposure.

- **Field-of-view (FOV)**: By using a very small FOV, i.e., a very strong detail magnification, as a general rule, a very sharp reconstruction filter has to be used. This is due to the fact that the magnification of the section under examination is typically, because a locally more detailed image needs to be analyzed. This immediately has consequences for the image noise, which can only be reduced with an increase in the mAs product and, therefore, an increase in the applied dose.

For practical assistance for dose estimation the calculation strategy of the effective dose from the mAs product is provided in Table 11.4.

Table 11.4. Calculation steps from the mAs product to the effective dose. The example is for a thorax study of an adult male patient on a Siemens SOMATOM 4 Plus scanner

Step	Variable	Calculation	Result	Example	Reference
1	Tube current	I		129 mA	Scan protocol
2	Exposition time	$\cdot t$		1 s	Scan protocol
		$= Q$	mAs product	129 mAs	(11.10)
3	Normalized $CTDI$ free-in-air	$\cdot_n CTDI_{air}$		0.18 mGy/mAs	Nagel (2002)
4	1st voltage correction	$\cdot k_{U,1}$		$(140/120)^2$	(11.20)
		$= CTDI_{air}$	CTDI free-in-air	31.605 mGy	(11.14)
5	Slice thickness	$\cdot d$		0.5 cm	Scan protocol
6	Number of slices	$\cdot n$		56	Scan protocol
		$= DLP_{air}$	Dose length product	884.94 mGy·cm	(11.17)
7	Conversion factor	$\cdot f_{average}$		0.0068	Table 11.3
8	System correction factor	$\cdot k_{CT}$		1	Nagel (2002)
9	2nd voltage correction	$\cdot k_{U,2}$		$(140/120)^2$	(11.21)
		$= E$	**Effective dose**	**8.19 mSv**	(11.19)

Finally, it shall be mentioned that the progress toward smaller detector elements also influences the dose. It has been derived in ▸ Sect. 9.8.3 that at a constant signal-to-noise ratio the dose increases to the power of four as soon as the detector element size is reduced. If in the formula of Brooks (cf. (11.18)) it is assumed that $b \approx d \approx \Delta \xi$ for the sampling distance in the rotation center and for the slice thickness respectively, then, in consideration of the definition for the signal-to-noise ratio $SNR = \mu/\sigma$, one immediately obtains the relation

$$D \propto \frac{(SNR)^2}{(\Delta \xi)^4} . \tag{11.22}$$

Obviously, this limits a further reduction of the size of the detector elements, since the dose may not be arbitrarily increased.

References

Abramowitz M, Stegun IA (1970) Handbook of mathematical functions with formulas, graphs and mathematical tables. National Bureau of Standards, Applied Mathematics Series 55, Washington

Agawal BK (1991) X-ray spectroscopy, Springer Series in Optical Sciences, vol 15. Springer, Berlin

Andia BI (2000) Nonlinear backprojection. Dissertation, University of Notre Dame, Indiana

Azizi SA (1987) Entwurf und Realisierung digitaler Filter. Oldenbourg, Munich

Barrett HH, Swindell W (1981) Radiological imaging: The theory of image formation, detection, and processing. Academic, San Diego

Berger MJ, Coursey JS, Zucker DS ESTAR, PSTAR, and ASTAR: Computer programs for calculating stopping-power and range tables for electrons, protons, and helium ions (version 1.2.2). National Institute of Standards and Technology, Gaithersburg; Available from http://physics.nist.gov/PhysRefData/Star/Text/contents.html

Bergmann L, Schäfer C (1999) Lehrbuch der Experimentalphysik, Elektromagnetismus, vol 2, de Gruyter, Berlin

Blanck C (1998) Understanding helical scanning. Williams and Wilkins, Baltimore

Bluestein LI (1970) A linear filtering approach to the computation of discrete Fourier transform. IEEE Trans Audio Electroacoust AU-18:451

Bouman CA, Sauer K (1996) A unified approach to statistical tomography using coordinate descent optimization. IEEE Trans Image Process 5:480

Bracewell RN (1965) The Fourier transform and its applications. McGraw-Hill, New York

Bracewell RN (2003) Fourier analysis and imaging. Kluwer Academic/Plenum, New York

Bronstein IN, Semendjajew KA (1979) Taschenbuch der Mathematik. Teubner, Leipzig

Brooks RA, DiChiro G (1976) Statistical limitations in X-ray reconstructive tomography. Med Phys 3:237–240

Brunst B (2002) High resolution digital flat panel detectors for X-ray applications – basics. In: Niederlag W, Lemke HU (eds) Medical imaging. Health Academy 02:63

Bushong SC (2000) Computed tomography. Essentials of Medical Imaging Series. McGraw-Hill, New York

Bushong SC (2001) Radiologic science for technologists. Mosby, St. Louis

Bushberg JT, Seibert JA, Leidholdt EM, Boone JM (2002) The essential physics of medical imaging. Lippincott, Williams and Wilkins, Philadelphia

Clack R, Defrise M (1994) Cone-beam reconstruction by the use of radon transform intermediate functions. J Opt Soc Am A 11:580

Cormack AM (1963) Representation of a function by its line integrals with some radiological applications I. J Appl Phys 34:2722

Cormack AM (1964) Representation of a function by its line integrals with some radiological applications II. J Appl Phys 35:195

Cormack AM (1982) Computed tomography: Some history and recent developments. Proc Symp Appl Math 27:35

Cunningham IA (2000) Computed tomography. In: Bronzino JD (ed) The biomedical engineering handbook, volume I. CRC, Boca Raton, pp 62–61

Curry TS III, Dowdey JE, Murry RC (1990) Christensen's physics of diagnostic radiology. Lippincott, Williams and Wilkins, Philadelphia

David B, Barschdorf H, Doormann V, Eckart R, Harding G, Schlomka J-P, Thran A, Bachmann P, Flisikowski P (2003) Liquid metal anode x-ray tube. In: Kyrala G, Gauthier J, McDonald C, Khounsary A (eds) Laser-generated and other laboratory X-ray and EUV sources, optics, and applications. SPIE 5196, pp 432–443

David B, Thran A, Eckart R (2004) Modeling and experimental investigation of X-ray spectra from a liquid metal anode X-ray tube. In: McDonald CA, Macrander AT, Ishikawa T, Morawe C, Wood JL (eds) X-ray sources and optics. SPIE 5537, pp 45–56

Deans SR (1983) The Radon transform and some of its applications. Wiley, New York

De Clerck NM, van Dyck D, Postnov AA (2003) Non-invasive high-resolution μCT of the inner structure of living animals. Microsc Anal 1:13

Defrise M, Clack R (1994) A cone-beam reconstruction algorithm using shift-variant filtering and cone-beam backprojection. IEEE Trans Med Imaging 13:186

Dempster AP, Laird NM, Rubin DB (1977) Maximum likelihood from incomplete data via the EM algorithm. J R Stat Soc B39:1

Demtröder W (2000) Experimentalphysik 3, Atome, Moleküle und Festkörper. Springer, Berlin

Dössel O (2001) Bildgebende Verfahren in der Medizin. Springer, Berlin

Dove EL (2001) Notes on computerized tomography, script 51:060. Bioimaging Fundamentals, The University of Iowa, College of Engineering, 2001.

Edholm PR (1977) Tomogram reconstruction using an opticophotographic method. Acta Radiol 18:126

Epstein CL (2003) Introduction to the mathematics of medical imaging. Pearson, Upper Saddle River

Feldkamp LA, Davis LC, Kress JW (1984) Practical cone-beam algorithm. J Opt Soc Am A 6:612

Fessler JA (1996) Mean and variance of implicitly defined biased estimators (such as penalized maximum likelihood). IEEE Trans Image Process 5:1346

Feynman R (1966) Lectures on physics, vol 1. Addison-Wesley, Massachusetts

Fichtenholz GM (1982) Differential- und Integralrechnung. VEB Deutscher Verlag der Wissenschaften, Berlin

Gay SB, Matthews AB (1998) Ten reasons why spiral CT is worth a million bucks. Diagn Imaging 20:111

Glover GH, Pelc NH (1981) An algorithm for the reduction of metal clip artifacts in CT reconstructions. Med Phys 8:799

Grangeat P (1990) Mathematical framework of cone-beam 3D reconstruction via the first derivative of the Radon transform. In: Herman GT, Louis AK, Natterer F (eds) Mathematical methods in tomography. Springer, Berlin, p 66

Grangeat P (1997) Indirect cone-beam three-dimensional image reconstruction. In: Roux C, Coatrieux J-L (eds) Contemporary perspectives in three-dimensional biomedical imaging. IOS, Amsterdam

Grass M, Köhler T, Proksa R (2000) 3D cone-beam CT reconstruction for circular trajectories. Phys Med Biol 45:329

Green PJ (1990) Bayesian reconstruction from emission tomography data using a modified EM algorithm. IEEE Trans Med Imaging 9:84

Grossmann G (1934) Procédé et dispositif pour la représentation radiographique des sections des corps. French Patent No. 771887

Hajnal JV, Hill DLG, Hawkes DJ (2001) Medical image registration. CRC, Boca Raton

Härer W, Lauritsch G, Mertelmeier T, Wiesent K (1999) Rekonstruktive Röntgenbildgebung. Phys Bl 55:37

Härer W, Lauritsch G, Mertelmeier T (2003) Tomographie – Prinzip und Potential der Schichtbildverfahren. In: Schmidt T (ed) Handbuch diagnostische Radiologie. Springer, Berlin

Harris FJ (1978) On the use of window functions for harmonic analysis with the discrete Fourier transform. Proc IEEE 66:51

Hebert T, Leahy R (1989) A generalized EM algorithm for 3D Bayesian reconstruction from Poisson data using Gibbs priors. IEEE Trans Med Imaging 8:194

Heinzerling J (1998) Röntgenstrahler. In: Ewen K (ed), Moderne Bildgebung. Thieme, Stuttgart, p 77

Helgason S (1999) The Radon transform, 2nd edn. Birkhäuser, Boston

Herman GT (1980) Image reconstruction from projections: The fundamentals of computerized tomography. Academic, New York

Heuser H (1992) Funktionalanalysis: Theorie und Anwendung. Teubner, Stuttgart

Hidajat N (2001) Bestimmung und Optimierung der Strahlendosis des Patienten bei der Computertomographie – Methoden, Probleme und Lösungsmöglichkeiten. Habilitationsschrift, Klinik für Strahlenheilkunde der Medizinischen Fakultät Charité der Humboldt-Universität zu Berlin

Hofer M (2000) CT-Kursbuch. Hofer, Düsseldorf

Horn BKP (1979) Fan-beam reconstruction methods. Proc IEEE 67:1616

Hounsfield GN (1973) Computerized transverse axial scanning (tomography). I. Description of system. Br J Radiol 46:1016

Hsieh J (2004) Computed tomography, principles, design, artifacts and recent advances. SPIE, Bellingham

Huesman RH, Gullberg GT, Greenberg WL, Budinger TF (1977) Users manual: Donner algorithms for reconstruction tomography. Berkeley Laboratory, University of California, http://cfi.lbl.gov/cfi_software.html

Huda W, Slone R (1995) Review of radiologic physics. Lippincott, Williams and Wilkins, Philadelphia

ICRP (International Commission on Radiological Protection) (1991) Publication 60, recommendations of the International Commission on Radiological Protection 60, Ann ICRP, vol 21/1–3

Ingerhed M (1999) Fast backprojection for computed tomography; implementation and evaluation. Linköping Studies in Science and Technology 759. Department of Electrical Engineering, Linköping University

Jacobson C (1996) Fourier methods in 3D reconstruction from cone-beam data, dissertation. Linköping Studies in Science and Technology 427. Institute of Technology, Linköping University

Jan J (2006) Medical image processing, reconstruction and restoration. CRC/Taylor and Francis, Boca Raton

Kachelrieß M, Schaller S, Kalender WA (2000) Advanced single-slice rebinning in cone-beam spiral CT. Med Phys 19:864

Kachelrieß M, Fuchs T, Schaller S, Kalender WA (2001) Advanced single-slice rebinning for tilted spiral cone-beam CT. Med Phys 28:1033

Kaczmarz S (1937) Angenäherte Auflösung von Systemen linearer Gleichungen. Bull Acad Pol Sci Lett A35:355

Kak AC, Slaney M (2001) Principles of computerized tomographic imaging. Classics in Applied Mathematics 33. IEEE, New York

Kalender WA (2000) Computertomographie. Publicis MCD, Munich

Kalender WA (2003) Der Einsatz von Flachbilddetektoren für die CT-Bildgebung. Radiologe 43:379

Kalender WA, Seissler W, Vock P (1989) Single-breath-hold spiral volumetric CT by continuous patient translation and scanner rotation. Radiology 173:414

Kamm K-F (1998) Grundlagen der Röntgenabbildung. In: Ewen K (ed) Moderne Bildgebung, Thieme, Stuttgart, p 45

Katsevich A (2001) Exact FBP-type inversion algorithm for spiral CT. 3D-2001. Proceedings of the Sixth International Meeting on Fully Three-Dimensional Image Reconstruction in Radiology and Nuclear Medicine. Asilomar, Pacific Grove, p 3

Kiencke U (1998) Signale und Systeme. Oldenbourg, Munich

Kimke T, Tuytschaever T (2006) Increasing the number of gray shades in medical display systems – how much is enough? J Digit Imaging 20:422

Klingen B (2001) Fouriertransformation für Ingenieur- und Naturwissenschaften, Springer, Berlin

Köhler T, Proksa R, Grass M (2000) A fast and efficient method for sequential cone-beam tomography. IEEE Trans Med Imaging 2:110

Köhler T, Proksa R, Bontus C, Grass M, Timmer J (2002) Artifact analysis of approximate helical cone-beam CT reconstruction algorithms. Med Phys 29:51

Kramers HA (1923) On the theory of X-ray absorption and of continuous X-ray spectrum, Philos Mag 46:836–871

Krestel E (ed) (1990) Imaging systems for medical diagnostics. Siemens, Berlin

Kudo H, Saito T (1994) Derivation and implementation of a cone-beam reconstruction algorithm for nonplanar orbits. IEEE Trans Med Imaging 13:196

Kudo H, Noo F, Defrise M (1998) Cone-beam filtered-backprojections algorithm for truncated helical data. Phys Med Biol 43:2885

Lange K, Fessler JA (1995) Globally convergent algorithm for maximum a posteriori transmission tomography. IEEE Trans Image Process 4:1430

Lange K, Bahn M, Little R (1987) A theoretical study of some maximum likelihood algorithms for emission and transmission tomography. IEEE Trans Med Imaging 6:106

Laubenberger T, Laubenberger J (1999) Technik der medizinischen Radiologie. Diagnostik, Strahlentherapie, Strahlenschutz. Deutscher Ärzte, Cologne

Lee SW, Cho G, Wang G (2002) Artifacts associated with implementation of the Grangeat formula. Med Phys 29:2871

Lehmann T, Oberschelp W, Pelikan E, Repges R (1997) Bildverarbeitung für die Medizin. Springer, Berlin

Leroy C, Rancoita P-G (2004) Principles of radiation interaction in matter and detection. World Scientific, London

Lewitt RM (1990) Multidimensional digital image representations using generalized Kaiser–Bessel window functions. J Opt Soc Am A, 7(10):1834–1846

Lewitt RM (1992) Alternatives to voxels for image representation in iterative reconstruction algorithms. Phys Med Biol 37(3):705–716

Lin W-T (1999) A computational-efficient cone-beam CT reconstruction algorithm using circle-and-line orbit. SPIE 3659:933

Lossau N (1995) Röntgen. Eine Entdeckung verändert unser Leben. vgs Verlagsgesellschaft, Cologne

Lüke HD (1999) Signalübertragung. Springer, Berlin

Magnusson M (1993) (Seger), linogram and other direct Fourier methods for tomography reconstruction, dissertation. Linköping Studies in Science and Technology 320. Institute of Technology, Linköping University

Messiah A (1981) Quantenmechanik. De Gruyter, Berlin

Morneburg H (ed) (1995) Bildgebende Systeme für die medizinische Diagnostik. Publicis MCD, Munich

Mudry KM (2000) X-ray. In: Bronzino J (ed) The biomedical engineering handbook, vol I. CRC, Boca Raton, p 61-1

Müller J (2006) Metal artefact reduction for X-ray computed tomography in combination with iterative reconstruction algorithms. Thesis, RheinAhrCampus Remagen

Nagel HD (ed) (2002) Strahlenexposition in der Computertomographie, 3rd edn. CTB, Hamburg

Natterer F (1999) Numerical methods in tomography. Acta Num

Natterer F (2001) The mathematics of computerized tomography. Classics in Applied Mathematics 32. IEEE, New York

Natterer F, Wübbeling F (2001) Mathematical methods in image reconstruction. SIAM Monographs on Mathematical Modelling and Computing, Philadelphia

Neitzel U (1998) Grundlagen der digitalen Bildgebung. In: Ewen K (ed) Moderne Bildgebung, Thieme, Stuttgart, p 63

Nilsson S (1999) Application of fast backprojection techniques for some inverse problems of integral geometry. Dissertation, Linköping Studies in Science and Technology, Thesis No. 499, Linköping University

Nuyts J, Michel C, Dupont P (2001) Maximum-likelihood expectation-maximization reconstruction of sinograms with arbitrary noise distribution using NEC-transformations. IEEE Trans Med Imaging 20:365–375

Oehler M, Buzug TM (2007) Statistical image reconstruction for inconsistent CT projection data. J Methods Inform Med 3:261

Oppenheim AV, Schafer RW (1999) Zeitdiskrete Signalverarbeitung. Oldenbourg, Munich

Orlov SS (1975) Theory of three dimensional reconstruction. I. Conditions for a complete set of projections. Sov Phys Crystallogr 20:312

Otendal M (2006) A compact high-brightness liquid-metal-jet X-ray source. Thesis, Stockholm

Papula L (2000) Mathematische Formelsammlung. Vieweg, Braunschweig

Parzen E (1961) Mathematical considerations in the estimation of spectra. Technometrics 3:167

Pfoh AH (2002) Volume computer tomography (VCT) – a new diagnostic imaging technique on the basis of high-resolution flat-panel detectors. In: Niederlag W, Lemke HU (eds) Medical imaging. Health Acad 02:7

Pol MJ, Rogers JV II, Kobayashi Y, Jacobs LL (2000) Computed tomography of an anolis lizard in Dominican amber: Systematic, taphonomic, biogeographic, and evolutionary implications. Palaeontol Electron 5:13

Press WH, Flannery BP, Teukolsky SA, Vetterling WT (1990) Numerical recipes in C: The art of scientific computing. Cambridge University Press, Cambridge

Proksa R, Köhler T, Grass M, Timmer J (2000) The n-Pi-method for helical cone-beam CT. IEEE Trans Med Imaging 19:848

Radon J (1917) Über die Bestimmung von Funktionen längs gewisser Mannigfaltigkeiten. Berichte der mathematisch-physikalischen Kl. Sächsischen Gesellschaft der Wissenschaften 59. Leipzig, p 262

Ramachandran GN, Lakshminarayanan AV (1971) Three-dimensional reconstruction from radiographs and electron micrographs. Proc Acad Sci USA 68:2236

Ramm AG, Katsevich AI (1996) The Radon transform and local tomography. CRC, Boca Raton

Romans LE (1995) Introduction to computed tomography. Lippincott, Williams and Wilkins, Philadelphia

Rosenfeld A, Kak AC (1982) Digital picture processing. Academic, New York

Ruhlmann J, Oehr P, Biersack HJ (1998) PET in der Onkologie – Grundlagen und klinische Anwendung. Springer, Heidelberg

Sassov A (1999) Desktop X-ray micro-CT. In: Computerized tomography for industrial applications and image processing in radiology. Proc DGZiP BB67-CD, Berlin, p 165

Sassov A (2002) Desktop X-ray micro-CT instruments. Proc SPIE 4503:282

Sassov A (2002) Comparison of fan-beam, cone-beam and spiral scan reconstruction in X-ray micro-CT. Proc SPIE 4503:124

Sauer K, Bouman C (1993) A local update strategy for iterative reconstructions from projections. IEEE Trans Signal Process 41:533

Schlegel W, Bille J (eds) (2002) Medizinische Physik, vol 2. Springer, Berlin

Schramm N (2001) Entwicklung eines hochauflösenden Einzelphotonen-Tomographen für kleine Objekte. Dissertation, Berichte des Forschungszentrums Jülich 3841

Schumacher H, Fischer B (2007) A new approach for motion correction in SPECT imaging. Proceedings of the BVM, Springer, Berlin, p 51

Seeram E (2001) Computed tomography. Saunders, Philadelphia

Sergé E (1965) Nuclei and particles. Benjamin, New York

Shepp LA, Logan BF (1974) The Fourier reconstruction of a head section. IEEE Trans Nucl Sci 21:21

Shepp LA, Vardi Y (1982) Maximum likelihood reconstruction for emission tomography. IEEE Trans Med Imaging 1:113

Stearns SD, Hush DR (1999) Digitale Verarbeitung analoger Signale. Oldenbourg, Munich

Sternberg S (2000) CT scans: A very high-dose diagnosis. USA Today, Nov. 20

Stevens GM (2000) Volumetric tomographic imaging. Dissertation, University of Stanford

Stierstorfer K, Flohr T, Bruder H (2002) Segmented multiple plane reconstruction – a novel approximate reconstruction scheme for multislice spiral CT. Phys Med Biol 47:2571–2581

Tam KC. Three-dimensional computerized tomography scanning method and system for large objects with small area detectors. US Patent 5,390,112

Thomsen D, Klein K, Oehler M, Pfinninger S, Reich F, Buzug TM (2003) Computertomographie in der Archäologie. In: Physikalische Methoden der Laser- und Medizintechnik. VDI Fortschritt-Bericht, Reihe 17: Biotechnik/Medizintechnik, Nr. 231. VDI, Düsseldorf, p 156

Tisson G (2006) Reconstruction from transversely truncated cone beam projections in micro-tomography. Thesis, University of Antwerp

Toft P (1996) The Radon transform. Theory and implementation. Thesis, Technical University of Denmark

Turbell H (1999) Three-dimensional image reconstruction in circular and helical computed tomography. Linköping Studies in Science and Technology. Thesis No. 760, Linköping University

Turbell H (2002) Cone-beam reconstruction using filtered backprojection. Linköping Studies in Science and Technology 672. Dissertation, Linköping University

Tuy H (1983) An inversion formula for cone-beam reconstruction. SIAM J Appl Math 43:546

Wang G, Lin TH, Cheng PC (1995) Error analysis on a generalized Feldkamp s cone-beam computed tomography algorithm. Scanning 17:361

Wang G, Zhao S, Cheng P-C (1998) Exact and approximate cone-beam X-ray microscopy. In: Cheng PC, Huang PP, Wu JL, Wang G, Kim HG (eds) Modern microscopes (I) – instrumentation and image processing. World Scientific, Singapore

Watson GN (1966) A treatise on the theory of Bessel functions, 2nd edn. Cambridge University Press, Cambridge

Webb S (1990) From the watching of shadows. Hilger, Bristol

Weisser G (2000) Technische Grundlagen der EBCT. In: Gaa J, Lehmann K-J, Georgi M (eds) MR-Angiographie und Elektronenstrahl-CT-Angiographie. Thieme, Stuttgart, p 145

Werner M (2000) Signale und Systeme. Vieweg, Braunschweig

Wilting JE (1999) Technical aspects of spiral CT. Medicamundi 43:34

Yang X, Horn BKP (2002) Cone-beam reconstruction – present and future

Yendiki A Fessler JA (2004) A comparison of rotation- and blob-based system models for depth-dependent detector response. Phys Med Biol 49(11):2157–2168

Zankl M, Panzer W, Drexler G (1991) The calculation of dose from external photon exposures using reference human phantoms and Monte Carlo methods. Part VI. Organ doses from computed tomographic examinations, GSF-Report 30/91, GSF Research Centre, Oberschleißheim

Zylka W, Wischmann HA (1996) On geometric distortions in CT images. Proceedings of the 18th International Conference of the IEEE. EMBS, Amsterdam

Subject Index

180°LI   315
360°LI   314

Abel transform   132, 133, 200
absorbed energy   486
absorption   31
  coefficient   33, 36
  edges   37
  probability   66
acceleration voltage   7, 16
acquisition time   500
ADC (analog–digital converter)   404
adjoint
  Radon transform   206, 207
  reconstruction problem   209
AEC (automatic exposure control)   493,
     494
afterglow   50, 73, 404
  artifact   437
algebraic reconstruction technique (ART)
     201, 202, 211, 216
algorithm layer   404
aliasing   136
  artifact   454
amalgam   42, 439
Ambrose, J.   6
Americium source   82
amplitude spectrum   141
analog tomography   77, 78
analog–digital converter (ADC)   404
angiographic CT (CTA)   481
angular dose modulation   495
annihilation   39, 96
  radiation   39
anode surface   25
antenna characteristic   26
anti-scatter grid   51, 404, 444, 445
anticathode   19

anvil   400
aperture angle   265, 266, 320
apparent spot size   25
approximate reconstruction   366
ART (algebraic reconstruction technique)
     201, 211, 215, 216, 218
artifact
  afterglow   437
  aliasing   435, 454
  beam-hardening   30, 46, 425
  cone-beam   458
  cupping   427
  electronic   435
  interpetrous lucency   430
  metal   438, 455
  motion   432, 450
  partial volume   423, 446
  ring   435
  sampling   435, 454
  scalloping   456
  scatter   443
  slice shearing   451
  spiral CT related   456
  staircasing   448
  streak   438
asymmetrical detector   301
attenuation coefficient   33
Auger
  electron   22, 36
  process   20, 22
autocorrelation function   141
automatic exposure control (AEC)   493,
     494
axial
  aperture   447
  slice   9
a priori model   235

backprojection operator   207
backscattering   25, 38
band-limited projection   259
Bayesian estimation   235
beam
    profile   26
    quality   24
beam collimation   404
beam-hardening   29, 201, 404
    artifact   426
    correction   73
beam-wise correction   211
beating heart   433
Bernoulli
    detectors   64
    distribution   61, 67, 69
Bessel function   126, 128, 131, 189, 221
Bethe–Bloch equation   22
binding energy   17
binomial
    coefficient   61
    distribution   63
    series   67
biograph   97
biological
    effect   486
    weighting   486
bismuth germanate   50
Blackman window   246
blank scan   73
blob   220, 221
blob-based reconstruction   219
Bluestein identity   149
blurring angle   78
Bockwinkel, H.B.A.   6
bone
    removal   481
    window   476, 478–480
bounded input–bounded output   106
bowtie filter   73, 497
Bracewell, R.N.   4, 6, 113
brain window   480
bremsstrahlung   20, 21, 26
Brooks' formula   469, 493, 502

C-arm   59
cadmium tungstate   50
calcium scoring   482
capacity   23

capture efficiency   48
cardiac imaging   499
cardio CT   482
cascaded Poisson process   65, 230
CAT (computerized axial tomography)   1, 79
cathode surface   17
Cauchy's fundamental theorem   133
causality   106
CBF (cerebral blood flow)   482
CBV (cerebral blood volume)   482
CCD (charge-coupled device)   94
central limit theorem   71
central section theorem   329, 331, 350
ceramic material   50
cerebral blood flow (CBF)   482
cerebral blood volume (CBV)   482
cesium iodide   50, 53, 58
characteristic line spectrum   20
charge-coupled device (CCD)   94
Chebyshev polynomials   198
chirp z-transform   147, 148, 171
circular   369
    harmonics   196, 198
    path   311
    trajectory   261, 366, 371, 393
Clack, R.   357, 387
clique   236
coherent scattering   34
coin-throwing experiment   61
coincidence
    detection   39
    measurement   96
collimator   51, 79, 84, 305, 316, 317
    lamella   51
colon peristalsis   433
comb function   136, 139
complementary
    rebinning   270, 315
    X-ray source   271
Compton
    collision   38
    electron   38
    scattering   31, 38, 40, 426, 430
Compton, A.H.   38
computed tomography dose index (CTDI)   489
computerized

axial tomography (CAT)   1, 79
   transaxial tomography (CTAT)   79
concentric squares   169
conditional
   distribution   236
   probability   64
cone angle   372, 388, 395
cone-beam
   CT   400
   detector   52
   geometry   53, 91, 336, 340, 345, 355, 366, 371, 395, 400, 458
conservation of energy   16
consistency condition   311
contrast   405, 409, 467
   reserve   501
convergence-generating function   183
conversion
   efficiency   21
   factor   491
convolution
   kernel   183, 245, 250, 252, 273, 278
   theorem   124, 126, 134, 188, 191, 356
Cormack transform   195, 198, 200
Cormack, A.M.   5
coronal slice   9
correction function   359
covariance matrix   238, 239
Cramer, H.   6
cross-section   38, 39
CT microscopy   93
CTA (angiographic CT)   481, 482
CTAT (computerized transaxial tomography)   79
CTDI (computed tomography dose index)   489
cupping artifact   427
curved detector array   272
cyclic path   371
cylindrical detector   388, 391

DAS (data acquisition system)   72
DAT (digital axial tomography)   79
data acquisition system (DAS)   72
deceleration cascade   21, 26
deconvolution   188–190
defective beams   436
Defrise, M.   357, 387
Delbrück scattering   40

δ-distribution   103, 109, 111, 112, 323
δ-line   114, 203
density function   72
dental
   fillings   438
   radiology   79
design matrix   205
detective quantum efficiency (DQE)   422
detector
   afterglow   437
   failure   437
   fan   89
   layer   404
   quarter shift   261
   sensitivity profile   309
   size   468, 469
   surface normal   330
DFT (discrete Fourier transform)   144
diagnostic energy window   39
digital axial tomography (DAT)   79
digitalization layer   404
dipole layer   17
Dirac comb   112
Dirac's delta impulse   109
Dirac, P.A.M.   39
direct
   inversion   210
   reconstruction   168
directional interpolation   441
discrete Fourier transform (DFT)   144
discretization   202, 243, 287, 403
display layer   404
divergent integral   122
DLP (dose length product)   491, 492
dose   422, 469, 485
   free-in-air   491
   length product (DLP)   491, 501
   plateau   489
DQE (detective quantum efficiency)   422
dual-energy method   430
dual-source CT   31, 434, 498, 499

EBCT (electron beam computerized tomography)   89, 434
ECG-trigger   316, 433, 495
edge
   filter   104
   preserving   237
Edholm, P.R.   83, 98

effective
  CTDI   490
  dose   486, 492
Ehrenfest, P.   6
eigenvalue problem   116
Einstein, A.   36
elastic scattering   34
electric dipole   19
electromagnetic spectrum   15
electron
  avalanche   72
  beam   25
  beam computerized tomography (EBCT)
    89, 434
  optics   17, 19
  velocity   16
electron–positron pair   39
electron-impact   15
  source   425
electron-nucleus collision   20
electronic
  artifact   435
  defect   435
  layer   404
EM (expectation maximation)   227
  algorithm   227
energy
  conservation of   16
  conversion   51
  density spectrum   141
  dose   486
equivalent dose   486
exact 3D reconstruction   357
expectation
  maximation (EM)   227
  value   62, 63, 67–69, 223, 225, 230, 231,
    421
exposure time   23

^{18}F-FDG   96
false alarm probability   423
fan-beam geometry   84, 85, 153, 262, 263,
  274, 281
fast Fourier transform (FFT)   168
FCD (focus center distance)   261
FDK reconstruction   372, 386–388
FDK-SLANT method   393, 394
Feldkamp, L.A.   371, 386
FFT (fast Fourier transform)   168

field-of-view (FOV)   501
filament   17
fill factor   52, 55, 56
filter kernel   501
filtered
  backprojection   179, 183, 185, 190, 191,
    194, 235, 257, 272, 326, 327, 357, 365,
    378, 386, 419, 424
  layergram   190
  projection   182, 243
fine structure   37
  constant   22
finite beam width   223
finite DFT   146
first generation CT   80, 241
first moment   68
fixpoint iteration   227, 233, 234
flat-panel detector   53, 57, 58, 93, 304, 337,
    345, 351, 373, 401
fluorescence time   437
flying focus   26, 242, 435
focal
  line   23
  spot   23, 30, 403, 412, 413, 446
focus
  center distance (FCD)   261
  diameter   25
  line   28
focusing cup   17
forward projection   211, 215, 216, 229
Fourier
  coefficients   196
  slice theorem   163, 165, 175, 181, 326, 327,
    333, 352
  transform   117–121, 123, 124, 132, 136,
    146, 164, 210, 242
Fourier–Abel–Hankel cycle   132, 133
Fourier–Bessel transform   128
fourth generation CT   87, 88, 241
fourth power law   469, 502
FOV (field-of-view)   501
frequency variable   166, 327
fundamental signals   102

gadolinium oxysulphide   50
gantry   90, 155
  tilt   7, 451, 474
gas detector   48
gate property   107

Gaussian
    blurring   407
    distribution   71
    function   102
    MRF   237
Geiger–Müller counter   48
generalized
    Gaussian MRF (GGMRF)   237
    Radon transform   331
    value   122
geometric
    efficiency   49, 51
    enlargement   86
geometrical design   241
German Employer's Liability Insurance
        Association   2
GGMRF (generalized Gaussian MRF)
        237
Gibbs
    distribution   236, 237
    phenomenon   252
    potential   236
grad operator   228, 234
gradient method   227
Grangeat method   350
Grangeat, P.   341, 345, 354
Gray   486
gray-value discrimination   476
great circle   335, 336
Grossmann tomograph   76, 77
Grossmann, G.   76

hammer   400
Hamming window   246
Hankel transform   128, 131, 132, 189, 199,
        200
Hanning window   246
head scanner   80
heat
    capacity   24
    load   23
    units   23
Heaviside step function   102
Heel effect   27
helical
    cone-beam reconstruction   394
    CT   90
    trajectory   311
Hertzian antenna   26

Hessian
    matrix   226, 232
    normal form   157, 176, 192
high-pass filter   126, 183
Hilbert transform   133, 134, 194
hot spot   97
Hough transformation   160
Hounsfield
    bars   430
    unit (HU)   475
Hounsfield, G.N.   5, 81
HU (Hounsfield unit)   475
human
    jaw   400
    visual system   476
hybrid Radon transform   331, 334, 335, 352
hyperplanes   212

ICRP (International Commission on
        Radiological Protection)   42
ideal low-pass filter   140
identity operator   133
ill-conditioned   209
    problem   205
ill-posed problems   235
ILST (iterative least squares techniques)
        211
image
    chain   405
    deterioration   412
    processing   404
imaging quality   410
impulse
    response   104, 106, 107, 115, 116, 178, 407,
        409, 413
    train   113
incoherent scattering   38
incomplete
    data   459
    Radon data   371, 387
inconsistency   311, 424, 427, 447, 456
inconsistent projection data   317
inherent beam weighting   201, 440
integrating detector   72
International Commission on Radiological
        Protection (ICRP)   42
interpolation   313
inverse
    fan   88

problem   3, 4, 234, 235
  Radon transformation   163, 169
  squares   170
inversion formula   329
irregular sampling   358
iso-dose line   41, 487
iso-surface   304, 460, 482
isotropic voxel   320
iterative
  ART   211
  fixpoint solution   237
  least squares techniques (ILST)   211

Jacobian   129, 179, 274, 281, 325, 326, 332, 340, 377
jinc function   128
jitter   136
joint probability   224, 230

k-space   117
Kaczmarz's method   211
Kaiser–Bessel window   220
Kalender, W.   91, 309
kerma   490
Klein–Nishina equation   38, 431
Korenblyum, B.I.   6
Kramers   21
Kudo, H.   357
Kuhn–Tucker conditions   226, 233

λ^3 dependence   36, 426
Lakshminarayanan, A.V.   244, 252, 253
Lambert–Beer's law   33, 46, 68, 230, 233, 425
Laplace transform   122
least squares minimum norm   206
likelihood function   225, 231
LIMAX (Liquid Metal Anode X-ray)   31
line
  pairs   404
  spread function   411
linear
  detector array   280
  frequency ramp   242, 248, 256, 278, 285, 354, 381
  interpolation   314
  system   104
  system of equations   201
  transmission   106

linear interpolation   417, 441
linogram   354
  method   170, 171
  sampling   148
liquid metal   24
Liquid Metal Anode X-ray (LIMAX)   30, 31
log likelihood function   225, 227, 232, 234, 235, 237, 239
Logan, B.F.   245, 249, 252, 253
longitudinal dose modulation   493, 494
Lorentz, H.A.   6
low-pass filter   126, 136, 248
LTI systems   105
lung emphysema   482

MAP (maximum *a posteriori*)   235
Markoff
  process   236
  random field (MRF)   236
mAs product   468, 499
mass attenuation coefficient   33, 37, 42
mass–energy equivalency   39
maximum *a posteriori* (MAP)   235
maximum likelihood
  expectation maximization (MLEM)   60
  method   223, 225, 228, 231
maximum-intensity projection (MIP)   481
mean   421
  transit time (MTT)   482
measurement field diameter (MFD)   259
measuring interval   269
meridian surface   348
metal
  artifact   438, 455
  artifact reduction   442
  filter   496
  shadow   438
MFD (measurement field diameter)   259
micro-CT   93, 337, 338
micro-focus tube   93
MIP (maximum-intensity projection)   481
mirror-image force   17
missing data   366, 440
MLEM (maximum likelihood expectation maximization)   60, 440
modulation   405
  transfer function (MTF)   25, 93, 296, 404, 406, 410, 413

Moiré
  effect   136
  pattern   437
Moore–Penrose inverse   206
Moseley's law   21
most probable solution   225, 231
motion
  artifact   84, 311, 433, 450
  correction   433
  parameter   358
MPR (multi-planar reformatting)   9, 11,
    310
MRF (Markoff random field)   236
MSAD (multiple scan average dose)   489
MSCT (multi-slice CT)   52, 337
MTF (modulation transfer function)   25,
    296, 404, 409, 410, 412, 466
MTT (mean transit time)   482
multi-line detector array   93, 319
multi-planar reformatting (MPR)   9, 11,
    310
multi-row arrays   52
multi-slice
  CT (MSCT)   52, 337
  detector   12
multiple scan average dose (MSAD)   489

N chooses *n*   61
n-PI method   398
NDT (nondestructive testing)   2
nearest neighbors method   208
NEC (noise-equivalent count)   73
  scaling   73
  shifting   73
neutrino   96
neutron   96
Newton–Raphson method   240
Nobel Prize   6, 15, 36, 38, 81
noise   235, 252, 422, 462, 466
noise-equivalent count (*NEC*)   73
non-Poisson statistics   72
non-radiative process   22
non-square integrable function   247
nondestructive testing (NDT)   2
normal distribution   71
normalized CTDI   491
nuclear resonance scattering   40
number of
  detectors   259

projections   259
Nyquist
  criterion   135, 258, 305
  frequency   136

object thickness   500
optical
  focus   25
  reconstruction method   98
  transfer function (OTF)   410
organ dose   491
Orlov's sufficiency condition   335
orthogonal
  matrix   207
  reformatting   10, 449
orthopan tomography   79
OTF (optical transfer function)   410
over-determined   208, 218
  system   205
overview scan   7, 9, 471, 472

pair production   31, 39, 40, 96
parallel-beam
  geometry   263, 269
  rebinning   265
paranasal sinuses   474
parenchymatous organs   476
Parseval's theorem   125
partial volume artifact   446, 456, 461
partition function   236, 237
Parzen window   246
Pascal's triangle   61
patient overview   9
penalty term   235
pencil-beam geometry   33, 79, 153, 258,
    262, 265, 272
penetration depth   26
penumbra fringe   24
perfusion   482
PET (positron emission tomography)   39,
    96, 204, 223
PET-CT   96
petrous bones   430
phase transfer function (PTF)   411
photo disintegration   40
photoelectric absorption   31, 36, 40, 426,
    430
photomultiplier   72
photon detector   50

photon–matter interaction   47, 430
physical layer   403
PI
  method   396
  window   395
pilot
  scan   472
  view   186
pitch   7, 316–318, 320, 395, 432, 449, 456, 489, 501
pixel   53, 218, 219
  noise   465
pixel-wise correction   211
planar detector   372, 390
planigraphy   78
planning overview   186
pleura window   478, 479
point-slope form   214, 329
point-spread function (PSF)   115–117, 178, 410
Poisson
  distribution   63, 65, 69, 70, 224, 230, 239, 421, 440, 462
  source   65
polar coordinates   129
polychromatic
  spectrum   201
  X-ray   29, 427
position invariance   105
positron   39, 96
  emission tomography (PET)   39, 96
  emitter   96
power
  of four law   469, 502
  theorem   125
practical CTDI   490
preprocessing chain   73
principal
  quantum number   21
  sections   9
prior   236
probability density function   63
projection
  integral   156, 323, 462
  pattern   187
  surface   322
pseudo
  inverse   207

sharpness   411
solution   206, 218
PSF (point-spread function)   178, 187, 410, 411
PTF (phase transfer function)   411

QED (quantum electrodynamics)   22
quantum
  efficiency   23, 49, 50, 57
  electrodynamics (QED)   22
  noise   462
quarter detector shift   242, 298, 435

radiation
  energy   16
  weighting factor   487
radiation-free transition   36
Radon
  inversion   328
  inversion formula   350, 353, 357, 399
  space   160, 161, 163, 184, 205, 264, 270, 290, 315, 369, 371, 459
  sphere   341–343, 348, 357, 360, 364
  transform   161, 213, 321, 323, 341, 343, 345, 348, 355
Radon's solution   151, 191, 327
Radon, J.   5, 7, 151, 193
Ramachandran, G.N.   244, 252, 253
ramp filter   279, 286, 289, 386, 389, 392, 419
rank   203, 208
ray-by-ray reconstruction   211
Rayleigh
  scattering   31, 34, 40
  theorem of   125, 465
rebinning   265, 266, 269, 272, 290, 390, 392
  of the fan beams   265
receiver operating characteristic (ROC)   422
reconstruction layer   404
rectangular function   102
reduced energy   38, 431
registration   96
regridding   168, 175
regular sequence   122, 123, 183
regularity conditions   151
regularization   209, 234–237, 247
  parameter   236, 237

relaxation parameter   215
residual inconsistency   441
respiration   433
Richardson–Dushman equation   19
ring artifact   86, 435
ROC (receiver operating characteristic)   422, 423
Röntgen, W.C.   5, 7
rotating
    anode disk   23
    unit vectors   157
rotation invariance   105

S-FDK (sequential FDK)   393
sagittal section   9
Saito, T.   357
sampling
    aperture   319
    artifact   435
    disk   155, 157
    theorem   135, 136, 260, 294–297, 306, 316, 435
    unit   90
scaling property   110
scalloping   456
scanogram   9, 186, 471
scatter
    coefficient   33
    diagram   41
scattering   31, 34
scintillator
    detector   50
    layer   55
scout view   9, 186, 471
second central moment   68
second generation CT   83, 241
secondary reconstruction   8, 304, 454
segmentation inaccuracy   459
self absorption   27
sensitivity   422
    profile   317, 318, 413, 423, 435
sequential FDK (S-FDK)   393
    method   393
seventh generation CT   92
shadow zone   369–371, 387, 390, 459
shah   113
Shannon's theorem   135, 242, 296
Shannon–Whittaker interpolation   140
shaped filter   497

Shepp, L.A.   245, 249, 252, 253
Shepp–Logan   288, 420
shock room   1
sidelobes   128, 245
Sievert   486
sift property   110, 114, 139, 160, 178, 194, 323
signal mean   68
signal transmission   403
signal-to-noise ratio (SNR)   68, 70, 421, 468, 493
signum function   121, 134
simple backprojection   175, 183, 185, 187, 207
sinc function   102, 121, 245
singular value   207, 209
    decomposition (SVD)   207
sinogram   160, 161, 205
sinusoidal trace   162
sixth generation CT   90
slice
    collimation   404
    direction   9
    sensitivity profile (SSP)   313, 318, 456
    thickness   7, 500
slip-ring   90, 91, 262, 311
slope-intercept form   213
SNR (signal-to-noise ratio)   421, 422, 469
sodium iodide   83
soft tissue window   476, 478
solid metal anode   16
solid-state detector   50
source trajectory   357, 358
sparse matrix   205, 210
spatial
    frequency   94, 117, 242
    resolution   93, 235, 252, 404, 466
    signal   102
specificity   422
specimen disk   93
SPECT (single photon emission computed tomography)   204, 223
spectral weighting function   242
spherical coordinates   325
spiral
    CT   90, 309, 312, 313, 317, 395, 456
    path   311
spiral-groove bearing   24
split detector   432

square integrable function   183
SSP (slice sensitivity profile)   318
staircasing artifact   449
standard
  deviation   68, 421, 463
  kernel   252
stationary detector ring   87
statistical
  photon model   60
  reconstruction   201, 225
step function   107, 108
Stirling equation   70
stirrup   400
stochastic
  fluctuation   462
  process   236
stratigraphy   78
Straton™ X-ray tube   31
sub-second scanners   90
sufficiency condition   336, 359, 371
supplementary helix   314, 315, 317
surface
  integral   321
  rendering   455, 459, 481
  segmentation   305
surrogate data   440
SVD (singular value decomposition)   207,
    209
symmetry properties   181
system
  matrix   205, 208, 210, 213, 218, 221
  of equations   208

T-FDK (tent-FDK) method   392
table deformation   454
Tam–Danielsson window   396
target area   25
temporal dose modulation   495, 497
tent-FDK (T-FDK) method   392
TFT (thin-film transistor)   53
theorem of Thales   160, 172, 342, 360, 367
thermal
  electron   16
  noise   462
thin-film transistor (TFT)   53
third generation CT   85, 87, 241, 261
Thomson
  cross-section   34
  scattering   34, 40

time to peak (TTP)   482
tissue
  mass   486
  weighting factor   487
tomography   78
tomosynthesis   77, 336
topogram   9, 186, 471
torus   369
tracer   96
translation invariance   105
transmission
  probability   66
  system   104
triangle function   102, 417
triangulation inaccuracy   459
triplet production   39
TTP (time to peak)   482
tube
  current   7, 467
  voltage   500
Tuy–Smith
  condition   369
  sufficiency condition   346, 368, 371, 387

unit
  of dose   486
  sphere   328, 335, 336
  vector   322, 350

variance   68, 69, 463
  of the origin   465
  of the Radon transform   462
  of the reconstruction   464
VCT (volume CT)   337
vessel analysis   482
virtual
  detector   281, 341, 373, 391, 396
  endoscopy   481
  linear detector   267, 280
volume
  CT (VCT)   337
  rendering   310, 449, 459, 479, 481
voxel   219, 221, 309, 310
voxel-based reconstruction   219

water correction   427
Wehnelt cylinder   17, 18
weighted
  CTDI   490

filtered backprojection   279
least squares   238, 239
MLEM   441
whole body scan   12
Wiener–Khintchine theorem   141
window
    function   244
    level   476
    width   476, 501
Wold, H.   6
wolfram target   90

X-ray   15
    anode   24
    detector   64

focus   24
quanta   230
shielding   33, 40
transform   339, 341, 345
tube   16
tube current   70
X-ray imaging layer   403
xenon   49

Yang Hui's triangle   61

Z^4 dependence   36, 438
z-transform   147
Ziedses des Plantes, B.G.   78

MIX
Papier aus verantwortungsvollen Quellen
Paper from responsible sources
FSC® C105338

FSC
www.fsc.org

If you have any concerns about our products,
you can contact us on
ProductSafety@springernature.com

In case Publisher is established outside the EU,
the EU authorized representative is:
Springer Nature Customer Service Center GmbH
Europaplatz 3, 69115 Heidelberg, Germany

Printed by Libri Plureos GmbH
in Hamburg, Germany